GENERALIZED RECURSION THEORY

STUDIES IN LOGIC

AND

THE FOUNDATIONS OF MATHEMATICS

VOLUME 79

NORTH-HOLLAND PUBLISHING COMPANY – AMSTERDAM • LONDON
AMERICAN ELSEVIER PUBLISHING COMPANY, INC. – NEW YORK

GENERALIZED RECURSION THEORY

PROCEEDINGS OF THE 1972 OSLO SYMPOSIUM

Edited by

J. E. FENSTAD
University of Oslo

and

P. G. HINMAN
University of Oslo
and
University of Michigan, Ann Arbor

1974

NORTH-HOLLAND PUBLISHING COMPANY – AMSTERDAM • LONDON
AMERICAN ELSEVIER PUBLISHING COMPANY, INC. – NEW YORK

Library of Congress Catalog Card Number 73-81531

North-Holland ISBN for the series 0 7204 22000
for this volume 0 7204 22760

American Elsevier ISBN 0 444 10545 X

Published by
North-Holland Publishing Company – Amsterdam
North-Holland Publishing Company, Ltd. – London

Sole distributors for the U.S.A. and Canada
American Elsevier Publishing Company, Inc.
52 Vanderbilt Avenue
New York, N.Y., 10017

PRINTED IN THE NETHERLANDS

PREFACE

The Symposium on Generalized Recursion Theory was held at the University of Oslo, June 12–16, 1972. The Symposium received generous financial support from the Norwegian Research Council and from the University of Oslo. About 50 persons attended the meeting.

This volume contains 12 of the papers presented at the meeting. Of the five remaining papers the contribution of Y.N. Moschovakis replaces the one originally presented, which will be published by North-Holland in 1973 under the title *Elementary Induction on Abstract Structures*. The paper "Post's problem for admissible sets" by S. Simpson is a later addition. The Editors asked K. Devlin to write a survey paper on the Jensen theory of the fine structure of the constructible hierarchy. The two remaining papers, by S. Aanderaa and L. Harrington, solve important problems left open at the end of the Symposium, and we are happy to include these papers in the Proceedings. We should finally note that the authors have been free to revise their papers after the Symposium, which in some cases has led to extensions of the results as originally reported.

We hope that the inclusion of a bibliography of papers on generalized recursion theory will increase the usefulness of the present volume. The participants of the Symposium agreed that a bibliography of the field would be useful, and the preparation of it was taken over by Gerald Sacks, who received extensive assistance from Leo Harrington. The Editors are grateful to them for their valuable work. The reader will note that the bibliography carries the disclaimer "uncritical". This is to emphasize that the purpose was not to present a comprehensive and scholarly bibliography of works relevant to generalized recursion theory, but to provide a useful list of some of the basic papers.

The Symposium was intended to present a broad view of methods and results in generalized recursion theory. We believe that the meeting achieved some measure of success toward this goal so that the published Proceedings also can serve as an introduction for the beginning research student who wants to specialize in this rich and fascinating branch of logic.

The Editors

CONTENTS

III. INDUCTIVE DEFINABILITY

IV. AXIOMATIC APPROACHES AND GENERAL DISCUSSION

V. A BIBLIOGRAPHY OF GENERALIZED RECURSION THEORY

PART I

RECURSION IN OBJECTS OF FINITE TYPE

J.E. Fenstad, P.G. Hinman (eds.), Generalized Recursion Theory

RECURSION IN THE SUPERJUMP

Peter ACZEL
University of Manchester

and

Peter G. HINMAN
University of Oslo
and University of Michigan

The *ordinary jump* operator of recursion theory is a function $oJ : {}^{\omega}\omega \to {}^{\omega}\omega$ defined by

$$oJ(\alpha)(m) = \begin{cases} 0, & \text{if } \{m\}(\alpha)\downarrow ; \\ 1, & \text{otherwise} . \end{cases}$$

By treating oJ as a (type-2) function: ${}^{\omega}\omega \times \omega \to \omega$ and coding the two arguments into one, we obtain from the schemata of [Kl] a notion of recursion relative to oJ. It is well-known that oJ is of the same degree as ${}^{2}E$, that a set A of natural numbers is recursive in oJ just in case it is hyperarithmetic, and that ω_1^{oJ}, the least ordinal not recursive in oJ, is just ω_1, the least non-recursive ordinal and the second admissible ordinal.

The same procedure can, of course, be applied to any function (jump) J. For example, hJ defined by

$$hJ(\alpha)(m) = \begin{cases} 0, & \text{if } \{m\}(\alpha, oJ)\downarrow ; \\ 1, & \text{otherwise,} \end{cases}$$

is recursively equivalent to the *hyperjump* and of the same degree as E_1. The sets of natural numbers recursive in hJ are the recursive analogue of the C-sets of descriptive set theory [Hi], and ω_1^{hJ} is the least recursively inaccessible ordinal.

The *superjump* S as defined in [Ga] is a type-3 jump defined by

$$S(J)(\alpha)(m) = \begin{cases} 0, & \text{if } \{m\}(\alpha, J)\downarrow: \\ 1, & \text{otherwise.} \end{cases}$$

In particular, $S(oJ) = hJ$. By coding arguments as above we may consider S as a function: $({}^{\omega}\omega)\omega \to \omega$. Thus from [Kl] we have a notion of recursion relative to S. In this paper we study some properties of this notion. In §§1 and 2 we discuss a hierarchy of jump operators, due to Platek, obtained by iterating S over a set of ordinal notations. §3 contains some results concerning the size of ω_1^S, the least ordinal not recursive in S. In §4 we extend Platek's hierarchy to one with the property that a set of natural numbers is recursive in S iff it is recursive in some jump operator occurring in the hierarchy. Finally, in §5 we discuss several other type-3 functionals which are in some sense equivalent to S.

In §§1 and 2 we assume familiarity with [Pl] and conform for the most part to his notation. The rest of the paper does not have this prerequisite, but we recommend to the reader the clear general discussion of [Pl, 257–263] as background.

§1. Platek's hierarchy

Modern mathematics must be considered more an art-form than a science, but it is perhaps a harder master than most of the arts: mathematics must not only be beautiful, it must also be correct. Sad to say, even the beautiful can be false and such an occurrance is the starting point of this paper. In [Pl], Platek constructs a hierarchy of jumps J_a^S indexed by elements a of a set $\mathcal{O}^S$ of ordinal notations. $J_1^S = oJ$, $J_2^S = hJ$, and, roughly speaking, the hierarchy is obtained by iterating S over the set of notations. The construction is closely parallel to that of Kleene's sets H_a for $a \in \mathcal{O}$, a set of notations for recursive ordinals, and even more closely parallel to that of Shoenfield's sets H_a^F for $a \in \mathcal{O}^F$, a set of notations for the ordinals recursive in the type-2 function F [Sh]. Since a set A of numbers is recursive in oJ iff it is recursive in some H_a $(a \in \mathcal{O})$, and, for any F in which oJ is recursive, A is recursive in F iff it is recursive in some H_a^F $(a \in \mathcal{O}^F)$, it would be elegant and satisfying if also A were recursive in S iff it is recursive in some J_a^S $(a \in \mathcal{O}^S)$. The main theorem of [Pl] assets that this is true – unfortunately it is not.

Before describing the counterexample to Platek's theorem, we point out where his argument breaks down. In the sentences beginning at the bottom of p. 265 of [Pl] and continuing at the top of p. 266 he applies the boundedness lemma to a function $\phi(\phi(a) \simeq (H(a))_0$, so that for $a \in \mathcal{O}^F$, H_a^F is recursive in $J_{\phi(a)}^S)$ to obtain a $d \in \mathcal{O}^S$ such that each H_a^F is recursive in J_d^S. The function H is defined by the recursion theorem over $\mathcal{O}^F$ and thus ϕ is partial recursive with domain including $\mathcal{O}^F$. The boundedness lemma applies to total functions $\psi \in \mathcal{R}$, so to use it here we would need to find such a ψ which agreed with ϕ on $\mathcal{O}^F$. The natural ψ to choose would be

$$\psi(a) = \begin{cases} \phi(a), & \text{if } a \in \mathcal{O}^F; \\ 0, & \text{otherwise.} \end{cases}$$

However, there is no apparent way to find a J-number for (the characteristic function of) $\mathcal{O}^F$ and thus show this $\psi \in \mathcal{R}$. In fact we shall see that for the functional K below, to which this step of the argument would have to apply, $\mathcal{O}^K$ does *not* have a J-number and there is no $d \in \mathcal{O}^S$ such that all H_a^K $(a \in \mathcal{O}^K)$ are recursive in J_d^S.

Our counterexample consists in defining a jump K with the following three properties:

(1) K is recursive in S;

(2) $a \in \mathcal{O}^S \to J_a^S$ is recursive in K;

(3) for any $A \subseteq \omega$, if A is recursive in K, then for some $a \in \mathcal{O}^S$, A is recursive in J_a^S.

From (1) follows that $S(K)$ is also recursive in S and hence that there exist $A \subseteq \omega$ recursive in S but not recursive in K – for example, $\{a : \{a\}(K)\downarrow\}$. Hence from (2) we have immediately that there are A recursive in S but in no J_a^S $(a \in \mathcal{O}^S)$. Property (3) completes the picture to show that the jumps J_a^S $(a \in \mathcal{O}^S)$ provide a natural hierarchy for the sets recursive in K. Of course, it also follows easily from (2) that $\mathcal{O}^K$ is not recursive in any J_a^S $(a \in \mathcal{O}^S)$ and that there is no $d \in \mathcal{O}^S$ such that all A recursive in K are recursive in J_d^S.

In terms of ordinals, the uniqueness theorem for $\mathcal{O}^S$ [Pl, p. 263] ensures that for every $\sigma < |\mathcal{O}^S| = \sup\{|a| : a \in \mathcal{O}^S\}$, σ is recursive in some J_a^S and thus by (2) recursive in K. Hence $|\mathcal{O}^S| \leq \omega_1^K$. Conversely, if σ is an ordinal recursive in K, then by (3) there is some $a \in \mathcal{O}^S$ such that $\sigma < \omega_1^{J_a^S} < |\mathcal{O}^S|$. The last inequality may be shown by constructing an order-preserving partial recursive map of $\mathcal{O}^{J_a^S}$ into $\mathcal{O}^S$ and obtaining an upperbound by use of J_{2a}^S. Thus $\omega_1^K = |\mathcal{O}^S|$. Of course $\mathcal{O}^K$ is recursive in $S(K)$ so $\omega_1^K < \omega_1^{S(K)} < \omega_1^S$.

We turn now to the definition of K and the proof of properties (1)–(3). K is based on the same idea as the jump T of [Pl, Th. VI].

Definition 1.1. (a) For any $\gamma : \omega \times \omega \to 2$ we write $p \leq_\gamma q$ if $\gamma(p,q) = 0$, let $\mathrm{Fld}(\gamma) = \{p : p \leq_\gamma p\}$ and let $p <_\gamma q$ if $p \leq_\gamma q$ and $p \neq q$;

(b) $\mathcal{W}$ is the set of $\gamma : \omega \times \omega \to 2$ such that $<_\gamma$ well-orders $\mathrm{Fld}(\gamma)$;

(c) for $\gamma \in \mathcal{W}$, $\|\gamma\|$ is the order type of $<_\gamma$;

(d) for any γ and $r \in \omega$,

$$(\gamma \upharpoonright r)(p,q) = \begin{cases} 0 & \text{if } p \leq_\gamma q \;\&\; q <_\gamma r; \\ 1 & \text{otherwise.} \end{cases}$$

Note that for $\gamma \in \mathcal{W}$ and $r \in \omega$, $\gamma \upharpoonright r \in \mathcal{W}$ and $\|\gamma \upharpoonright r\| = 0$ if $r \notin \mathrm{Fld}(\gamma)$ while $\|\gamma \upharpoonright r\| < \|\gamma\|$ if $r \in \mathrm{Fld}(\gamma)$.

Definition 1.2. (a) for $\gamma \in \mathcal{W}$ and $\|\gamma\| = 0$, $K_\gamma = oJ$;

(b) for $\gamma \in \mathcal{W}$ and $\|\gamma\| > 0$, $K_\gamma = \lambda\alpha \cdot S(K_{\gamma \upharpoonright \alpha(0)})(\alpha^+)$, where $\alpha^+(m) = \alpha(m+1)$;

(c) $$K(\gamma, \alpha) = \begin{cases} K_\gamma(\alpha), & \text{if } \gamma \in \mathcal{W}; \\ \lambda m \cdot 0, & \text{otherwise.} \end{cases}$$

Theorem 1.3. *K is recursive in S.*

Proof. Since $\mathcal{W}$ is recursive in hJ, it is recursive in S. Hence if

$$G(e,m,\gamma,\alpha) \simeq \begin{cases} oJ(\alpha)(m), & \text{if } \gamma \in \mathcal{W} \text{ and } \|\gamma\| = 0; \\ S(\lambda\beta\lambda n \cdot \{e\}^S(n, \gamma \upharpoonright \alpha(0), \beta))(\alpha^+)(m), & \text{if } \gamma \in \mathcal{W} \text{ and } \|\gamma\| > 0; \\ 0, & \text{otherwise,} \end{cases}$$

then G is partial recursive in S. By the recursion theorem there exists an $\bar{e}$ such that $G(\bar{e},m,\gamma,\alpha) \simeq \{\bar{e}\}^S(m,\gamma,\alpha)$. If $F = \{\bar{e}\}^S$, we claim that for all α, γ, and m:

$$K(\gamma,\alpha)(m) = F(m,\gamma,\alpha) .$$

If $\gamma \notin \mathcal{W}$ or $\|\gamma\| = 0$ this is evident, so we proceed by induction on $\|\gamma\|$. Suppose $\|\gamma\| > 0$. Then for each γ and β, $F(n,\gamma \restriction \alpha(0),\beta) = K(\gamma \restriction \alpha(0),\beta)(n)$ so $K_\gamma(\alpha)(m) = S(\lambda\beta\lambda n \cdot K(\gamma \restriction \alpha(0),\beta)(n))(\alpha^+)(m) = F(m,\gamma,\alpha)$. □

Note that the only properties of S used in the preceding theorem are that oJ and hJ are recursive in S. The next lemma records the other (very general) properties of S that are needed in this section. Thus our methods apply in many other situations. For example, they provide an alternative proof to the result of [Mo, § 1] that Kleene's proposed hierarchy for ${}^3E$ fails to exhaust even the sets of numbers recursive in ${}^3E$.

Lemma 1.4. (a) *There exists a primitive recursive ϕ such that for all J and α, $J(\alpha) = \phi(\alpha,S(J))$;*

(b) *there exists a primitive recursive f such that for any e, J, and J', if J is recursive in J' with index e, then $S(J)$ is recursive in $S(J')$ with index $f(e)$.*

Proof. (a) is proved in two lines just as the corresponding fact one type down with oJ in place of S.

For (b), suppose that for all β and q

$$J(\beta)(q) \simeq \{e\}(q,\beta,J') .$$

It suffices to show that there is a primitive recursive g such that for any a, $\boldsymbol{m} = (m_0, ..., m_{k-1})$, and $\alpha \in {}^\omega\omega$,

$$\{a\}(\boldsymbol{m},\alpha,J) \simeq \{g(a,e)\}(\boldsymbol{m},\alpha,J') .$$

as then

$$S(J)(\alpha)(a) = S(J')(\alpha)(g(a,e)) .$$

g is defined by the recursion theorem and by cases depending on the index a.

The only difficult case is when

$$\{a\}(\boldsymbol{m},\alpha,J) \simeq J(\lambda p \cdot \{b\}(p,\boldsymbol{m},\alpha,J))(q)\,,$$

where b and q are coded into a. Here we define $g(a,e)$ to be the "natural" index such that

$$\{g(a,e)\}(\boldsymbol{m},\alpha,J') \simeq 0 \cdot J'(\lambda p \cdot \{g(b,e)\}(p,\boldsymbol{m},\alpha,J'))(q)$$

$$+ \{e\}(q,\lambda p \cdot \{g(b,e)\}(p,\boldsymbol{m},\alpha,J'),J')\,.$$

It is straightforward to show for this case that

$$\{a\}(\boldsymbol{m},\alpha,J) \simeq n \to \{g(a,e)\}(\boldsymbol{m},\alpha,J') \simeq n\,.$$

Conversely, if $\{g(a,e)\}(\boldsymbol{m},\alpha,J') \simeq n$, then by virtue of the first term in the definition, $\lambda p \cdot \{g(b,e)\}(p,\boldsymbol{m},\alpha,J')$ is total and all computations of its values are subcomputations of the computation of $\{g(a,e)\}(\boldsymbol{m},\alpha,J')$. Hence, the induction hypothesis guarantees that $\lambda p \cdot \{b\}(p,\boldsymbol{m},\alpha,J)$ is the same total function and thus that $\{a\}(\boldsymbol{m},\alpha,J)$ is defined with value n.

In the case $\{a\}(b,\boldsymbol{m},\alpha,J) \simeq \{b\}(\boldsymbol{m},\alpha,J)$ we take $g(a,e)$ to be the "natural" index c, computed from an index for g, such that

$$\{c\}(b,\boldsymbol{m},\alpha,J') \simeq \{g(b,d)\}(\boldsymbol{m},\alpha,J')\,.\ \Box$$

Corollary 1.5. *There exist primitive recursive f, g, and h such that for any $\gamma \in \mathcal{W}$ and any p, q:*
(a) *If $\|\gamma\| > 0$ then $S(K_{\gamma\restriction p})$ is recursive in K_γ with index $f(p)$;*
(b) *$p <_\gamma q \to K_{\gamma\restriction p}$ is recursive in $K_{\gamma\restriction q}$ with index $g(p,q)$;*
(c) *if $\|\gamma\| = \|\gamma\restriction p\| + 1$, then K_γ is recursive in $S(K_{\gamma\restriction p})$ and γ with index $h(p)$.*

Proof. (a) is immediate from the definition. When $p <_\gamma q$, $\gamma\restriction p = (\gamma\restriction q)\restriction p$ so (b) follows from (a) and 1.4(a). Suppose $\|\gamma\| = \|\gamma\restriction p\| + 1$. If $r <_\gamma p$, then $K_{\gamma\restriction r}$ may be computed from $K_{\gamma\restriction p}$ by (b). If $r \not<_\gamma p$, then either $r = p$, so that $K_{\gamma\restriction r}$ is trivially recursive in $K_{\gamma\restriction p}$, or $\|\gamma\restriction r\| = 0$ and $K_{\gamma\restriction r} = oJ$ so again is recursive in $K_{\gamma\restriction p}$ (with index computable from γ). Hence using γ we may compute each $K_{\gamma\restriction r}$ uniformly from $K_{\gamma\restriction p}$. By 1.4(b) the same is true of $S(K_{\gamma\restriction r})$

and $S(K_{\gamma \upharpoonright p})$ and from the definition of K_γ it is clear that this is sufficient to establish (c). □

To obtain result (2) it is much more convenient to work with a hierarchy slightly different from that of [Pl] obtained by introducing an ordering relation $<_S$ and at limit stages $3^a \cdot 5^e$ requiring that $\lambda m \cdot \{e\}(m, J_a^S)$ ascend in the ordering $<_S$. The construction is entirely parallel to that of [Sh]. Let $\bar{O}^S$ and $\bar{J}_a^S$ denote the set of notations and jumps thus obtained. The new system is a subsystem of Platek's and it is an easy exercise to prove.

Lemma 1.6. *There exist partial recursive f and g such that for all $a \in O^S$, $f(a) \in \bar{O}^S$ and J_a^S is recursive in $\bar{J}_{f(a)}^S$ with index $g(a)$.*

For each $a \in \bar{O}^S$, let γ_a be defined by

$$\gamma_a(p,q) = \begin{cases} 0, & \text{if } p \leq_S q <_S a; \\ 1, & \text{otherwise.} \end{cases}$$

Then $\gamma_a \in \mathcal{W}$ and $\|\gamma_a\| = |a|$. Note that if $b <_S a$, then $\gamma_a \upharpoonright b = \gamma_b$.

Lemma 1.7. *There exist partial recursive f and g such that for all $a \in \bar{O}^S$.*
(a) *γ_a is recursive in $\bar{J}_a^S$ with index $f(a)$;*
(b) *$\bar{J}_a^S$ is recursive in K_{γ_a} with index $g(a)$.*

Proof. (a) is proved by a straightforward induction over $\bar{O}^S$. At limit stages one uses the fact that oJ is recursive in $\bar{J}_a^S$. For (b) we define g by the recursion theorem over $\bar{O}^S$ as follows. For $a = 1$, $\bar{J}_a^S = K_{\gamma_a} = oJ$. If $a = 2^b$, the induction hypothesis yields that J_b^S is recursive in $K_{\gamma_b} = K_{\gamma_a \upharpoonright b}$ with index $g(b)$. Then by using 1.5(a) and 1.4(b) we may compute an index $g(a)$ of $\bar{J}_a^S = S(\bar{J}_b^S)$ from K_{γ_a}. If $a = 3^b \cdot 5^e$, let $\phi(m) = \{e\}(m, \bar{J}_b^S)$. Since $b <_S a$ we can find as in the previous case an index of $\bar{J}_b^S$ and hence of ϕ from $K_{\gamma_a \upharpoonright b}$ and then by 1.5(a) from K_{γ_a}. Similarly, for each m we can compute an index of $\bar{J}_{\phi(m)}^S$ from K_{γ_a}. Putting these together, we obtain an index $g(a)$ for $\bar{J}_a^S$ from K_{γ_a}. □

Theorem 1.8. *There is a partial recursive h such that for all $a \in \bar{O}^S$, $\bar{J}_a^S$ is recursive in K with index $h(a)$.*

Proof. We again define h by the recursion theorem over $\bar{\mathcal{O}}^S$. For $a = 1$, $\bar{J}_a^S = oJ = \lambda\alpha \cdot K(\lambda m \cdot 1, \alpha)$, so it is clear how to pick $h(1)$. For $a = 2^b$, using $h(b)$ and (a) of the previous lemma we can compute an index for γ_b from K. By 1.7(b) we have an index of $\bar{J}_a^S$ from K and γ_a. But γ_a is obviously recursive in γ_b so we have one from K alone. For $a = 3^b \cdot 5^e$, let $\phi(m) = \{e\}(m, \bar{J}_b^S)$. Using $h(b)$ we can find an index of ϕ from K and using $h(\phi(m))$ we can find an index of $\bar{J}_{\phi(m)}^S$ from K. Hence we can find an index of $\bar{J}_a^S$ from K. □

Result (2) now follows immediately from 1.6 and 1.8.

Lemma 1.9. *There exist primitive recursive f and g such that for any* $d \in \mathcal{O}^S$ *and any* $\gamma \in \mathcal{W}$ *recursive in* J_d^S *with index e,* $f(d,e) \in \mathcal{O}^S$ *and* K_γ *is recursive in* $J_{f(d,e)}^S$ *with index* $g(d,e)$.

Proof. Let d, e, and γ be as described. We first define functions ϕ and ψ recursive in J_d^S, with indices from J_d^S depending only on e, such that for all p, $\phi(p) \in \mathcal{O}^S$ and $K_{\gamma \restriction p}$ is recursive in $J_{\phi(p)}^S$ with index $\psi(p)$. Once this is done we can compute from the index of ϕ from $\bar{J}_d^S$ by the boundedness lemma a $c \in \mathcal{O}^S$ such that $|d| \leq |c|$ and $\forall p \cdot |\phi(p)| < |c|$. Then using 1.4(b) and the uniqueness lemma, for all p, $S(K_{\gamma \restriction p})$ is recursive in J_c^S with index computable from $\psi(p)$. Since ψ is also recursive in J_c^S it follows that K_γ is recursive in J_c^S. Hence we may take $f(d,e) = c$ and $g(d,e)$ an index formalizing the preceding labyrinthine computation.

It remains to define ϕ and ψ. Given p, we first check from γ and oJ, hence from J_d^S, whether $p \in \mathrm{Fld}(\gamma)$ and $\|\gamma \restriction p\| > 0$. If not, set $\phi(p) = d$ (a trick useful below) and $\psi(p)$ any index of $K_{\gamma \restriction p} = oJ$ from J_d^S. If so, we assume as induction hypothesis that for q and r such that $\|\gamma \restriction q\| < \|\gamma \restriction r\| < \|\gamma \restriction p\|$, $\phi(q)$ and $\phi(r)$ are defined and $|\phi(q)| < |\phi(r)|$. If p has an immediate predecessor q in the ordering $<_\gamma$, this can be determined and q computed from J_d^S. Then by 1.5(c) we may compute an index of $K_{\gamma | p}$ from $S(K_{\gamma | q})$ and γ and thus, using the induction hypothesis and 1.4(b), from $S(J_{\phi(q)}^S)$ and γ. Since γ is recursive in J_d^S and $|d| \leq |\phi(q)|$, it suffices to take $\phi(p) = 2^{\phi(q)}$ and $\psi(p)$ an appropriate index.

Finally, suppose $\|\gamma \restriction p\|$ is limit ordinal. Let θ be defined by: $\theta(0)$ = the unique $q \in \mathrm{Fld}(\gamma)$ such that $\|\gamma \restriction q\| = 0$;

$$\theta(m+1) = \text{least } q \cdot \theta(m) <_\gamma q <_\gamma p\,.$$

It is clear that θ is recursive in γ and oJ, hence in J^S_d, and that θ defines a strictly increasing cofinal sequence in $<_{\gamma\restriction p}$. Let $\eta(m) = \phi(\theta(m))$ and let a be an index of η from J^S_d. Then $3^d \cdot 5^a \in \mathcal{O}^S$ and we claim that $K_{\gamma\restriction p}$ is recursive in $J^S_{3^d\cdot 5^a}$. It suffices to show that for all r, $S(K_{(\gamma\restriction p)\restriction r})$ is recursive in $J^S_{3^d\cdot 5^a}$ (uniformly in r, of course!). If $r \not<_\gamma p$, then $K_{(\gamma\restriction p)\restriction r} = oJ$. If $r <_\gamma p$, we compute the least m such that $r <_\gamma \theta(m)$. Then $|\phi(r)| < |\phi(\theta(m))| = |\eta(m)|$ so by 1.4(b) and the uniqueness lemma, $S(K_{(\gamma\restriction p)\restriction r})$ is recursive in $J^S_{\eta(m)}$ which in turn is recursive in $J^S_{3^d\cdot 5^a}$. Thus we take $\phi(p) = 3^d \cdot 5^a$. □

Theorem 1.10. *There exist primitive recursive f and g such that for any a,* $\boldsymbol{m}$, *and n:*

$$\{a\}(\boldsymbol{m},K) \simeq n \rightarrow f(a,\langle \boldsymbol{m}\rangle) \in \mathcal{O}^S$$

and

$$\{g(a)\}(\boldsymbol{m}, J^S_{f(a,\langle \boldsymbol{m}\rangle)}) \simeq n\ .$$

Proof. We define f and g by recursion over computations in K. Most cases are straightforward – we consider only the case where K is applied. Suppose

$$\{a\}(\boldsymbol{m},K) \simeq K(\lambda p \cdot \{b\}(p,\boldsymbol{m},K), \lambda q \cdot \{c\}(q,\boldsymbol{m},K))\ .$$

By the induction hypothesis, $\forall p \cdot f(b,\langle p,\boldsymbol{m}\rangle) \in \mathcal{O}^S$ so by the boundedness lemma we can compute $d \in \mathcal{O}^S$ such that $\forall p \cdot |f(b,\langle p,\boldsymbol{m}\rangle)| < |d|$. Hence $\gamma = \lambda p \cdot \{b\}(p,\boldsymbol{m},K)$ is recursive in J^S_d with index (say) e, computable from $g(b)$. Then appropriate values for $f(a,\langle \boldsymbol{m}\rangle)$ and $g(a)$ can be computed using the functions of Lemma 1.9. □

Corollary 1.11. *For any $A \subseteq \omega$, if A is recursive in K, then for some $a \in \mathcal{O}^S$, A is recursive in J^S_a.*

Proof. Immediate from 1.10 and the boundedness lemma. □

§2. The ordinals $\leq |\mathcal{O}^S|$

In this section we shall characterize the ordinals $\omega_1^{J^S_a}$ (for $a \in \mathcal{O}^S$) and $|\mathcal{O}^S|$.

Definition 2.1. For any ordinals σ and τ:

(a) σ is *0-recursively inaccessible* iff $\sigma > \omega$ and σ is admissible;

(b) σ is $\tau+1$-recursively inaccessible iff σ is τ-recursively inaccessible and a limit of τ-recursively inaccessibles.

(c) if τ is a limit ordinal, σ is *τ-recursively inaccessible* iff σ is ρ-recursively inaccessible for all $\rho < \tau$.

Our aim is to prove:

Theorem 2.2. (a) *For all $a \in \bar{O}^S$, $\omega_1^{J_a^S}$ is the least σ which is $|a|$-recursively inaccessible*;

(b) $|\bar{O}^S|$ *is the least σ such that σ is σ-recursively inaccessible.*

This follows easily from the following two results:

(1) For any jump J and any $a \in \bar{O}^S$, if J_a^S is recursive in J, then ω_1^J is $|a|$-recursively inaccessible;

(2) for any ordinal σ and any $a \in \bar{O}^S$ if σ is $|a|$-recursively inaccessible, then $\omega_1^{J_a^S} \leq \sigma$.

Proof of Theorem 2.2. By (1) $\omega_1^{J_a^S}$ is $|a|$-recursively inaccessible and by (2) it must be the least such. By (2) of § 1, J_a^S is recursive in K for all $a \in \bar{O}^S$ so by (1), $\omega_1^K = |\bar{O}^S|$ is $|a|$-recursively inaccessible for all $a \in \bar{O}^S$. As $|\bar{O}^S|$ is a limit ordinal, $|\bar{O}^S|$ is thus $|\bar{O}^S|$-recursively inaccessible. If $\sigma < |\bar{O}^S|$, $\sigma = |a|$ for some $a \in \bar{O}^S$. Since $|a| < \omega_1^{J_a^S}$, it follows from (2) that σ is not σ-recursively inaccessible. □

We now turn to the proof of (1). For any jump J, let

$$\Omega(J) = \{\omega_1^F : F \text{ is type 2 and } J \text{ is recursive in } F\}.$$

Lemma 2.3. *For any jump J and ordinal σ, if $\sigma \in \Omega(S(J))$, then $\sigma \in \Omega(J)$ and σ is a limit of ordinals in $\Omega(J)$.*

Proof. Let $\sigma = \omega_1^F$ with $S(J)$ recursive in F. By 1.4(a), J is recursive in F so $\sigma \in \Omega(J)$. For any $\tau < \sigma$ there exists a set $A \subseteq \omega$ such that A is recursive in F and $\tau < \omega_1^A$ (for example, A codes a well-ordering of ω in type τ). Then

$$\tau < \omega_1^A \leq \omega_1^{J,A} < \omega_1^{S(J),A} \leq \omega_1^F = \sigma .$$

But clearly $\omega_1^{J,A} \in \Omega(J)$. Hence σ is a limit of ordinals in $\Omega(J)$. □

Theorem 2.4. *For any $a \in \mathcal{O}^S$ and any ordinal σ, if $\sigma \in \Omega(J_a^S)$, then σ is $|a|$-recursively inaccessible.*

Proof. We proceed by induction on $|a|$. If $|a| = 0$, then $J_a^S = oJ$ and it is well-known that any $\sigma \in \Omega(oJ)$ is admissible and greater than ω, hence 0-recursively inaccessible. If $a = 2^b$, $\Omega(J_a^S) = \Omega(S(J_b^S))$. By the induction hypothesis, elements of $\Omega(J_b^S)$ are $|b|$-recursively inaccessible so any $\sigma \in \Omega(J_a^S)$ is $|a|$-recursively inaccessible by Lemma 2.3. If $a = 3^b \cdot 5^e$ and $\phi(n) = \{e\}(n, J_b^S)$, then any $\sigma \in \Omega(J_a^S)$ is clearly also in $\Omega(J_{\phi(n)}^S)$ for each n and thus by the induction hypothesis is $|\phi(n)|$-recursively inaccessible. Since $|a| = \sup\{|\phi(n)| : n \in \omega\}$ and is a limit ordinal, σ is $|a|$-recursively inaccessible. □

This establishes result (1) and we now turn to (2). Our proof will require some detailed information about recursion on ordinals. As there is at present no good reference for this material, we digress here to state the facts that we need. First we note that the class of primitive ordinal recursive functions has closure properties similar to the class of ordinary primitive recursive functions; in particular, if g is primitive ordinal recursive and

$$f(\rho, \boldsymbol{\mu}) \simeq \sup_{\pi<\rho} g(\pi, \boldsymbol{\mu}),$$

then so is f. With each admissible ordinal κ and each $a < \kappa$ is associated a k-ary partial function $\{a\}_\kappa$ on κ for a natural number k "decodable" from a. A function f on κ is κ-partial recursive just in case $f = \{a\}_\kappa$ for some $a < \kappa$. By allowing indices from κ (not just ω), we include all constant functions with values $< \kappa$ among the κ-recursive functions.

(A) (Uniform Normal Form Theorem) *For each $k > 0$ there exists a primitive ordinal recursive relation T_k such that for any κ*

(i) *if κ is admissible, then for any a, $\boldsymbol{\mu} = (\mu_0, \ldots, \mu_{k-1})$, and $\nu < \kappa$,*

$$\{a\}_\kappa(\boldsymbol{\mu}) \simeq \nu \leftrightarrow \exists v_{v<\kappa}[T_k(a, \boldsymbol{\mu}, v) \wedge (v)_0 = \nu];$$

(ii) *κ is admissible iff for any a and $\mu < \kappa$,*

$$\forall \pi_{\pi<\mu} \exists v_{v<\kappa} T_1(a, \pi, v) \to \exists \rho_{\rho<\kappa} \forall \pi_{\pi<\mu} \exists v_{v<\rho} T_1(a, \pi, v).$$

(B) (Uniform Iteration Theorem). *For each $i \in \omega$ there is a primitive ordinal recursive function* $\mathbf{Sb}_i$ *such that for any admissible κ and any a, $\boldsymbol{\mu} < \kappa$*

$$\{\mathbf{Sb}_i(a, \mu_0, \ldots, \mu_i)\}_\kappa(\mu_{i+1}, \ldots, \mu_{k-1}) \simeq \{a\}_\kappa(\boldsymbol{\mu}) .$$

(C) (Recursion Theorem). *For any admissible κ and any κ-partial recursive function f, there exists an $\bar{e}$ such that*

$$\{\bar{e}\}_\kappa(\boldsymbol{\mu}) \simeq f(\bar{e}, \boldsymbol{\mu}) .$$

As corollaries of (A) we have:

(D) $\{\kappa : \kappa$ is admissible$\}$ is primitive ordinal recursive.

(E) For any admissible κ and λ and any $a, \boldsymbol{\mu}, \nu < \lambda$,

$$\lambda < \kappa \wedge \{a\}_\lambda(\boldsymbol{\mu}) \simeq \nu \rightarrow \{a\}_\kappa(\boldsymbol{\mu}) \simeq \nu .$$

(F) For any 1-recursively inaccessible κ and any a, $\boldsymbol{\mu} < \kappa$,

(i) $\{a\}_\kappa(\boldsymbol{\mu})\downarrow \leftrightarrow \exists\lambda_{\lambda<\kappa}$ [λ is admissible $\wedge \{a\}_\lambda(\boldsymbol{\mu})\downarrow$];

(ii) $\forall p_{p<\omega} \{a\}_\kappa(p)\downarrow \leftrightarrow \exists\lambda_{\lambda<\kappa}$ [λ is admissible $\wedge \forall p_{p<\omega} \{a\}_\lambda(p)\downarrow$].

Lemma 2.5. *For any admissible $\kappa > \omega$, any κ-recursive well-ordering of ω has order-type $< \kappa$.*

Proof. Let R be a κ-recursive well-ordering of ω of type λ. By the Recursion Theorem (C) there exists a κ-partial recursive function f such that for all $m \in \omega$,

$$f(m) \simeq \text{least } \sigma_{\sigma<\kappa}[\forall n_{n<\omega}(R(n,m) \rightarrow f(n) < \sigma)] .$$

Then $\lambda = \sup_{m<\omega} f(m) + 1$ so $\lambda < \kappa$ by the admissibility of κ. □

The next definition is the key to our proof of (2):

Definition 2.6. For any jump J, any $\kappa > \omega$, and any $e < \kappa$, J is *κ-effective* *with index e* iff κ is admissible and for any $a < \kappa$ such that $\{a\}_\kappa \restriction \omega \in {}^\omega\omega$,

$$J(\{a\}_\kappa \restriction \omega) = \lambda m_{m<\omega} \{e\}_\kappa(a,m) = \{\mathbf{Sb}_1(e,a)\}_\kappa \restriction \omega .$$

J is κ*-effective* iff J is κ-effective with some index. $\mathbf{Ef}_{J,e}$ is the set of all κ such that J is κ-effective with index e.

Now (2) follows immediately from:

Theorem 2.7. *For any jump J and any κ,*
(a) *if J is κ-effective, then $\omega_1^J \leq \kappa$;*
(b) *If $a \in \bar{O}^S$ and κ is $|a|$-recursively inaccessible, then J_a^S is κ-effective.*

This, in turn, will follow from some lemmas.

Lemma 2.8. *There exists a primitive recursive g such that for any κ and e such that J is κ-effective with index e, any a, any c such that $\{c\}_\kappa \restriction \omega \in {}^\omega\omega$ and any $\boldsymbol{m} = m_0, \ldots, m_{k-1} < \omega$*

$$\{a\}(\boldsymbol{m}, \{c\}_\kappa \restriction \omega, J) \simeq \{g(a,e)\}_\kappa(\boldsymbol{m}, c) .$$

Proof. The proof is very similar to that of Lemma 1.4(b). g is defined as there except for two cases:
(i) if $\{a\}(\boldsymbol{m}, \alpha, J) \simeq \alpha(m_i)$, then $g(a,e)$ is chosen so that

$$\{g(a,e)\}_\kappa(\boldsymbol{m}, c) \simeq \{c\}_\kappa(m_i) ;$$

(ii) if $\{a\}(\boldsymbol{m}, \alpha, J) \simeq J(\lambda p \cdot \{b\}(p, \boldsymbol{m}, \alpha, J))(q)$, we choose $g(a,e)$ to be the "natural" index such that

$$\{g(a,e)\}_\kappa(\boldsymbol{m}, c) \simeq 0 \cdot \sup_{p<\omega} \{g(b,e)\}_\kappa(p, \boldsymbol{m}, c) + \{e\}_\kappa(g'(b,e,\boldsymbol{m},c), q) ,$$

where

$$\{g'(b,e,\boldsymbol{m},c)\}_\kappa(p) \simeq \{g(b,e)\}(p, \boldsymbol{m}, c) .$$

The proof that g is as required is essentially identical to that of Lemma 1.4(b). □

Part (a) of Theorem 2.7 follows immediately from Lemmas 2.5 and 2.8. Part (b) will follow from the next three lemmas, which are the induction steps in the definition of a recursive function f such that if $a \in \mathcal{O}^S$ and κ if $|a|$-recursively inaccessible then J_a^S is κ-effective with index $f(a)$.

Lemma 2.9. *There exists a $c_0 < \omega$ such that oJ is κ-effective with index c_0 for every admissible $\kappa > \omega$.*

Proof. Let R be a primitive recursive relation such that

$$\{m\}(\alpha)\downarrow \leftrightarrow \exists p_{p<\omega} R(\bar{\alpha}(p), m) .$$

For any admissible κ, let

$$f_\kappa(a,m) = (0, \text{ if } \exists p_{p<\omega} R(\overline{\{a\}_\kappa}(p), m) ; \quad 1, \text{ otherwise}) .$$

Since the definition of f_κ is uniform with respect to κ, there exists an index c_0 such that $f_\kappa = \{c_0\}_\kappa$ for every admissible κ. Then oJ is κ-effective with index c_0. □

Lemma 2.10. *There exists an primitive ordinal recursive f_1 such that for any $\kappa > \omega$, any jump J, and any e, if $\kappa \in \mathbf{Ef}_{J,e}$ and κ is a limit of ordinals $\lambda \in \mathbf{Ef}_{J,e}$, then $\kappa \in \mathbf{Ef}_{S(J),f_1(e)}$. If $e < \omega$, then also $f_1(e) < \omega$.*

Proof. Let

$$h_\kappa(e,a) \simeq \text{least } \lambda_{\lambda<\kappa} [\lambda \in \mathbf{Ef}_{J,e} \wedge \{\dot{a}\}_\lambda \restriction \omega \in {}^\omega\omega] ,$$

and

$$f_\kappa(e,a,m) \simeq \begin{cases} 0, \text{ if } \exists v_{v<h_\kappa(e,a)} T_1(g(m,e),a,v) ; \\ 1, \text{ otherwise.} \end{cases}$$

We claim first that for κ as in the hypothesis of the lemma, if $\{a\}_\kappa \restriction \omega \in {}^\omega\omega$, then for all $m \in \omega$,

$$S(J)(\{a\}_\kappa \restriction \omega)(m) = f_\kappa(e,a,m) .$$

Since κ is 1-recursively inaccessible, it follows from (F)(ii) that $h_\kappa(e,a)\downarrow$ so that $f_\kappa(e,a,m)$ is also defined with value 0 or 1. Then

$$\begin{aligned} S(J)(\{a\}_\kappa \upharpoonright \omega)(m) = 0 &\leftrightarrow \{m\}(\{a\}_\kappa \upharpoonright \omega, J)\downarrow \\ &\leftrightarrow \{m\}(\{a\}_{h_\kappa(e,a)} \upharpoonright \omega, J)\downarrow \\ &\leftrightarrow \{g(m,e)\}_{h_\kappa(e,a)}(a)\downarrow \\ &\leftrightarrow f_\kappa(e,a,m) = 0 \,. \end{aligned}$$

The second equivalence uses (E) and the third Lemma 2.8.

To complete the proof, it suffices to show that f_κ is uniformly κ-partial recursive, that is, that for some fixed index c_1, $f_\kappa = \{c_1\}_\kappa$ for all κ satisfying the hypothesis. Given this, $f_1(e) = \mathbf{Sb}_1(c_1, e)$ is the required function. Let

$$\begin{aligned} \mathbf{Ef}_e = \{\lambda : \lambda \text{ is admissible} \wedge \omega, e < \lambda \\ \wedge \forall a_{a<\lambda} [\{a\}_\lambda \upharpoonright \omega \in {}^\omega\omega \rightarrow \{\mathbf{Sb}_1(e,a)\}_\lambda \upharpoonright \omega \in {}^\omega\omega]\} \,. \end{aligned}$$

It follows from (A), (D), and closure under bounded quantification that $\mathbf{Ef}_e$ is primitive ordinal recursive. We claim that for $\kappa \in \mathbf{Ef}_{J,e}$,

$$\forall\lambda_{\lambda<\kappa} [\lambda \in \mathbf{Ef}_{J,e} \leftrightarrow \lambda \in \mathbf{Ef}_e] \,.$$

The implication $(\rightarrow)$ is immediate from the definition of $\mathbf{Ef}_{J,e}$. Suppose $\lambda \in \mathbf{Ef}_e$ and $\{a\}_\lambda \upharpoonright \omega \in {}^\omega\omega$. Then by (E), $\{a\}_\lambda \upharpoonright \omega = \{a\}_\kappa \upharpoonright \omega$ so that

$$J(\{a\}_\lambda \upharpoonright \omega) = J(\{a\}_\kappa \upharpoonright \omega) = \{\mathbf{Sb}_1(e,a)\}_\kappa \upharpoonright \omega = \{\mathbf{Sb}_1(e,a)\}_\lambda \upharpoonright \omega \,.$$

Now in the definition of h_κ we may replace $\mathbf{Ef}_{J,e}$ by $\mathbf{Ef}_e$ without altering the value when $\kappa \in \mathbf{Ef}_{J,e}$. Thus h_κ and f_κ are uniformly κ-partial recursive for κ satisfying the hypothesis of the lemma. □

Lemma 2.11. *There exists a primitive ordinal recursive f_2 such that for any $\kappa > \omega$, any b, and any indexed set of jumps $\{J_n : n \in \omega\}$ such that for all n, J_n is κ-effective with index $\{b\}_\kappa(n)$, if $J = \lambda\alpha m \cdot J_{(m)_0}(\alpha)((m)_1)$, then J is κ-effective with index $f_2(b)$. If $b < \omega$, then also $f_2(b) < \omega$.*

Proof. If $\{a\}_\kappa \upharpoonright \omega \in {}^\omega\omega$, then

$$\begin{aligned} J(\{a\}_\kappa \upharpoonright \omega)(m) &= J_{(m)_0}(\{a\}_\kappa \upharpoonright \omega)((m)_1) \\ &= \{\{b\}_\kappa((m)_0)\}_\kappa(a,(m)_1) \\ &= \{c_2\}_\kappa(b,a,m) \end{aligned}$$

for an appropriate c_2. Thus it suffices to set $f_2(b) = \mathbf{Sb}_1(c_2, b)$. □

Proof of Theorem 2.7. (a) follows immediately from Lemmas 2.5 and 2.7. For (b), let $\bar{f}_i = f_i \upharpoonright \omega$. Then $\bar{f}_1$ and $\bar{f}_2$ are (ordinary) primitive recursive. By the (ordinary) recursion theorem there exists a primitive recursive function f with index $\hat{f}$ such that:

$$\begin{aligned} &f(1) = c_0\ ; \\ &f(2^a) = \bar{f}_1(f(a))\ ; \\ &f(3^a \cdot 5^e) = \bar{f}_2(h(\hat{f}, a, e))\ , \end{aligned}$$

where h is a primitive recursive function such that for any admissible κ

$$\{h(\hat{f}, a, e)\}_\kappa(n) \simeq f(\{g(f(a), e)\}_\kappa(n))\ ,$$

with g as in Lemma 2.8. It is now straightforward to prove by induction on $\mathcal{O}^S$ that for $a \in \mathcal{O}^S$, if κ is $|a|$-recursively inaccessible, then J_a^S is κ-effective with index $f(a)$. □

§3. S and the first recursively Mahlo ordinal

In this section we extend the ideas of §2 to obtain a bound for ω_1^S, the least ordinal not recursive in S.

Definition 3.1. (i) For any ordinal κ, κ is *recursively Mahlo* iff κ is admissible and for any κ-recursive function f from κ to κ, there is an admissible $\lambda < \kappa$ which is closed under f;

(ii) ρ_0 is the least recursively Mahlo ordinal.

It is easy to see that ρ_0 is ρ_0-recursively inaccessible and is not the first such ordinal. Hence $|\bar{O}^S| < \rho_0$. We shall show here that $\omega_1^S \leq \rho_0$. It has recently been shown by Leo Harrington that $\omega_1^S = \rho_0$; his proof appears in his contribution to this volume. That $\omega_1^S = \rho_0$ appeared as a theorem in [Ac], but its purported proof there depended on the fallacious results of [Pl].

Definition 3.2. For any $\kappa > \omega$ and any $e < \kappa$, S is *κ-effective with index e* iff κ is admissible and for any J and d such that J is κ-effective with index d, $S(J)$ is κ-effective with index $\mathbf{Sb}_1(e,d)$. S is *κ-effective* iff S is κ-effective with some index.

The next lemma is analogous to a weak form of Lemma 2.8.

Lemma 3.3. *There exists a primitive ordinal recursive function g such that for any κ and e such that S is κ-effective with index e, any $a < \omega$ and $\boldsymbol{m} = m_0, \ldots, m_{k-1} < \omega$, any $\boldsymbol{c} = c_0, \ldots, c_{l-1} < \kappa$ and $\boldsymbol{\alpha} = \alpha_0, \ldots, \alpha_{l-1} \in {}^{\omega}\omega$ such that $\alpha_i = \{c_i\}_\kappa \restriction \omega$ for $i < l$, if $\{a\}(\boldsymbol{m}, \boldsymbol{\alpha}, S) \simeq n$ then $\{g(a,e)\}_\kappa(\boldsymbol{m}, \boldsymbol{c}) \simeq n$.*

Proof. This is similar to that of Lemma 2.8 and we treat only the case

$$\{a\}(\boldsymbol{m}, \boldsymbol{\alpha}, S) \simeq S(J)(\beta)(p)\,,$$

where $J = \lambda\gamma q \cdot \{b_0\}(q, \boldsymbol{m}, \boldsymbol{\alpha}, \gamma, S)$ and $\beta = \lambda r \cdot \{b_1\}(r, \boldsymbol{m}, \boldsymbol{\alpha}, S)$, with b_0, b_1, and p coded into a. Given that $\{a\}(\boldsymbol{m}, \boldsymbol{\alpha}, S) \simeq n$, it follows that J and β are both total objects with all of their computations "preceding" the given one. By the induction hypothesis we can compute from a, $\boldsymbol{m}$ and $\boldsymbol{c}$ an index d, such that J is κ-effective with index d, and b_2 such that $\beta = \{b_2\}_\kappa \restriction \omega$. Then

$$S(J)(\beta)(p) \simeq \{\mathbf{Sb}_1(e,d)\}_\kappa(b_2, p) \simeq \{e\}_\kappa(d, b_2, p)\,.$$

Hence it suffices to choose $g(a,e)$ such that

$$\{g(a,e)\}_\kappa(\boldsymbol{m}, \boldsymbol{c}) \simeq \{e\}_\kappa(d, b_2, p)\,. \square$$

Corollary 3.4. *For any κ, if S is κ-effective, then $\omega_1^S \leq \kappa$.*

Proof. If S is κ-effective, then by the preceding lemma any well-ordering of ω recursive in S is also κ-recursive, hence of order-type $<\kappa$ by Lemma 2.5. □

Lemma 3.5. *For any J, e, and κ, if κ is recursively Mahlo and $\kappa \in \mathbf{Ef}_{J,e}$, then κ is also a limit of ordinals $\lambda \in \mathbf{Ef}_{J,e}$.*

Proof. Suppose κ is recursively Mahlo, $\kappa \in \mathbf{Ef}_{J,e}$ and let $\mathbf{Ef}_e$ be as in the proof of Lemma 2.10. Then for any admissible λ with $\omega, e < \lambda < \kappa$, $\lambda \in \mathbf{Ef}_{J,e} \leftrightarrow \lambda \in \mathbf{Ef}_e$. With the help of the uniform normal form theorem (A) we can define a primitive ordinal recursive relation R such that for such λ,

$$\lambda \in \mathbf{Ef}_{J,e} \leftrightarrow \forall a_{a<\lambda} [\exists \sigma_{\sigma<\lambda} R(a,\sigma) \rightarrow \exists \tau_{\tau<\lambda} R(\mathbf{Sb}_1(e,a),\tau)] .$$

Now let λ_0 be any ordinal $<\kappa$ and set

$$f(a,\sigma) \simeq \text{least } \rho_{\rho<\kappa} [R(a,\sigma) \rightarrow \exists \tau_{\tau<\rho} R(\mathbf{Sb}_1(e,a),\tau) \wedge \omega, \lambda_0 < \rho] .$$

Clearly for any admissible λ,

$$\lambda \text{ is closed under } f \leftrightarrow \omega, \lambda_0 < \lambda \text{ and } \lambda \in \mathbf{Ef}_{J,e} .$$

Since κ is recursively Mahlo and f is κ-recursive, there is an admissible $\lambda < \kappa$ such that λ is closed under f and hence a $\lambda \in \mathbf{Ef}_{J,e}$ with $\lambda_0 < \lambda < \kappa$. □

Theorem 3.6. *For any κ, if κ is recursively Mahlo, then S is κ-effective.*

Proof. Suppose κ is recursively Mahlo and let e be an index such that

$$\{e\}_\kappa(d,a,m) \simeq \{f_1(d)\}(a,m) ,$$

where f_1 is the function of Lemma 2.10. We claim that S is κ-effective with index e. Let J be any jump which is κ-effective with index d : $\kappa \in \mathbf{Ef}_{J,d}$. By the preceding lemma, κ is also a limit of ordinals $\lambda \in \mathbf{Ef}_{J,d}$, so by Lemma 2.10 $S(J)$ is κ-effective with index $f_1(d)$. But then $S(J)$ is also κ-effective with index $\mathbf{Sb}_1(e,d)$ as required. □

Corollary 3.7. $\omega_1^S \leq \rho_0$.

For ordinals κ which are projectible to ω, the converse of Theorem 3.6 holds also:

Theorem 3.8. *For any κ which is projectible to ω, κ is recursively Mahlo iff S is κ-effective.*

Proof. Suppose S is κ-effective, say with index e, and let h be a $1-1$ κ-recursive function which projects κ into ω. We define a hierarchy of jumps J_σ for $\sigma < \kappa$ as follows:

$$J_0 = {}^oJ\ ;$$

$$J_{\sigma+1} = S(J_\sigma)\ ;$$

and for limit σ

$$J_\sigma(\alpha)(\langle m,n\rangle) = \begin{cases} J_{h^{-1}(m)}(\alpha)(n), & \text{if } m \in \text{image}\,(h \restriction \sigma); \\ 0, & \text{otherwise.} \end{cases}$$

We define, by use of the recursion theorem, a κ-partial recursive function g such that for all $\sigma < \kappa$, J_σ is κ-effective with index $g(\sigma)$:

$$g(0) = c_0\ ; \qquad (c_0 \text{ from Lemma 2.8})$$

$$g(\sigma+1) = \mathbf{Sb}_1(e, g(\sigma))\ ;$$

and for limit σ, $g(\sigma)$ is an index b such that

$$\{b\}_\kappa(a,\langle m,n\rangle) \simeq \begin{cases} \{g(h^{-1}(m))\}_\kappa(a,n), & \text{if } m \in \text{image}\,(h \restriction \sigma); \\ 0, & \text{otherwise.} \end{cases}$$

Now let f be any κ-recursive function. We aim to find an admissible $\lambda < \kappa$ such that λ is closed under f. Let T be the jump defined by

$$T(\gamma,\alpha)(m) = \begin{cases} J_{f(\|\gamma\|)+1}(\alpha)(m), & \text{if } \gamma \in \mathcal{W} \text{ and } \|\gamma\| < \kappa\ ; \\ 0, & \text{otherwise.} \end{cases}$$

Using the function g and the fact that $\mathcal{W}$ is recursive in S, it is easy to see that T is κ-effective. Hence $S(T)$ is also κ-effective so

$$\lambda = \omega_1^T < \omega_1^{S(T)} \leq \kappa .$$

Then λ is admissible and it remains to show that λ is closed under f. If $\sigma < \lambda$, then $\sigma = \|\gamma\|$ for some $\gamma \in \mathcal{W}$ and recursive in T. Hence $J_{f(\sigma)+1} = \lambda\alpha \cdot T(\gamma, \alpha)$ is recursive in T and $\omega_T^{J_{f(\sigma)+1}} < \lambda$. Note that for any $\sigma < \tau < \lambda$, $S(J_\sigma)$ is recursive in J_τ so $\omega_1^{J_\sigma} < \omega_1^{J_\tau}$. Hence $f(\sigma) \leq \omega_1^{J_{f(\sigma)}} < \omega_1^{J_{f(\sigma)+1}} < \lambda$. □

Corollary 3.9. *ρ_0 is the least κ such that S is κ-effective.*

§4. A countable hierarchy for 1–SC (S)

We next define an extension of Platek's hierarchy and show that a set of natural numbers is recursive in S iff it appears in our hierarchy. To guide the reader through the inevitable forest of indices and recursions we sketch first the main points of argument. We first define a set of notations $\bar{O}$, and for each $u \in \bar{O}$ a jump J_u, by adjoining the method of [Ri] to Platek's hierarchy – that is, whenever Platek's inductive definition grinds to a halt but the set of notations is closed under some new primitive recursive function $\{d\}$, we add 7^d to the set of notations, collect all previous J_u's to form J_{7^d}, and go on. With relatively minor alterations to the proofs of [Sh] and [Pl] we establish uniqueness and boundedness lemmas.

We aim to show that computations relative to S can be done relative to the various J_u ($u \in \bar{O}$) as in Theorem 1.10. The difficult point, which failed for Platek's hierarchy, is to show that from such a method for computing F we can find one for $S(F)$. Here we accomplish this roughly as follows (the details differ somewhat from this sketch). From the instructions for F we can compute a primitive recursive $\xi : \bar{O} \to \bar{O}$ such that for any $u \in \bar{O}$ and any α recursive in J_u we can compute the value $F(\alpha)$ from $J_{\xi(u)}$. If d is a primitive recursive index for ξ, then $7^d \in \bar{O}$. A computation similar to that of the proof of 1.10 shows that computations from F can be done relative to J_{7^d} and hence that the "diagonal set" $\{a : \{a\}(F)\downarrow\}$ is recursive in $S(J_{7^d})$. Then to compute $S(F)(\alpha)$ for any α computable from the J_u's we apply this procedure to a recursive join of F and α.

Definition 4.1. We define by ordinal recursion a sequence of sets $\mathcal{O}^\sigma$ and, for $u \in \mathcal{O}$, an ordinal $|u|$ and a jump J_u as follows:

(a) Let $\mathcal{O}^{(\sigma)} = \bigcup\{\mathcal{O}^\tau : \tau < \sigma\}$ and for $u \in \mathcal{O}^{(\sigma)}$ let $[a,u]$ denote the function $\lambda m \cdot \{a\}(m, J_u)$.

(b) Let $\mathcal{O}_1^\sigma$ be the smallest set with the following properties:
  (i) $\mathcal{O}^{(\sigma)} \subseteq \mathcal{O}_1^\sigma$;
  (ii) $1 \in \mathcal{O}_1^\sigma$;
  (iii) $\forall v[v \in \mathcal{O}^{(\sigma)} \to 2^v \in \mathcal{O}_1^\sigma]$;
  (iv) for any $v \in \mathcal{O}^{(\sigma)}$ and any b, if $[b,v]$ is a total function γ such that $\gamma(0) = v$ and for all m, $\gamma(m) \in \mathcal{O}^{(\sigma)}$ and $|\gamma(m)| < |\gamma(m+1)|$, then $3^v \cdot 5^b \in \mathcal{O}_1^\sigma$.

(c) Let $\mathcal{O}_2^\sigma$ be the smallest set with the following properties:
  (i) $\mathcal{O}^{(\sigma)} \subseteq \mathcal{O}_2^\sigma$;
  (ii) for any primitive recursive index d, if $\forall v[v \in \mathcal{O}^{(\sigma)} \to \{d\}(v) \in \mathcal{O}^{(\sigma)}]$ and $\forall v \forall v'[|v| \leq |v'| < \sigma \to |\{d\}(v)| \leq |\{d\}(v')| < \sigma]$ then $7^d \in \mathcal{O}_2^\sigma$.

Then
$$\mathcal{O}^\sigma = \begin{cases} \mathcal{O}_1^\sigma, & \text{if } \mathcal{O}^{(\sigma)} \neq \mathcal{O}_1^\sigma; \\ \mathcal{O}_2^\sigma, & \text{otherwise.} \end{cases}$$

For $u \in \mathcal{O}^\sigma \sim \mathcal{O}^{(\sigma)}$, $|u| = \sigma$ and we define J_u as follows:

$$u = 1 \;:\; J_u = {}^oJ;$$

$$u = 2^v : J_u = S(J_v);$$

$$u = 3^v \cdot 5^b : J_u(\alpha)(\langle m,n\rangle) = J_{[b,v](m)}(\alpha)(n);$$

$$u = 7^d : J_u(\alpha)(\langle m,n\rangle) = \begin{cases} J_m(\alpha)(n), & \text{if } m \in \mathcal{O}^{(\sigma)}; \\ 0, & \text{otherwise.} \end{cases}$$

(d) Let $|\mathcal{O}| = \text{least } \sigma[\mathcal{O}^{(\sigma)} = \mathcal{O}^\sigma]$ and $\mathcal{O} = \mathcal{O}^{|\mathcal{O}|}$. For $u \notin \mathcal{O}$, $|u| = |\mathcal{O}|$.

Lemma 4.2. *For any a, b, d, u, and v:*
  (a) $|u| < |3^v \cdot 5^b| \to |2^u| < |3^v \cdot 5^b|$;
  (b) $|u| < |7^d| \to |2^u| < |7^d|$;
  (c) $|u| < |7^d|$ *and* $3^u \cdot 5^a \in \mathcal{O}$ *and* $\forall m(|[a,u](m)| < |7^d|)$ $\to |3^u \cdot 5^a| < |7^d|$;

(d) *d is a primitive recursive index and* $\forall v[v \in \bar{O} \to \{d\}(v) \in \bar{O}]$ *and* $\forall v \forall v'[|v| \leq |v'| < |\bar{O}| \to |\{d\}(v)| \leq |\{d\}(v')| < |\bar{O}|] \to 7^d \in \bar{O}$;

(e) $7^d \in \bar{O}$ *and* $|v| < |7^d| \to |\{d\}(v)| < |7^d|$;

(f) *d is a primitive recursive index and* $|7^e| < |7^d| \to \exists v(|v| < |7^e| \leq |\{d\}(v)|)$.

Proof. Most are immediate from the definition and are included to facilitate understanding the nature of $\bar{O}$. (c) depends on the fact that if $|7^d| = \sigma$, then $\bar{O}^{(\sigma)} = \bar{O}_1^\sigma$. For (f), if there were no such v, then for $\sigma = |7^e|$, $7^d \in \bar{O}_2^\sigma = \bar{O}^\sigma$ so $|7^d| \leq |7^e|$. □

Lemma 4.3 (Uniqueness). *There exist partial recursive f and g such that for all $v \in \bar{O}$:*

(a) $[f(v), v]$ *is the characteristic functions of* $\{u : |u| < |v|\}$;

(b) *for any u, if $|u| \leq |v|$, then J_u is recursive in J_v with index $g(u, v)$.*

Proof. The functions f and g are defined by the recursion theorem over $\bar{O}$ much as in [Sh, p. 104–106]. Our equivalent of [Sh, (1)–(3) p. 104] is Lemma 1.4 and the fact that for all $v \in \bar{O}, oJ$ is recursive in J_v with index computable from v. Thus all cases not involving a notation of the form 7^d may be treated as they are there and we consider here only the cases that are new.

(1) $u = 7^d$ and $v = 2^w$:

(a) $|u| < |v| \leftrightarrow |u| < |w|$ & $\exists e(w = 7^e$ *and* $|u| \leq |w|)$. The first clause can be checked from J_w by the induction hypothesis. By 4.2(f) the second holds just in case

$$\forall w'(|w'| < |w| \to |\{d\}(w')| < |w|)$$

which again can be checked from J_w.

(b) $|u| \leq |v| \to |u| \leq |w|$ so J_u is recursive in J_w with index $g(u, w)$, hence in J_v with index computable from 1.4(a).

(2) $u = 7^d$ and $v = 3^w \cdot 5^a$: exactly as in [Sh].

(3) $v = 7^d$:

(a) $|u| < |v| \leftrightarrow \exists n \cdot J_v(\lambda p \cdot 0)(\langle u, n \rangle) \neq 0$.

(b) $|u| \leq |v| \to |u| < |v|$ or $\exists e\,(u = 7^e$ *and* $|u| = |v|)$. Given $|u| \leq |v|$ we can, by (a), decide which is the case. If $|u| < |v|$, then $J_u(\alpha)(n) = J_v(\alpha)(\langle u, n \rangle)$. If $|u| = |v|$, $J_u = J_v$. □

Corollary 4.4. *For any* $u \in \mathcal{O}$, $\omega_1^{J_u} > |u|$.

We shall now prove that for all $u \in \mathcal{O}$, J_u is recursive in S. In earlier instances of this type of hierarchy (e.g. [Sh]) the analogous result was nearly trivial. This is not quite so here and in fact the proof eluded the authors until Leo Harrington provided the key idea of Theorem 4.6 below.

Lemma 4.5. *There exists a functional F recursive in S such that for all u, m, α, and γ,*

$$F(u,m,\alpha,\gamma) = \begin{cases} J_u(\alpha)(m) + 1, & \text{if } \gamma \in \mathcal{W} \wedge u \in \mathcal{O}^{\|\gamma\|}; \\ 0, & \text{otherwise.} \end{cases}$$

Proof. The definition of F is similar to that of the corresponding functional F in the proof of Theorem 1.3. We shall describe intuitively how F is computed and leave it to the reader to define the appropriate auxiliary functional G and apply the recursion theorem. The proof that F is as required will be by induction on $\|\gamma\|$ for $\gamma \in \mathcal{W}$, so to compute $F(u,m,\alpha,\gamma)$ we may assume $\gamma \in \mathcal{W}$ and that for all $p \in \mathrm{Fld}(\gamma)$, $F(u,m,\alpha,\gamma \restriction p)$ is the correct value. Hence we may decide recursively in S, whether or not a given ν belongs to $\mathcal{O}^{(\|\gamma\|)}$ by computing whether or not $\exists p[F(\nu,m,\alpha,\gamma \restriction p) > 0]$, and if the answer is yes we may compute the unique p_ν such that $|\nu| = \|\gamma \restriction p_\nu\|$.

Now we proceed by cases on u.

$u = 1$: $\quad F(u,m,\alpha,\gamma) = oJ(\alpha)(m)$;

$u = 2^\nu$:

$$F(u,m,\alpha,\gamma) = \begin{cases} S(\lambda\beta n[F(\nu,n,\beta,\gamma \restriction p_\nu) - 1])(\alpha)(m) + 1, & \text{if } p_\nu \text{ exists;} \\ 0, & \text{otherwise.} \end{cases}$$

$u = 3^\nu \cdot 5^b$: if the following two conditions are satisfied:

(i) p_ν exists;

(ii) if $f(m) \simeq \{b\}(m, \lambda\beta n[F(\nu,n,\beta,\gamma \restriction p_\nu) - 1])$, then for all m, $p_{f(m)}$ exists and $p_{f(m)} <_\gamma p_{f(m+1)}$ then

$$F(u,\langle m,n\rangle,\alpha,\gamma) \simeq F(f(m),n,\alpha,\gamma \restriction p_{f(m)}) ;$$

otherwise,

$$F(u,\langle m,n\rangle,\alpha,\gamma) = 0\,.$$

$u = 7^d$: if p_u exists, $F(u,\langle m,n\rangle,\alpha,\gamma) = F(u,\langle m,n\rangle,\alpha,\gamma\restriction p_u)$; otherwise, if the following three conditions are satisfied:

(i) d is a primitive recursive index;

(ii) $\forall v[p_v \text{ exists} \to p_{\{d\}(v)} \text{ exists}]$;

(iii) $\forall v \forall v'[p_v \text{ and } p_{v'} \text{ exists} \wedge p_v \leq_\gamma p_{v'} \to p_{\{d\}(v)} \leq_\gamma p_{\{d\}(v')}]$;

then

$$F(u,\langle m,n\rangle,\alpha,\gamma) = \begin{cases} F(m,n,\alpha,\gamma\restriction p_m), & \text{if } p_m \text{ exists;} \\ 0, & \text{otherwise;} \end{cases}$$

otherwise, $F(u,\langle m,n\rangle,\alpha,n) = 0$.

u is not of one of these forms: $F(u,m,\alpha,\gamma) = 0$. □

Theorem 4.6. *There exists a functional G partial recursive in S such that for any u, m, and α,*

$$u \in \mathcal{O} \to G(u,m,\alpha) \simeq J_u(\alpha)(m)\,.$$

Proof. We define G by recursion over $\mathcal{O}$. The clauses for $u = 1, 2^v$, or $3^v \cdot 5^b$ are routine (in the sense that they differ little from the corresponding clauses in [Sh] or [Pl]) and we omit them. If $u = 7^d$, let H be defined by:

$$H(v,m,\alpha,\gamma) \simeq \begin{cases} G(\{d\}(v),m,\alpha), & \text{if } \gamma \in \mathcal{W} \wedge v \in \mathcal{O}^{(\|\gamma\|)} u \notin \mathcal{O}^{(\|\gamma\|)}; \\ 0, & \text{otherwise.} \end{cases}$$

By the preceding lemma H is partial recursive in S. Furthermore, the conditions of the first clause imply $|v| < |u| = |7^d|$ and hence $|\{d\}(v)| < |u|$, so that by the induction hypothesis for this stage, $G(\{d\}(v),m,\alpha)$ is defined for all m and α. Thus H is a total functional. Let I be a jump recursive in S such that both F and H are recursive in I (F being as in Lemma 4.5).

We claim that $\omega_1^I \geq |u|$. If not, then as ω_1^I is a limit ordinal one of the following must hold:

(i) $\omega_1^I = |3^\nu \cdot 5^b| < |u|$: then $|\nu| < \omega_1^I$ so that there exists $\gamma \in \mathcal{W}$, γ recursive in I, such that $|\nu| \leq \|\gamma\|$. Then

$$J_\nu(\alpha)(m) = F(\nu, m, \alpha, \gamma) .$$

So J_ν is recursive in I. Similarly, for all m, $J_{\{b,\nu\}(m)}$ is recursive in I with index computable from m. Hence $J_{3^\nu \cdot 5^b}$ is recursive in I so by Corollary 4.4, $|3^\nu \cdot 5^b| < \omega_1^I$, contrary to assumption.

(ii) $\omega_1^I = |7^e| < |u|$: then by Lemma 4.2(f) there exists a ν such that $|\nu| < \omega_1^I \leq |\{d\}(\nu)|$. Choose $\gamma \in \mathcal{W}$, γ recursive in I, such that $|\nu| < \|\gamma\|$. Then

$$J_{\{d\}(\nu)}(\alpha)(m) = H(\nu, m, \alpha, \gamma) ,$$

so $J_{\{d\}(\nu)}$ is recursive in I, which again leads to a contradiction by use of Corollary 4.4.

Hence, $\omega_1^I \geq |u|$.

There exists a fixed index $\bar{a}$ such that for any jump J, $\lambda p \cdot \{\bar{a}\}(p, S(J))$ codes a well-ordering of type ω_1^J. Hence it suffices to set

$$G(7^d, m, \alpha) = F(7^d, m, \alpha, \lambda p \cdot \{\bar{a}\}(p, S(I))) . \quad \square$$

Corollary 4.7. (a) *For all* $u \in \mathcal{O}$, J_u *is recursive in* S;
(b) $|\mathcal{O}| \leq \omega_1^S$.

We now turn to proving that every set of natural numbers recursive in S is recursive in some J_u ($u \in \mathcal{O}$). To this end we need a boundedness lemma for $\mathcal{O}$. As in [Sh] we shall derive this through the use of a partial recursive "addition" function $\oplus$ on $\mathcal{O}$. Apparently it is not possible to define such a function with the property $|u \oplus \nu| = |u| + |\nu|$. Fortunately, some weaker properties suffice.

Definition 4.8. An ordinal σ is *superadmissible* iff $\sigma = |\mathcal{O}|$, or for some $7^d \in \mathcal{O}$, $\sigma = |7^d|$.

Lemma 4.9. *There exists a partial recursive function* $\oplus$ *such that for any* $u, \nu \in \mathcal{O}$:

(a) *$u \oplus v$ is defined and* $u \oplus v \in \bar{O}$;
(b) *for any superadmissible* σ, *if* $|u|, |v| < \sigma$, *then* $|u \oplus v| < \sigma$;
(c) $|u \oplus v| \geq \max\{|u|, |v|\}$;
(d) $v \neq 1 \rightarrow |u| < |u \oplus v|$;
(e) $|u'| \leq |u| \rightarrow |u' \oplus v| \leq |u \oplus v|$.

Proof. We define $\oplus$ by recursion on $v \in \bar{O}$ (much as in [Sh]) as follows:
(1) $u \oplus 1 = u$;
(2) $u \oplus 2^v = 2^{u \oplus v}$;
(3) $u \oplus 3^v \cdot 5^b = 3^{u \oplus v} \cdot 5^c$, where $[c, u \oplus v](m) = u \oplus [b, v](m)$. Such a c exists and can be easily computed by the uniqueness lemma and the induction hypothesis that $|v| \leq |u \oplus v|$;
(4) $u \oplus 7^d = 3^{7^d} \cdot 5^c$, where, if $\bar{0} = 1$ and $\overline{m+1} = 2^{\bar{m}}$,

$$[c, 7^d](m) = \begin{cases} 7^d \oplus \bar{m}, & \text{if } |u| < |7^d|\,; \\ 7^d, & \text{if } |u| \geq |7^d| \text{ and } m = 0\,; \\ u \oplus \bar{m}, & \text{if } |u| \geq |7^d| \text{ and } m > 0\,. \end{cases}$$

By (a) of the uniqueness lemma, the cases are recursive in J_{7^d} so such a c exists.

We prove (a)–(e) by induction on $v \in \bar{O}$. Cases (1)–(3) pose no unusual problems – for (b) we use 4.2(b),(c). For case (4), let $\gamma = [c, 7^d]$. Then clearly $\gamma(0) = 7^d$ and for all m, $\gamma(m+1) = 2^{\gamma(m)}$ so $|\gamma(m)| < |\gamma(m+1)|$. Hence $u \oplus 7^d \in \bar{O}$. Properties (b), (c) and (d) are obvious. Suppose $|u'| \leq |u|$. If $|u| < |7^d|$, then also $|u'| < |7^d|$ so $u' \oplus v = u \oplus v$ and (e) is satisfied. Otherwise, by the induction hypothesis, for all m, $|u' \oplus \bar{m}| \leq |u \oplus \bar{m}|$ so again $|u' \oplus v| \leq |u \oplus v|$. □

Lemma 4.10 (Boundedness). *There exists a primitive recursive ζ such that for any superadmissible σ and any a, u such that $|u| < \sigma$, $[a,u]$ is total, and* $\forall m(|[a,u](m)| < \sigma)$:
(a) $\operatorname{Sup}\{|[a,u](m)| : m \in \omega\} < |\zeta(a,u)| < \sigma$;
(b) *If* $|v| \leq |u|$ *and* $[b,v]$ *is a total function such that for every* m, $[b,v](m)$ *is equal either to* 1 *or to* $[a,u](m)$, *then* $\zeta(b,v) \leq \zeta(a,u)$.

Proof. Let $\zeta(a,u) = 3^u \cdot 5^c$, where $[c,u](0) \simeq u$ and $[c,u](m+1) \simeq [c,u](m) \oplus 2^{[a,u](m)}$. Then (a) is obvious from 4.9(a)–(d) and 4.2(a).

Suppose b and ν are as in the hypothesis of (b), and $\zeta(b,\nu) = 3^\nu \cdot 5^d$. It suffices to show

$$\forall m(|[d,\nu](m)| \leq |[c,u](m)|).$$

For $m = 0$ this is just the hypothesis $|\nu| \leq |u|$. Assume it is true for m. If $[b,\nu](m) = 1$, then by 4.9(d)

$$[d,\nu](m+1) = 2^{[d,\nu](m)} \leq 2^{[c,u](m)} \leq [c,u](m+1).$$

If $[b,\nu](m) = [a,u](m)$, the result follows by 4.9(e). □

To provide a convenient setting for discussing computations relative to various initial segments of the J_u's, we next define a sequence of computation systems similar to $\mathscr{A}$ of [Ri, 283–284]. The intuition is that for a given σ, a function is σ – computable iff it can be computed effectively except for references to "indexed oracles" for all $J_u(|u|<\sigma)$.

Definition 4.11. For every $\sigma \leq |\mathcal{O}|$:

(a) the relation $[\![x]\!]^\sigma(\boldsymbol{m}) \simeq n$ is defined inductively as in [Kl] for type-2 with a clause S8 for functional application of the form:

$$u \in \mathcal{O}^{(\sigma)} \text{ and } J_u([\![y]\!]^\sigma)(m) \simeq n \to [\![\langle 8,u,y\rangle]\!]^\sigma(m) \simeq n.$$

(b) $\mathcal{R}_1^\sigma$ is the class of partial functions ϕ such that for some x, $\phi(\boldsymbol{m}) \simeq [\![x]\!]^\sigma(\boldsymbol{m})$. Such a ϕ is called *σ-computable with σ-index x.*

(c) $\mathcal{R}_2^\sigma$ is the class of partial functionals F which are σ computable on $\mathcal{R}_1^\sigma$ – that is, F is defined on all (total) $\alpha \in \mathcal{R}_1^\sigma$ and there exists a σ-computable ϕ such that for all x, if $[\![x]\!]^\sigma$ is a total unary function, then $\phi(x) \simeq F([\![x]\!]^\sigma)$. A σ-index for ϕ is also called a *σ-index for F.*

When $\sigma = |\mathcal{O}|$, we shall usually omit the superscript σ.

Lemma 4.12. *There exist primitive recursive f and ρ such that for any x, $\boldsymbol{m}$, and n, and any $\nu \in \mathcal{O}$, if $\sigma = |\nu|$:*

(a) $[\![x]\!]^\sigma(\boldsymbol{m}) \simeq [f(x,\nu),\nu](\boldsymbol{m})$;

(b) *if σ is superadmissible,* $[\![x]\!]^\sigma(\boldsymbol{m})\downarrow \leftrightarrow \rho(x,\langle\boldsymbol{m}\rangle) \in \mathcal{O}^{(\sigma)}$, *and if so* $\tau = |\rho(x,\langle\boldsymbol{m}\rangle)|$, *and* $[\![x]\!]^\sigma(\boldsymbol{m}) \simeq [\![x]\!]^\tau(\boldsymbol{m})$.

Proof. f and ρ are defined by straightforward recursions and the properties established by corresponding inductions. For (b) we use the boundedness properties of superadmissibles. □

Corollary 4.13. *There exists a primitive recursive η such that for any superadmissible σ and any x, if $[\![x]\!]^\sigma$ is total, then $\eta(x) \in \mathcal{O}^{(\sigma)}$ and $[\![x]\!]^\sigma = [\![x]\!]^{|\eta(x)|}$. In particular, $[\![x]\!]^\sigma$ is recursive in $J_{\eta(x)}$. Hence the total σ-computable functions are just those recursive in some $J_u(|u| < \sigma)$.*

Proof. If $[\![x]\!]^\sigma$ is total, take $\eta(x)$ to be an element of $\mathcal{O}^{(\sigma)}$ such that for all $\boldsymbol{m}$, $|\rho(x, \langle \boldsymbol{m} \rangle)| < |\eta(x)|$. Such an η is easily defined by use of Lemma 4.10. □

The next lemma is the key to the succes of our method. For application to the main Theorem it is most convenient to have it formulated in terms of functionals, but the intuition behind it is in terms of jumps: for every $J \in \mathcal{R}_2$ there exists a superadmissible $\sigma < |\mathcal{O}|$ such that for all $\alpha \in \mathcal{R}_1^\sigma$, $J(\alpha) \in \mathcal{R}_1^\sigma$. This form can be obtained from the version below by an application of the boundedness lemma.

Lemma 4.14. *There exists a primitive recursive θ such that for any $F \in \mathcal{R}_2$ with $|\mathcal{O}|$-index x, $\theta(x) \in \mathcal{O}$, $\sigma = |\theta(x)|$ is superadmissible, and $F \in \mathcal{R}_2^\sigma$ with σ-index x.*

Proof. Since F has $|\mathcal{O}|$-index x, we have from 4.12(b) that for total $[\![y]\!]$, $F([\![y]\!]) = [\![x]\!](y) = [\![x]\!]^\tau(y)$ for $\tau = |\rho(x,y)|$. For each $\nu \in \mathcal{O}$, let

$$A_\nu = \{y : [\![y]\!]^{|\nu|} \text{ is total}\}.$$

It is obvious that $|\nu| \leq |\nu'| \to A_\nu \subseteq A_{\nu'}$ and it follows from 4.12(a) that A_ν is semi-recursive in J_ν and hence recursive in J_{2^ν}. Hence there exists an index c computable from x such that for each $\nu \in \mathcal{O}$ and all y:

$$[c, 2^\nu](y) = \begin{cases} \rho(x,y), & \text{if } y \in A_\nu\,; \\ 1, & \text{otherwise.} \end{cases}$$

Let $\xi(\nu) = \zeta(c, 2^\nu)$ (ζ from 4.10) and set $\theta(x) = 7^d$ where d is a primitive recursive index for ξ.

It is clear that ξ maps $\bar{O}$ into $\bar{O}$, so to show $7^d \in \bar{O}$ we need only check the monotonicity condition of 4.2(d). But if $|v| \leq |w|$, then clearly the functions $[c, 2^v]$ and $[c, 2^w]$ satisfy the hypothesis of 4.10(b) so $|\xi(v)| = |\zeta(c, 2^v)| \leq |\zeta(c, 2^w)| = |\xi(w)|$.

The ordinal $\sigma = |\theta(x)| = |7^d|$ is superadmissible by definition and it remains to show that x is a σ-index for F. For any y such that $[\![y]\!]^\sigma$ is total, $F([\![y]\!]^\sigma) = F([\![y]\!]) = [\![x]\!](y) = [\![x]\!]^\tau(y)$ for $\tau = |\rho(x,y)|$. By 3.9, $[\![y]\!]^\sigma = [\![y]\!]^{|\eta(y)|}$ so that $y \in A_{\eta(y)}$ and $\tau = |\rho(x,y)| \leq \zeta(c, 2^{\eta(y)}) = \xi(\eta(y))$. Since $|\eta(y)| < \sigma$, by 4.2(e) so is $\xi(\eta(y)) < \sigma$, hence $\tau < \sigma$. Thus $F([\![y]\!]^\sigma) = [\![x]\!]^\sigma(y)$ as required. □

Lemma 4.15. *There exists a primitive recursive f such that for any superadmissible σ and any $F \in \mathcal{R}_2^\sigma$ with σ-index x, for all a and* $\boldsymbol{m}$

$$\{a\}(\boldsymbol{m},F) \simeq [\![f(a,x)]\!]^\sigma(\boldsymbol{m}) .$$

Proof. The definition of f is by a standard application of the recursion theorem similar to those in the proofs of Lemmas 1.4(b) and 2.8 and we consider only the case

$$\{a\}(\boldsymbol{m},F) \simeq F(\lambda p \cdot \{b\}(p,\boldsymbol{m},F)) .$$

By the induction hypothesis

$$\{a\}(\boldsymbol{m},F) \simeq F(\lambda p \cdot [\![f(b,x)]\!]^\sigma(p,\boldsymbol{m}))$$

$$\simeq F([\![g(b,x,\boldsymbol{m})]\!]^\sigma)$$

for some primitive recursive (substitution) function g. Then we pick $f(a,x)$ to be an index such that

$$[\![f(a,x)]\!]^\sigma(\boldsymbol{m}) \simeq 0 \cdot J_1([\![g(b,x,\boldsymbol{m})]\!]^\sigma) + [\![x]\!]^\sigma(g(b,x,\boldsymbol{m})) .$$

The first term is, of course, to ensure that $\lambda p \cdot \{b\}(p,\boldsymbol{m},F)$ is a total function. □

Corollary 4.16. *For any superadmissible* $\sigma \leq |\bar{O}|$ *and any* $F \in \mathcal{R}_2^\sigma$, 1-sc$(F)$ $\subseteq \mathcal{R}_1^\sigma$.

In the introduction we defined S as a function on jumps. It is obviously equivalent to consider it as a function on functionals defined by

$$S(F)(\alpha) = \begin{cases} 0, & \text{if } \{\alpha(0)\}(\alpha^+, F)\downarrow ; \\ 1, & \text{otherwise.} \end{cases}$$

This form is more convenient for treating computations in S.

Lemma 4.17. *There exists a primitive recursive* π *such that for any total* F *and any* x, *if* $F \in \mathcal{R}_2$ *with* $|\bar{O}|$*-index* x, *then* $S(F) \in \mathcal{R}_2$ *with* $|\bar{O}|$*-index* $\pi(x)$.

Proof. Suppose $F \in \mathcal{R}_2$ with $|\bar{O}|$-index x. Let f be a primitive recursive function such that for any total $[\![y]\!] \in \mathcal{R}_2$, $f(x,y)$ is an $|\bar{O}|$-index for a recursive join F_y of F and $[\![y]\!]$. By Lemma 4.14, $\theta(f(x,y)) \in \mathcal{R}$, $\sigma(y) = |\theta(f(x,y))|$ is superadmissible, and $F_y \in \mathcal{R}_2^{\sigma(y)}$. By 4.15 and 4.12(a) we can define a primitive recursive g such that

$$\{a\}(\boldsymbol{m}, F_y) \simeq [g(a,x,y), \theta(f(x,y))](\boldsymbol{m}) ,$$

and thus a primitive recursive h such that

$$\{a\}([\![y]\!], F) \simeq [h(a,x,y), \theta(f(x,y))](0) .$$

Thus the relation $\{a\}([\![y]\!],F)\downarrow$ is uniformly semicomputable from $J_{\theta(f(x,y))}$ and hence computable from $S(J_{\theta(f(x,y))})$ which is $J_{2\theta(f(x,y))}$. From this it is easy to compute a $|\bar{O}|$-index for $S(F)$. □

Theorem 4.18. *There exists a primitive recursive* f *such that for any* $a < \omega$, $\boldsymbol{m} = m_0, \ldots, m_{k-1} < \omega$, $\boldsymbol{x} = x_0, \ldots, x_{l-1} < \omega$ *and* $\boldsymbol{\alpha} = \alpha_0, \ldots, \alpha_{l-1} \in {}^\omega\omega$ *such that* $\alpha_i = [\![x_i]\!] \in \mathcal{R}_1$ for $i < l$, *if* $\{a\}(\boldsymbol{m}, \boldsymbol{\alpha}, S) \simeq n$ *then* $[\![f(a)]\!](\boldsymbol{m}, \boldsymbol{x}) \simeq n$.

Proof. This is similar to the proof of Lemma 3.3, and we again proceed immediately to the main case where

$$\{a\}(\boldsymbol{m},\boldsymbol{\alpha},S) \simeq S(F)(\alpha)$$

with $F = \lambda\beta \cdot \{b\}(\boldsymbol{m},\boldsymbol{\alpha},\beta,S)$ and $\alpha = \lambda q \cdot \{c\}(q,\boldsymbol{m},\boldsymbol{\alpha},S)$. The assumption that $\{a\}(\boldsymbol{m},\boldsymbol{\alpha},S)\downarrow$ guarantees that F and α are total. Furthermore, from the induction hypothesis it is clear that there exists primitive recursive g and h such that $F \in \mathcal{R}_2$ with $|\bar{O}|$–index $g(b,\boldsymbol{m},\boldsymbol{x})$ and $\alpha \in \mathcal{R}_1$ with $|\bar{O}|$-index $h(c,\boldsymbol{m},\boldsymbol{x})$. Hence by 4.17 $S(F)(\alpha) = [\![\pi(g(b,\boldsymbol{m},\boldsymbol{x}))]\!](h(c,\boldsymbol{m},\boldsymbol{x}))$. Then it suffices to let $f(a)$ be an $|\bar{O}|$-index for this as a function of $\boldsymbol{m}$ and $\boldsymbol{x}$. □

Corollary 4.19. *For any ϕ and F:*

(a) *ϕ partial recursive in $S \rightarrow \phi \in \mathcal{R}_1$;*

(b) *F partial recursive in S and defined on all $\alpha \in \mathcal{R}_1 \rightarrow F \in \mathcal{R}_2$.*

Corollary 4.10. (a) For any α, α is recursive in S iff for some $u \in \bar{O}$, α is recursive in J_u;

(b) $|\bar{O}| = \omega_1^S$.

Proof. (a) is immediate from 4.19(a) and 4.13; (b) follows from (a) and 4.7. □

§5. Functionals equivalent to S

Along with the superjump S, Gandy introduced in [Ga] two other functionals and sketched proofs that all three generate the same class of type-1 functions. In [Pl], Platek made use of some slightly altered versions of Gandy's functionals. As most details of the proofs of equivalence were not included in either article, we thought it worthwhile to present here (reasonably) complete proofs. To avoid repetition we shall treat in detail only Platek's versions of the functionals and mention at the end how the proofs may be altered to handle also Gandy's original versions. For completeness' sake we also include the equivalence of a fourth functional. The main ideas of this section are due to Gandy.

In order to formulate our results we shall need to extend the interpretation of Kleene's schemes S0 – S9 in [Kl]. He only considers computations for partial functionals $\lambda\mathfrak{a}\,\varphi(\mathfrak{a})$ where $\mathfrak{a}$ is a sequence of total objects. Also he only defines relative recursion relative to such functionals. We wish to consider these notions for partial functionals where some of the arguments may range over *partial* type 2 objects. We shall continue to use $F, G, H, \ldots$ to denote total type 2 objects while we shall use $\dot{F}, \dot{G}, \ldots$ to denote partial type 2 objects. Kleene's scheme S8.2 must be supplemented by a new scheme S8.$\dot{2}$ introducing the new type of variable:

$$\text{S8.}\dot{2} \qquad \varphi(\dot{F}, \mathfrak{b}) \simeq \dot{F}(\lambda t \chi(\dot{F}, t, \mathfrak{b}))\,;$$

with appropriate modifications in the indexing to take care of the new scheme and the new type of variable. By using the scheme S0 as in [Kl] we may relativise to a sequence $\Psi_1, \ldots, \Psi_l$ of functionals, some of whose arguments may be partial. But it will be necessary to insist that $\Psi_1, \ldots, \Psi_l$ are *consistent* in the following sense: if $\Psi_i(\mathfrak{a})\downarrow$ and $\mathfrak{a}'$ results from $\mathfrak{a}$ by extending some of the partial type 2 objects occuring in $\mathfrak{a}$ then $\Psi_i(\mathfrak{a}') \simeq \Psi_i(\mathfrak{a})$. Without this restriction the inductive definition for the graph of the enumerating functional (see 3.8 of [Kl]) would no longer be monotone, so that a crucial ingredient of [Kl] would be missing. It is not hard to see that if φ is partial recursive in $\Psi_1, \ldots, \Psi_l$ where $\Psi_1, \ldots, \Psi_l$ are consistent then φ is also consistent.

It is important to distinguish between variables F and $\dot{F}$ even in some contexts where this might not appear necessary. For example let ${}^3 0 = \lambda F.0$ and ${}^3\dot{0} = \lambda \dot{F}.0$. Clearly these are distinct and both are partial recursive, using essentially the same schemes. But what about ${}^3\overline{0} = \lambda \dot{F}.{}^3 0(\lambda \alpha \dot{F}(\alpha))$? ${}^3\overline{0}$ is extensionally identical with ${}^3 0$, but it is not partial recursive, and hence must be distinguished from ${}^3 0$ on the grounds that ${}^3\overline{0}$ and ${}^3 0$ have arguments of different type.

In the next two definitions we introduce the three functionals we shall compare with S.

Definition 5.1. For any total F,

$$\mathcal{E}(F) = \begin{cases} 0, & \text{if } \exists\alpha[\alpha \text{ recursive in } {}^2E, F \text{ and } F(\alpha) = 0]; \\ 1, & \text{otherwise.} \end{cases}$$

We recall from [Pl] that a partial functional $\dot{F}$ is called *acceptable* iff the domain of $\dot{F}$ includes all α recursive in ${}^2E, \dot{F}$. If $\dot{F}$ is acceptable and $\dot{F} \subseteq \dot{G}$, then $\dot{G}$ has the same 1-section as $\dot{F}$.

Definition 5.2. For any acceptable $\dot{F}$,

$$S^+(a,\dot{F}) \simeq \begin{cases} 0, & \text{if } \{a\}(\dot{F})\downarrow; \\ 1, & \text{otherwise.} \end{cases}$$

$$\mathcal{E}^+(\dot{F}) \simeq \begin{cases} 0, & \text{if } \exists\alpha\,[\alpha \text{ recursive in } {}^2E, \dot{F} \text{ and } \dot{F}(\alpha) \simeq 0]; \\ 1, & \text{otherwise.} \end{cases}$$

Both S^+ and $\mathcal{E}^+$ are undefined for non-acceptable F.

We note that S^+ does not seem to be naturally interpreted as a jump since the jump of an acceptable functional would not in general be itself acceptable. Computations relative to $\mathcal{E}^+$ are the same as what are called "weak computations from $\mathcal{E}$" in [Pl]. Of course the proof in [Pl, Corollary II] that the same α are weakly computable from $\mathcal{E}$ as are strongly computable from $\mathcal{E}$ depends on the incorrect hierarchy result.

Note also that both S^+ and $\mathcal{E}^+$ are *consistent*. This follows from the fact that if $\dot{F}$ is acceptable and $\dot{F} \subseteq \dot{G}$, then $\dot{G}$ is also acceptable, for any a and m, $\{a\}(m, {}^2E, \dot{F}) \simeq \{a\}(m, {}^2E, \dot{G})$ (because only values $\dot{F}(\alpha)$ for α recursive in ${}^2E, \dot{F}$ are used in the computation), and $1\text{-sc}({}^2E, \dot{F}) = 1\text{-sc}({}^2E, \dot{G})$.

Lemma 5.3. (a) $\mathcal{E}$ *is recursive in* $\mathcal{E}^+$;
(b) S *is recursive in* S^+;
(c) $\mathcal{E}$ *is recursive in* S;

Proof. (a) is trivial as $\mathcal{E}(F) \simeq \mathcal{E}^+(\lambda\alpha F(\alpha))$. For (b) note that $S(F)(\alpha)(a) \simeq S^+(p(a), \langle F, \alpha\rangle)$ where p is a primitive recursive function such that $\{p(a)\}(\langle F, \alpha\rangle) \simeq \{a\}(\alpha, F)$. For (c) note that

$$\mathcal{E}(F) = 0 \Leftrightarrow \exists a\{b\}(a, {}^2E, F) \simeq 0\,,$$

where $b \in \omega$ such that $\{b\}(a, {}^2E, F) \simeq F(\lambda m\{a\}(m, {}^2E, F))$. Hence

$$\&(F) = 0 \Leftrightarrow {}^2E(\lambda a\{b\}(a, {}^2E, F)) \simeq 0$$

$$\Leftrightarrow \{e\}(\lambda t.0, \langle {}^2E, F\rangle)\downarrow ,$$

where $e \in \omega$ such that $\{e\}(\alpha, \langle {}^2E, F\rangle) \simeq 0$ if ${}^2E(\lambda a\{b\}(a, {}^2E, F)) \simeq 0$, and is undefined otherwise. Then $\&(F) \simeq S(\langle {}^2E, F\rangle)(\lambda t.0)(e)$. □

Our aim now is to show that $\&^+$ and S^+ are both partial recursive in $\&$. From this together with Lemma 5.3 and the appropriate transitivity results it follows that $\&, S, \&^+, S^+$ are all partial recursive in each other and hence have the same 1-sections.

We shall use the notion of a computation tree relative to ${}^2E, \dot{F}$. Our notion differs somewhat from that introduced in [Kl]. Let $\Omega[\dot{F}]$ be the inductively defined set of all tuples $\langle a, \boldsymbol{m}, n\rangle$ such that $\{a\}(\boldsymbol{m}, {}^2E, \dot{F}) \simeq n$. Let $\preceq^{\dot{F}}$ be the well-founded partial ordering of $\Omega[\dot{F}]$ corresponding to the relation 'precedes or equals as subcomputation' among computations. For example, if a is the index for the composition of $\{b\}$ and $\{c\}$, $\{c\}(\boldsymbol{m}, {}^2E, \dot{F}) \simeq n$, then for some p, $\langle c, \boldsymbol{m}, p\rangle$ and $\langle b, p, \boldsymbol{m}, n\rangle$ are the unique immediate $\preceq^{\dot{F}}$-predecessors of $\langle a, \boldsymbol{m}, n\rangle$. Of course, the precise definition depends on a particular indexing, but we shall not delve into this level of detail.

Let $T^{\dot{F}}_{a,\boldsymbol{m},n}$ be the restriction of $\preceq^{\dot{F}}$ to $\{s : s \preceq^{\dot{F}} \langle a, \boldsymbol{m}, n\rangle\}$. Formally, of course $T^{\dot{F}}_{a,\boldsymbol{m},n}$ is most easily defined by recursion on computations. Note that for any $\langle a, \boldsymbol{m}, n\rangle \in \Omega[\dot{F}]$, $T^{\dot{F}}_{a,\boldsymbol{m},n}$ is a well-founded partial ordering with largest element $\langle a, \boldsymbol{m}, n\rangle$ and no limit points. Any such ordering T has a natural ordinal $\|T\|$ associated with it. If $T \restriction u = \{\langle s, t\rangle : T(s,t)$ and $T(t,u)\}$, then for any u other than the T-largest element, $\|T \restriction u\| < \|T\|$.

Lemma 5.4. *For any* $\langle a, \boldsymbol{m}, n\rangle \in \Omega[\dot{F}]$, $T^{\dot{F}}_{a,\boldsymbol{m},n}$ *is recursive in* ${}^2E, \dot{F}$.

Proof. Construct in the usual way a primitive recursive f such that for $\langle a, \boldsymbol{m}, n\rangle \in \Omega[\dot{F}]$, $f(\langle a, \boldsymbol{m}, n\rangle)$ is an index of $T^{\dot{F}}_{a,\boldsymbol{m},n}$ from ${}^2E, \dot{F}$. □

Definition 5.5. (a) $\mathcal{S}^{\dot{F}}(T) \leftrightarrow$ for some $\langle a, \boldsymbol{m}, n\rangle \in \Omega[\dot{F}]$, $T = T^{\dot{F}}_{a,\boldsymbol{m},n}$;

(b) if $\mathcal{S}^{\dot{F}}(T)$, a_T, $\boldsymbol{m}_T$, and n_T are the unique a, $\boldsymbol{m}$, and n such that $T = T^{\dot{F}}_{a,\boldsymbol{m},n}$.

Lemma 5.6. *There exists a function* Φ *partial recursive in* E_1 *such that*

(a) $\dot{F}$ *is acceptable* $\Leftrightarrow \lambda T \Phi(T, \dot{F})$ *is total*;

(b) *If $\dot{F}$ is acceptable then $\lambda T \Phi(T, \dot{F})$ is the characteristic function of $\mathcal{S}^{\dot{F}}$.*

Proof. Let $\mathcal{S}'(T)$ be the conjunction of the following clauses:
(a) T is a well-founded partial order with largest element but no limit points;
(b) all elements of the field T are of the form $\langle a, \boldsymbol{m}, n \rangle$;
(c) for all u other than the T-largest element, $\mathcal{S}'(T \upharpoonright u)$;
(d) for all $t = \langle a, \boldsymbol{m}, n \rangle$ in the field of T, *one* of the following holds:
(1) a is an index for an initial function and $T = \{\langle t, t \rangle\}$;
(2) for some b and c, $\{a\}(\boldsymbol{m}, {}^2E, \dot{F}) \simeq \{b\}(\{c\}(\boldsymbol{m}, {}^2E, \dot{F}), \boldsymbol{m}, {}^2E, \dot{F})$ and for some p, $s_1 = \langle c, \boldsymbol{m}, p \rangle$ and $s_2 = \langle b, p, \boldsymbol{m}, n \rangle$ are the unique immediate T-predecessors of t;
(3) for some b, $\{a\}(\boldsymbol{m}, {}^2E, \dot{F}) \simeq {}^2E(\alpha)$ or $\dot{F}(\alpha)$ where $\alpha = \lambda p \cdot \{b\}(p, \boldsymbol{m}, {}^2E, \dot{F})$ and $\forall p \exists ! q$ such that $\langle b, p, \boldsymbol{m}, q \rangle$ is an immediate T-predecessor of t and all immediate T-predecessors of t are of this form; further, if ${}^2E$ was applied then $q = 0$ or 1 depending on whether or not one of the immediate T-predecessors of t is of the form $\langle b, p, \boldsymbol{m}, 0 \rangle$;
(4) other cases similarly.

$\mathcal{S}'$ is well-defined by recursion on $\| T \|$, and a proof that it is recursive in E_1 follows the lines of the proof of 1.3. Of course, E_1 is needed for (a) and for the number quantifiers.

For any T such that $\mathcal{S}'(T)$ and any $t = \langle a, \boldsymbol{m}, n \rangle$ in the field of T which describes application of $\dot{F}$ as in (3), let $\beta_t = \lambda p \cdot$ unique q [$\langle b, p, \boldsymbol{m}, q \rangle$ is an immediate T-predecessor of t]. Then set

$$\Phi(T, \dot{F}) \simeq \begin{cases} 0, & \text{if } \mathcal{S}'(T) \text{ and for all } t = \langle a, \boldsymbol{m}, n \rangle \in \mathrm{Fld}(T) \text{ which describe} \\ & \text{application of } \dot{F}, \dot{F}(\beta_t) = n; \\ 1, & \text{if either} \sim \mathcal{S}'(T) \text{ or } \mathcal{S}'(T) \text{ but for some } t = \langle a, \boldsymbol{m}, n \rangle \in \mathrm{Fld}(T) \\ & \text{which describes application of } \dot{F}, \exists n' \, (\dot{F}(\beta_t) \simeq n' \text{ and } n' \neq n). \end{cases}$$

Thus Φ is partial recursive in E_1 and it suffices to prove for any acceptable $\dot{F}$
(i) $\langle a, \boldsymbol{m}, n \rangle \in \Omega[\dot{F}] \to \Phi(T^{\dot{F}}_{a, \boldsymbol{m}, n}, \dot{F}) \simeq 0$,
(ii) $\forall T [\Phi(T, \dot{F}) \downarrow$ *and* $(\Phi(T, \dot{F}) \simeq 0 \to \mathcal{S}^{\dot{F}}(T))]$, and for any $\dot{F}$
(iii) if $\lambda T \Phi(T, \dot{F})$ is total then $\dot{F}$ is acceptable.

The proof of (i) is a straightforward induction over $\Omega[\dot{F}]$. For (ii) if $\sim \mathcal{S}'(T)$, $\Phi(T, \dot{F}) \simeq 1$ so assume $\mathcal{S}'(T)$. We now proceed by induction on $\| T \|$

and by cases according to the T-largest element $t = \langle a, \boldsymbol{m}, n\rangle$. We consider only the case where t describes an application of $\dot{F}$. By the induction hypothesis $\forall p \cdot \Phi(T \upharpoonright \langle b, p, \boldsymbol{m}, \beta_t(p)\rangle, \dot{F})\downarrow$. If any of these has value 1, then clearly also $\Phi(T, \dot{F}) \simeq 1$. If they all have value 0, then $\forall p \cdot \mathcal{S}^{\dot{F}}(T \upharpoonright \langle b, p, \boldsymbol{m}, \beta_t(p)\rangle)$ which implies $\forall p \cdot \beta_t(p) \simeq \{b\}(p, \boldsymbol{m}, {}^2E, \dot{F})$. Hence β_t is recursive in ${}^2E, \dot{F}$, so, as $\dot{F}$ is acceptable, $\dot{F}(\beta_t)\downarrow$ and thus $\Phi(T, \dot{F})\downarrow$. If $\Phi(T, \dot{F}) \simeq 0$, then $\dot{F}(\beta_t) \simeq n$ and thus $\mathcal{S}^{\dot{F}}(T)$.

For (iii) let $\lambda T \Phi(T, \dot{F})$ be total and let $\alpha = \lambda q \{e\}(q, {}^2E, \dot{F})$. We must show that $\dot{F}(\alpha)\downarrow$. For each $q \in \omega$ $T^{\dot{F}}_{e,q,\alpha(q)} \in \mathcal{S}'$. Let b describe the application of $\dot{F}$ to α; i.e. $\{b\}({}^2E, \dot{F}) \simeq \dot{F}(\lambda q\{e\}(q, {}^2E, \dot{F}))$. Now let T be the unique tree in $\mathcal{S}'$ with T-largest element $\langle b, 0\rangle$ and $T^{\dot{F}}_{e,q,\alpha(q)}$, for each $q \in \omega$, as immediate subtrees, i.e. $T(s,t)$ iff either $\exists q T^{\dot{F}}_{e,q,\alpha(q)}(s,t)$ or $(s = t = \langle b, 0\rangle)$ or $(\exists q T^{\dot{F}}_{e,q,\alpha(q)}(s,s)$ and $t = \langle b, 0\rangle)$. If $t = \langle a, \boldsymbol{m}, n\rangle \in \mathrm{Fld}(T)$ describes application of $\dot{F}$ then either $t = \langle b, 0\rangle$, when $\beta_t = \alpha$, or else $t \in \mathrm{Fld}(T^{\dot{F}}_{e,q,\alpha(q)})$ for some q, when $\dot{F}(\beta_t) \simeq n$. Hence

$$\Phi(T, \dot{F}) \simeq 0 \Leftrightarrow \dot{F}(\alpha) \simeq 0\,;$$

and

$$\Phi(t, \dot{F}) \simeq 1 \Leftrightarrow \exists n' \dot{F}(\alpha) \simeq n' \;\&\; n' \neq 0\,.$$

So if $\lambda T \Phi(T, \dot{F})$ is total it follows that $F(\alpha)\downarrow$. □

Definition 5.7. For any $\dot{F}$ and T:

$$\dot{F}^*(T) \simeq \begin{cases} 0, & \text{if } \Phi(T,\dot{F}) \simeq 0 \textit{ and } a_T \text{ is an index describing application of } \dot{F} \textit{ and } n_T = 0; \\ 1, & \text{if } \Phi(T,\dot{F}) \simeq 0 \textit{ but } (a_T \text{ is not an index describing application of } \dot{F} \textit{ or } n_T \neq 0); \\ 2, & \text{if } \Phi(T,\dot{F}) \simeq 1. \end{cases}$$

Lemma 5.8. *For any* $\dot{F}$

(a) $\dot{F}$ *is acceptable* $\Leftrightarrow \dot{F}^*$ *is total; and*

(b) *if* $\dot{F}$ *is acceptable then* $1\text{-sc}\,({}^2E, F) \subseteq 1\text{-sc}\,(\dot{F}^*)$.

Proof. (a) follows immediately from the definition of $\dot{F}^*$ and Lemma 5.6. For (b) we define a primitive recursive f such that

$$\{a\}(\boldsymbol{m}, {}^2E, \dot{F}) \simeq n \rightarrow \lambda r \cdot \{f(a,\langle \boldsymbol{m}\rangle)\}(r, \dot{F}^*) \text{ is the characteristic function of } T^{\dot{F}}_{a,\boldsymbol{m},n}.$$

The definition of f for most cases is routine. We treat only the case when a describes an application of $\dot{F}$, say

$$\{a\}(\boldsymbol{m}, {}^2E, \dot{F}) \simeq \dot{F}(\beta) \simeq n \text{ where } \beta = \lambda p \cdot \{b\}(p, \boldsymbol{m}, {}^2E, \dot{F}).$$

For any q, let T_q be the tree with largest element $\langle a, \boldsymbol{m}, q\rangle$ and all $T^{\dot{F}}_{b,p,\boldsymbol{m},\beta(p)}$ as immediate subtrees. Then $\mathcal{S}^{\dot{F}}(T_q)$ iff $q = n$. From the induction hypothesis we can define a primitive recursive g such that for all q, $\lambda r \cdot \{g(a,\boldsymbol{m},q)\}(r, \dot{F}^*)$ is the characteristic function of T_q. Then $n = \text{least } q\ [\dot{F}^*(T_q) \leq 1]$ and $f(a,\boldsymbol{m}) \simeq g(a,\boldsymbol{m},n)$ is a proper value.

Finally for any a, $\boldsymbol{m}$ such that $\{a\}(\boldsymbol{m}, {}^2E, \dot{F})\downarrow$,

$$\{a\}(\boldsymbol{m}, {}^2E, \dot{F}) \simeq \text{least } n \cdot \{f(a,\boldsymbol{m})\}(\langle a, \boldsymbol{m}, n\rangle, \dot{F}^*) = 0. \quad \square$$

Theorem 5.9 (Gandy). $\mathcal{E}^+(\dot{F}) \simeq \mathcal{E}(\dot{F}^*)$; *hence* $\mathcal{E}^+$ *is partial recursive in* $\mathcal{E}$.

Proof. Suppose first $\mathcal{E}^+(\dot{F}) = 0$ so for some b, $\dot{F}(\lambda p \cdot \{b\}(p, {}^2E, \dot{F})) \simeq 0$. Then if a is an index such that $\{a\}(\dot{F}) \simeq \dot{F}(\lambda p \cdot \{b\}(p, {}^2E, \dot{F}))$, $\dot{F}^*(T^{\dot{F}}_{a,\phi,0}) = 0$. By 5.4, $T^{\dot{F}}_{a,\phi,0}$ is recursive in ${}^2E, \dot{F}$ and thus by 5.8 recursive in $\dot{F}^*$. Hence $\mathcal{E}(\dot{F}^*) = 0$.

Conversely, if $\mathcal{E}(\dot{F}^*) = 0$, say $\dot{F}^*(T) = 0$, then $\Phi(T, \dot{F}) \simeq 0$ and for some b, $\{a_T\}(\boldsymbol{m}_T, {}^2E, \dot{F}) \simeq \dot{F}(\lambda p \cdot \{b\}(p, \boldsymbol{m}_T, {}^2E, \dot{F}))$. But then $\mathcal{S}^{\dot{F}}(T)$ and this is a correct computation from ${}^2E, \dot{F}$. Hence $\lambda p \cdot \{b\}(p, \boldsymbol{m}_T, {}^2E, \dot{F})$ is a function β recursive in ${}^2E, \dot{F}$ and $\dot{F}(\beta) \simeq n_T = 0$. Thus $\mathcal{E}^+(\dot{F}) = 0$.

Lemma 5.8(a) shows that $\mathcal{E}^+(\dot{F})\downarrow$ iff $\mathcal{E}(\dot{F}^*)\downarrow$, so that the first part is proved. The second part follows from the fact that E_1 is clearly recursive in $\mathcal{E}$ and the passage from $\dot{F}$ to $\dot{F}^*$ is uniform. $\square$

Definition 5.10. Let $\bar{\mathcal{S}}'$ be defined like $\mathcal{S}'$ except that no provision is made for application of ${}^2E$. Then for any $\dot{F}$, T, and a,

$$\dot{F}^{**}_a(T) \simeq \begin{cases} 0, \text{ if } \Phi(T,\dot{F}) \simeq 0 \textit{ and } \bar{\mathcal{S}}'(T) \textit{ and } a_T = a \text{ and } \boldsymbol{m}_T = \phi; \\ 1, \text{ if } \Phi(T,\dot{F}) \simeq 0 \textit{ but } (\sim \bar{\mathcal{S}}'(T) \textit{ or } a_T \neq a \textit{ or } \boldsymbol{m}_T \neq \phi); \\ 2, \text{ if } \Phi(T,\dot{F}) \simeq 1. \end{cases}$$

Lemma 5.11. *For any a and* $\dot{F}$
(a) $\dot{F}$ *is acceptable* $\Leftrightarrow \dot{F}_a^{**}$ *is total; and*
(b) *if* $\dot{F}$ *is acceptable then* 1-sc $({}^2E, \dot{F}) \subseteq$ 1-sc $(\dot{F}_a^{**})$.

Proof. As for 5.8. □

Theorem 5.12. $S^+(a, \dot{F}) \simeq \&(\dot{F}_a^{**})$; *hence,* S^+ *is partial recursive in* &.

Proof. Suppose first $S^+(a, \dot{F}) \simeq 0$, so for some n, $\{a\}(\dot{F}) \simeq n$. Hence $\dot{F}_a^{**}(T^{\dot{F}}_{a,\phi,n}) = 0$. By 5.4 $T^{\dot{F}}_{a,\phi,n}$ is recursive in ${}^2E, \dot{F}$ and thus by 5.11 recursive in $\dot{F}_a^{**}$. Hence $\&(\dot{F}_a^{**}) = 0$. Conversely, if $\&(\dot{F}_a^{**}) = 0$, say $\dot{F}_a^{**}(T) = 0$, then $\Phi(T, \dot{F}) \simeq 0$ and $\mathcal{S}'(T)$, so T is a correct computation tree involving only $\dot{F}$. Hence $\{a_T\}(F) \simeq n_T$ so $S^+(a, \dot{F}) \simeq 0$. Lemma 5.11(a) shows that $S^+(a, \dot{F})\downarrow$ iff $\&(\dot{F}_a^{**})\downarrow$, so that the first part is proved. The second part follows just as in 5.9. □

Thus from 5.3 and 5.12 we obtain the equivalence of all four functionals and, in particular, the identity of their 1-sections. Gandy's original functional was

$$\bar{\&}(F) = \begin{cases} 0, & \text{if } \exists\alpha\,[\alpha \text{ recursive in } F \textit{ and } F(\alpha) = 0]\,; \\ 1, & \text{otherwise,} \end{cases}$$

and its natural extension $\bar{\&}^+$ defined for all *recursively satisfactory* $\dot{F}$, that is, $\dot{F}$ such that the domain of $\dot{F}$ includes all α recursive in $\dot{F}$. A minor modification of 5.3(c) shows $\bar{\&}$ is recursive in S and the proof of 5.12 actually shows S^+ partial recursive in $\bar{\&}$. $\bar{\&}^+$ is an extension of $\bar{\&}$ so $\bar{\&}$ is partial recursive in $\bar{\&}^+$. For the converse of this, let $\bar{\Phi}$ be defined from $\mathcal{S}'$ as Φ was from $\mathcal{S}'$ and define $\bar{F}^*$ from $\bar{\Phi}$ in such a way as to ensure ${}^2E$ is recursive in $\bar{F}^*$. Then for any recursively satisfactory $\dot{F}$, $\bar{\&}^+(\dot{F}) = \bar{\&}(\bar{F}^*)$. Hence $\bar{\&}$ and $\bar{\&}^+$ also have the same 1-sections as S. We leave it to the reader to check that $\bar{S}^+$, the extension of S^+ to all recursively satisfactory $\dot{F}$, also has the same 1-section.

Bibliography

[Ac] P. Aczel, The ordinals of the superjump and other related functionals, abstract, Conference in Mathematical Logic-London 1970, Lecture Notes In Mathematics 255, Ed. Wilfrid Hodges (Springer, Berlin, 1971) 336–337.

[Ga] R.O. Gandy, General recursive functionals of finite type and hierarchies of functions, paper given at the Symposium on Mathematical Logic, University of Clermont-Ferrand, June 1962 (mimeographed).

[Hi] P.G. Hinman, Hierarchies of effective descriptive set theory, Trans. Amer. Math. Soc. 142 (August 1969) 111–140.

[Kl] S.C. Kleene, Recursive functionals and quantifiers of finite types I, Trans. Amer. Math. Soc. 91 (1959) 1–52.

[Mo] Y.N. Moschovakis, Hyperanalytic predicates, Trans. Amer. Math. Soc. 129 (1967) 249–282.

[Pl] R.A. Platek, A countable hierarchy for the superjump, Logic Colloquium '69, edited by R.O. Gandy and C.E.M. Yates (North-Holland, Amsterdam, 1971). 257–271.

[Ri] W. Richter, Recursively Mahlo ordinals and inductive definitions, Logic Colloquium '69, edited by R.O. Gandy and C.E.M. Yates (North-Holland, Amsterdam, 1971) 273–288.

[Sh] J.R. Shoenfield, A hierarchy based on a type-two object, Trans. Amer. Math. Soc. 134 (October, 1968) 103–108.

J.E.Fenstad. P.G.Hinman (eds.), Generalized Recursion Theory

THE SUPERJUMP AND THE FIRST RECURSIVELY MAHLO ORDINAL

Leo HARRINGTON
Massachusetts Institute of Technology

The purpose of this paper is to continue the investigation of the superjump, which was first defined by Gandy in [Ga], which was studied despite adversity in [Pl] and [Ac], and which has recently been examined with notable success in [A–H]. The main result of this paper is that ω_1^S, the first ordinal not recursive in the superjump, is the first recursively Mahlo ordinal. This result was announced in [Ac], but its proof hinged on some corollaries of a fallacious theorem from [Pl]. These corollaries turn out to be correct.

As prerequisites, a working knowledge of Kleene's (or some equivalent) formulation of recursion on higher type functionals, [Kl], would be desirable. The reader would also be advised to acquaint himself with the discussions in [Pl] and [A–H], and most particularly with §3 and §4 of [A–H].

I should acknowledge here that my debt to [Pl], [Ri] and [A–H] is even greater than the derivative nature of this paper would suggest.

§0

A few definitions seem to be in order.

Definition. The superjump, S, is a type 3 functional defined by

$$S(F,e) = \begin{cases} 0 & \text{if } \{e\}^F\downarrow \\ 1 & \text{otherwise} \end{cases}$$

(where F is any type 2 functional).

^2E is a type 2 functional defined by

$$^2E(X) = \begin{cases} 0 & \text{if } X = \emptyset \\ 1 & \text{otherwise} \end{cases}$$

(where X is any real).

Let F be a fixed type 2 functional. F will stay more or less fixed till further notice. The rest of this paper will be developed relative to F. This will seemingly increase the generality of the results while in no way increasing the difficulty of their proofs. In fact F will in general be tacitly ignored.

Definition. $1-\mathrm{Sc}(F, {}^2E)$, $[1-\mathrm{Sc}(F,S)]$, is the collection of reals recursive in $F, {}^2E$ [recursive in F, S].

Let σ be an ordinal. Define $L_\sigma(F)$ by: $L_\sigma(F) = \bigcup_{\delta<\sigma} \{x \subseteq L_\delta(F); x$ is first order definable with parameters over $\langle L_\delta(F), \epsilon, F \upharpoonright (2^\omega \cap L_\delta(F))\rangle\}$. Let $M_\sigma(F)$ denote the structure $\langle L_\sigma(F), \epsilon, F \upharpoonright (2^\omega \cap L_\sigma(F))\rangle$.

σ is an *F-admissible* ordinal if $M_\sigma(F)$ is an admissible structure.

σ is *F-recursively Mahlo* if σ is F-admissible and if for any function $f : \sigma \to \sigma$, Δ_1 over $M_\sigma(F)$, there is an F-admissible $\beta < \sigma$ such that $f''\beta \subseteq \beta$.

$\omega_1^F(\rho_0^F)$ is the first F-admissible (F-recursively Mahlo) ordinal.

In [Sh], Shoenfield constructed a hierarchy for $1-\mathrm{Sc}(F, {}^2E)$ which could be used to show that $1-\mathrm{Sc}(F, {}^2E) = 2^\omega \cap L_{\omega_1^F}(F)$. In §1 Shoenfield's hierarchy will be extended to a hierarchy for $2^\omega \cap L_{\rho_0^F}(F)$. This is done in a way totally analogous to the extension in [A–H] of the hierarchy from [Pl]. The technique of [Ri] is the guiding influence in both cases. This new hierarchy will then be employed in §2 to demonstrate that $1-\mathrm{Sc}(F,S) = 2^\omega \cap L_{\rho_0^F}(F)$. In §3 the possibility of finding a reasonable notion of partial recursiveness in S is discussed.

§1

In this section the hierarchy for $2^\omega \cap L_{\rho_0^F}(F)$ is defined, and the crucial lemma concerning this hierarchy is proven.

Definition. A set of notations, $\mathcal{N}^F$, a function $||^F : \mathcal{N}^F \to$ Ordinals, and for each $n \in \mathcal{N}^F$ a real, H_n^F, are defined by induction as follows

(i) $1 \in \mathcal{N}^F$, $|1|^F = 0$, $H_1^F = \omega$.

(ii) If $x \in \mathcal{N}^F$ then: $2^x \in \mathcal{N}^F$, $|2^x|^F = |x|^F + 1$, and $H_{2^x}^F = \{\langle e, a\rangle; \{e\}^{H_x^F}$ is total and $F(\{e\}^{H_x^F}) = a\}$

(iii) If $n \in \mathcal{N}^F$ and if for all $i \in \omega \, \{e\}^{H_n^F}(i) \in \mathcal{N}^F$ then: $3^n \cdot 5^e \in \mathcal{N}^F$, $|3^n \cdot 5^e|^F$ = the 1st limit ordinal greater than $|n|^F$ and $|\{e\}^{H_n^F}(i)|^F$ for all $i \in \omega$, and $H_{3^n \cdot 5^e} = \{\langle m, 0\rangle; |m|^F \not< |3^n \cdot 5^e|^F\} \cup \{\langle m, a+1\rangle; |m|^F < |3^n \cdot 5^e|^F$ and $a \in H_m^F\}$.

(iv) If there is an ordinal σ such that (a) σ is a limit ordinal; (b) $\sigma \neq |3^a \cdot 5^b|^F$ for all a, b; and (c) for all n, if $|n|^F < \sigma$ then $|\{e\}(n)|^F < \sigma$; then: $7^e \in \mathcal{N}^F$, $|7^e|^F$ = the 1st ordinal satisfying (a), (b) and (c), and $H_{7^e}^F = \{\langle m, 0\rangle; |m|^F \not< |7^e|^F\} \cup \{\langle m, a+1\rangle; |m|^F < |7^e|^F$ and $a \in H_m^F\}$.

(It should be noted here that this hierarchical definition has some atypical features when compared with corresponding definitions in [Sh], [Pl] or [A–H]. The most glaring dissimilarity is that $H_n^F = H_m^F$ if $|n|^F = |m|^F$.)

For an ordinal σ, let $\mathcal{N}_\sigma^F = \{n \in \mathcal{N}^F; |n|^F < \sigma\}$. For $n \notin \mathcal{N}^F$ let $|n|^F = \infty$. Let $|\mathcal{N}^F| = \sup\{|n|^F; n \in \mathcal{N}^F\}$.

Till further notice the superscript F will be systematically deleted.

Definition. For an ordinal σ, σ is *$\mathcal{N}$-admissible* if $\sigma = |7^e|$ for some $7^e \in \mathcal{N}$; σ is *$\mathcal{N}$-inaccessible* if σ is $\mathcal{N}$-admissible and a limit of $\mathcal{N}$-admissibles.

Notice that for any $n \in \mathcal{N}$ there is an $\mathcal{N}$-inaccessible ordinal greater than $|n|$.

Remarks. (1) For $n, m \in \mathcal{N}$, if $|n| \leq |m|$ then $\mathcal{N}_n$ and H_n are both recursive in H_m uniformly from n, m; if $|n| < |m|$ then the Turing jump of H_n is recursive in H_m uniformly from n, m.

(2) For $n \in \mathcal{N}$, if n is $\mathcal{N}$-admissible (or even if $|n| = \omega \cdot |n|$) then there is a real I_n, recursive in H_n, which codes $M_{|n|}(F)$. The I_n's are in fact recursive in the H_n's uniformly.

(3) For any ordinal σ, $\{\langle n, a\rangle; n \in \mathcal{N}_\sigma$ and $a \in H_n\}$ is Σ_1 over $M_\sigma(F)$.

By remarks (2) and (3) we have: $\{X; X \leq H_n$ for some $n \in \mathcal{N}\} = 2^\omega \cap L_{|\mathcal{N}|}(F)$.

It should be clear that ρ_0^F cannot receive a notation, and so $|\mathcal{N}| \leq \rho_0^F$. To demonstrate now that our hierarchy is in fact a hierarchy for $2^\omega \cap L\rho_0^F(F)$, we need only show that $|\mathcal{N}|$ is F-recursively Mahlo. The key lemma is the following.

Lemma. *There is a recursive function p such that if at least one of* m_0, m_1 *is in* $\mathcal{N}$, *and if* β *is the first* $\mathcal{N}$*-inaccessible ordinal greater than* $\min(|m_0|, |m_1|)$, *then* $p(m_0, m_1) \in \mathcal{N}_\beta$, *and* $|p(m_0, m_1)| \geq \min(|m_0|, |m_1|)$.

Proof. The proof is of course by effective transfinite induction on $\mu = \min(|m_0|, |m_1|)$, so assume that $p(m'_0, m'_1)$ is defined and works whenever $\min(|m'_0|, |m'_1|) < \mu$. The only novel cases are when one of m_0, m_1 is of the form 7^e. (As may have been noticed, throughout this paper generalized recursive functions have been preferred to primitive recursive ones. When dealing with potential notations of the form 7^e or $3^n \cdot 5^e$, this could be a cause for worry as $\{e\}$ or $\{e\}^{H_n}$ need not be total. This difficulty is easily handled since an oracle for H_2 or H_{2n} can always be presumed, and so $\{e\}$ or $\{e\}^{H_n}$ may be converted into a total function. For the remainder of this proof, this annoyance will be ignored.)

As a notational convenience define, given a set of ordinals A, the *strict limit sup* of A, $\operatorname{sls} A$, to be the 1st limit ordinal greater than every member of A.

Case 1: $m_0 = 7^{e_0}, m_1 = 2^x$

Find an e such that for all $n \in \mathcal{N}$

$$|\{e\}(n)| = \begin{cases} \operatorname{sls}\{|n|\} & \text{if } |n| \geq \mu \\ \operatorname{sls}\{|n|, |p(\{e_0\}(n), x)|\} & \text{if } |n| < \mu \end{cases}$$

(Such an e can be found: By remark (1) there is a recursive function k such that for $n \in \mathcal{N}$ $\{k(n)\}^{H_{2n}} = \mathcal{N}_{|n|+1}$. Let h be a recursive function such that for all integers n and all reals X

$$\{h(n)\}^X(i) = \begin{cases} 1 & \text{if either } m_0 \text{ or } m_1 \text{ is in } \{k(n)\}^X \\ p(\{e_0\}(n), x) & \text{otherwise}. \end{cases}$$

Get e so that $\{e\}(n) = 3^{2n} \cdot 5^{h(n)}$.) Notice that this respects the induction hypothesis as $|x| \geq \mu \Rightarrow \mu = |7^{e_0}|$ and so $|n| < \mu \Rightarrow |\{e_0\}(n)| < |7^{e_0}| = \mu$. Let $p(m_0, m_1) = 7^e$.

$|p(m_0, m_1)| < \beta$: By the induction hypothesis $n \in \mathcal{N}_\beta \Rightarrow \{e\}(n) \in \mathcal{N}_\beta$. Let $\sigma = \operatorname{sls}(\{|\{e\}(n)|; n \in \mathcal{N}_\mu\} \cup \{\mu\})$. So $\sigma \leq \beta$. But σ has a notation of the form $3^m \cdot 5^-$ (where $|m| = \mu$), and so $\sigma < \beta$ since β has no such notation. But for $n \in \mathcal{N}_\beta \sim \mathcal{N}_\mu$, $|\{e\}(n)| = \operatorname{sls}\{|n|\} = |n| + \omega$. Thus $|7^e| \leq$ the 1st $\mathcal{N}$-admissible ordinal greater than σ, which is less than β since β is $\mathcal{N}$-inaccessible. Therefore $|7^e| < \beta$.

$|p(m_0, m_1)| \geq \mu$: Suppose not. Then $|n| < |7^e| \Rightarrow |n| < \mu$ and so

$|\{e\}(n)| > |p(\{e_0\}(n),x)|$, and $\{e_0\}(n),x$ satisfies the I.H. (induction hypothesis). But $|n| < |7^e| \Rightarrow |\{e\}(n)| < |7^e| \leq |x|$ and therefore $|\{e_0\}(n)| \leq |p(\{e_0\}(n),x)| < |\{e\}(n)|$. But then $n \in \mathcal{N}_{|7^e|} \Rightarrow |\{e_0\}(n)| < |\{e\}(n)|$, and so $|7^{e_0}| \leq |7^e| < \mu$.

Case 2: $m_0 = 7^{e_0}, m_1 = 7^{e_1}$

In a way similar to case 1, using the uniformities mentioned in remark (1), find an e such that for $n \in \mathcal{N}$

$$|\{e\}(n)| = \begin{cases} \text{sls}\{|n|\} & \text{if } |n| \geq \mu \\ \text{sls}(\{|n|\} \cup \{|p(\{e_0\}(n_0), \{e_1\}(n_1))|; n_0, n_1 \in \mathcal{N}_{|n|}\}) & \text{if } |n| < \mu. \end{cases}$$

Notice that $n_0, n_1 \in \mathcal{N}_\mu \Rightarrow \min(|\{e_0\}(n_0)|, |\{e_1\}(n_1)|) < \min(|7^{e_0}|, |7^{e_1}|) = \mu$ and so the I.H. is maintained. Let $p(m_0, m_1) = 7^e$.

By an argument entirely similar to that in case 1, we get $|p(m_0, m_1)| < \beta$.

$p(m_0, m_1)| \geq \mu$: Suppose not. Then $|7^e| < |7^{e_0}|$ and so there is $n_0 \in \mathcal{N}_{|7^e|}$ such that $|\{e_0\}(n_0)| \geq |7^e|$. Similarly there is $n_1 \in \mathcal{N}_{|7^e|}$ such that $|\{e_1\}(n_1)| \geq |7^e|$. But pick $n \in \mathcal{N}_{|7^e|}$ so that $|n| > |n_0|, |n_1|$. Then $|7^e| > |\{e\}(n)| > |p(\{e_0\}(n_0), \{e_1\}(n_1))| \geq \min(|\{e_0\}(n_0)|, |\{e_1\}(n_1)|) \geq |7^e|$.

Case 3: $m_0 = 7^{e_0}, m_1 = 3^{n_1} \cdot 5^{e_1}$

Claim. *There is a recursive function f such that given q, if $|q| < \mu$ or if $|m_0| = \mu$, then $f(q) \in \mathcal{N}_\beta$ and $|f(q)| \geq \min(|m_0|, |q|)$.*

Proof. Find a recursive function e such that for all $n \in \mathcal{N}$

$$|\{e(q)\}(n)| = \begin{cases} \text{sls}\{|n|\} & \text{if } |n| \geq \min(|m_0|, |q|) \\ \text{sls}\{|n|, |p\{e_0\}(n), q)|\} & \text{otherwise} \end{cases}$$

Notice that either $|q| < \mu$, or $|m_0| = \mu = \min(|m_0|, |q|)$ in which case $|n| < |m_0| \Rightarrow |\{e_0\}(n)| < |m_0| = \mu$. So if $|n| < \min(|m_0|, |q|)$ then one of q, $\{e_0\}(n)$ is in $\mathcal{N}_\mu$, and thus the I.H. is preserved.

Let $f(q) = 7^{e(q)}$. As usual $|f(q)| < \beta$.

$|f(q)| \geq \min(|m_0|, |q|)$: Suppose not. Then $|n| < |f(q)| \Rightarrow |n| < \min(|m_0|, |q|) \Rightarrow$ the I.H. applies to $\{e_0\}(n), q$ and $|\{e(q)\}(n)| > |p(\{e_0\}(n), q)|$. But $|n| < |f(q)| \Rightarrow |\{e(q)\}(n)| < |7^{e(q)}| < |q|$, and so $|\{e_0\}(n)| \leq |p(\{e_0\}(n), q)| < |\{e(q)\}(n)|$. Thus $|7^{e(q)}| \geq |7^{e_0}|$.

Now to finish case 3. Notice that by the claim $f(n_1) \in \mathcal{N}_\beta$ and $|f(n_1)| \geq \min(|m_0|, |n_1|)$. Let x_i denote the possibility imaginary object

$\{e_1\}^{H_{n_1}}(i)$. If $n_1 \in \mathcal{N}$ then the x_i's have substance, and the claim implies that $f(x_i) \in \mathcal{N}_\beta$ and $f(x_i) \geq \min(|m_0|, |x_i|)$.

Let $m = 2^{f(n_1)}$. Find e such that

$$|\{e\}^{H_m}(i)| = \begin{cases} \text{sls}\,\{|m|\} & \text{if } |f(n_1)| \geq |m_0| \\ \text{sls}\,\{|m|, |f(x_i)|\} & \text{otherwise.} \end{cases}$$

Notice that $|f(n_1)| < |m_0| \Rightarrow |n_1| \leq |f(n_1)|$ and thus the x_i's have substance, and therefore the I.H. is not violated. Let $p(m_0, m_1) = 3^m \cdot 5^e$. As always $|p(m_0, m_1)| < \beta$.

$|p(m_0, m_1)| \geq \mu$: If $|3^m \cdot 5^e| \not\geq |m_0|$, then $|m_0| > |f(n_1)| \geq |n_1|$, and thus $|\{e\}^{H_m}(i)| > |f(x_i)| \geq |x_i|$. So $|3^m \cdot 5^e| \geq |3^{n_1} \cdot 5^{e_1}|$.

§2

It is now possible to show that $|\mathcal{N}|$ is F-recursively Mahlo. The proof will only be sketched as it is precisely the same as Richter's argument in [Ri]. In fact the $\mathcal{N}$ hierarchy (for F trivial) is essentially just Richter's system of notations for ρ_0 with the onerous $3^{\langle x,y \rangle}$ clause eliminated.

Our first job is to show that $|\mathcal{N}|$ and the $\mathcal{N}$-admissible ordinals are in fact F-admissible. This should be clear for $\mathcal{N}$-admissible ordinals which are not $\mathcal{N}$-inaccessible – the first $\mathcal{N}$-admissible greater than $|n|$ is just ω_1^{F,H_n}. So let β be either an $\mathcal{N}$-inaccessible ordinal or $|\mathcal{N}|$.

We will place a recursion theoretic structure on the hierarchy restricted to $\mathcal{N}_\beta$ as follows.

Definition. $[e]^{\mathcal{N}_\beta}(x) \cong y$ means $e = \langle e_0, e_1 \rangle$, $\{e_0\}(x) \in \mathcal{N}_\beta$, and $\{e_1\}^{H_{\{e_0\}(x)}}(x) \cong y$.

$[e]^{\mathcal{N}_\beta}(x)\downarrow$ means $[e]^{\mathcal{N}_\beta}(x) \cong y$ for some y. If $[e]^{\mathcal{N}_\beta}(x)\downarrow$ then $|[e]^{\mathcal{N}_\beta}(x)| = |\{e_0\}(x)|$, otherwise $|[e]^{\mathcal{N}_\beta}(x)| \models \infty$.

A partial function φ is partial $\mathcal{N}_\beta$-recursive if for some e and all x, $\varphi(x) \cong [e]^{\mathcal{N}_\beta}(x)$.

We now give a series of theorems which together imply that β is F-admissible.

Lemma 1. *If φ is a partial $\mathcal{N}_\beta$-recursive function, and if the range of φ is a* *subset of* $\mathcal{N}_\beta$, *then there is an e such that for all x in the domain of* φ, $|\varphi(x)| \leq |\{e\}(x)| < \beta$.

Theorem 1 (Bounding). *If* φ *is a total* $\mathcal{N}_\beta$*-recursive function, and if the range of* φ *is a subset of* $\mathcal{N}_\beta$, *then* $\sup\{|\varphi(x)|; x \in \omega\} < \beta$.

Lemma 2. *There is a partial* $\mathcal{N}_\beta$*-recursive function* θ *such that*

$$\theta(e_0, x_0, e_1, x_1) = \begin{cases} 0 & \text{if } [e_0]^{\mathcal{N}_\beta}(x_0)\downarrow \text{ and } |[e_0]^{\mathcal{N}_\beta}(x_0)| \leq |[e_1]^{\mathcal{N}_\beta}(x_1)| \\ 1 & \text{if } |[e_1]^{\mathcal{N}_\beta}(x_1)| < |[e_0]^{\mathcal{N}_\beta}(x_0)| \\ \uparrow & \textit{otherwise} \end{cases}$$

Lemma 2 should be an immediate corollary of the Lemma in §1.

Theorem 2 (Selection). *There is a partial* $\mathcal{N}_\beta$*-recursive function* ψ *such that for all e, if* $[e]^{\mathcal{N}_\beta}(x)\downarrow$ *for some x, then* $[e]^{\mathcal{N}_\beta}(\psi(e))\downarrow$.

Theorem 2 is a soft consequence of Lemma 2. For an actual proof of a similar theorem in a different setting, see [Mo] or [Gr]. [Ga] is the prototype.

Theorem 3. *Let f be* Σ_1 *over* $M_\beta(F)$, *and assume that f maps a subset of* β *into* β. *Then there is a partial* $\mathcal{N}_\beta$*-recursive function* φ *such that for all n,* $\varphi(n)\downarrow$ *iff* $|n| \in \operatorname{dom} f$, *in which case* $f(|n|) = |\varphi(n)|$.

Theorem 3 can be proven by combining remark (2) of §1 with a few applications of Theorem 2.

Theorems 1 and 3 should now make it clear that β is F-admissible. But this, together with Theorem 3 and Lemma 1 for the case $\beta = |\mathcal{N}|$, should make it equally clear that $|\mathcal{N}|$ is F-recursively Mahlo. So to summarize.

Theorem 4 $|\mathcal{N}| = \rho_0^F$. *The* $\mathcal{N}$*-admissible ordinals are just the F-admissible ordinals which are less than* ρ_0^F. *The partial* $\mathcal{N}$*-recursive functions are just the partial functions which are* Σ_1 *over* $M_{\rho_0^F}(F)$.

It is much easier to show that our hierarchy is also a hierarchy for $1-\mathrm{Sc}(F, S)$. This can actually be seen from its definition (the Lemma of §1 is not needed). Arguments similar to those in §4 of [A–H] give:

Theorem 5. *There are recursive functions f and g such that*
(a) $[e]^{\mathcal{N}}(x) \simeq y$ *iff* $\{f(e)\}^{S,F}(x) \simeq y$
(b) *If* $\{e\}^{S,F}(x) \simeq y$ *then* $[g(e)]^{\mathcal{N}}(x) \simeq y$.

(a) can be proven fairly directly, as in §4 of [A–H]. (b) is best proven in the following form: there is a recursive function h such that, if $[e_0]^{\mathcal{N}}$ is total and if $\{e_1\}^{S,F}([e_0]^{\mathcal{N}}, x) \simeq y$, then $[h(e_0, e_1)]^{\mathcal{N}}(x) \simeq y$. h should be constructed by induction on the length, in the sense of [Ga] or [Gr], of the computation associated with $\{e_1\}^{S,F}([e_0]^{\mathcal{N}}, x)$.

Theorems 5 and 4 immediately yield that $1-\mathrm{Sc}(F,S) = 2^{\omega} \cap L_{\rho_0^F}(F)$.

By letting F become unfixed, it is possible to catch a glimpse of the type 3 functionals recursive in S. They are just the total functionals of the form $\lambda F[[e]^{\mathcal{N}^F}(0)]$, or equivalently of the form $\lambda F[\mu x(M_{\rho_0^F}(F) \models \varphi(x))]$ where $\varphi(x)$ is a uniformized Σ_1 formula in the language appropriate to the structures $M_{\rho_0^F}(F)$.

§3

In §2 we were able to characterize the reals recursive in S. The next natural step would be to investigate the reals r.e. in S. Unfortunately the fact that S is of type 3 swamps all more delicate considerations. The reals r.e. in S are just the $\mathbf{\Pi}_2^1$ reals, which is definitely more than $1-\mathrm{Sc}(S)$ warrants – there are plenty of reals which are r.e. and co-r.e. in S and yet not recursive in S. This seems to arise as a consequence of the ability of a computation from S to diverge for totally gratuitous, rather than inherent or ineluctable, reasons. (See [Pl] pp. 262–263.) If these gratuitously diverging computations are systematically excluded, then S is seen to have a much more natural r.e. structure, as for example Platek's weak computability [Pl]. Before trying to justify such systematic exclusions of divergent computations, it would be best to make this notion precise.

Definition. Let G be a higher type functional of unspecified type. A recursive function f is a *type n reindexing* for G if for any type n a and any e, if $\{e\}^G(a) \simeq x$ then $\{fe\}^G(a) \simeq x$. (Notice that $\{fe\}^G(a)\downarrow$ is possible even if $\{e\}^G(a)\uparrow$.) f is a *minimal* type n reindexing for G if, for any other type n

reindexing g, there is a recursive function h such that for all type n a and all e, $\{fe\}^G(a) \cong \{g(he)\}^G(a)$.

If f is a type n reindexing for G then f gives rise to a natural enumeration of type $n+1$ partial functionals which are recursive in G: the eth partial functional is just $\lambda a[\{fe\}^G(a)]$. All total type $n+1$ functionals recursive in G appear in this enumeration, but some partial functionals recursive in G may be omitted. Any partial functional omitted in this way may be viewed as being gratuitously recursive in G. (For example: for many functionals G, it is possible to find $H \leq G$ such that there are more type 1 partial functionals recursive in $H \vee G$ than there are recursive in G. These partial functionals have little right to be recursive in $H \vee G$, and this intuition is upheld by the fact that there is a reindexing for $H \vee G$ which omits them.)

Call a partial functional *strongly recursive* in G if no reindexing for G excludes it. This may seem a stringent requirement, but for the least pathological higher type functionals, the functionals of type n in which ${}^nE$ is recursive, this is no restriction since the identity type n reindexing is minimal. It is possible to have minimal reindexings different than the identity. The situation seems to be the following:

Definition. f is an *acceptable* type n reindexing for G if there is a type $n+1$ partial functional p recursive in G such that for any type n a_0, a_1 and any e_0, e_1

(i) if $p(a_0, e_0, a_1, e_1) \cong i$ then $i = 0$ or 1 and $\{fe_i\}^G(a_i)\downarrow$, and

(ii) if $\{f(e_i)\}^G(a_i)\downarrow$ for some i, $i = 0$ or 1, then $p(a_0, e_0, a_1, e_1)\downarrow$.

An acceptable reindexing supports a very reasonable recursion theory, which in some cases is more reasonable than the usual one. Type 0 selection, (the analogue of Theorem 2 of §2, see [Ga] or [Gr]), though not necessarily true of the identity reindexing, is true for acceptable reindexings. Acceptable reindexings also possess some degree of naturalness:

Theorem. *An acceptable type n reindexing for G is minimal. If f and g are two acceptable type n reindexings for G then there is a recursive permutation h such that* $\{fe\}^G(a) \cong \{g(he)\}^G(a)$ *for all type n a.*

Returning to S, Theorem 5 of §2 may now be summarized by saying that

S has an acceptable type 2 reindexing, which may now be viewed as giving S a natural partial recursive structure. The type 3 partial functionals which are strongly recursive in S are the partial functionals mentioned at the end of §2.

The results of this paper can be generalized to other type levels. Define the type $n+3$ superjump, ${}^{n+3}S$, by

$$ {}^{n+3}S(G,e) = \begin{cases} 0 & \text{if } \{e\}^G\downarrow \\ 1 & \text{otherwise} \end{cases} $$

(where G is any type $n+2$ functional). There is an acceptable type $n+2$ reindexing for ${}^{n+3}S$, and this reindexing satisfies Grilliot's formulation of type $n-1$ selection (see [Gr] and [Ma]). There seems to be no model theoretic characterization of $(n+1)-\mathrm{Sc}\,({}^{n+3}S)$, but it is possible to characterize $(n+1)-\mathrm{Sc}^n({}^{n+3}S) = \bigcup_{a \text{ of type } n}(n+1)-\mathrm{Sc}\,(a, {}^{n+3}S)$ along lines similar to the characterization of $(n+1)-\mathrm{Sc}^n({}^{n+2}E)$ in Chapter 3 of [Ma]. There is also a hierarchy for $(n+1)-\mathrm{Sc}\,({}^{n+3}S)$ which may be gotten by appropriately extending a hierarchy similar to the one for ${}^{n+2}E$ described in [Sa]. In all this ${}^{n+3}S$ behaves remarkably like a type $n+2$ object.

References

[Ac] P. Aczel, The ordinal of the superjump and related functionals, in: Conference in Mathematical Logic – London '70 (Springer, Berlin, 1972) pp. 336–337.

[A–H] P. Aczel and P.G. Hinman, Recursion in the Superjump, this volume.

[Ga] R.O. Gandy, General recursive functionals of finite type and hierarchies of functions (A paper given at the Symposium on Mathematical Logic held at the University of Clermont Ferrand, June 1962).

[Gr] T.J. Grilliot, Selection functions for recursive functionals, Notre Dame Journal of Formal Logic 10 (1969) 225–234.

[Kl] S.C. Kleene, Recursive functionals and quantifiers of finite type, Trans. Am. Math. Soc. 91 (1959) 1–52; 108 (1963) 106–142.

[Pl] R.A. Platek, A countable hierarchy for the Superjump, in: R.O. Gandy, C.E.M. Yates (eds.) Logic Colloquium '69 (North-Holland, Amsterdam, 1971) pp. 257–271.

[Ma] D.B. MacQeen, Post's problem for recursion in higher types, Ph.D. dissertation, MIT (1972).

[Mo] Y.N. Moschovakis, Hyperanalytic predicates, Trans. Am. Math. Soc. 129 (1967) 249–282.

[Ri] W. Richter, Recursively Mahlo ordinals, in: R.O. Gandy, C.E.M. Yates (eds) Logic Colloquium '69 (North-Holland, Amsterdam, 1971) pp. 273–288.

[Sa] G.E. Sacks, The 1-Section of a type n object, this volume.

[Sh] J.R. Shoenfield, A hierarchy based on a type 2 object, Trans. Am. Math. Soc. 134 (1968) 103–108.

J.E.Fenstad, P.G.Hinman (eds.), Generalized Recursion Theory

STRUCTURAL CHARACTERIZATIONS OF CLASSES OF RELATIONS

Yiannis N. MOSCHOVAKIS [1]
University of California, Los Angeles

For each object U of finite type over the integers and for each $k \geq 1$, put

$$
\begin{aligned}
{}_k\text{en}(U) &= \textit{the } k\text{-envelope of } U \\
&= \textit{the set of all relations with arguments of type} < k \\
&\quad \textit{which are semirecursive in } U.
\end{aligned}
$$

We will give a fairly simple, indexing-free characterization of ${}_k\text{en}(U)$ when U is of type $m \geq 2$, $m - 1 \leq k \leq m + 1$ and U is *normal*, i.e. the existential quantifier ${}^mE$ over the objects of type $m - 2$ is recursive in U.

Our method will also give simple structural characterizations of some other interesting classes of relations, e.g. the $\boldsymbol{\Sigma}^1_2$ and the (positive) second order inductively definable relations with arguments of type 0 and 1. But our main aim is to make a start in the study of structural properties of envelopes. There is evidence that this study will increase our understanding of recursion in higher types. For example, we will also show that *the* 1-*envelope of a normal object of type greater than* 2 *is not the* 1-*envelope of any normal type*-2 *object*. Also, *the class of* $\boldsymbol{\Pi}^1_n$ $(n \geq 2)$ *or* $\boldsymbol{\Sigma}^1_m$ $(m \geq 1)$ *relations on the integers is not the* 1-*envelope of any normal type*-2 *object* (this result is independently due to A.S. Kechris). Thus the 1-envelope of a normal object U codes substantially more information about U than the 1-section of U which according to Sacks [1973] fails to determine whether *type*$(U) = 2$ or *type*$(U) > 2$.

[1] During the preparation of this paper the author was partially supported by a Sloan Foundation Fellowship and by a grant from the National Science Foundation.

1. Preliminaries

The set $T^{(j)}$ of objects of type j over the integers is defined by the induction

$$T^{(0)} = \omega = \{0, 1, 2, ...\},$$

$$T^{(j+1)} = \textit{all unary functions on } T^{(j)} \textit{ to } \omega .$$

A *point*

$$x = (x_1, ..., x_n)$$

is a tuple of objects of finite type and *the type of* x is the largest j such that some x_i is of type j. A (product) *space* is a cartesian product

$$\mathcal{X} = X_1 \times ... \times X_n$$

where each X_i is some $T^{(j)}$ and *the type of* $\mathcal{X}$ is the largest j such that some X_i is $T^{(j)}$. If $x = (x_1, ..., x_n)$ and $y = (y_1, ..., y_m)$, we write $(x,y) = (x_1, ..., x_n, y_1, ..., y_m)$. Similarly, if $\mathcal{X} = X_1 \times ... \times X_n$ and $\mathcal{Y} = Y_1 \times ... \times Y_m$, then $\mathcal{X} \times \mathcal{Y} = X_1 \times ... \times X_n \times Y_1 \times ... \times Y_m$.

A *relation of type* $k > 0$ is any subset $R \subseteq \mathcal{X}$ of a space of type $k - 1$. We also call these *pointsets of type* k and we write interchangeably

$$R(x) \Leftrightarrow x \in R .$$

Following Kleene, we let $\alpha^j, \beta^j, \gamma^j$ vary over $T^{(j)}$. It is also customary to reserve the variables n, k, m, ... for naming integers and the unsuperscripted Greek letters α, β, γ, ... for naming objects of type 1, or *reals*.

We assume that the reader is familiar with at least the basic definitions and facts of Kleene [1959]. In particular, it is defined there what it means for a (total) function

$$f : \mathcal{X} \to \omega$$

to be *primitive recursive*. This is by means of eight natural schemes and involves no indexing. A function

$$f : \mathcal{X} \to T^{(j+1)}$$

is primitive recursive if there is some primitive recursive

$$g : T^{(j)} \times \mathcal{X} \to \omega$$

such that

$$f(x) = \lambda\alpha^j g(\alpha^j, x) .$$

Similarly,

$$f : \mathcal{X} \to \mathcal{Y} = Y_1 \times \ldots \times Y_m$$

is primitive recursive, if there exist primitive recursive functions

$$f_i : \mathcal{X} \to Y_i$$

such that

$$f(x) = (f_1(x), \ldots, f_m(x)) .$$

If $\mathcal{X}$ and $\mathcal{Y}$ are of type ≤ 1, then these definitions agree with the classical definitions of primitive recursive functions and functionals.

A *partial function*

$$f : \mathcal{X} \to \omega$$

is a function with domain some $D \subseteq \mathcal{X}$ which takes values in ω. We write

$$f(x) = n$$

to indicate that *f is defined at x and has value n.* We can think of a (total) function $f : \mathcal{X} \to \omega$ as a partial function which happens to be totally defined on .

The definition of *recursive partial functions* in Kleene [1959] adds only one more scheme to those needed for primitive recursion but is substantially more complicated in its interpretation. An inductive definition is given of a set of triples

$$\mathcal{K} = \{(e, \mathfrak{a}, n) : \{e\}(\mathfrak{a}) = n\} ,$$

where e, n vary over ω and $\mathfrak{a}$ varies over tuples of arbitrary length of objects of arbitrary finite types. A partial function $f: \mathcal{X} \to \omega$ with domain (a subset of) some fixed space $\mathcal{X}$ is then *recursive*, if there is some integer e (an index of f) such that for all $x \in \mathcal{X}$ and all $n \in \omega$,

$$f(x) = n \Longleftrightarrow (e, x, n) \in \mathcal{K} .$$

A relation $R \subseteq \mathcal{X}$ is *recursive* if its characteristic function

$$\chi_R(x) = \begin{cases} 0 & \textit{if } R(x) , \\ 1 & \textit{if } \neg R(x) \end{cases}$$

is recursive and *semirecursive* (or recursively enumerable) if there is a recursive partial $f: \mathcal{X} \to \omega$ such that

$$R = \textit{Domain}\,(f) = \{x : f(x) \textit{ is defined}\} .$$

These notions relativize to a fixed object U of type $m > 0$ by substitution, i.e. a partial function $f: \mathcal{X} \to \omega$ is *recursive in U* if

$$f(x) = g(U, x)$$

with some recursive partial $g: T^{(m)} \times \mathcal{X} \to \omega$. Similarly, a relation R is *recursive in U* if χ_R is recursive in U, and it is *semirecursive in U* if there is a partial function $f: \mathcal{X} \to \omega$, recursive in U, such that $R = \textit{Domain}\,(f)$.

It is simple to check that a total function $f: \mathcal{X} \to \omega$ is recursive (in U) if and only if the graph of f,

$$\textit{Graph}\,(f) = \{(x, n) : f(x) = n\}$$

is recursive (in U) as a relation. This is not always true for partial functions, which is why it is not convenient in this context to identify partial functions with their graphs.

Since an object F of type $k > 0$ is a function of type k, it makes sense to ask whether *F is recursive in U*. This relation is transitive. We agree that the objects of type 0, the integers, are recursive in every object. If F is recursive in G and G is recursive in F, we say that F and G are (recursively) *equivalent* or that they have the same *Kleene degree*. It follows from one of Kleene's basic substitution results that if U and V are equivalent, then for every pointset R,

$$R \textit{ is recursive in } U \Longleftrightarrow R \textit{ is recursive in } V .$$

Also, if U and V are equivalent and of the same type, then

$$R \textit{ is semirecursive in } U \Longleftrightarrow R \textit{ is semirecursive in } V .$$

Following Kleene [1963], we define for each object U and each $k \geq 1$, *the k-section of U*,

$$_k\mathrm{sc}(U) = \{R : R \textit{ is a pointset of type} \leq k \textit{ and } R \textit{ is recursive in } U\}.$$

We also define *the k-envelope of U* as in the introduction,

$$_k\mathrm{en}(U) = \{R : R \textit{ is a pointset of type} \leq k \textit{ and } R \textit{ is semirecursive in } U\}.$$

The object ^mE $(m \geq 2)$ representing quantification over $T^{(m-2)}$ is defined by

$$^mE(\alpha^{m-1}) = \begin{cases} 0 \textit{ if } (\exists\beta^{m-2})[\alpha^{m-1}(\beta^{m-2}) = 0], \\ 1 \textit{ otherwise.} \end{cases}$$

We call an object U of type m *normal* if ^mE is recursive in U. (All objects of type 0 or 1 are normal.) Almost all reasonable structure results that are known about recursion relative to an object U have been proved on the hypothesis that U is normal, and many are known to fail if U is not normal.

Kleene's inductive definition of recursion naturally assigns to each pair $e, \mathfrak{a}$ such that $\{e\}(\mathfrak{a})$ is defined an ordinal $|e, \mathfrak{a}|$, *the stage* of the induction at which we first recognized that $(e, \mathfrak{a}, n) \in \mathfrak{X}$, for some n. The following theorem is the correct general version of Theorem 6 of Moschovakis [1967] and can be established by a variation of the proof given there:

The Stage Comparison Theorem for Kleene Recursion. Let x, y vary over the spaces $\mathcal{X}, \mathcal{Y}$ respectively, both of type $\leq m$ $(m \geq 2)$. There is a recursive partial function $f(\alpha^m, e, x, z, y)$ such that

$$\{e\}(x) \textit{ is defined and } |e,x| \leq |z,y| \Rightarrow f({}^mE, e, x, z, y) = 0\,,$$

$$\{z\}(y) \textit{ is defined and } |z,y| < |e,x| \Rightarrow f({}^mE, e, x, z, y) = 1.$$

(Here $|e,x| \leq |z,y|$ is true if $\{z\}(y)$ is not defined.)

This result is due to Gandy [1962] for $m = 2$ and to Platek [1966] for $m \geq 3$, independently of Moschovakis [1967]. (The proof in Moschovakis [1967] is given only for $m = 3$ and Grilliot [1967] extended it to all m.) It is the basic theorem about recursion relative to normal objects.

One of the easy consequences of the Stage Comparison Theorem is the existence of *selection operators*: *for each recursive partial function*

$$f : \omega \times \mathcal{X} \to \omega$$

with *type* $(\mathcal{X}) \leq m$ $(m \geq 2)$ *there is a recursive partial function* $g(\alpha^m, x)$ *such that*

$$(\exists n)[f(n,x) \textit{ is defined}] \Leftrightarrow g({}^mE, x) \textit{ is defined},$$

$$(\exists n)[f(n,x) \textit{ is defined}] \Rightarrow f(g({}^mE, x), x) \textit{ is defined}.$$

From this follows immediately that *a partial function* $f : \mathcal{X} \to \omega$ *with type* $(\mathcal{X}) \leq m$ *is recursive in a normal object U of type m if and only if the graph of f is semirecursive in U*. In these circumstances we often identify a partial function with its graph.

A *pointclass* is any collection of pointsets. We call a pointclass Γ *of type k*, or a *k-pointclass*, if every pointset in Γ is of type $\leq k$, e.g. if Γ is a k-section or a k-envelope.

A pointclass Γ *is closed under* $\exists^j$ if whenever $R \subseteq T^{(j)} \times \mathcal{X}$ is in Γ, so is P defined by

$$P(x) \Leftrightarrow (\exists \alpha^j) R(\alpha^j, x)\,.$$

Closure under $\forall^j$, & and $\vee$ is defined similarly.

A k-pointclass Γ *is closed under primitive recursive substitution* if it contains all primitive recursive pointsets of type $\leq k$ and if whenever $\mathcal{X}$, $\mathcal{Y}$ are of type $< k$ $R \subseteq \mathcal{Y}$ is in Γ and $f: \mathcal{X} \to \mathcal{Y}$ is primitive recursive, then $f^{-1}[R] = \{x : f(x) \in R\}$ is also in Γ.

Using the basic definitions and the Stage Comparison Theorem, one easily proves that *if U is normal of type* $m \geq 2$ *and* $1 \leq k \leq m+1$, *then* ${}_k\text{en}(U)$ *is closed under primitive recursive substitution*, &, $\vee$, $\exists^0$ *and* $\forall^j$ *for every* $j \leq m-2$.

A k-pointclass Γ is *ω-parametrized* if for every space $\mathcal{X}$ of type $< k$ there is a relation $G \subseteq \omega \times \mathcal{X}$ in Γ such that for every $R \subseteq \mathcal{X}$,

$$R \text{ is in } \Gamma \Leftrightarrow \text{for some } e \in \omega, R = G_e = \{x \in \mathcal{X} : G(e,x)\}.$$

It is immediate from the definitions that *for each object U and each* $k \geq 1$, ${}_k\text{en}(U)$ *is ω-parametrized.*

Finally we look at a more complicated structural property that pointclasses may posses. A *norm* on a pointset R is any function

$$\sigma : R \to \mathit{Ordinals}$$

which assigns ordinals to the members of R. (Sometimes it is convenient to insist that σ is onto an initial segment of the ordinals, but we do not do this here.) With each norm σ on R there are naturally associated two relations,

$$x \leq^*_\sigma y \Leftrightarrow R(x) \,\&\, [\neg R(y) \vee \sigma(x) \leq \sigma(y)],$$

$$x <^*_\sigma y \Leftrightarrow R(x) \,\&\, [\neg R(y) \vee \sigma(x) < \sigma(y)].$$

If R is in a pointclass Γ, we call σ a *Γ-norm* if both $\leq^*_\sigma$ and $<^*_\sigma$ are in Γ. Notice that both $\leq^*_\sigma$ and $<^*_\sigma$ are relations of the same type as R.

A pointclass Γ is *normed* (or satisfies the *Prewellordering Property*) if every pointset in Γ admits a Γ-norm.

The Stage Comparison Theorem implies immediately that *if U is normal of type m and* $k \leq m+1$, *then* ${}_k\text{en}(U)$ *is normed*. This is the *Prewellordering Theorem* for semirecursion relative to a normal object and implies directly most of the known structural properties of envelopes and sections, e.g. the existence or hierarchies on sections.

2. The Main Lemma

The key to our results is the closure of reasonably rich ω-parametrized, normed classes under *monotone inductive definability*.

Let

$$\Phi : Power(\mathcal{X}) \to Power(\mathcal{X})$$

be an operator on subsets of $\mathcal{X}$ which is *monotone*, i.e.

$$S \subseteq T \Rightarrow \Phi(S) \subseteq \Phi(T) .$$

Using the notation for induction of Moschovakis [1973], we define for each ordinal ξ *the stage* I_Φ^ξ by the recursion

$$I_\Phi^\xi = \Phi(\bigcup_{\eta<\xi} I_\Phi^\eta) .$$

It follows that

$$\eta \leq \xi \Rightarrow I_\Phi^\eta \subseteq I_\Phi^\xi$$

and that the set

$$I_\Phi = \bigcup_\xi I_\Phi^\xi$$

is the least (under inclusion) fixed point of Φ.

A k-pointclass Γ is *uniformly closed under* a monotone operator

$$\Phi : Power(\mathcal{X}) \to Power(\mathcal{X}) ,$$

with $type(\mathcal{X}) < k$, if for every space $\mathcal{Y}$ of type $< k$ and every $P \subseteq \mathcal{X} \times \mathcal{Y}$ in Γ, the relation $Q \subseteq \mathcal{X} \times \mathcal{Y}$ defined by

$$Q(x,y) \Leftrightarrow x \in \Phi(\{x' : P(x',y)\})$$

is also in Γ. This says that if for each $y \in \mathcal{Y}$ we define

$$P_y = \{x' : P(x',y)\}$$

by fixing a parameter y in a Γ-set P, then each set $\Phi(P_y)$ is also obtained by fixing y in some Γ-set Q,

$$\Phi(P_y) = \{x : Q(x,y)\} .$$

It may happen that

$$\Phi : Power(\mathcal{X} \times \omega) \to Power(\mathcal{X} \times \omega)$$

preserves single-valuedness, i.e. whenever $R \subseteq \mathcal{X} \times \omega$ is (the graph of) a partial function, then so is $\Phi(R)$. Since the empty set $\emptyset$ is a partial function, it follows in this case that I_Φ is a partial function too. For such operations Φ, we say that a k-pointclass Γ is *uniformly closed for partial functions under* Φ, if for each space $\mathcal{Y}$ of type $<k$, whenever $P \subseteq \mathcal{X} \times \omega \times \mathcal{Y}$ is in Γ and for every $y \in \mathcal{Y}$, the set $P_y = \{(x',n') : P(x',n',y)\}$ is a partial function and $Q \subseteq \mathcal{X} \times \omega \times \mathcal{Y}$ is defined by

$$Q(x,n,y) \Leftrightarrow (x,n) \in \Phi(\{(x',n') : P(x',n',y)\}) ,$$

then Q is also in Γ.

Main Lemma. *Let Γ be a k-pointclass which is ω-parametrized, normed and closed under primitive recursive substitution, let $\mathcal{X}$ be a space of type $<k$ and assume that Γ is uniformly closed under the monotone operator*

$$\Phi : Power(\mathcal{X}) \to Power(\mathcal{X}) .$$

Then I_Φ is in Γ.

Similarly, if Γ is ω-parametrized, normed and closed under &, $\vee$, $\forall^0$ and primitive recursive substitution, if type$(\mathcal{X}) < k$ and

$$\Phi : Power(\mathcal{X} \times \omega) \to Power(\mathcal{X} \times \omega)$$

preserves single-valuedness and if Γ is uniformly closed for partial functions under Φ, then the partial function I_Φ is in Γ.

Proof. Choose $G \subseteq \omega \times \omega \times \mathcal{X}$ in Γ which parametrizes the subsets of

$\omega \times \mathcal{X}$ in Γ, let

$$\sigma : G \to \textit{Ordinals}$$

be a Γ-norm on G and put

$$Q(m,x) \Leftrightarrow x \in \Phi(\{x' : (m,m,x') <_\sigma^* (m,m,x)\}) .$$

The relation

$$P(x',m,y) \Leftrightarrow (m,m,x') <_\sigma^* (m,m,y)$$

is in Γ by closure under primitive recursive substitution, so by the uniform closure under Φ, the relation

$$Q_1(x,m,y) \Leftrightarrow x \in \Phi(\{x' : P(x',m,y)\})$$

is in Γ and hence Q is in Γ, since

$$Q(m,x) \Leftrightarrow Q_1(x,m,x) .$$

It follows that for a fixed $e^* \in \omega$,

$$Q(m,x) \Leftrightarrow G(e^*,m,x) .$$

Put

$$R(x) \Leftrightarrow Q(e^*,x) \Leftrightarrow G(e^*,e^*,x)$$

and define a norm τ on R by

$$\tau(x) = \sigma(e^*,e^*,x) .$$

Immediately from the definitions, we have the equivalence

(∗) $$R(x) \Leftrightarrow x \in \Phi(\{x' : x' <_\tau^* x\}) .$$

We now show that (∗) characterizes I_Φ, i.e. any relation R which admits a norm τ satisfying (∗) must be I_Φ. This will complete the proof, since we have found a relation in Γ which satisfies (∗).

Step 1. *If R satisfies* $(*)$, *then* $R(x) \Rightarrow x \in I_\Phi$.

Proof of Step 1 is by induction on $\tau(x)$. By $(*)$, from $R(x)$ we get $x \in \Phi(\{x' : x' <_\tau^* x\})$ and by the induction hypothesis, $\{x' : x' <_\tau^* x\} \subseteq I_\Phi$, so by the monotonicity of Φ, $x \in \Phi(I_\Phi)$. But $\Phi(I_\Phi) = I_\Phi$, since I_Φ is a fixed point of Φ.

Step 2. *If R satisfies* $(*)$, *then* $x \in I_\Phi \Rightarrow R(x)$.

Proof of Step 2. We show by induction on ξ, that

$$x \in I_\Phi^\xi \Rightarrow R(x) .$$

For this, it will be enough to obtain a contradiction from the hypotheses

$$x \in I_\Phi^\xi , \qquad \bigcup_{\eta<\xi} I_\Phi^\eta \subseteq R , \qquad \neg R(x) .$$

From the definition of $<_\tau^*$, if $\neg R(x)$, then

$$\{x' : x' <_\tau^* x\} = R .$$

From the definition of I_Φ^ξ, we have

$$x \in \Phi(\bigcup_{\eta<\xi} I_\Phi^\eta) ,$$

and from $\neg R(x)$ and $(*)$, we have

$$x \notin \Phi(\{x' : x' <_\tau^* x\}) = \Phi(R) .$$

Finally, from the induction hypothesis we have

$$\bigcup_{\eta<\xi} I_\Phi^\eta \subseteq R ,$$

which together with the two preceeding displayed formulas directly contradicts the monotonicity of Φ.

To prove the second part of the theorem, we first show that under the hypotheses on Γ, for each $\mathcal{Y}$ of type $< k$ there is a partial function on $\omega \times \mathcal{Y}$,

$$G \subseteq \omega \times \mathcal{Y} \times \omega$$

which is in Γ and which parametrizes the partial functions in Γ, on $\mathcal{Y}$. (We could have substituted this in the hypotheses instead of the extra closure properties of Γ.)

Choose

$$H \subseteq \omega \times \mathcal{Y} \times \omega$$

in Γ which parametrizes the Γ-subsets of $\mathcal{Y} \times \omega$ and take

$$\sigma : H \to \mathit{Ordinals}$$

to be a Γ-norm on H. Put

$$\begin{aligned} G(m,y,n) \Leftrightarrow {} & H(m,y,n) \\ & \& \ \sigma(m,y,n) = \mathit{infimum}\{\sigma(m,y,n') : H(m,y,n')\} \\ & \& \ n = \mathit{infimum}\{n' : H(m,y,n') \ \& \ \sigma(m,y,n') = \sigma(m,y,n)\}. \end{aligned}$$

It is immediate that G parametrizes the partial functions on $\mathcal{Y}$ which are in Γ. That G itself is in Γ, follows from the equivalence

$$\begin{aligned} G(m,y,n) \Leftrightarrow {} & H(m,y,n) \\ & \& \ (\forall n')[(m,y,n) \leq_\sigma^* (m,y,n')] \\ & \& \ (\forall n')[(m,y,n) <_\sigma^* (m,y,n') \vee n \leq n']. \end{aligned}$$

Using this, choose $G \subseteq \omega \times \omega \times \mathcal{X} \times \omega$ in Γ, parametrizing the Γ-partial functions on $\omega \times \mathcal{X}$, choose a Γ-norm σ on G, put

$$Q(m,x,n) \Leftrightarrow (x,n) \in \Phi(\{(x',n') : (m,m,x',n') <_\sigma^* (m,m,x,n)\})$$

and argue exactly as in the proof of the first part.

This proof is a small variation on the proof of Theorem (9A.2) in Moschovakis [1973], which in turn was patterned on the proof of the First Recursion Theorem of Moschovakis [1971].

As a first and quite typical application of this Lemma, consider the following two simple but useful results.

Proposition 1. *If a k-pointclass Γ is ω-parametrized, normed and closed under $\vee$ and primitive recursive substitution, then Γ is also closed under $\exists^0$.*

Proof. Suppose $R \subseteq \omega \times \mathcal{X}$ is in Γ, define Φ by

$$(m,x) \in \Phi(S) \Longleftrightarrow R(m,x) \vee (m+1,x) \in S .$$

Clearly Γ is uniformly closed under Φ, hence by the Main Lemma I_Φ is in Γ. From the definition of Φ it is immediate that

$$R(m,x) \Rightarrow (m,x) \in I_\Phi ,$$

$$(m+1,x) \in I_\Phi \Rightarrow (m,x) \in I_\Phi ;$$

hence

$$(\exists m) R(m,x) \Rightarrow (0,x) \in I_\Phi .$$

Conversely, a very simple induction on ξ shows that for all n,

$$(n,x) \in I_\Phi^\xi \Rightarrow (\exists m) R(m,x) ,$$

so that, in particular,

$$(0,x) \in I_\Phi \Rightarrow (\exists m) R(m,x) .$$

Thus

$$(\exists m) R(m,x) \Longleftrightarrow (0,x) \in I_\Phi$$

and Γ is closed under $\exists^0$.

Proposition 2. *If a k-pointclass Γ is ω-parametrized, normed and closed under primitive recursive substitution, $\vee$ and &, then Γ is also closed under bounded universal number quantification, i.e. if $R \subseteq \omega \times \mathcal{X}$ is in Γ, so is $P(m,x)$ defined by*

$$P(m,x) \Longleftrightarrow (\forall n \leq m) R(n,x) .$$

Proof. Given $R \subseteq \omega \times \mathcal{X}$ in Γ, define Φ by

$$(m,x) \in \Phi(S) \Longleftrightarrow [m = 0 \;\&\; R(0,x)]$$
$$\vee\; [m > 0 \;\&\; R(m,x) \;\&\; (m-1,x) \in S] \;.$$

It is easy to check that

$$(m,x) \in I_\Phi \Longleftrightarrow (\forall n \leq m) R(n,x) \;.$$

3. Characterizations by closure under logical operations

We start with a very simple characterization of the $\mathbf{\Sigma}^0_1$ (recursively enumerable) relations.

(I) *For* $k = 1, 2$, *the class of* Σ^0_1 *k-pointsets is the smallest k-pointclass which is* ω*-parametrized, normed and closed under primitive recursive substitution and* $\vee$.

Proof. It is well known that the class of $\mathbf{\Sigma}^0_1$ k-pointsets ($k = 1,2$) has all these properties, e.g. to prove that it is normed, write each $\mathbf{\Sigma}^0_1$ relation R in the form

$$R(x) \Longleftrightarrow (\exists m) P(m,x)$$

with P primitive recursive and define σ on R by

$$\sigma(x) = \textit{least } m \textit{ such that } P(m,x) \;.$$

Conversely, if Γ satisfies all these conditions, then Γ is closed under $\exists^0$ by Proposition 1, and since Γ contains all primitive recursive relations, it must contain all $\mathbf{\Sigma}^0_1$ relations too.

The next characterization is equally simple but more interesting.

(II) *For* $k = 1, 2$, *the class of* $\mathbf{\Pi}^1_1$ *k-pointsets is the smallest k-pointclass which is* ω*-parametrized, normed and closed under primitive recursive substitution,* $\vee$ *and* $\forall^0$.

Proof. It is again well known that the class of $\mathbf{\Pi}^1_1$ k-pointsets ($k = 1,2$) has all these properties. Here, the Prewellordering Property is a nontrivial classical result and it is proved using the fact that every $\mathbf{\Pi}^1_1$ relation.is of the form

$$(*) \qquad R(x) \Longleftrightarrow (\forall\alpha)(\exists m)P(\bar{\alpha}(m),x)\,,$$

with P primitive recursive. (We are using the standard notation,

$$\bar{\alpha}(0) = \langle\ \rangle = 1\,,$$

$$\bar{\alpha}(m) = \langle\alpha(0), \ldots, \alpha(m-1)\rangle$$

$$= 2^{\alpha(0)+1}, \ldots, (\textit{the } m\textit{'th prime})^{\alpha(m-1)+1}\,.)$$

In view of the Main Lemma, the converse is just the classical theorem of Kleene that every $\mathbf{\Pi}^1_1$ relation is inductively definable, see Spector [1961]. Briefly, to prove that every k-pointclass Γ satisfying the conditions of the theorem must contain all $\mathbf{\Pi}^1_1$ relations, assume that R satisfies $(*)$ with a primitive recursive P and put

$$(u,x) \in \Phi(S) \Longleftrightarrow P(u,x) \vee (\forall t)(u^\frown\langle t\rangle,x) \in S\,,$$

where $u^\frown v$ is the obvious primitive recursive concatenation function on sequence codes. Clearly Γ is uniformly closed under.Φ, so by the Main Lemma I_Φ is in Γ. Now verify

$$R(x) \Longleftrightarrow (\langle\ \rangle,x) \in I_\Phi\,,$$

by showing directly from the definition of Φ that

$$(\langle\ \rangle,x) \notin I_\Phi \Rightarrow (\exists\alpha)(\forall m)\neg P(\bar{\alpha}(m),x)$$

and then proving

$$(\forall m)\neg P(\bar{\alpha}(m),x) \Rightarrow (\forall m)(\bar{\alpha}(m),x) \notin I_\Phi$$

by a reductio ad absurdum.

From this we get immediately the next result.

(III) *The class of* $\mathbf{\Sigma}^1_2$ *pointsets of type* 2 *is the smallest* 2*-pointclass which is* ω*-parametrized, normed and closed under primitive recursive substitution,* $\vee$, $\forall^0$ *and* $\exists^1$.

Proof. That the class of $\mathbf{\Sigma}^1_2$ pointsets has these properties is again well known – for the Prewellordering Property see Rogers [1967] or Moschovakis [1973], section 9B. Conversely, if Γ has all these properties, it must contain all $\mathbf{\Pi}^1_1$ 2-pointsets by (II), and hence it must contain all $\mathbf{\Sigma}^1_2$ 2-pointsets by closure under $\exists^1$.

We do not know a structural characterization of this sort for the class of Σ^1_2 relations on ω.

(I)–(III) and the trivial closure properties of $\mathbf{\Sigma}^0_1, \mathbf{\Pi}^1_1$ and $\mathbf{\Sigma}^1_2$ determine completely a 2-pointclass Γ on the hypothesis that it is ω-parametrized, normed, closed under primitive recursive substitution and $\vee$ and also closed under the operations in a set

$$K \subseteq \{\&, \exists^0, \forall^0, \exists^1\};$$

in fact Γ must be $\mathbf{\Sigma}^0_1, \mathbf{\Pi}^1_1$ or $\mathbf{\Sigma}^1_2$. To study closure under the full set of second order operations on ω, including $\forall^1$, we now consider induction in analysis.

Let $\mathcal{L}$ be the customary two-sorted language of analysis (or second order number theory), with variables $n, k, m, \ldots$ over $\omega, \alpha, \beta, \gamma, \ldots$ over $T^{(1)} = {}^\omega\omega$, with prime formulas

$$m = n, \;\; \alpha = \beta, \;\; m + n = k, \;\; m \cdot n = k, \;\; \alpha(m) = n,$$

the logical connectives $\&, \vee, \neg, \rightarrow$ and the quantifiers $\exists^0, \forall^0, \exists^1, \forall^1$. We take it here with the standard interpretation. For each space $\mathcal{X}$ of type 1, we obtain the extension $\mathcal{L}(S)$ by adding a relation variable S to $\mathcal{L}$ and the prime formulas $S(x)$, where x is the proper sequence of variables so it denotes a point of $\mathcal{X}$. An operator

$$\Phi : Power(\mathcal{X}) \rightarrow Power(\mathcal{X})$$

is *positive, second order definable* if there is a formula $Q(x,S)$ of $\mathcal{L}(S)$ with only positive occurrences of S such that for all $S \subseteq \mathcal{X}$, in the standard interpretation,

$$\Phi(S) = \{x : Q(x,S)\} .$$

Such operators are monotone. We call $R \subseteq \mathcal{X}$ *absolutely second order inductive* if there is some positive, second order definable operator

$$\Phi : Power(\omega^k \times \mathcal{X}) \to Power(\omega^k \times \mathcal{X})$$

and integers $n_1, \ldots, n_k$, such that

$$R(x) \Longleftrightarrow (n_1, \ldots, n_k, x) \in I_\Phi .$$

These are the *absolutely semihyperprojective* relations of Moschovakis [1969]. In Moschovakis [1973] we restricted ourselves for simplicity to induction on one-sorted structures and we allowed substitution of arbitrary constants from the domain of that structure in the language. One verifies trivially that a 2-pointset R is *inductive on the structure of analysis* in the sense of Moschovakis [1973] if and only if there is an absolutely second order inductive relation P in the present sense and a real α_0, such that

$$R(x) \Longleftrightarrow P(\alpha_0, x) .$$

(IV) *The class of absolutely second order inductive pointsets of type* ≤ 2 *is the smallest 2-pointclass which is* ω*-parametrized, normed and closed under primitive recursive substitution,* $\vee$, $\exists^1$ *and* $\forall^1$.

Proof. That the class of absolutely second order inductive pointsets of type 2 has these properties follows easily by the "lightface versions" of theorems (1D.1), (3A.3) and (5D.2) of Moschovakis [1973].

To prove the converse, we show first that if Γ satisfies the hypotheses, then Γ is also closed under &, $\exists^0$, $\forall^0$. This is completely trivial for $\exists^0$ and $\forall^0$. For &, let

$$P(x) \Longleftrightarrow G(e_0, x) ,$$

$$Q(x) \Longleftrightarrow G(e_1, x) ,$$

where e_0, e_1 are fixed integers and $G \subseteq \omega \times \mathcal{X}$ parametrizes the subsets of $\mathcal{X}$ in Γ and put

$$f(\alpha,x) = \begin{cases} (e_0,x) & \textit{if} \quad \alpha(0) = 0\,, \\ (e_1,x) & \textit{if} \quad \alpha(0) > 0\,. \end{cases}$$

Then

$$P(x) \;\&\; Q(x) \Longleftrightarrow (\forall\alpha)\, G(f(\alpha,x))\,,$$

so that the conjunction of P and Q is in Γ.

It follows that a pointclass Γ with all these properties is uniformly closed under every positive, second order definable operator Φ, so by the Main Lemma Γ must contain every I_Φ and hence every absolutely second order definable pointset.

The results in (I)–(IV) can be summarized in the table below where each line is to be read as follows: *the smallest 2-pointclass which is ω-parametrized, normed and closed under primitive recursive substitution and the operations in the first column is the pointclass in the second column; it is also closed under the operations in the third column.*

	Operations	*Smallest Class*	*Additional Closure*
(I)	$\vee$	$\boldsymbol{\Sigma}^0_1$	$\vee, \&, \exists^0, \exists^1$
(II)	$\vee, \forall^0$	$\boldsymbol{\Pi}^1_1$	$\vee, \&, \exists^0, \forall^0, \forall^1$
(III)	$\vee, \forall^0, \exists^1$	$\boldsymbol{\Sigma}^1_2$	$\vee, \&, \exists^0, \forall^0, \exists^1$
(IV)	$\vee, \exists^1, \forall^1$	*abs. second order inductive*	$\vee, \&, \exists^0, \forall^0, \exists^1, \forall^1$

Using this table and very trivial arguments we can easily compute the smallest ω-parametrized, normed 2-pointclass which is closed under primitive recursive substitution, $\vee$ and any set of operations $K \subseteq \{\&, \exists^0, \forall^0, \exists^1, \forall^1\}$.

It is obvious from the proofs that the hypothesis of closure under primitive recursive substitution can be substantially relaxed in these characterizations. One can in fact put down characterizations that are almost algebraic in flavor, but which are naturally a bit more complicated.

4. Characterization of envelopes

We now turn to the characterization of envelopes promised in the introduction.

(V) *Let U be a normal object of type $m \geq 2$ and let $m-1 \leq k \leq m+1$. Then the k-envelope of U is the smallest k-pointclass Γ which is ω-parametrized, normed, closed under primitive recursive substitution, $\vee$ and & and which has the following two additional closure properties*:

(1) *If $2 \leq j < k$ and $g(\alpha^j, \beta^{j-2}, x)$ is a partial function with graph in Γ, then the partial function*

$$f(\alpha^j, x) = \alpha^j(\lambda\beta^{j-2} g(\alpha^j, \beta^{j-2}, x))$$

also has graph in Γ.

(2) *If $g(\alpha^{m-2}, x)$ is a partial function with graph in Γ, then the partial function*

$$f(x) = U(\lambda\alpha^{m-2} g(\alpha^{m-2}, x))$$

also has graph in Γ.

Proof. That ${}_k\mathrm{en}(U)$ has all these properties under the hypotheses on U, k is immediate from the discussion in section 1.

To prove the converse, assume that Γ satisfies all the conditions of the theorem. By Propositions 1 and 2 we also know that Γ is closed under $\exists^0$ and bounded universal number quantification.

Let $\langle\alpha_0^j, ..., \alpha_n^j\rangle$, $(\alpha^j)_i$ be the primitive recursive tuple-coding functions defined in Kleene [1959], and relativizing Kleene's $\{e\}[\alpha^0, \alpha^1, ..., \alpha^{k-1}]$, put

$$\{e\}^U[\alpha^0, \alpha^1, ..., \alpha^{k-1}] = \{e\}(U, (\alpha^0)_0, ..., (\alpha^0)_{n_0}, ..., (\alpha^{k-1})_0, ..., (\alpha^{k-1})_{n_{k-1}}),$$

where the number n_i of variables in each type $i < k$ is recoverable primitively recursively from the index e in the coding used by Kleene. The inductive definition of the relation $\{e\}(U, \mathfrak{a}) = n$ determines a monotone operator

$$\Phi : Power(\omega \times \mathcal{X} \times \omega) \Rightarrow Power(\omega \times \mathcal{X} \times \omega),$$

where

$$\mathcal{X} = T^{(0)} \times T^{(1)} \times \dots \times T^{(k-1)} ,$$

such that

$$\{e\}^U[\alpha^0, \dots, \alpha^{k-1}] = n \Leftrightarrow (e, \alpha^0, \dots, \alpha^{k-1}, n) \in I_\Phi .$$

The definition of $(e,x,n) \in \Phi(S)$ is a disjunction of ten clauses, S1 – S9 and S8(U), the last corresponding to the clause for $\{e\}(U, \mathfrak{a}) = n$ which applies U,

S8(U) $\qquad f(\mathfrak{b}) = U(\lambda\alpha^{m-2} g(\alpha^{m-2}, \mathfrak{b}))$.

Certainly Φ preserves single-valuedness. Now the closure properties of Γ imply that Γ is uniformly closed under Φ: we use $\exists^0$ to express clause S4 for composition we use both $\exists^0$ and bounded universal number quantification to express clause S5 for primitive recursion, we use (1) to express clause S8 and (2) to express clause S8(U) and of course we use closure under primitive recursive substitution all over. The explicit computation is a bit messy but direct and we omit it.

By the second part of the Main Lemma then, I_Φ is in Γ, hence the relation $\{e\}^U[\alpha^0, \dots, \alpha^{k-1}] = n$ is in Γ, hence by closure under primitive recursive substitution, every pointset of type $\leq k$ which is semirecursive in U is in Γ.

It is perhaps worth putting down explicitly the result for $m = 3$, $k = 2$. (The statement is very similar for the 1-envelope of a type-2 object.)

The 2-envelope of a normal type-3 object U is the smallest class Γ of pointsets of type ≤ 2, which is ω-parametrized, normed, closed under primitive recursive substitution, $\vee$ and & and such that whenever $g(\alpha, x)$ is a partial function in Γ, so is $f(x)$, defined by

$$f(x) = U(\lambda\alpha g(\alpha, x)) .$$

The Main Lemma does not yield a characterization of this sort for the 1-envelope of a normal type-3 object.

The special case $U = {}^3E$ makes precise the difference between the 2-envelope of ${}^3E$, the class of semihyperanalytical relations of Kleene and the class of absolutely second order inductive relations which we characterized in (IV). It is trivial that ${}_2\mathrm{en}({}^3E)$ is closed under $\forall^1$. Hence the only difference between these two classes is that the collection of absolutely second order inductive pointsets is closed under $\exists^1$, while ${}_2\text{en}\,({}^3E)$ is only closed under a very restricted kind of existential quantification on $T^{(1)}$; if $g(\alpha, x)$ is a partial function (with graph) in ${}_2\text{en}\,({}^3E)$, then

$$R(x) \Longleftrightarrow (\forall\alpha)[g(\alpha,x) \textit{ is defined}] \;\&\; (\exists\alpha)[g(\alpha,x) = 0]$$

is also in ${}_2\text{en}\,({}^3E)$. This is a kind of "restricted" $\exists^1$. It is known from Corollary 10.2 of Moschovakis [1967] that ${}_2\text{en}\,({}^3E)$ is not closed under $\exists^1$.

The hypothesis that U is normal is essential in (V), as otherwise ${}_k\text{en}\,(U)$ need not be normed or closed under $\vee$. However, a similar characterization of the class of partial functions in ${}_k\text{en}\,(U)$ for an arbitrary U of type $m \geq 2$ $(m-1 \leq k \leq m+1)$ can be read off the results in Moschovakis [1971]. (To obtain such a characterization was one of the motivations for introducing axiomatic computation theories.) Briefly, the key property of these partial functions is that for each $\mathfrak{X}$ of type $< k$ there is an ω-parametrization $G: \omega \times \mathfrak{X} \to \omega$ of those with domain in $\mathfrak{X}$, which is in ${}_k\text{en}\,(U)$ and which admits a norm $\sigma: G \to \textit{Ordinals}$ having the natural properties of a *measure of complexity* (or *length*) *of computations*, as expressed in the axioms for a computation theory. This length function need not satisfy any definability conditions.

5. The type of a 1-envelope

We show here that for sufficiently rich 2-pointclasses Γ, the class of 1-pointsets in Γ cannot be the envelope of a type-2 normal object.

As usual, we let $\neg\Gamma$ be *the dual class* of Γ,

$$\neg\Gamma = \{\mathfrak{X} - R : R \subseteq \mathfrak{X}, R \textit{ is in } \Gamma\}$$

and we put

$$\boldsymbol{\Delta} = \Gamma \cap \neg\Gamma .$$

We say that $\boldsymbol{\Delta}$ is closed under $\exists^1 \cap \forall^1$, if whenever $P(\alpha, x)$, $Q(\beta, x)$ are in $\boldsymbol{\Delta}$ and for all x,

$$(\exists\alpha)P(\alpha,x) \Longleftrightarrow (\forall\beta)\,Q(\beta,x)\,,$$

then the relation

$$R(x) \Longleftrightarrow (\exists\alpha) P(\alpha, x)$$

is also in $\mathbf{\Delta}$.

For each real α put

$$\leq_\alpha = \{(n,m) : \alpha(\langle n,m\rangle) = 0\}$$

and let WO be the set of reals which code wellorderings of ω,

$$\alpha \in \mathrm{WO} \Longleftrightarrow \leq_\alpha \textit{ is a wellordering with field } \omega\,.$$

If $\alpha \in \mathrm{WO}$, we let $|t|_\alpha$ be the rank of t in the wellordering $\leq \alpha$.

The tool for the proof of the theorem is the following representation theorem for the 1-section of a type-2 object proved in section 2 of Moschovakis [1967] (see also Corollary 7.1 of [1967] and section 8).

Let U be an object of type 2 and define a set $N \subseteq \omega$ and for each $z \in N$ a function $f_z : \omega \to \omega$ by the following simultaneous induction:

(1) *For each integer q, $\langle 1,q\rangle \in N$ and for all t, $f_{\langle 1,q\rangle}(t) = q$. (Introduction of constants.)*

(2) *If $w \in N$, then $\langle 2,w\rangle \in N$ and for all t, $f_{\langle 2,w\rangle}(t) = U(f_w)$. (Introduction of U.)*

(3) *If $w \in N$ and if the (absolutely) recursive partial function $\{e\}(\alpha,t)$ is totally defined when $\alpha = f_w$ and if for all t,*

$$\{e\}(f_w, t) = e(w,t) \in N\,,$$

then $\langle 3,w,e\rangle \in N$ and for all t, $f_{\langle 3,w,e\rangle}(t) = f_{e(w,t)}(t)$. (Diagonalization.)

Then, a unary function is recursive in U if and only if it is f_z for some $z \in N$.

Moreover, if U is normal, then N is semirecursive in U and the partial function

$$u(z) = n \Longleftrightarrow z \in N \;\&\; U(f_z) = n$$

is recursive in U.

We understand the definition (1)–(3) as a transfinite recursion in the usual way, so that

$$N = \bigcup_{\xi} N(\xi)\,,$$

where for each ordinal ξ, $N(\xi)$ consists of the integers put in N in $\leq \xi$ steps. The induction clearly closes at some countable ordinal.

Theorem. *Let Γ be a 2-pointclass which is closed under primitive recursive substitution, $\vee$, &, $\exists^0$, $\forall^0$, and assume further that the set* WO *of codes of wellorderings is in $\mathbf{\Delta} = \Gamma \cap \neg\Gamma$ and that $\mathbf{\Delta}$ is closed under $\exists^1 \cap \forall^1$. Then the class of* 1*-pointsets in Γ is not the* 1*-envelope of any normal type-*2 *object.*

Proof. Assume towards a contradiction that U is normal of type 2 and that ${}_1\mathrm{en}(U)$ consists precisely of the relations on ω in Γ, let N, $\{f_z\}_{z\in N}$ be the canonical hierarchy of the 1-section of U as described above. Put

$$\begin{aligned} P(\alpha,\beta,\gamma) \Leftrightarrow{} & \alpha \in \mathrm{WO} \\ & \& \; (\forall x)(\forall z)\{[\beta(\langle x,z\rangle) = 0 \Leftrightarrow z \in N(|x|_\alpha)] \\ & \& \; [\beta(\langle x,z\rangle) = 0 \Rightarrow (\gamma)_z = f_z]\}\,, \end{aligned}$$

intuitively,

$$P(\alpha,\beta,\gamma) \Leftrightarrow \alpha \in \mathrm{WO} \;\&\; (\beta,\gamma) \textit{ code the definition of } N, \{f_z\}_{z\in N} \textit{ along } \leq_\alpha .$$

It will be easy to complete the proof of the theorem if we can show that P is in $\mathbf{\Delta}$.

Put

$$P_1(z,q) \Leftrightarrow z = \langle 1,q\rangle\,,$$

$$P_2(z,w,\alpha,\beta,x) \Leftrightarrow z = \langle 2,w\rangle \;\&\; (\exists y <_\alpha x)[\beta(\langle y,w\rangle) = 0]\,,$$

$$\begin{aligned} P_3(z,w,e,\alpha,\beta,\gamma,x) \Leftrightarrow{} & z = \langle 3,w,e\rangle \;\&\; (\exists y <_\alpha x)[\beta(\langle y,w\rangle) = 0] \\ & \& \; \lambda t\{e\}((\gamma)_w,t) \textit{ is total } \& \; (\forall t)(\exists y <_\alpha x)[\beta(\langle y,\{e\}((\gamma)_w,t)\rangle) = 0] \end{aligned}$$

and then set

$$\begin{aligned}
Q(\alpha,\beta,\gamma,x) \Leftrightarrow \alpha \in \mathrm{WO} \\
& \& (\forall z)\{\beta(\langle x,z\rangle) = 0 \Leftrightarrow [(\exists q)P_1(z,q) \\
& \quad \vee (\exists w)P_2(z,w,\alpha,\beta,x) \vee (\exists w)(\exists e)P_3(z,w,e,\alpha,\beta,\gamma,x)]\} \\
& \& (\forall z)\{\beta(\langle x,z\rangle) = 0 \Rightarrow [(\exists q)[P_1(z,q) \& (\forall t)[(\gamma)_z(t) = q]] \\
& \quad \vee (\exists w)[P_2(z,w,\alpha,\beta,x) \& (\forall t)[(\gamma)_z(t) = U((\gamma)_w)]] \\
& \quad \vee (\exists w)(\exists e)[P_3(z,w,e,\alpha,\beta,\gamma,x) \& (\forall t)[(\gamma)_z(t) = (\gamma)_{e(\gamma,w,t)}(t)]]]\} ,
\end{aligned}$$

where

$$e(\gamma,w,t) = \{e\}((\gamma)_w,t) .$$

Intuitively,

$Q(\alpha,\beta,\gamma,x) \Leftrightarrow \alpha \in \mathrm{WO}$ & (β,γ) *locally seem to code the definition of* N,

$$\{f_z\}_{z\in N} \text{ at } |x|_\alpha .$$

It is trivial to check by transfinite induction along $\leq_\alpha$ the two implications establishing

$$P(\alpha,\beta,\gamma) \Leftrightarrow (\forall x)\, Q(\alpha,\beta,\gamma,x) .$$

We now substitute for Q in this equivalence two approximations in Γ and $\neg\Gamma$.

Define $Q_1(\alpha,\beta,\gamma,x)$ by replacing

$$(\gamma)_z(t) = U((\gamma)_w)$$

in the definition of Q by

$$(\gamma)_z(t) = u(w) ,$$

where $u(w)$ is the partial function, recursive in U, which is associated with the canonical hierarchy on ${}_1\mathrm{sc}(U)$. Clearly Q_1 is in Γ and again it is trivial to check by transfinite induction along $\leq_\alpha$ the two implications establishing

$$P(\alpha,\beta,\gamma) \Leftrightarrow (\forall x)\, Q_1(\alpha,\beta,\gamma,x) ,$$

so that P is in Γ. Now define $Q_2(\alpha,\beta,\gamma,x)$ by replacing

$$(\gamma)_z(t) = U((\gamma)_w)$$

in the definition of Q by

$$(\forall n)[u(w) = n \Rightarrow (\gamma)_z(t) = n] .$$

Clearly Q_2 is in $\neg\Gamma$ and the proof of

$$P(\alpha,\beta,\gamma) \Longleftrightarrow (\forall x)\, Q_2(\alpha,\beta,\gamma,x)$$

is exactly as before, completing the proof that P is in $\mathbf{\Delta}$.

To complete the proof of the theorem, put

$$R(\alpha,\beta,\gamma,x) \Longleftrightarrow P(\alpha,\beta,\gamma)$$
$$\& \; (\forall z)[\beta(\langle x,z\rangle) = 0 \Rightarrow (\exists y <_\alpha x)\beta(\langle y,z\rangle) = 0] \; ;$$

intuitively,

$$R(\alpha,\beta,\gamma,x) \Longleftrightarrow P(\alpha,\beta,\gamma) \;\&\; \textit{the recursion defining } N,\ \{f_z\}_{z\in N} \textit{ closes by } |x|_\alpha .$$

Clearly R is in $\mathbf{\Delta}$ and

$$z \in N \Longleftrightarrow (\exists\alpha)(\exists\beta)(\exists\gamma)(\exists x)[R(\alpha,\beta,\gamma,x) \;\&\; \beta(\langle x,z\rangle) = 0]$$
$$(\forall\alpha)(\forall\beta)(\forall\gamma)(\forall x)[R(\alpha,\beta,\gamma,x) \Rightarrow \beta(\langle x,z\rangle) = 0] ,$$

so by the closure of $\mathbf{\Delta}$ under $\exists^1 \cap \forall^1$, N is in $\mathbf{\Delta}$. A similar trivial argument shows that the function

$$g(z) = \begin{cases} f_z(z)+1 & \textit{if } z \in N , \\ 0 & \textit{otherwise} \end{cases}$$

has graph in $\mathbf{\Delta}$, which is a contradiction since g is different from all the f_z's, $z \in N$, and hence not recursive in U.

Corollary 1. *If U is normal of type $m \geq 3$, then* ${}_1\text{en}(U)$ *is not the* 1*-envelope of any normal type-*2 *object.*

Corollary 2. *The class of* $\mathbf{\Sigma}^1_n$ $(n \geq 1)$ *or* $\mathbf{\Pi}^1_m$ $(m \geq 2)$ *relations on* ω *is not the* 1*-envelope of any normal type-*2 *object.*

Both corollaries are immediate, except for the case of $\mathbf{\Sigma}^1_1$ in the second, where the result is well known, e.g. it follows from the fact that $\mathbf{\Sigma}^1_1$ is not normed, while the 1-envelope of every normal type-2 object is normed. The second Corollary was independently formulated and proved by A.S. Kechris.

Several interesting problems suggest themselves. We state two of them as conjectures, for sharpness, although we do not have now any hard evidence that these statements rather than their negations hold.

Conjecture 1. If U is normal of type $m \geq 3$, then the 1-envelope of U is not the 1-envelope of any normal object of type less than m.

Conjecture 2. The class of $\mathbf{\Sigma}^1_2$ relations on ω is not the 1-envelope of any object U of finite type such that ${}^2E$ is recursive in U.

Of course it is quite probable that there are perverse, forcing-type counterexamples to these conjectures. We would consider positive answers (to these or similarly motivated questions) very interesting, as they would tend to show that there is a real notion of "type" in Kleene semirecursion, even if the types are lost when one looks only at the recursive objects.

Bibliography

Gandy, R.O.
[1962] General recursive functionals of finite type and hierarchies of functions, Proceedings of Logic Colloquium Clermont Ferrand, 1962, 5–24.

Grilliot, T.J.
[1967] Recursive functions of finite higher types, Ph.D. Thesis, Duke University, 1967.

Kleene, S.C.
[1959] Recursive functionals and quantifiers of finite types I, Trans. Amer. Math. Soc. 91 (1959) 1–52.

[1963] Recursive functionals and quantifiers of finite types II, Trans. Amer. Math. Soc. 108 (1963) 106–142.

Moschovakis, Y.N.
[1967] Hyperanalytic predicates, Trans. Amer. Math. Soc. 129 (1967) 249–282.
[1969] Abstract first order computability I, and II, Trans. Amer. Math. Soc. 138 (1969) 427–464.and 465–504.
[1971] Axioms for computation theories – first draft, in: R.O. Gandy, C.E.M. Yates (eds) Logic Colloquium '69 (North-Holland, Amsterdam, 1971) 199–255.
[1973] Elementary Induction on Abstract Structures (North-Holland, Amsterdam, 1973).

Platek, R.
[1966] Foundations of recursion theory, Ph.D. Thesis, Stanford University, 1966.

Sacks, G.E.
[1973] The 1-section of a type-*n* object, this volume.

Spector, C.
[1961] Inductively defined sets of natural numbers, Infinitistic Methods (Pergamon, New York, 1961) 97–102.

Added in proof. I thank D. Normann, A.S. Kechris and L. Harrington for noticing that Conjecture 2 was formulated in the first draft of this paper so that it was trivially false.

L. Harrington and J. Moldestad recently disproved Conjecture 1 by showing (independently) that the 1-envelope of every normal object U of type ≥ 3 is also the 1-envelope of some normal type-3 object. (July 9, 1973.)

J.E.Fenstad, P.G.Hinman (eds.), Generalized Recursion Theory
© North-Holland Publ. Comp., 1974

THE 1-SECTION OF A TYPE n OBJECT

Gerald E. SACKS [1]
Harvard University, Massachusetts Institute of Technology

1. Introduction

This paper is a puffing up of the proof of the plus-one theorem for the case $k = 1$. Let ${}^nE$ be the representing function of the equality predicate for all objects X, Y of type less than n : ${}^nE(X, Y) = 0$ if $X = Y$, and $= 1$ if $X \neq Y$.

Plus-One Theorem. *Let U be of type n. Suppose ${}^nE$ is recursive in U and $k < n$. Then there exists a V of type $k+1$ such that the objects of type k recursive in V are the same as those recursive in U. Furthermore ${}^{k+1}E$ is recursive in V.*

Recursion in objects of finite type was discovered by Kleene [2]. An equivalent formulation, needed for the forcing argument of section 4, is given in section 2. The proof of the plus-one theorem for the case $k > 1$ will be given in [3]. It is largely a consequence of a stability lemma described in section 5.

${}_k\mathrm{sc}U$ is the set of all objects of type k recursive in U, and is called the k-section of U. The plus-one theorem states that all k-sections generated by finite type objects (in which the appropriate equality predicates are recursive) are generated by type $k+1$ objects. The first result on k-sections was Kleene's [2]: ${}_1\mathrm{sc}{}^2E$ is the set of hyperarithmetic reals. He asked if the $\mathbf{\Delta}^1_2$ reals constituted the 1-section of some type 2 object. They do by virtue of the complete characterization of 1-sections of type 2 objects (in which ${}^2E$ is recursive) developed in section 4.

[1] The preparation of this paper was partially supported by NSF contract GP-29079. Its principal results were announced in [1]. The author is grateful to T. Grilliot for steering him towards k-section problems.

Section 2 redoes some of the elements of recursion in objects of finite type. Section 3 introduces the notion of abstract 1-section and shows that many familiar collections of reals, among then the 1-section of ${}^nE$ ($n \geq 2$) and the set of lightface $\mathbf{\Delta}^i_j$ reals ($\min(i,j) \geq 1$), are countable abstract 1-sections. Section 4 proves every countable abstract 1-section is the 1-section of some type 2 object in which ${}^2E$ is recursive by means of a forcing argument of the sort associated with generic classes rather than sets. Section 5 describes some further results based on the technique of section 4, and speculates on the nature of abstract k-sections when $k > 1$.

2. Recursion in higher types

The objects of type 0 are the nonnegative integers. An object of type $n > 0$ is a total function whose arguments range over all objects of type $< n$ and whose values are objects of type $< n$. Any object of type i can be inflated to an equivalent object of type $j > i$ by adding dummy arguments. Any object of type $n > 0$ is equivalent to one of type n whose values are either 0 or 1; e.g. the function $F(x)$ is equivalent to the representing function of the predicate $F(x) = y$. Any finite sequence of objects is equivalent to a single object; e.g. the pair $F_0(x), F_1(x)$ is equivalent to $H(n,x)$, where $H(n,x) = F_n(x)$ for $n < 2$, and = 0 otherwise. The previous three sentences should make the ambiguities in what follows tolerable.

Fix $n \geq 0$, and let $F, G, \ldots$ be objects of type $n+2$ called *functionals*. The definition of $G \leq F, {}^{n+2}E$ (read G is recursive in $F, {}^{n+2}E$) is given by means of a hierarchy inspired by those of Shoenfield [4] and Grilliot [5]. D. MacQueen (cf. [6]) has shown that $G \leq F, {}^{n+2}E$ if and only if G is recursive in F, ${}^{n+2}E$ in the sense of Kleene [2]. MacQueen's argument is not unlike that of Grilliot [5]. The hierarchical approach via total objects is equivalent to Kleene's schematic approach via partial objects because of the presence of ${}^{n+2}E$.

Functions $f, g, h, \ldots$ and *sets* $R, S, T, \ldots$ are objects of type $n+1$. *Individuals* $a, b, c, \ldots$ are objects of type at most n. *Subindividuals* $r, s, t, \ldots$ are objects of type less than n when $n > 0$ and are nonnegative integers when $n = 0$.

Fix F and g. The hierarchy $\{S_\sigma\}$ of sets, defined by induction on σ, is designed to define "G is recursive in $F, {}^{n+2}E, g$ via e" by induction on n.

Stage 0. S_0 contains $\langle 1, a\rangle$ for every a, and $\langle 1, a, m\rangle$ whenever $ga = m$. $\langle 1, a\rangle$ is an index for S_0 for every a.

Stage $\sigma+1$. $\langle 2^e, a\rangle$ is an index for $S_{\sigma+1}$ if (i) $\langle 2^e, a\rangle \notin S_\sigma$, (ii) $\langle m, a\rangle$ is an index for S_σ for some integer m, and (iii) there is a function f recursive in S_σ, ${}^{n+1}E, a$ via e.

Clause (iii) makes sense by induction on n. If $n = 0$ the clause states f is Turing reducible to S via e. The f of clause (iii) is denoted by $\lambda b\{e\}^{S_\sigma, {}^{n+1}E, a}(b)$.

$S_{\sigma+1}$ is S_σ augmented by: (1) all indices for $S_{\sigma+1}$; (2) all quadruples $\langle 3^e, a, b, m\rangle$ if $\langle 2^e, a\rangle$ is an index for $S_{\sigma+1}$ and

$$\{e\}^{S_\sigma, {}^{n+1}E, a}(b) = m\ ;$$

and (3) all triples $\langle 5^e, a, m\rangle$ if $\langle 2^e, a\rangle$ is an index for S_σ and

$$F(\lambda b\{e\}^{S_\sigma, {}^{n+1}E, a}(b)) = m\ .$$

$|\langle e, a\rangle| = \sigma$ means $\langle e, a\rangle$ is an index for S_σ.

Stage λ, *where* λ *is a limit*. $\langle 7^e, a\rangle$ is an index for S_λ if $\langle 2^e, a\rangle$ is an index for some $S_{\delta+1} < \lambda$ and $\lambda b\{e\}^{S_\sigma, {}^{n+1}E, a}(b)$ is the characteristic function of a set T of indices such that

$$\lambda = \sup\{|b|\ |\ b \in T\}\ .$$

S_λ is $\bigcup\{S_\delta | \delta < \lambda\}$ augmented by all indices for S_λ.

f is said to be recursive in $F, {}^{n+2}E, g, a$ via e, written

$$f = \{e\}^{F, {}^{n+2}E, g, a}$$

if $\langle 2^e, a\rangle$ is an index for some $S_{\sigma+1}$ and f is $\lambda b\{e\}^{S_\sigma, {}^{n+1}E, a}(b)$. G is said to be recursive in $F, {}^{n+2}E, g$ via e, written

$$G = \{e\}^{F, {}^{n+2}E, g}$$

if $G(h)$ is $\{e\}^{F, {}^{n+2}E, g, h}(0)$ for all h.

R is recursive in $F, {}^{n+2}E$ if its characteristic function is. ${}_k\mathrm{sc}(F, {}^{n+2}E)$ is the set of all objects of type k recursive in $F, {}^{n+2}E$. If ${}^{n+2}E$ is recursive in F in the sense of Kleene [2] then ${}_k\mathrm{sc}(F, {}^{n+2}E) = {}_k\mathrm{sc}F$.

R is recursively enumerable in $F, {}^{n+2}E$ via e if $R = \{a | \langle 2^e, a\rangle$ is an index$\}$. This notion of recursive enumerability coincides with Kleene's [2]. Thus R is recursive in $F, {}^{n+2}E$ if and only if both R and its complement are recursively enumerable in $F, {}^{n+2}E$.

The principal fact needed in the next section is the existence of a selection operator discovered by R. Gandy [7], who had to wrestle with Kleene's definition. His result follows more readily from the hierarchial definition of this section. [2]

Gandy's Selection Operator.
There exists a recursive function $\lambda e | e^*$ such that if T is a nonempty subset of ω recursively enumerable in $F, {}^{n+2}E$ via e, then $\{e^*\}^{F, {}^{n+2}E}(0)$ is defined and belongs to T.

3. Abstract 1-sections

Let A be a nonempty transitive set. A is said to be an abstract 1-section if it is closed under the operations of pairing and union, and satisfies axiom (1) and schemas (2)–(3), where $a \in A$ and $\mathcal{G}(y)$ and $\mathcal{G}(x,y)$ are $\mathbf{\Delta}_0$ formulas of **ZF** (i.e. formulas whose quantifiers are restricted, cf. Levy [10]) with parameters in A.

(1) Local countability: (x) (x is countable).

(2) $\mathbf{\Delta}_0$ separation:

$$(\mathsf{E}x)(y)[y \in x \leftrightarrow y \in a \ \& \ \mathcal{G}(y)].$$

(3) $\mathbf{\Delta}_0$ dependent choice:

$$(x)(\mathsf{E}y)\mathcal{G}(x,y) \rightarrow (\mathsf{E}h)(n)\mathcal{G}(hn, h(n+1)),$$

where $h\colon \omega \rightarrow A$.

If A is an abstract 1-section, then A is an admissible set as defined by Platek [8], and every member of A is hereditarily countable.

Each hereditarily countable set x can be encoded by a real number Y. Let m be the function that takes a code Y to the set mY encoded by Y. The relation

$$Y \text{ is a code } \& \ mY = x$$

[2] Not surprising, since his argument is based on a shadowy hierarchy of computations arising from the Kleene schemas of recursion. The existence of Gandy's selection operator was proved by him when $n = 0$, by Moschovakis [16] when $n = 1$, and by Platek [8] when $n > 1$.

is defined by induction on the rank of x: Y is a code for $\{m(Y_n) | n<\omega\}$, where $Y_n = \{a | 2^n \cdot 3^a \in Y\}$. The set of all codes is $\mathbf{\Pi}_1^1$, since a code is little more than a wellfounded relation on ω.

Proposition 3.1. *Let d be a code and $\mathcal{G}(x)$ a $\mathbf{\Delta}_0$ formula of* **ZF**. *The predicate $P(e)$, defined by*

$$(\mathsf{E}n)[e=d_n] \ \& \ \mathcal{G}(me) \, ,$$

is recursive in $d, {}^2E$.

Proof. If sets correspond to codes, then restricted set quantifiers correspond to number quantifiers.

For each $K \subset 2^\omega$ let $\mathcal{M}K$ be the set of all sets with codes in K. It is convenient to ignore the distinction between K and $\mathcal{M}K$ (e.g. to say K is an abstract 1-section instead of $\mathcal{M}K$ is an abstract 1-section) when every member of K is coded by some member of K. The next proposition extends a result of Platek [8] to the effect that the least ordinal not in $\mathcal{M}({}_1\mathrm{sc}^n E)$ is $\mathbf{\Sigma}_1$ admissible.

Proposition 3.2. *Suppose $n>1$ and U is a type n object in which ${}^nE$ is recursive. Then the 1-section of U is a countable abstract 1-section.*

Proof. It is immediate that $\mathcal{M}({}_1\mathrm{sc}\,U)$ is closed under the operations of pairing and union. To check $\mathbf{\Delta}_0$ separation let $\mathcal{G}(y)$ be a $\mathbf{\Delta}_0$ formula and $d \in {}_1\mathrm{sc}\,U$ be a code. It suffices to find a code $c \in {}_1\mathrm{sc}\,U$ such that
$(y)[y \in mc \leftrightarrow y \in md \ \& \ \mathcal{G}(y)]$ holds in $\mathcal{M}({}_1\mathrm{sc}\,U)$. The predicate $P(e)$, defined by

$$(\mathsf{E}n)[e=d_n] \ \& \ \mathcal{G}(me) \, ,$$

is recursive in U by Proposition 3.1. The desired c is such that $\{c_n | n<\omega\}$ is an enumeration of all the e's that satisfy $P(e)$.

To check $\mathbf{\Delta}_0$ dependent choice let $\mathcal{G}(x,y)$ be a $\mathbf{\Delta}_0$ formula such that

$$(x)(\mathsf{E}y)\mathcal{G}(x,y)$$

holds in $\mathcal{M}({}_1\text{sc}\, U)$. Suppose $\{p\}^U$ is a code. Let Q_p be the set of all n such that

$$\{n\}^U \text{ is a code } \& \ \mathcal{J}(m\{p\}^U, m\{n\}^U).$$

Q_p is recursively enumerable in U (uniformly in p). Gandy's selection operator yields a partial function t recursive in[3] U such that $tp \in Q_p$ whenever $\{p\}^U$ is a code. Define g recursively in U by:

$$g0 = e_0 \quad (\{e_0\}^U \text{ is a code for } 0),$$

$$g(n+1) = t(gn).$$

Then $(n)\mathcal{J}(m\{gn\}^U, m\{g(n+1)\}^U)$.

For each ordinal α let L_α be the set of sets constructible in the sense of Gödel [11] via ordinals less than α. α is said to be $\mathbf{\Sigma}_1$ admissible if L_α satisfies the $\mathbf{\Sigma}_1$ replacement axiom schema of **ZF**.

Proposition 3.3. *If α is a $\mathbf{\Sigma}_1$ admissible ordinal, then $L_\alpha \cap 2^\omega$ is an abstract* 1-*section.*

Proof. Gödel [11] shows $L \cap 2^\omega = L_{\omega_1} \cap 2^\omega$, where ω_1 is the least ordinal not countable in L. His argument restricted to L_α shows

$$L_\alpha \cap 2^\omega = L_{\omega_1^\alpha} \cap 2^\omega,$$

where ω_1^α is the least ordinal not countable in L_α. It follows $L_{\omega_1^\alpha}$ satisfies local countability. Gödel's wellordering of L restricted to $L_{\omega_1^\alpha}$ is $\mathbf{\Sigma}_1$ over $L_{\omega_1^\alpha}$.

Let $\mathbf{\Delta}_j^i$ be the set of all lightface $\mathbf{\Delta}_j^i$ reals.

Proposition 3.4. *If* $\min(i,j) \geq 1$, *then $\mathbf{\Delta}_j^i$ is a countable abstract* 1-*section.*

Proof. If $i = j = 1$, then the proposition follows from Spector's boundedness theorem [9] for $\mathbf{\Sigma}_1^1$ subsets of Kleene's O and Kreisel's selection operator [12] for $\mathbf{\Pi}_1^1$ predicates of numbers.

[3] I.e. the graph of t is recursively enumerable in U.

Assume max $(i,j) > 1$. Let **HC** be the set of hereditarily countable sets. The Kondo–Addison uniformization of $\mathbf{\Pi}_1^1$ predicates of reals implies **HC** satisfies $\mathbf{\Delta}_0$ dependent choice. Consequently **HC** is an abstract 1-section. It suffices to show $\mathcal{M}(\mathbf{\Delta}_j^i)$ is a $\mathbf{\Sigma}_1$ substructure of **HC**. Let $\mathcal{G}(x)$ be a $\mathbf{\Delta}_0$ formula with parameters in $\mathcal{M}(\mathbf{\Delta}_j^i)$ such that

$$\mathbf{HC} \models (\mathbf{E}x)\mathcal{G}(x) .$$

There exists an arithmetic predicate $A(Y)$ such that

$$[Y \text{ is a code } \ \& \ A(Y)] \leftrightarrow \mathbf{HC} \models \mathcal{G}(mY) .$$

The set parameters occurring in $\mathcal{G}(x)$ correspond to $\mathbf{\Delta}_j^i$ codes occurring in $A(Y)$. The Kondo–Addison uniformization supplies a code Z such that Z satisfies $A(Y)$ and is $\mathbf{\Delta}_2^1$ in the $\mathbf{\Delta}_j^i$ codes occurring in $A(Y)$. Since max $(i,j) > 1$, $Z \in \mathbf{\Delta}_j^i$. Hence $\mathcal{M}(\mathbf{\Delta}_j^i) \models (\mathbf{E}x)\mathcal{G}(x)$.

4. Generic type 2 objects

Let K be a countable abstract 1-section. Suppose F maps ω^ω into ω and is 0 off K. If F is generic in the sense of the following forcing relation, then the 1-section of $(F, {}^2E)$ is K.

Let p be a partial function from ω^ω into ω. p generates a hierarchy $\{T_\sigma\}$ of reals as defined below. If p is total, then the T_σ's are equivalent to the S_σ's of section 2 when $n = 0$ and $F = p$. If p is not total, then there may be a σ such that $T_{\sigma+1}$ has an index but is not total.

Stage 0. $T_0 = \{1\}$. 1 is an index for T_0. T_0 is total.

Stage $\sigma+1$. 2^e is an index for $T_{\sigma+1}$ if T_σ has an index, T_σ is total, $2^e \notin T_\sigma$, and $\{e\}^{T_\sigma}(m)$ is defined for all m. ($\{e\}^{T_\sigma}$ is the unique partial function from ω into ω recursive in T_σ via Gödel number e.)

$T_{\sigma+1}$ is total if it has an index and $p(\lambda m\{e\}^{T_\sigma}(m))$ is defined whenever 2^e is an index for $T_{\sigma+1}$.

$T_{\sigma+1}$ is T_σ augmented by: all indices for $T_{\sigma+1}$; all triples $\langle 3^e, m, n\rangle$ such that $\{e\}^{T_\sigma}(m) = n$ and 2^e is an index for $T_{\sigma+1}$; and all pairs $\langle 5^e, n\rangle$ such that $p(\lambda m\{e\}^{T_\sigma}(m)) = n$ and 2^e is an index for $T_{\sigma+1}$.

$$|m| = \sigma \text{ means } m \text{ is an index for } T_\sigma .$$

Stage λ (limits). 7^e is an index for T_λ if 2^e is an index for $T_{\delta+1}$ for some $\delta < \lambda$ and $\lambda m\{e\}^{T_\delta}(m)$ is the characteristic function of a set R of indices such that

$$\lambda = \sup \{|m| \mid m \in R\}.$$

T_λ is total if it has an index. T_λ is $\bigcup \{T_\delta | \delta < \lambda\}$ augmented by all indices for T_λ.

p is said to *generate* T_σ if T_σ has an index and is total. Fact H is easily proved by induction on σ.

Fact H. If $\sigma < \gamma$ and p generates T_σ and T_γ, then T_σ has lower Turing degree than T_γ.

Suppose p is total and S is a real. The arguments of Shoenfield [4] show S is Turing reducible to some T_σ generated by p if and only if S is recursive in p, ${}^2E$ in the sense of Kleene [2].

If $T_{\sigma+1}$ has an index but is not total, then $T_{\sigma+1}$ is said to be the *maximum* of p. p is a *forcing condition* if it meets two requirements.

(1) $p \in \mathfrak{M}K$ and has a maximum.

(2) X is in the domain of p if and only if X is Turing reducible to T_δ for some $\delta < \sigma$, where $T_{\sigma+1}$ is the maximum of p.

Requirement (2) is not as limiting as it may appear, because Fact H implies that the generation of T_σ by p utilizes the value of $p(X)$ only if X is Turing reducible to T_δ for some $\delta < \sigma$.

From this point on $p, q, r, \ldots$ denote forcing conditions. p is extended by q (in symbols $p \supset q$) if the graph of p is contained in the graph of q.

The language $\mathcal{L}(K)$ will be used to define the desired generic F's. The individual constants of $\mathcal{L}(K)$ are: $\underline{m}$ for each $m \in \omega$; $\underline{f}$ for each $f \in \omega^\omega \cap \mathfrak{M}K$; $\underline{\sigma}$ and $T_{\underline{\sigma}}$ for each ordinal $\sigma \in \mathfrak{M}K$; and $\underline{S}$ for each $S \in 2^\omega \cap \mathfrak{M}K$. The variables of $\mathcal{L}(K)$ are: $x, y, \ldots$ (numbers); $\mu, \nu, \ldots$ (ordinals); and T_μ, $T_\nu, \ldots$ (sets). The atomic formulas of $\mathcal{L}(K)$ are of the form: $|x| = \mu$, $|fx| = \mu$, $\mu < \nu$, and $\underline{S} \leq T_\mu$. The sentences of $\mathcal{L}(K)$ are built up from the atomic formulas by substitution of appropriate individual constants for variables and by application of propositional connectives (& and $\sim$) and existential number and ordinal quantifiers.

$\mathcal{J}$ is a *ranked* sentence of rank σ if $\mathcal{J}$ contains no ordinal quantifiers and σ is the least ordinal greater than every ordinal occurring in $\mathcal{J}$. Such an $\mathcal{J}$ is *true* in $\{T_\delta | \delta < \sigma\}$ if it is true when $\underline{\delta} < \underline{\gamma}$ is interpreted as $\delta < \gamma$, $|\underline{m}| = \underline{\delta}$ as

m is an index for T_δ, and $\underline{S} \leq \underline{T_\delta}$ as S is Turing reducible to T_δ. The forcing relation $\Vdash$ is defined inductively.

(i) $p \Vdash \mathcal{J}$ if $\mathcal{J}$ is of rank σ, p generates T_δ for all $\delta < \sigma$, and $\mathcal{J}$ is true in $\{T_\delta \mid \delta < \sigma\}$.

(ii) $p \Vdash (\mathrm{E}x)\mathcal{J}(x)$ if $p \Vdash \mathcal{J}(\underline{m})$ for some m.

(iii) $p \Vdash (\mathrm{E}\mu)\mathcal{J}(\mu)$ if $p \Vdash \mathcal{J}(\underline{\sigma})$ for some σ.

(iv) $p \Vdash \mathcal{J} \;\&\; \mathcal{G}$ if $p \Vdash \mathcal{J}$ and $p \Vdash \mathcal{G}$.

(v) $p \Vdash \sim \mathcal{J}$ if $(q)_{p \supset q} \sim [q \Vdash \mathcal{J}]$ and $\mathcal{J}$ is not ranked.

A sentence $\mathcal{G}$ is said to be $\mathbf{\Sigma}_1$ if it is in prenex normal form and contains no universal quantifiers.

Proposition 4.1. *The relation $p \Vdash \mathcal{G}$, restricted to $\mathbf{\Sigma}_1$ $\mathcal{G}$'s, is $\mathbf{\Sigma}_1$ over $\mathcal{M}K$.*

Proof. Suppose $w \in \mathcal{M}K$ and is a partial function from ω^ω into ω. The set of T_σ's generated by w is $\mathbf{\Sigma}_1$ over $\mathcal{M}K$ uniformly in w. (This last is a consequence of the $\mathbf{\Sigma}_1$ admissibility of $\mathcal{M}K$ and the autonomous fashion in which indices are assigned to T_λ when λ is a limit ordinal.) Thus if w has a maximum, then that maximum belongs to $\mathcal{M}K$. It follows that the set of forcing conditions is $\mathbf{\Sigma}_1$ over $\mathcal{M}K$.

Let F map ω^ω into ω and be 0 off K. F satisfies p (in symbols $F \in p$) if the graph of p is contained in the graph of F. F is *generic* if for each sentence $\mathcal{G}$ of the language $\mathcal{L}(K)$ there is a p such that $F \in p$ and either $p \Vdash \mathcal{G}$ or $p \Vdash \sim \mathcal{G}$. Generic F's exist because there are only countably many sentences to be forced. Standard arguments [15] show: if F is generic, then each true statement about F (expressible in the language $\mathcal{L}(K)$) is forced by some p satisfied by F.

Lemma 4.2. *If F is generic, then $K \subset {}_1\mathrm{sc}(F, {}^2E)$.*

Proof. Suppose S: $\omega \to \omega$ belongs to K. Fix p; since F is generic it suffices to find a $q \subset p$ such that

$$q \Vdash \underline{S} \leq \underline{T_\sigma}$$

for some σ. Since p has a maximum there is an e and a γ such that p generates T_γ, 2^e is an index for $T_{\gamma+1}$ and

$$p(\lambda m\{e\}^{T_\gamma}(m))$$

is undefined. Let $\lambda n | e_n$ be a recursive function such that

$$\{e_n\}^X \simeq n + \{e\}^X$$

for all $X \subset \omega$ and $n \in \omega$. Clearly $\{e_n\}^X$ is total if and only if $\{e\}^X$ is. It follows 2^{e_n} is an index for $T_{\gamma+1}$ because 2^e is. In addition

$$p(\lambda m\{e_n\}^{T_\gamma}(m))$$

is undefined, because the domain of p is an initial segment of Turing degrees.

Choose $q \subset p$ so that the domain of q consists of all functions Turing reducible to T_γ, and so that

$$q(\lambda m\{e_n\}^{T_\gamma}(m)) = Sn$$

for all $n \in \omega$. Thus q generates $T_{\gamma+1}$ but not $T_{\gamma+2}$, since $q(T_{\gamma+1})$ is undefined. And S is Turing reducible to $T_{\gamma+1}$ since

$$Sn = r \leftrightarrow \langle 5^{e_n}, r\rangle \in T_{\gamma+1}$$

for all n and r.

Lemma 4.3. *If F is generic, then ${}_1\mathrm{sc}(F, {}^2E) \subset K$.*

Proof. Suppose $S \in {}_1\mathrm{sc}(F, {}^2E) - K$ for the sake of a reductio ad absurdum. Then S is Turing reducible to some T_σ generated by F but not in $\mathcal{M}K$; $\sigma \notin \mathcal{M}K$ since F is generic. Let α be the least ordinal not in $\mathcal{M}K$. Then F generates some T_α with index 7^e. Thus 2^e is an index for some $T_{\delta+1}$ generated by F in $\mathcal{M}K$, and $\{e\}^{T_\delta}$ is the characteristic function of a set R such that

$$\alpha = \sup\{|n| \mid n \in R\}.$$

Let $f \in \mathcal{M}K$ enumerate R. Since F is generic there is a p satisfied by F such that:

(a) $p \Vdash (x)(\mathsf{E}\mu)[|\underline{fx}| = \mu]$;
(b) $p \Vdash (\mu)(\mathsf{E}x)(\mathsf{E}\nu)[\mu < \nu \ \& \ |\underline{fx}| = \nu]$;
(c) $p \Vdash \underline{R} \leq \underline{T_\delta}$;
(d) $p \Vdash \sim(\mathsf{E}\mu)[|\underline{7}^e| = \mu]$.

(a) is equivalent to

(a*) $(m)(q)_{p\supset q}(\mathsf{E}r)_{q\supset r}(\mathsf{E}\sigma)[r \Vdash |\underline{fm}| = \underline{\sigma}]$,

and (b) is equivalent to

(b*) $(\sigma)(q)_{p\supset q}(\mathsf{E}r)_{q\supset r}(\mathsf{E}m)(\mathsf{E}\gamma)_{\sigma<\gamma}[r \Vdash |\underline{fm}| = \underline{\gamma}]$.

It follows from (a*), (b*), Proposition 4.1 and the validity of the $\mathbf{\Sigma}_1$ dependent choice schema in $\mathcal{M}K$ that there exist functions $\lambda m | p_m$ and $\lambda m | \sigma_m$ in $\mathcal{M}K$ such that

$$p \supset p_m \supset p_{m+1}, \qquad p_m \Vdash |\underline{fm}| = \underline{\sigma_m}$$

and $(m)(\mathsf{E}n)[\delta < \sigma_m < \sigma_n]$. Let $\lambda = \sup\{\sigma_m | m \in \omega\}$ and $w =$ $\bigcup\{p_m | m \in \omega\}$. Clearly $\lambda, w \in \mathcal{M}K$. w generates $\underline{T_{\delta+1}}$ with index 2^e thanks to (c) and the fact that $\delta < \lambda$. Consequently w generates some T_λ with index 7^e. None of the p_m's generate T_λ by (d). It follows from requirement (2) of the definition of forcing condition and Fact H that $w(T_\lambda)$ is undefined. Thus w has a maximum (namely $T_{\lambda+1}$) and so is a forcing condition. But then $w \Vdash |7^e| = \underline{\lambda}$, an impossibility according to (d).

Theorem 4.4. *K is a countable abstract 1-section if and only if K is the 1-section of some type 2 object in which ^2E is recursive.*

Proof. If G is a type 2 object in which ^2E is recursive, then ${}_1\mathrm{sc}\,G$ is a countable abstract 1-section by Lemma 3.2.

Suppose K is a countable abstract 1-section. Let F be generic in the sense of Lemma 4.2, and let G be the recursive join of F and ^2E. Then $K = {}_1\mathrm{sc}\,G$ by Lemmas 4.2 and 4.3.

The following three corollaries of Theorem 4.4 are consequences of Propositions 3.2–3.4.

Corollary 4.5. *Suppose $n > 2$ and U is a type n object in which ^nE is recursive. Then there exists a type 2 object V such that*

$$_1\mathrm{sc}\,U = {}_1\mathrm{sc}\,V$$

and ^2E *is recursive in* V.

Corollary 4.6. *Suppose* α *is a countable* Σ_1 *admissible ordinal. Then there exists a type* 2 *object* V *such that*

$$L_\alpha \cap 2^\omega = {}_1\mathrm{sc}\,V$$

and ^2E *is recursive in* V.

Corollary 4.7. *If* $\min(i,j) \geq 1$, *then the set of all lightface* $\mathbf{\Delta}^i_j$ *reals is the* 1-*section of some type* 2 *object in which* ^2E *is recursive.*

5. Further results

The method of section 4 is applicable to the study of Gandy's superjump [7]. Theorems 5.1 and 5.2 are typical results of [14] and were inspired by some questions raised by P. Hinman at the 1969 Manchester Logic Colloquium. Let F and G be objects of type 2, G' the superjump of G, and E_1 the superjump of ^2E. $_1\mathrm{sc}\,G$ is said to be closed under hyperjump if

$$_1\mathrm{sc}(E_1, X) \subset {}_1\mathrm{sc}\,G$$

for every $X \in {}_1\mathrm{sc}\,G$.

Theorem 5.1. *Suppose* $_1\mathrm{sc}\,G$ *is closed under hyperjump. Then there exists an* F *such that*

$$_1\mathrm{sc}\,G = {}_1\mathrm{sc}(F')\,.$$

Theorem 5.2. (Assume $2^\omega = \omega_1$.) *There exists an* H *such that* $(G)(\mathbf{E}F)[H \leq G \to F' \equiv G]$.

The method of section 4 does not appear to suffice for the proof of the plus-one theorem when $k > 1$. A stability result is needed to overcome problems caused by gaps in the hierarchy of section 2, gaps that fall between objects recursive in $F, {}^{n+2}E$ when $n > 0$. Call R subrecursive in $F, {}^{n+2}E$ if R is recursive in some S_σ (as defined in section 2) with an index of the form $\langle 2^e, r\rangle$, where r is a subindividual. The stability result in question says: each nonempty recursively enumerable (in $F, {}^{n+2}E$) collection of subrecursive (in $F, {}^{n+2}E$) sets must have a recursive (in $F, {}^{n+2}E$) set among its members.

At this writing it is not known if there exists a decent notion of abstract k-section when $k > 1$. Decency requires that Theorem 4.4 remain true when "1-section" is replaced by "k-section" and "2" by "k+1".

References

[1] G.E. Sacks, Recursion in objects of finite type, Proceedings of the International Congress of Mathematicians 1 (1970) 251–254.
[2] S.C. Kleene, Recursive functionals and quantifiers of finite type, Trans. Amer. Math. Soc. 91 (1959) 1–52; 108 (1963) 106–142.
[3] G.E. Sacks, The k-section of a type n object, to appear.
[4] J. Shoenfield, A hierarchy based on a type 2 object, Trans. Amer. Math. Soc. 134 (1968) 103–108.
[5] T. Grilliot, Hierarchies based on objects of finite type, Jour. Symb. Log. 34 (1969) 177–182.
[6] G.E. Sacks, Higher Recursion Theory, Springer Verlag, to appear.
[7] R. Gandy, General recursive functionals of finite type and hierarchies of functions, University of Clermont-Ferrand (1962).
[8] R. Platek, Foundations of Recursion Theory, Ph.D. Thesis, Stanford (1966).
[9] C. Spector, Recursive well-orderings, Jour. Symb. Log. 20 (1955) 151–163.
[10] A. Levy, A hierarchy of formulas in set theory, Memoirs of the American Mathematical Society, Number 57 (1965).
[11] K. Gödel, The Consistency of the Axiom of Choice and of the Generalized Continuum Hypothesis (Princeton University Press, Princeton, 1966).
[12] G. Kreisel, Set theoretic problems suggested by the notion of potential totality, in: Infinitistic Methods (Pergamon Press, Oxford, and PWN, Warsaw, 1961) pp. 103–140.
[13] J. Shoenfield, The problem of predicativity, Essays on the Foundations of Mathematics (Magnes Press, Jerusalem, 1961 and North-Holland, Amsterdam, 1962).
[14] G.E. Sacks, Inverting the superjump, to appear.
[15] S. Feferman, Some applications of the notion of forcing and generic sets, Fund. Math. 56 (1965) 325–345.
[16] Y. Moschovakis, Hyperanalytic predicates, Trans. Amer. Math. Soc. 129 (1967) 249–282.

PART II

SETS AND ORDINALS

J.F.Fenstad, P.G.Hinman (eds.), Generalized Recursion Theory

ADMISSIBLE SETS OVER MODELS OF SET THEORY

K. Jon BARWISE [1]
University of Wisconsin, Madison and Stanford University

§1. Introduction

The addition of urelements gives a new dimension to the theory of admissible sets, a dimension which has applications in several parts of logic. To see why this addition is an obvious step to take we begin by reviewing the development of Zermelo–Fraenkel set theory, **ZF**, as it is usually presented (see for example Shoenfield [1967], §9.1).

The fundamental tenet of set theory is that given a collection of mathematical objects, subcollections are themselves perfectly reasonable mathematical objects, as are collections of these new objects, and so on. Thus we begin with a collection M of objects called *urelements* which we think of as being given outright. We construct sets on the collection M in stages. At each stage α, we have available all urelements and all sets constructed at previous stages. A collection is a *set* if it is formed at some stage in this construction; the collection of all sets built on M is denoted by V_M.

Now it turns out that *if* we allow strong enough principles of construction at each stage α, and *if* we assume that there are enough stages, then the urelements become redundant in that all ordinary mathematical objects occur, up to isomorphism, in V, i.e. in V_M for the empty collection M. It is for this reason that the axioms of **ZF** explicitly rule out the existence of urelements; the combination of the power set and replacement axioms are so strong as to make urelements unnecessary.

So formulated, **ZF** provides us with an extremely elegant way to organize existing mathematics. It does this at a cost, though. The principle of parsimony, historically of great importance in mathematics, is violated at almost every

[1] Research for and preparation of this paper were supported by NSF GP 27633 and NSF GP 34091X, respectively.

turn. And one of the main advantages of the axiomatic method is lost since **ZF** has so few recognizable models in which to interpret its theorems. For these reasons, and others familiar to anyone versed in generalized recursion theory, it eventually becomes profitable to look at set theories weaker than **ZF**, weaker in the principles of set existence which they allow us to use. The theory we have in mind here is the Kripke–Platek theory **KP** for admissible sets.

We now come to the main point. As we weaken the principles allowed in the construction of sets (in the move from **ZF** to **KP**) we destroy the earlier justification for throwing out urelements. In this paper we put them back in by "weakening" **KP** to a theory **KPU** which does not rule out the existence of urelements. **KP** will be equivalent to the theory **KPU** + "there are no urelements". The result is worth the trouble.

There is a great deal of folk literature about admissible sets as well as about admissible sets with urelements. A large portion of our talk at Oslo was devoted to a review of this folk material. When it came to writing it soon became obvious that neither time nor space would permit a complete treatment in this paper. We are currently at work on such a treatment, however, and plan to publish it as a textbook on admissible sets.

In this paper, then, we abandon once again any reader ignorant of the basics of admissible sets, and discuss the material from the last third of our Oslo talk: admissible covers of nonstandard models. We have chosen this topic because it offers nice examples of the new degree of freedom afforded by urelements in admissible sets, examples in recursion theory and in the model theory of set theory. Proofs not given here will be found in the book referred to above.

§2. The axioms of KPU

Let L be a first order language and let $\mathfrak{M} = \langle M, ... \rangle$ be a structure for the language L. We wish to form admissible sets which have M as a collection of urelements; these admissible sets are the intended models of the theory **KPU**, other models of **KPU** being so called non-standard admissible sets on M.

The theory **KPU** is formulated relative to a language $L^* = L(\in, ...)$ which extends L by adding a membership symbol $\in$ and, possibly, other function, relation and constant symbols. Rather than describe L^* precisely, we describe its class of structures, leaving it to the reader to formalize L^* in a way that suits his prejudices.

2.1. Definition. A *structure* for L^* consists of

(1) a structure $\mathfrak{M} = \langle M, \ldots\rangle$ for the language L, $M = \emptyset$ being kept open as a possibility,

(2) a nonempty set A disjoint from M.

(3) a relation $E \subseteq (M \cup A) \times A$ which interprets the symbol $\in$,

(4) other function, relation and constants on $M \cup A$ to interpret any other symbols in $L(\in, \ldots)$.

We denote such a structure by $\mathfrak{A}_{\mathfrak{M}} = (\mathfrak{M}\,; A, E, \ldots)$.

We use variables of L^* subject to the following conventions: Given a structure $\mathfrak{A}_{\mathfrak{M}} = (\mathfrak{M}; A, E, \ldots)$ for L^*,

$p, q, r, p_1, \ldots$	range over M	(urelements)
$a, b, c, d, a_1, \ldots$	range over A	(sets)
$x, y, z, \ldots$	range over $M \cup A$.	

We use u, v, w to denote any kind of variable. This notation gives us an easy way to assert that something holds of sets, or of urelements. For example, $\forall p\, \exists a\, \forall x\, (x \in a \leftrightarrow x = p)$ asserts that $\{p\}$ exists for any urelement p, where as $\forall p\, \exists a\, \forall q (q \in a \leftrightarrow a = p)$ asserts that there is a set a whose intersection with the class of all urelements is $\{p\}$.

The axioms of **KPU** are of three kinds. The axioms of extensionality and foundation concern the basic nature of sets. The axioms of pair, union and $\mathbf{\Delta}_0$-separation deal with the principles of set construction available to us. The most powerful axiom, $\mathbf{\Delta}_0$-collection, guarantees that there are enough stages in our construction process. In order to state the latter two axioms we need to define the notion of $\mathbf{\Delta}_0$-formula of $L(\in, \ldots)$, due to Lévy [1965].

2.2. Definition. The collection of $\mathbf{\Delta}_0$-formulas of a language $L(\in, \ldots)$ is the smallest collection Y containing the atomic formulas of L^* and closed under

(1) if ϕ is in Y then so is $\neg\phi$

(2) if ϕ, ψ are in Y so are $(\phi \wedge \psi)$ and $(\phi \vee \psi)$

(3) if ϕ is in Y then so are

$$\forall u \in v \phi \quad \text{and} \quad \exists u \in v \phi$$

for all variables, u and v.

The importance of Δ_0-formulas rests in the fact that many useful predicates can be defined by Δ_0-formulas and that any predicate defined by a Δ_0-formula is very absolute.

2.3. Definition. The theory **KPU** (relative to a language L(∈, ...)) consists of the universal closures of the following formulas:

Extensionality: $\forall x\,(x \in a \leftrightarrow x \in b) \rightarrow a = b$

Foundation: $\exists a\, \phi(a) \rightarrow \exists a [\phi(a) \wedge \forall b \in a \neg \phi(b)]$ for all formulas $\phi(a)$ in which b does not occur free.

Pair: $\exists a\,(x \in a \wedge y \in a)$

Union: $\exists b \forall y \in a \forall x \in y (x \in b)$

Δ_0-Separation: $\exists b \forall x\,(x \in b \leftrightarrow x \in a \wedge \phi(x))$ for all Δ_0 formulas in which b does not occur free.

Δ_0-Collection: $\forall x \in a \exists y\, \phi(x,y) \rightarrow \exists b \forall x \in a \exists y \in b\, \phi(x,y)$ for all Δ_0 formulas in which b does not occur free.

2.4. Definition. 1. $\mathbf{KPU}^+$ is **KPU** plus the axiom:

$$\exists a \forall x\, [x \in a \leftrightarrow \exists p\,(x = p)]$$

which asserts that there is a set of all urelements.

2. **KP** is **KPU** plus the axiom

$$\forall x\, \exists a\,(x = a)$$

which asserts that there are no urelements.

One word of caution. There are some axioms built into our definition of structure for L(∈, ...). For example, the sentence

$$\forall p \forall x\,(x \notin p)$$

follows from 2.1.3, and

$$\forall p \forall a\,(p \neq a)$$

follows from 2.1.2.

In a systematic treatment one would now develop axiomatically a large part of elementary set theory in **KPU**. It is done almost exactly as it is for **KP**, the only trouble being that there is no such axiomatic development in print for **KP**. We must therefore leave it to the reader to work most of this out for himself. In particular, he should verify that the following are provable in **KPU**.

"There is a unique set a with no elements"

"Given a, there is a unique set $b = \bigcup a$ such that $x \in b$ iff $\exists y \in a\,(x \in y)$."

"Given a, b there is a unique set $c = a \cup b$ such that $x \in c$ iff $x \in a$ or $x \in b$."

"Given a, b there is a unique set $c = a \cap b$ such that $x \in c$ iff $x \in a$ and $x \in b$."

We define, as usual, the ordered pair of x,y by

$$\langle x,y \rangle = \{\{x\}, \{x,y\}\}$$

and prove that $\langle x,y \rangle = \langle z,w \rangle$ iff $x = z$ and $y = w$, and then prove in **KPU** that

"for all a, b there is a set $c = a \times b$, the Cartesian product of a and b, such that

$$c = \{\langle x,y \rangle : x \in b \text{ and } y \in b\}."$$

§3. Some useful principles provable in KPU

A Σ_1 formula is one of the form $\exists u\, \phi(u)$ where ϕ is a Δ_0-formula. It turns out that a wide class of formulas are equivalent to Σ_1 formulas in **KPU**.

3.1. Definition. The class of Σ formulas is the smallest class of formulas Y containing the Δ_0 formulas and closed under 2.2.2, 2.2.3 and

$$\text{if } \phi \text{ is in } Y \text{ so is } \exists u \phi, \text{ for all variables } u\,.$$

Thus, for example, the predicate, "x is a set of urelements" can be written

$$\exists a\,(x = a) \wedge \forall u \in x\, \exists p\,(x = p)$$

which is Σ but, as written, is not Σ_1. We will show, however, that for every

Σ formula ϕ there is a Σ_1 formula ϕ' with the same free variables such that **KPU** $\vdash \phi \leftrightarrow \phi'$.

Given a formula ϕ and a variable w we write $\phi^{(w)}$ for the result of replacing each unbounded quantifier in ϕ by a bounded quantifier:

$$\exists u \quad \text{by} \quad \exists u \in w$$

$$\forall u \quad \text{by} \quad \forall u \in w$$

for all variables u. Thus $\phi^{(w)}$ is a Δ_0 formula. If ϕ is Δ_0 then $\phi^{(w)} = \phi$, since there are no unbounded quantifiers ϕ.

3.2. Lemma. *For each Σ formula ϕ the following are logically valid (i.e. true in all structures $\mathfrak{A}_{\mathfrak{M}}$):*

$$\phi^{(u)} \wedge u \subseteq v \rightarrow \phi^{(v)}$$

$$\phi^{(u)} \rightarrow \phi$$

where $u \subseteq v$ abbreviates the formula $\forall x\, [x \in u \rightarrow x \in v]$.

3.3. Σ reflection principle. *For all Σ formulas ϕ the following is a theorem of* **KPU**:

$$\phi \leftrightarrow \exists a \phi^{(a)}\, .$$

(We assume a is any set variable not occuring free in ϕ, and stop making such assumptions explicit in the remainder of this paper.)

Proof. We know from the previous lemma that $\exists a \phi^{(a)} \rightarrow \phi$ is just plain valid, so the axioms of **KPU** come in only in showing $\phi \rightarrow \exists a \phi^{(a)}$. The proof is by induction on ϕ, the case for Δ_0 formulas being trivial. We take the three most interesting cases, leaving the other two to the reader.
Case (i). ϕ is $\psi \wedge \theta$. Assume

$$\textbf{KPU} \vdash \psi \leftrightarrow \exists a\, \psi^{(a)}$$

and

$$\mathsf{KPU} \vdash \theta \leftrightarrow \exists a \theta^{(a)}$$

as induction hypothesis and prove

$$\mathsf{KPU} \vdash (\psi \wedge \theta) \to \exists a [\psi \wedge \theta]^{(a)} .$$

Let us work in **KPU**, assuming $\psi \wedge \theta$ and proving $\exists a [\psi^{(a)} \wedge \theta^{(a)}]$. Now there are a_1, a_2 such that $\psi^{(a_1)}, \theta^{(a_2)}$ so let $a = a_1 \cup a_2$. Then $\phi^{(a)}$ and $\psi^{(a)}$ hold by the above lemma.

Case (ii). ϕ is $\forall u \in v \psi(u)$. Assume that

$$\mathsf{KPU} \vdash \psi \leftrightarrow \exists a \, \psi^{(a)} .$$

Again, working in **KPU** assume $\forall u \in v \, \psi(a)$ and prove $\exists a \forall u \in v \, \psi(u)^{(a)}$. For each $u \in v$ there is an b such that $\psi(u)^{(b)}$, so by $\mathbf{\Delta}_0$-collection there is an a_0 such that $\forall u \in v \exists b \in a_0 \, \psi(u)^{(b)}$. Let $a = \bigcup a_0$. Now for every $u \in v \, \exists b \subseteq a \, \psi(u)^{(b)}$ so $\forall u \in v \, \psi(u)^{(a)}$ by the above lemma.

Case (iii). ϕ is $\exists u \, \psi(u)$. Assume $\psi(u) \leftrightarrow \exists b \, \psi(u)^{(b)}$ proved and suppose $\exists u \, \psi(u)$ true. We need an a such that $\exists u \in a \, \psi(u)^{(a)}$. If $\psi(u)$ holds, pick b so that $\psi(u)^{(b)}$ and let $a = b \cup \{u\}$. Then $u \in a$ and $\psi(u)^{(a)}$ by the above lemma. □

In his original development of admissible sets, Platek took the $\mathbf{\Sigma}$ reflection principles as one of the axioms, since it is more useful than $\mathbf{\Delta}_0$-collection. $\mathbf{\Delta}_0$-collection, however, is usually easier to verify in a particular admissible set. We list below some of the consequences of the $\mathbf{\Sigma}$ reflection principle.

3.4. $\mathbf{\Sigma}$-collection. *For every $\mathbf{\Sigma}$ formula ϕ the following is a theorem of* **KPU**: *If $\forall x \in a \, \exists y \, \phi(x,y)$ then there is a set b such that*

$$\forall x \in a \, \exists y \in b \, \phi(x,y)$$

and

$$\forall y \in b \, \exists x \in a \phi(x,y) .$$

3.5. $\mathbf{\Delta}$-separation. *For any two $\mathbf{\Sigma}$ formulas $\phi(x), \psi(x)$, the following is a* *theorem of* **KPU**: *If for* $x \in a$, $\phi(x) \leftrightarrow \neg\psi(x)$ *then there is a set* b,

$$b = \{x \in a : \phi(x)\} .$$

3.6. Σ-replacement. *For each* Σ *formula* $\phi(x,y)$ *the following is a theorem of* **KPU**: *If* $\forall x \in a\, \exists! y\, \phi(x,y)$ *then there is a function* f *with domain* a *such that* $\forall x \in a\, \phi(x, f(x))$.

The above is sometimes unusable because of the uniqueness condition in the hypothesis. In these situations it is 3.7 that often comes to the rescue.

3.7. Strong Σ-replacement. *For each* Σ *formula* $\phi(x,y)$ *the following is a theorem of* **KPU**: *If* $\forall x \in a\, \exists y\, \phi(x,y)$ *then there is a function* f *with domain* a *such that for every* $x \in a$

$$f(x) \neq 0$$

$$\forall y \in f(x)\, \phi(x,y) .$$

A set a is *transitive* if for all $x \in a$ and all $y \in x$, $y \in a$. Thus if a is a set of urelements it is transitive. An urelement is never transitive since only sets are transitive. We can prove in **KPU** that for every x there is a unique transitive set a with $x \subseteq a$ such that if b is any other transitive set containing x, then $a \subseteq b$. This set a is called the *transitive closure* of x, $\mathrm{TC}(x)$. Using TC one can go on to justify recursive definitions over $\in$. For example, the support function can be defined by

$$\mathrm{Sp}(p) = \{p\}$$

$$\mathrm{Sp}(a) = \bigcup_{x \in a} \mathrm{Sp}(x) .$$

Thus, $\mathrm{Sp}(a) = \{p : p \in \mathrm{TC}(a)\}$. A *pure set* is a set a with empty support, $\mathrm{Sp}(a) = 0$.

We will also need the second recursion theorem which for **KPU** takes the following form.

3.8. Second recursion theorem for KPU. *Let $\phi(\boldsymbol{x}, \boldsymbol{y}, R)$ be a Σ formula of* $\mathrm{L}(\in, ..., R)$, *where $\boldsymbol{x} = x_1 ... x_n$, $\boldsymbol{y} = y_1 ... y_k$ and R is an n-ary relation symbol occuring positively in ϕ. There is Σ_1 formula $\psi(\boldsymbol{x}, \boldsymbol{y})$ of* $\mathrm{L}(\in, ...)$ *such that*

$$\mathsf{KPU} \vdash \forall \boldsymbol{x} \forall \boldsymbol{y} [\psi(\boldsymbol{x}, \boldsymbol{y}) \leftrightarrow \phi(\boldsymbol{x}, \boldsymbol{y}, \psi(\cdot, \boldsymbol{y}))] .$$

Proof. We are using $\phi(\boldsymbol{x}, \boldsymbol{y}, \psi(\cdot, \boldsymbol{y}))$ to denote the result of replacing $R(z_1 ... z_n)$ by $\psi(z_1 ... z_n, y_1 ... y_k)$ wherever it occurs in ϕ. To simplify notation we consider the case where $n = k = 1$. Let $\theta(x, y, z)$ be

$$\exists a [\phi(x, y, \mathrm{Sat}(z, \cdot, y, z))]^{(a)}$$

where $\mathrm{Sat}(z, u_1 u_2 u_3)$ is the Σ_1 satisfaction relation for Σ_1 formulas z of $3 = n + k + 1$ variables (cf. Lévy [1965]). Let m be the Gödel number of this formula $\theta(x, y, z)$ and let $\psi(x, y)$ be $\theta(x, y, m)$, or rather the Σ_1 formula equivalent to it where m has been replaced by its definition. Then

$$\begin{aligned} \psi(x,y) &\leftrightarrow \theta(x, y, m) \\ &\leftrightarrow \exists a [\phi(x, y, \mathrm{Sat}(m, \cdot, y, m))]^{(a)} \\ &\leftrightarrow \phi(x, y, \mathrm{Sat}(m, \cdot, y, m)) \\ &\leftrightarrow \phi(x, y, \theta(\cdot, y, m)) \\ &\leftrightarrow \phi(x, y, \psi(\cdot, y)) . \quad \square \end{aligned}$$

§4. Admissible sets over M

It facilitates matters if we fix notation and let V_M denote the most generous possible universe of sets built on M so that our admissible sets on M will be substructures of V_M. Thus, we define

$$\begin{aligned} &V_M(0) = 0 \\ &V_M(\alpha+1) = \text{Power set}\,(V_M(\alpha) \cup M) \\ &V_M(\lambda) = \textstyle\bigcup_{\alpha<\lambda} V_M(\alpha) \quad \text{if } \lambda \text{ is a limit ordinal} \end{aligned}$$

and let

$$V_M = \bigcup_\alpha V_M(\alpha)$$

where the latter union is taken over all ordinals. If we need to keep things straight for some reason we subscript notions with an M to denote their interpretations in V_M. For examples, $\in_M$ denotes the membership relation of V_M where each $p \in M$ is taken as having no elements (even though in some other context M might be a set of sets) and "a is transitive$_M$" means "$x \in_M y \in_M a$ implies $x \in_M a$".

4.1. Definition. A structure $\mathfrak{A}_\mathfrak{M} = (\mathfrak{M}\,; A, E, ...)$ for $L(\in, ...)$ is *admissible* if $\mathfrak{A}_\mathfrak{M}$ is a model of **KPU**, if A is a transitive$_M$ subset of V_M (where $\mathfrak{M} = \langle M, ...\rangle$) and E is the restriction of $\in_M$ to $M \cup A$.

If ordinary admissible sets are pictured as in fig. 1a, as they often are in informal discussions, then admissible sets with urelements should be pictured somewhat as in fig. 1b.

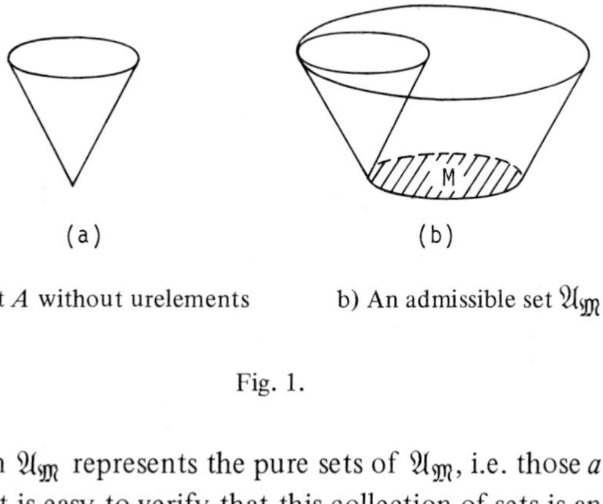

a) An admissible set A without urelements b) An admissible set $\mathfrak{A}_\mathfrak{M}$ over M

Fig. 1.

The small cone in $\mathfrak{A}_\mathfrak{M}$ represents the pure sets of $\mathfrak{A}_\mathfrak{M}$, i.e. those $a \in A_\mathfrak{M}$ with empty support. It is easy to verify that this collection of sets is an admissible set (without urelements).

4.2. *Example*. For any infinite cardinal κ define $H(\kappa)_\mathfrak{M} = (\mathfrak{M}; A, \in)$ where $A = \{a \in V_M : \mathrm{TC}_M(a)$ has cardinality $|\mathrm{TC}_M(a)| < \kappa\}$. For any such κ, $H(\kappa)_\mathfrak{M}$ is admissible. $H(\kappa)_\mathfrak{M}$ is a model of **KPU**$^+$ iff $\kappa > |M|$. We usually denote $H(\omega)_\mathfrak{M}$ by $\mathrm{HF}_\mathfrak{M}$ since it consists of the hereditarily finite sets relative to $\mathfrak{M}$.

A subset R of the admissible $\mathfrak{A}_{\mathfrak{M}}$ is Σ_1 if it is definable on $\mathfrak{A}_{\mathfrak{M}}$ by a Σ_1 formula with parameters from $A \cup M$. R is Δ_1 if both R and its complement $(A \cup M) - R$ are Σ_1 on $\mathfrak{A}_{\mathfrak{M}}$.

4.3. *Example.* If $\mathfrak{M}$ is an acceptable structure then a subset of $\mathfrak{M}$ is semi-search computable on $\mathfrak{M}$ iff it is Σ_1 on $\mathrm{HF}_{\mathfrak{M}}$, the notations of acceptable and semi-search computable being those of Moschovakis. The result is due to Gordon.

The *ordinal of an admissible set* is the least ordinal not in the admissible set.

4.4. *Example.* Given any structure $\mathfrak{M}$ there is a smallest admissible set over which is a model of **KPU**$^+$, i.e. where the set M itself is an element of the admissible set. Denote this admissible set by $\mathrm{HYP}_{\mathfrak{M}}$. The proofs in Barwise–Gandy–Moschovakis [1971], if carried out in this setting, show that if $\mathfrak{M}$ is acceptable then a relation S is inductive on $\mathfrak{M}$ iff it is Σ_1 on $\mathrm{HYP}_{\mathfrak{M}}$ and is hyperelementary on $\mathfrak{M}$ iff it is an element of $\mathrm{HYP}_{\mathfrak{M}}$. The ordinal of $\mathrm{HYP}_{\mathfrak{M}}$ is the closure ordinal of the class of first order positive inductive definitions over $\mathfrak{M}$. (We are using inductive and hyperelementary, as in Moschovakis [1973], for what was called semi-hyperprojective and hyperprojective in Barwise–Gandy–Moschovakis [1971].)

4.5. *Example.* The results in infinitary logic of Barwise [1969a] all go through in this more general setting without change. To see how this may nevertheless be a significant extension, suppose $\mathfrak{A}_{\mathfrak{M}} = (\mathfrak{M}; A, \in_M, ...)$ is countable and admissible with ordinal ω. (We will see many examples of such $\mathfrak{A}_{\mathfrak{M}}$ in the following sections.) Let $\mathbf{T}_1, \mathbf{T}_2$ be theories of the admissible fragment L_A of $\mathrm{L}_{\omega_1\omega}$, Σ_1 on $\mathfrak{A}_{\mathfrak{M}}$, such that every $\phi \in \mathbf{T}_2$ is a pure set (and hence finitary). If every *finite* subset of $\mathbf{T}_2$ is consistent with $\mathbf{T}_1$ then $\mathbf{T}_1 \cup \mathbf{T}_2$ is consistent, even though $\mathbf{T}_1$ may have infinitary sentences in it. The proof is a simple consequence of the compactness theorem for L_A.

§5. Properties of the admissible cover

Admissible sets $\mathfrak{A}_{\mathfrak{M}} = (\mathfrak{M}; A, \in, ...)$ embody certain principles of set contribution, the ordinals α in A give us the stages, the sets of rank α the principles of set formation available at stage α. What then is to be made of nonstandard models of **KPU**, **KP** or **ZF**. The results we discuss here shows that there is a hard core of admissibility in the heart of even the most non-standard models.

5.1. Definition. Given a model $\mathfrak{M} = \langle M, E, \ldots\rangle$, where E is binary, the *covering function* C_E for $\mathfrak{M}$ is the function which assigns to each $x \in M$ the set

$$x_E = \{y \in M : yEx\}.$$

The name "covering function" is new but the function itself is basic in the study of models of set theory. For example, if $\mathfrak{M} = \langle M, E\rangle$ is a substructure of $\mathfrak{N} = \langle N, F\rangle$ then an $x \in M$ is *fixed* by $\mathfrak{N}$ if $x_E = x_F$; otherwise x is *enlarged* by $\mathfrak{N}$. If $x_E = x_F$ for all $x \in M$ then $\mathfrak{N}$ is an end extension of $\mathfrak{M}$, written $\mathfrak{M} \subseteq_{\text{end}} \mathfrak{N}$.

The covering function for $\mathfrak{M} = \langle M, E\rangle$ maps elements of M to subsets of M hence to elements of V_M. If $\mathfrak{M} \models$ Extensionality then this function is one-one. There are many admissible $\mathfrak{A}_{\mathfrak{M}}$ which are admissible with respect to the covering function for $\mathfrak{M}$. For example, any $H(\kappa)_{\mathfrak{M}}$ for $\kappa > |M|$. If $\mathfrak{M}$ satisfies enough axioms of set theory, however, there is one admissible $\mathfrak{A}_{\mathfrak{M}}$ which really lives over $\mathfrak{M}$. This set is called the admissible cover of $\mathfrak{M}$ and is the object of study of this paper.

5.2. Theorem. *Let* **T** *be some set theory containing* **KP** *and let* $\mathfrak{M} = \langle M, E\rangle$ *be a model of* **T**, *standard or nonstandard, with covering function* C_E. *There is an admissible set* $\mathfrak{A}_M$ *over* $\mathfrak{M}$, *called the* admissible cover *of* $\mathfrak{M}$, *with Properties I–IX listed below.*

Property I. C_E *maps* M *into* $\mathfrak{A}_M$ *and* $\mathfrak{A}_M$ *is admissible with respect to* C_E; *i.e.* $\mathfrak{A}_M = (M; A_M, \in \restriction A_M, C_E)$ *is admissible.*

This is equivalent to saying that $(\mathfrak{M}; A_M, \in_M, C_E)$ is admissible since E can be recovered from C_E.

Property II. *There is a function* $*: A_M \cup M \to M$ *satisfying*:

$$p^* = p \quad \text{for all} \quad p \in M$$

$$(a^*)_E = \{b^* : b \in a\} \quad \text{for} \quad a \in A_M.$$

One might call $*$ an $\in$-retraction of $\mathfrak{A}_M$ onto $\mathfrak{M}$. It is Property II which insures that the admissible cover of $\mathfrak{M}$ really lives over $\mathfrak{M}$. To be more precise:

Property III. *$\mathfrak{A}_M$ is uniquely determined by Properties I and II. In fact, $\mathfrak{A}_M$ is contained in any admissible $\mathfrak{B}_M$ satisfying I and contains any admissible $\mathfrak{B}_M$ satisfying II.*

Since I and II characterize the admissible cover of $\mathfrak{M}$, all other properties could be derived from them, but such a procedure would cause us to duplicate many steps in the proofs of I and II. For example, the following is obvious from the proof of II.

Property IV. *The cardinality of $\mathfrak{A}_M$ is the same as that of M.*

The well founded part of a model $\mathfrak{M} = \langle M, E\rangle$ of **KP**, $\mathbf{WF}(\mathfrak{M})$, is the transitive set (in the sense of V) which is the range of the following collapsing function clpse. The domain of clpse is the set of all $x \in M$ for which E is well founded on $\mathrm{TC}^{\mathfrak{M}}(x)$, so that the following makes sense:

$$\mathrm{clpse}(x) = \{\mathrm{clpse}(y) : y \in x_E\} .$$

If $\mathfrak{M}$ is well founded then clpse: $\mathfrak{M} \cong \langle \mathbf{WF}(\mathfrak{M}), \in\rangle$.

Property V. *The pure sets of the admissible cover of $\mathfrak{M}$ are exactly those sets in $\mathbf{WF}(\mathfrak{M})$, the well founded part of $\mathfrak{M}$. In particular, the ordinal of A_M is just the ordinal of the well founded part of $\mathfrak{M}$.*

We will attempt a picture of $\mathfrak{M}$ and its admissible cover at this point. The dotted line in $\mathfrak{M}$ represents the point at which $\mathfrak{M}$ becomes non-standard, the lower portion being isomorphic to $\mathbf{WF}(\mathfrak{M})$.

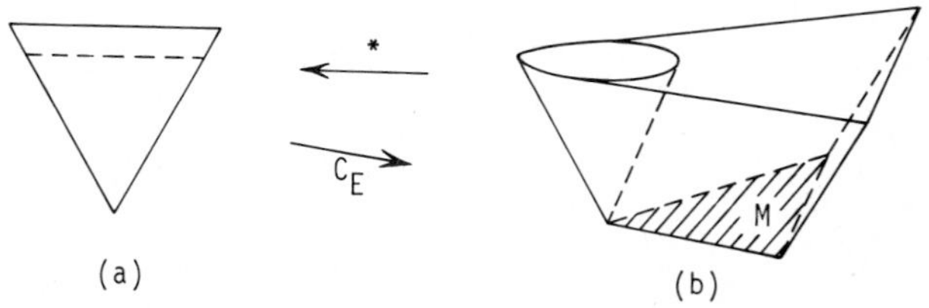

a) $\mathfrak{M} = \langle M, E\rangle$, a model of set theory b) The admissible cover $\mathfrak{A}_M$ of $\mathfrak{M}$

Fig. 2.

This gives a hint of the new dimension now available to us. Consider, for example, a model $\mathfrak{M}$ of **ZF** with nonstandard integers. The admissible cover of $\mathfrak{M}$ has many infinite sets in it (a_E for any infinite integer a, for example), but only the natural numbers for ordinals. This is in stark contrast to old fashioned admissible sets where if $A \neq \mathrm{HF}$ then $\omega \in A$.

The next three properties are of a more recursion theoretic nature.

We say that an inductive definition Γ on $\mathfrak{M}$ is a Σ inductive definition if the clause $\boldsymbol{x} \in \Gamma(R)$ is given by a Σ formula $\phi(\boldsymbol{x},R)$ with R occuring positively in ϕ:

$$\boldsymbol{x} \in \Gamma(R) \leftrightarrow \phi(\boldsymbol{x},R)\,.$$

We use I_ϕ for the fixed point of Γ, as in Moschovakis [1973]. A relation S of n variables is Σ inductively definable on $\mathfrak{M}$ if there is a Σ inductive definition I_ϕ of $n+k$ variables, $k \geq 0$, and $x_1 \ldots x_k$ in $\mathfrak{M}$ such that

$$R(y_1 \ldots y_n) \Leftrightarrow (y_1 \ldots y_n, x_1 \ldots x_k) \in I_\phi\,.$$

For example, the domain of the function clpse is Σ inductively defined on $\mathfrak{M}$ by the formula $\phi(x,R)$:

$$\forall y \in x\, R(y)\,.$$

A theorem of Gandy shows that if $\mathfrak{M}$ is an admissible set then any Σ inductively definable set on $\mathfrak{M}$ is Σ_1 on $\mathfrak{M}$. If $\mathfrak{M}$ is nonstandard, however, I_ϕ need not be even first order definable over $\mathfrak{M}$, as the domain of the function clpse shows.

Property VI. *A relation S on $\mathfrak{M}$ is Σ inductively definable on $\mathfrak{M}$ if and only if it is Σ_1 on the admissible cover of $\mathfrak{M}$. The closure ordinal of a Σ inductive definition is at most the ordinal of A_M.*

In Barwise [1969b] we showed that the strict-$\mathbf{\Pi}_1^1$ relations on an admissible set A coincide with the s.i.i.d. relations of Kunen [1968], and hence if A is countable the strict $\mathbf{\Pi}_1^1$ relations coincide with the Σ_1 relations. The proofs of these results carry over verbatim to admissible sets with urelements.

In Aczel [1970] it was announced that if $\mathfrak{M}$ is a countable nonstandard model of **KP** then the $s-\mathbf{\Pi}_1^1$ relations on $\mathfrak{M}$ coincide, not with the $\mathbf{\Sigma}_1$ relations, but with the $\mathbf{\Sigma}$ inductively definable relations on $\mathfrak{M}$. This follows from Properties VI and VII, and the above paragraph.

Property VII. *A relation S on $\mathfrak{M}$ is $s-\mathbf{\Pi}_1^1$ on $\mathfrak{M}$ iff it is $s-\mathbf{\Pi}_1^1$ on the admissible cover of $\mathfrak{M}$.*

The non-trivial half of VII and half of III can be derived from VIII.

Property VIII. *The admissible cover of $\mathfrak{M}$ is the $\in$-hard core (in the sense of Kreisel) of the class of all models of* **KPU** *of the form*

$$(\mathfrak{N}\,; B, E, C_F)$$

where $\mathfrak{N} = \langle N, F\rangle$ is an end extension of $\mathfrak{M}$ satisfying **KP**.

In the next section we are going to discuss the use of admissible covers in constructing models of set theory. The first application uses only properties I and II. In some applications, however, we need the following. Given $\mathfrak{M} \subseteq \mathfrak{N}$ and $M_0 \subseteq M$ we write

$$\mathfrak{M} \prec_1 \mathfrak{N} \ \ [\text{wrt}\, M_0]$$

to indicate that every $\mathbf{\Sigma}_1$ sentence with parameters from X has the same truth value in $\mathfrak{M}$ as in $\mathfrak{N}$. If $M_0 = M$ we write $\mathfrak{M} \prec_1 \mathfrak{N}$ and if $X = \emptyset$ we write $\mathfrak{M} \equiv_1 \mathfrak{N}$.

Let $\mathfrak{M} = \langle M, E\rangle$ and $\mathfrak{N} = \langle N, F\rangle$ be models of **KP** with admissible covers $\mathfrak{A}_M$ and $\mathfrak{A}_N$. It follows from VIII that $\mathfrak{M} \subseteq_{\text{end}} \mathfrak{N}$ iff $\mathfrak{A}_M \subseteq \mathfrak{A}_N$. In this case if $x \in \mathfrak{A}_M$ then x^* has the same value regardless of whether the $\in$-retraction $*$ is taken in the sense of $\mathfrak{A}_M$ or $\mathfrak{A}_N$.

Property IX. *Given $\mathfrak{M}$ and $\mathfrak{N}$ and $\mathfrak{N}$ as above with $\mathfrak{M} \subseteq_{\text{end}} \mathfrak{N}$ and $M_0 \subseteq M$, let $A_0 = \{x \in A_M : x^* \in M_0\}$. Then*

$$\mathfrak{M} \prec_1 \mathfrak{N} \ \ [\text{wrt}\, M_0]$$

if and only if

$$\mathfrak{A}_M \prec_1 \mathfrak{A}_N \ [\text{wrt}\, A_0]\ .$$

In particular, if $\mathfrak{M} \equiv_1 \mathfrak{N}$ then $\mathfrak{A}_M \equiv_1 \mathfrak{A}_N$ and if $\mathfrak{M} \prec_1 \mathfrak{N}$ then $\mathfrak{A}_M \prec_1 \mathfrak{A}_N$. There are a number of other relations (e.g. $\equiv, \prec$) which lift up from $(\mathfrak{M}, \mathfrak{N})$ to $(\mathfrak{A}_M, \mathfrak{A}_N)$ but the above is what is needed in the applications we have in mind.

We could have considered admissible covers of models of **KPU**, rather than **KP** and all of the properties would go through with slight modification. We restricted ourselves to models $\mathfrak{M}$ of **KP** for pedagogical reasons only.

§6. Infinitary methods in the model theory of set theory, revisited: Applications of the admissible cover

We first came across the admissible cover in trying to understand what was behind some of the proofs in Barwise [1970]. The general situation was this. We had a countable model $\mathfrak{M} = \langle M, E \rangle$ of set theory and we wanted to construct an end extension of $\mathfrak{M}$ with certain properties. If M was a transitive set with $E = \in \restriction M$ then we were able to exploit the completeness and compactness theorems for L_M, the admissible fragment of $L_{\omega_1\omega}$ associated with M. If $\mathfrak{M}$ was nonstandard, however, we were forced into some strange considerations. It is the admissible cover which unifies these two cases and allows the simpler proofs to go through whether $\mathfrak{M}$ is standard or not.

We give two examples of this here. In the first we extend a theorem of Friedman's from standard $\mathfrak{M}$ to arbitrary $\mathfrak{M}$. This result uses only Properties I and II of admissible covers, I in order to form sentences like

$$\forall x (x \in \bar{a} \leftrightarrow \mathbf{V}_{b \in a_E} x = \bar{b})$$

in A_M and II to know that if a theory T is an element of A_M then the set of $a \in M$ with $\bar{a}$ mentioned in T is a set of the form d_E, $d \in M$, and hence is bounded in $\mathfrak{M}$.

Friedman's result is the following: Let $\mathfrak{M} = \langle A, \in \rangle$ be a countable admissible set and let **T** be a theory of L_A which is Σ_1 on A, contains **KP**, and is true in some end extension of $\mathfrak{M}$. (For example, **T** might be **ZF** and it might be false in $\mathfrak{M}$.) The conclusion is that $\mathfrak{M}$ has an end extension $\mathfrak{N}$ with new ordinals which is a model of **T** but where all the new ordinals are nonstandard. We can extend his result as follows.

6.1. Theorem. *Let* $\mathfrak{M} = \langle M, E\rangle$ *be a countable model of* **KP** *with admissible cover* $\mathfrak{A}_M$. *Let* **T** *be a (finitary or infinitary) theory* Σ_1 *definable on* $\mathfrak{A}_M$ *which is true in* $\mathfrak{M}$ *or some end extension of* $\mathfrak{M}$. *Then* **T** *has a model* $\mathfrak{N}$ *which is an end extension of* $\mathfrak{M}$ *with new ordinals but such that no ordinal of* $\mathfrak{N}$ *is the least upper bound of the ordinals of* $\mathfrak{M}$.

Proof. Friedman's proof used "supercomplete" theories and the infinitary compactness theorem. We could give an extension of Friedman's proof, but for variety we replace supercomplete theories by the forcing version of the omitting types theorem, Theorem 2.2 in Keisler [1973]. Let L_{A_M} be the admissible fragment of $L_{\omega_1\omega}$ associated with $\mathfrak{M}_M$ and let L_B be the smallest fragment containing L_{A_M} and the sentence ψ:

$$\forall x\,[\theta(x) \to \exists y\,(x > y \wedge \theta(y))]$$

where $\theta(x)$ is

$$\bigwedge_{a \in \mathrm{Ord}(\mathfrak{M})} x > \bar{a}\ ,$$

$\mathrm{Ord}(\mathfrak{M}) = \{a \in M\colon \mathfrak{M} \models a \text{ is an ordinal}\}$, and $\bar{a}$ is a constant symbol in A_M used to denote a. Let Φ be the class of formulas of L_B which are either in L_{A_M} or of the form $\theta(x)$ or $\theta(y)$. Let $\mathcal{M}$ be the class of all models of **T** which are end extensions of $\mathfrak{M}$ with new ordinals; i.e. models of **T**$'$:

$$\mathbf{T}$$

$$\forall x\,[x \in \bar{a} \leftrightarrow \bigvee_{b \in a_E} x = \bar{b}] \qquad \text{for all } a \in M$$

$$\exists x\,\theta(x)$$

T$'$ is nonempty by the compactness theorem for L_{A_M}. The sentence ψ can be written as an $\forall\bigvee\exists(\Phi)$ sentence:

$$\forall x\,[\bigvee_{a \in \mathrm{Ord}(\mathfrak{M})} x \not> \bar{a} \vee \exists y\,[x > y \wedge \theta(y))]\ .$$

So by the version of the omitting types theorem given in Theorem 2.2 of Keisler [1973], we will know that ψ is true in some model $\mathfrak{N} \in \mathcal{M}$ if we can show that: if $p(c_1 \ldots c_n)$ is a finite subset of $\Phi(C)$ satisfiable in $\mathcal{M}$ then for each $c_k \in C$ one of the following is satisfiable in $\mathcal{M}$:

$$p'(a) = p(c_1 \dots c_n) \cup \{c_k \not> \bar{a}\} \quad \text{for some } a \in \mathrm{Ord}(\mathfrak{M})$$

$$p'' = p(c_1 \dots c_n) \cup \{\exists y\, [c_k > y \wedge \theta(y)]\}\,.$$

Assume that none of the $p'(a)$ are satisfiable in $\mathcal{M}$ so that $\mathbf{T}' \cup p(c_1 \dots c_n) \models \theta(c_k)$. We need to show that $\mathbf{T}' \cup p''$ is consistent. It suffices to show that the following $\boldsymbol{\Sigma}_1$ theory of L_{A_M} is consistent:

$$\begin{array}{ll} \mathbf{T} & \\ \forall x\, [x \in \bar{a} \leftrightarrow \bigvee_{b \in a_E} x = \bar{b}] & \text{for all } a \in M \\ c_k > \bar{a} & \text{all } a \in \mathrm{Ord}(\mathfrak{M}) \\ \phi(c_1 \dots c_n) & \text{all } \phi \in p(c_1 \dots c_n) \cap A_M \\ c_i > \bar{a} & \text{all } a \in \mathrm{Ord}(\mathfrak{M}) \\ & \text{if } \theta(c_i) \in p(c_1 \dots c_n)\,. \\ c_k > d & \\ d > \bar{a} & \text{all } a \in \mathrm{Ord}(\mathfrak{M}) \end{array}$$

where d is a new constant symbol in C, different from $c_1 \dots c_n, c_k$. But any A_M-finite subset of this theory is clearly satisfiable in any model of $\mathbf{T}' \cup p(c_1 \dots c_n)$, since then we need only have $d > \bar{a}$ for a in some set $X \in A_M, X \subseteq \mathrm{Ord}(\mathfrak{M})$ and any such set is bounded in $\mathfrak{M}$. □

For our second example we modify (by strengthening both the hypothesis and conclusion) one of the results in Barwise [1970]. This result shows how property IX comes into play. A subset of an admissible set is $\boldsymbol{\Sigma}_1^-$ if it is definable by a $\boldsymbol{\Sigma}_1$ formula without parameters.

6.2. Theorem. *Let $\mathfrak{M}$ be a countable model of $\mathbf{T}$, where $\mathbf{T}$ is a theory containing* **ZFC** *which is $\boldsymbol{\Sigma}_1^-$ on the admissible cover $\mathfrak{A}_M$ of $\mathfrak{M}$. Let κ be a cardinal of $\mathfrak{M}$, regular or not, such that*

$$\lambda < \kappa \rightarrow 2^\lambda \leq \kappa$$

holds in $\mathfrak{M}$. There is a model $\mathfrak{N}$ of $\mathbf{T}$ which is an end extension of $\mathfrak{M}$ such that:

1) *no subsets of any* $\lambda < \kappa$ *are added in* $\mathfrak{N}$ (*hence all cardinals* $\lambda < \kappa$ *of* $\mathfrak{M}$ *are still cardinals in* $\mathfrak{N}$), *and*
2) *all cardinals of* $\mathfrak{M}$ *greater than* κ *have cardinality* κ *in* $\mathfrak{N}$ (*hence many new subsets of* κ *are added by* $\mathfrak{N}$).

Proof. Let $\mathfrak{M} = \langle M, E\rangle$ and let $\mathfrak{M}_0 = \langle M_0, E \restriction M_0\rangle$ where $M_0 = \{x \in M\colon \mathfrak{M} \models$ "TC(x) has power $\leq \kappa$"$\}$. Then $\mathfrak{M}_0 \prec_1 \mathfrak{M}$ by Lévy [1965]. Since $\mathfrak{M}_0$ is transitive in $\mathfrak{M}$, $\mathfrak{M}_0 \subseteq_{\text{end}} \mathfrak{M}$. Let $\mathfrak{A}_M$ and $\mathfrak{A}_{M_0}$ be the admissible covers of $\mathfrak{M}$ and $\mathfrak{M}_0$ respectively, so that $\mathfrak{A}_{M_0} \prec_1 \mathfrak{A}_M$ by Property IX. For $\lambda < \kappa$ in $\mathfrak{M}$ we let $p(\lambda)$ denote the power set of λ in the sense of $\mathfrak{M}$. We need to show that the following theory $\mathbf{T}'$ is consistent:

i) $\mathbf{T}$

ii) $\forall x (x \in \bar{a} \leftrightarrow \bigvee_{b \in a_E} x = \bar{b})$ for all $a \in M$

iii) $\bigwedge_{\lambda \in \kappa_E} \forall x (x \subseteq \bar{\lambda} \to \bigvee_{a \in p(\lambda)_E} x = \bar{a})$

iv) $|c| = \bar{\kappa}$

v) $\bar{a} \in c$ all $a \in M$.

The theory $\mathbf{T}'$ is $\mathbf{\Sigma}_1$ definable using as parameter the sequence

$$\langle p(\lambda) : \lambda < \kappa\rangle$$

in the $\mathbf{\Sigma}_1$ definition (in order to write the sentence (iii)) and this set is an element of M_0 by the hypothesis $\lambda < \kappa \Rightarrow 2^\lambda \leq \kappa$. If $\mathbf{T}'$ is inconsistent there is a proof in A_M of a contradiction from $\mathbf{T}'$, by the completeness theorem for countable admissible fragments. This $\mathbf{\Sigma}_1$ property of A_M is thus true in A_{M_0}, so there is proof in A_{M_0} of a contradiction. This proof can use only $\leq \kappa$ axioms from $\mathbf{T}'$ and each constant $\bar{a}$ in these axioms has $a \in M_0$. Thus there is an inconsistent subset $\mathbf{T}_0 \subseteq \mathbf{T}'$ which asserts (v) only for $\leq \kappa$ sets a, each of them in M_0. But this is absurd since $\mathfrak{M}$ itself can be made into a model of any such $\mathbf{T}_0$ by the proper assignment of an element of M_0 to the constant symbol c. □

Actually we see that the theory $\mathbf{T}$ can be $\mathbf{\Sigma}_1$ using parameters from M_0 and the proof still goes through. We could combine the above two theorems to make $\mathfrak{N}$ have no least upper bound for the ordinals in $\mathfrak{N}$. We can also combine Theorem 6.2 with a theorem of Keisler and Morley (Corollary A on page 137 of Keisler [1971]) to get

$$\mathfrak{N} \models \text{"}\lambda \text{ is a cardinal} > \kappa\text{"}$$

implies

$$|\lambda_E| = \aleph_1$$

so that $\aleph_1$ subsets of κ are added by $\mathfrak{N}$ without adding new subsets to any $\lambda < \kappa$.

We should point out that some hypothesis like our

$$\mathfrak{M} \models (\lambda < \kappa \rightarrow 2^\lambda < \kappa)$$

is needed for the result. For example, if the sentence

$$2^{\aleph_0} = \aleph_2$$

is in **T** and $\kappa = \aleph_1^{\mathfrak{M}}$ then any end extension of $\mathfrak{M}$ which is a model of **T** and satisfies

$$|\aleph_2^{\mathfrak{M}}| = \aleph_1$$

must add new sets of integers (the old power set of ω is enumerated by a sequence of length an ordinal α with $|\alpha| = \aleph_1$ true in $\mathfrak{N}$).

§7. The construction of the admissible cover and related admissible sets with urelements

In this section we will sketch a construction which gives the admissible cover, but we shall not verify all of its properties. On the other hand, we shall carry out the construction in a general setting which gives many other examples of admissible sets with urelements.

For a simple such example, consider an admissible set A without urelements, with a $\boldsymbol{\Sigma}_1$ subset M of A. There is a natural admissible set A_M over M with A as the collection of pure sets of A_M. The general situation needs the following definition.

7.1. Definition. A structure $\mathfrak{N} = \langle N, R_1 \dots R_l \rangle$ is $\mathbf{\Sigma}_1$ on a structure $\mathfrak{B}_{\mathfrak{M}} = (\mathfrak{M}; B, F, \dots)$ if N is $\mathbf{\Sigma}_1$ on $\mathfrak{B}_{\mathfrak{M}}$ and if there are $\mathbf{\Sigma}_1$ relations $R_1^+ \dots R_l^+, R_1^- \dots R_l^-$ on $\mathfrak{B}_{\mathfrak{M}}$ such that for all $i \leq l$ and all $x_1 \dots x_{n_i} \in N$

$$R_i(x_1 \dots x_{n_i}) \Leftrightarrow R_i^+(x_1 \dots x_{n_i})$$

$$\Leftrightarrow \neg R_i^-(x_1 \dots x_{n_i})$$

where R_i is n_i-ary.

7.2. Theorem. *Let $\mathfrak{B}_{\mathfrak{M}} = (\mathfrak{M}; B, F, \dots)$ be a model of* **KPU** *and let $\mathfrak{N} = \langle N, R_1 \dots R_l \rangle$ be $\mathbf{\Sigma}_1$ on $\mathfrak{B}_{\mathfrak{M}}$. There is a largest admissible set $\mathfrak{A} = (\mathfrak{N}; A, \in_N)$ such that the map $*$: $\mathfrak{A}_{\mathfrak{N}} \to \mathfrak{B}_{\mathfrak{M}}$ is everywhere defined:*

$$p^* = p \quad for \quad p \in N$$

$$(a^*)_F = \{b^* : b \in_N a\} .$$

For $b \in N$, if $b_F \subseteq N$ then $b_F \in A$.

We shall sketch a proof of this for the case where there is just one binary relation R (i.e. $l = 1, n_1 = 2$).

The structure $\mathfrak{B}_{\mathfrak{M}}$ is a model of **KPU** as formulated in some language $\mathrm{L}(\in, \dots)$ (where $\mathfrak{M}$ is an L structure) whereas $\mathfrak{A}_{\mathfrak{N}}$ will be a model of **KPU** as formulated in $\mathrm{L}_0(\in)$, where L_0 has just the binary symbol R. To keep things straight we denote this second **KPU** by $\mathbf{KPU}_0$.

Choose $\mathbf{\Sigma}_1$ formulas (of $\mathrm{L}(\in, \dots)$) $\sigma(x)$, $\phi(x,y)$, $\psi(x,y)$ so that for all $x \in M \cup B$

$$x \in N \Leftrightarrow \mathfrak{B}_{\mathfrak{M}} \models \sigma(x)$$

and for $x, y \in N$

$$R(x,y) \Leftrightarrow \mathfrak{B}_{\mathfrak{M}} \models \phi(x,y)$$

$$\Leftrightarrow \mathfrak{B}_{\mathfrak{M}} \models \neg \psi(x,y) .$$

Let $\mathbf{KPU}_1$ be **KPU** plus the following axiom:

$$\forall x \forall y [\sigma(x) \wedge \sigma(y) \to (\sigma(x,y) \leftrightarrow \neg \psi(x,y))]$$

which, of course, is true in $\mathfrak{B}_{\mathfrak{M}}$. The heart of the proof of the theorem consists of constructing a suitable interpretation I of $\mathbf{KPU}_0$ in the theory $\mathbf{KPU}_1$. Toward this end define $\boldsymbol{\Sigma}_1$ predicates of $\mathrm{L}(\in, ...)$ as follows, using $\hat{x}$ for $\langle 0,x\rangle$ and $\check{x}$ for $\langle 1,x\rangle$.

$$\begin{aligned}
&\mathrm{Point}(a) \Longleftrightarrow \exists x[\sigma(x) \wedge a = \hat{x}]\\
&\mathrm{Set}(a) \Longleftrightarrow \exists b\,[a = \check{b} \wedge \forall x \in b(\mathrm{Set}(x) \vee \mathrm{Point}(x))]\\
&x \mathbin{\&} y \Longleftrightarrow \exists z\,(y = \check{z} \wedge x \in z)\\
&R^{\bullet}(a,b) \Longleftrightarrow \exists x\,\exists y(a = \hat{x} \wedge b = \hat{y} \wedge \phi(x,y))\\
&R^{\bullet\bullet}(a,b) \Longleftrightarrow \exists x\,\exists y\,(a = \hat{x} \wedge b = \hat{y} \wedge \psi(x,y))
\end{aligned}$$

The predicate $\mathrm{Set}(a)$ needs the second recursion theorem for **KPU** in its definition, the predicate $x \mathbin{\&} y$ can be written as a $\boldsymbol{\Delta}_0$ formula. The interpretation I of $\mathrm{L}_0(\in)$ into $\mathbf{KPU}_1$ is given by the following instructions: replace each

positive occurrence of R		by $R^{\bullet}$
negative occurrence of R		by $R^{\bullet\bullet}$
occurrence of $x \in y$		by $x \mathbin{\&} y$
quantifier over urelements	$\forall p(...)$	by $\forall a[\mathrm{Point}(a) \to (...)]$
	$\exists p(...)$	by $\exists a[\mathrm{Point}(a) \wedge (...)]$
quantifier over sets	$\forall a(...)$	by $\forall a[\mathrm{Set}(a) \to (...)]$
	$\exists a(...)$	by $\exists a[\mathrm{Set}(a) \wedge (...)]$
undetermined quantifiers	$\forall x(...)$	by $\forall a[\mathrm{Point}(a) \vee \mathrm{Set}(a) \to (...)]$
	$\exists x(...)$	by $\exists a[(\mathrm{Point}(a) \vee \mathrm{Set}(a)) \wedge (...)]$.

Equality and the propositional operations are not altered. One must carry this out in a sensible way so that no clashes of variables occur. Denote the translation of a sentence ϕ of $\mathrm{L}_0(\in)$ under the above by ϕ^I.

7.3. Lemma. *For each axiom ϕ of $\mathbf{KPU}_0$, ϕ^I is a theorem of $\mathbf{KPU}_1$.*

Proof. We run quickly through the axioms.

Extensionality: The translation reads: for all Sets a and b, if $x \mathbin{\&} a \leftrightarrow x \mathbin{\&} b$ for all Points and Sets x, then $a = b$. Now if $\mathrm{Set}(a)$, $\mathrm{Set}(b)$ then $a = \check{a}_0$, $b = \check{b}_0$, every $x \in a_0$ is a Point or a Set and $x \mathbin{\&} a \leftrightarrow x \in a_0$; similarly for b and b_0. Hence $a_0 = b_0$ so $a = b$.

Foundation: If there is a Set a such that $\phi^I(a)$ then there is a Set a so that $\phi^I(a)$ but for all Sets $b \mathrel{\&} a$, $\neg\phi^I(b)$. Note that $x \mathrel{\&} y \to \mathrm{rk}(x) < \mathrm{rk}(y)$ so choose a of least rank.

Pair: Let $c_0 = \{a,b\}$ and $c = \check{c}_0$. If a, b are Points or Sets then Set(c) and $x \mathrel{\&} c$ iff $x = a \vee x = b$.

Union: Given a Set a let $b = \{z : \exists y \mathrel{\&} a, z \mathrel{\&} y\}$, by $\boldsymbol{\Delta}_0$-separation, and let $c = \check{b}$. Then $z \mathrel{\&} c$ iff $z \mathrel{\&} y \mathrel{\&} a$ for some y.

$\boldsymbol{\Delta}_0$-separation: We wish to show that given a Set a there is a Set b such that for all x,

$$x \mathrel{\&} b \Leftrightarrow x \mathrel{\&} a \wedge \phi^I(x) .$$

If $a = \check{a}_0$ let $b_0 = \{x \in a_0 : \phi^I(x)\}$ by 3.5 and let $b = \check{b}_0$. (The point here is that ϕ^I is only a $\boldsymbol{\Sigma}$ formula but $(\neg\phi)^I$ is also a $\boldsymbol{\Sigma}$ formula.)

$\boldsymbol{\Delta}_0$-collection: Assume Set(a) and for every $x \mathrel{\&} a$ there is a y, either a Point or a Set such that $\phi^I(x,y)$, where ϕ is $\boldsymbol{\Delta}_0$ in $\mathrm{L}_0(\in)$. Now the whole statement ψ

$$\forall x \mathrel{\&} a \, \exists y \, [(\mathrm{Point}(y) \vee \mathrm{Set}(y)) \wedge \phi^I(x,y)]$$

is a $\boldsymbol{\Sigma}$ formula (since $x \mathrel{\&} a \to x \in \mathrm{TC}(a)$) so, by $\boldsymbol{\Sigma}$-reflection, there is a w, $a \subseteq w$, such that $\psi^{(w)}$ holds. Let

$$b_0 = \bigcup_{x \mathrel{\&} a} \{y \in w : (\mathrm{Point}^{(w)}(y) \vee \mathrm{Set}^{(w)}(y)) \wedge (\phi^I)^{(w)}(x,y)\}$$

and $b = \check{b}_0$. One easily checks that Set(b) and $\forall x \mathrel{\&} a \, \exists y \mathrel{\&} b \, \phi^I(x,y)$. □

Now when one has an interpretation I of one theory $\mathbf{T}_0$ in another theory $\mathbf{T}_1$ then any model $\mathfrak{B}$ of $\mathbf{T}_1$ has associated with it a model $\mathfrak{B}^{-I}$ of $\mathbf{T}_0$ in the usual way. Applying this to our situation the structure $\mathfrak{B}_{\mathfrak{M}}$ gives rise via I to a structure $\mathfrak{B}_{\mathfrak{M}}{}^{-I}$ which is a model of $\mathbf{KPU}_0$. The structure $\mathfrak{B}_{\mathfrak{M}}{}^{-I}$ has the form

$$\mathfrak{B}_{\mathfrak{M}}{}^{-I} = (\mathfrak{N}'; A', E')$$

where $\mathfrak{N}'$ is isomorphic to $\mathfrak{N}$ via the map $\hat{a} \to a$, A' is the set of $b \in B$ such that

$$\mathfrak{B}_{\mathfrak{M}} \models \mathrm{Set}(b)$$

and $E' = \mathcal{E} \upharpoonright A'$. If $\mathfrak{B}_{\mathfrak{M}}$ is admissible then E' is well founded and $\mathfrak{B}_{\mathfrak{M}}^{-I}$ is isomorphic to the desired admissible set. In general, however, $\mathfrak{B}_{\mathfrak{M}}$ need not be admissible but only a model of **KPU** in which case E' need not be well founded. Thus, we need to apply the following lemma to $\mathfrak{B}_{\mathfrak{M}}^{-I}$.

7.4. Lemma. *The well founded part of a model of* **KPU** *is again a model of* **KPU**.

The proof is like that given in A6 of Barwise [1972]. Combining these two lemmas, with a natural isomorphism η ($\eta(\hat{a}) = a$ and extend η: $V_{N'} \to V_N$ via $\eta(a) = \{\eta(b): b \in a\}$) gives an admissible set $_{\mathfrak{N}}$. The mapping $*$ arises as follows:

Let us work again in $\mathbf{KPU}_1$. Define

$$a^* = a_0 \text{ if Point}(a) \text{ and } \hat{a}_0 = a$$

$$a^* = \{b^* : b \mathrel{\mathcal{E}} a\} \text{ if Set}(a)$$

using the second recursion theorem for **KPU**. This mapping, interpreted, maps $\mathfrak{B}_{\mathfrak{M}}^{-I} \to \mathfrak{B}_{\mathfrak{M}}$, maps $\mathfrak{N}'$ to $\mathfrak{N}$ and satisfies

$$(a^*)_F = \{b^* : bE'a\} .$$

Restricting this map to the well founded part of $\mathfrak{B}_{\mathfrak{M}}^{-I}$ and then mapping it over to $\mathfrak{A}_{\mathfrak{N}}$ via the natural isomorphism η gives the conclusion of the theorem.

To see that $\mathfrak{A}_{\mathfrak{N}}$ is the largest admissible set with such a map $*$, suppose $\mathfrak{A}'_{\mathfrak{N}}$ where some other admissible over $\mathfrak{N}$ with a map $*$: $\mathfrak{A}'_{\mathfrak{N}} \to \mathfrak{B}_{\mathfrak{M}}$

$$p^* = p \quad \text{for} \quad p \in N$$

$$(a^*)_F = \{b^* : b \in_N a\} .$$

Consider the map $''$ defined in $\mathfrak{N}'_{\eta}$ by recursion over $\in_N$:

$$p'' = \langle 0, p \rangle$$

$$a'' = \langle 1, \{b'' : b \in a\} \rangle .$$

If one chases an x the long way around the following diagram, one sees by induction on $\in_N$ that all the maps are defined and that $\bar{x}$, the result of all the composite functions, is x for all $x \in \mathfrak{A}'_{\mathfrak{N}}$.

$$\begin{array}{ccc} \mathfrak{A}'_{\mathfrak{N}} & \xrightarrow{\ ''\ } & \mathfrak{A}'_{\mathfrak{N}} \\ \vdots & & \downarrow \\ \vdots \quad \mathrm{Id} & & \mathfrak{B}_{\mathfrak{N}}^{-I} \\ \downarrow & & \downarrow \\ \mathfrak{A}_{\mathfrak{N}} & \xleftarrow{\ \eta\ } & \mathrm{WF}\,(\mathfrak{B}_{\mathfrak{N}}^{-I}) \end{array}$$

Hence $\mathfrak{A}'_{\mathfrak{N}} \subseteq \mathfrak{A}_{\mathfrak{N}}$.

Finally, if $b \in N$ and $b_F \subseteq N$ then let $a = \langle 1, \{\langle 0, c\rangle : cFb\}\rangle$ in $\mathfrak{B}_{\mathfrak{M}}$.

This concludes the proof of the theorem. When one applies this to obtain the admissible cover, one has $\mathfrak{B}_{\mathfrak{M}} = \mathfrak{N}$ = a model of **KP**. Most of the properties of the admissible cover should be at least plausible from the above proof. Note that in order to show that the admissible cover is admissible with the covering function, one has to have $\mathbf{KPU}_0$ formulated in $L_0(\in, f)$ where f is a 1-ary function symbol. In the interpretation I one then has to replace $f(x) = y$ by $\dot{f}(x) = y$ where $\dot{f}$ is defined in $\mathbf{KPU}_1$ by

$$\dot{f}(x) = \{\hat{y} : y \in x\}^{\vee}.$$

References

Aczel, P.
[1970] Implicit and inductive definability (abstract), J. Symbolic Logic 35, p. 599.

Barwise, K.J.
[1969a] Infinitary logic and admissible sets, J. Symbolic Logic, 34, pp. 226–252.
[1969b] Applications of strict Π_1^1 predicates to infinitary logic, J. Symbolic Logic 34, pp. 409–423.
[1971] Infinitary methods in the model theory of set theory, Logic Colloquium '69, North-Holland, 1971, pp. 53–66.
[1972] Absolute logics and $L_{\infty\omega}$, Annals of Mathematical Logic 4, pp. 309–340.

Barwise–Gandy–Moschovakis
[1971] The next admissible set, J. Symbolic Logic 36, pp. 108–120.

Friedman, H.
[1972] Countable models of set theories, to appear.

Keisler, H.J.
[1971] Model Theory for Infinitary Logic (North-Holland, Amsterdam).

[1973] Forcing and the omitting types theorem, to appear.

Kunen, K.
[1968] Implicit definability and infinitary languages, J. Symbolic Logic 33, 446–451.

Lévy, A.
[1965] A hierarchy of formulas in set theory, Memoirs of Amer. Math. Soc. No. 57.

Moschovakis, Y.N.
[1973] Elementary induction on abstract structures (North-Holland, Amsterdam)

Shoenfield, J.
[1967] Mathematical Logic (Addison-Wesley, Reading, Mass.)

J.E.Fenstad, P.G.Hinman (eds.), Generalized Recursion Theory
©North-Holland Publ. Comp., 1974

AN INTRODUCTION TO THE FINE STRUCTURE OF THE CONSTRUCTIBLE HIERARCHY

(Results of Ronald Jensen)

Keith J. DEVLIN
University of Oslo

§0. Introduction

We shall work in Zermelo–Fraenkel set theory (including the axiom of choice) throughout, and denote this theory by ZFC. We shall adopt the usual, well-known, notations and conventions of contemporary set theory (e.g. an ordinal is defined to be the set of all smaller ordinals, cardinals are initial ordinals, etc.).

The paper is entirely self-contained, but some familiarity with the usual definition of the constructible universe, L, in terms of definability, and the proof that L is a model of ZFC + GCH + V = L, will be helpful.

The exposition is based, with permission, very strongly on a set of notes [1] written by *Ronald Jensen* and entitled "The Fine Structure of the Constructible Hierarchy". Except where otherwise stated, the results are entirely those of Professor Jensen. It is convenient at this point for us to express our appreciation of several illuminating discussions with Professor Jensen on his work in general.

[1] Since we wrote this paper, a slightly revised version of these notes has been published as a research paper. See R.B. Jensen, "The Fine Structure of the Constructible Hierarchy", Annals of Mathematical Logic, Vol. 4 [1972], p. 229. The present paper represents a lengthy discourse on an expansion of the earlier parts of Jensen's paper, and it is hoped that the somewhat more leisurely pace we adopt here (as opposed to Jensen's paper) will be of benefit to those not predominantly interested in the set theoretical consequences of the Fine Structure Theory. For those who are so inclined, the notation we use is almost identical to that of Jensen, so this paper should provide a good introduction to Jensen's.

Previously, Jensen worked, as did most other people, with the usual "constructible hierarchy". Thus, one defines, inductively, sets L_α, $\alpha \in OR$, by setting $L_0 = \emptyset$, $L_\lambda = \bigcup_{\alpha<\lambda} L_\alpha$ if $\lim(\lambda)$, and $L_{\alpha+1}$ = the set of all $x \subset L_\alpha$ such that for some $\in$-formula φ and some $a_1, ..., a_n \in L_\alpha$, $x = \{z \in L_\alpha | L_\alpha \models \varphi[z, a_1, ..., a_n]\}$. One then defines the *constructible universe* as the class $L = \bigcup_{\alpha \in OR} L_\alpha$. Now, the important facts concerning this definition which one uses when studying L, are, firstly, that the construction is (in a strong way, to be made precise later) Σ_1-definable, and thus has certain absoluteness properties, and, secondly, that $L_{\alpha+1}$ contains all and only those subsets of L_α which are L_α-definable (and which, therefore, must be in L if L is to be a model of ZFC). But if, indeed, these are the only conditions which we require (and loosely speaking they are), then it is clear that our above definition is unnecessarily restrictive. For instance, there are many simply definable functions on sets under which L must be closed, but which increase rank – and these functions will lead out of the sets L_α. For instance, unless $\lim(\alpha)$, L_α will not be closed under the formation of ordered pairs. Since this function plays a central role in even the most elementary parts of set theory, we see that this defect becomes quite important (though not unavoidable) when we try to study the fine structure of L rather than L itself. So, following Jensen, we define a new hierarchy of "constructible sets", which is sufficiently like the L-hierarchy to preserve the two properties mentioned above, but which has the extra property that each level in the hierarchy is closed under ordered pairs, etc. More precisely, we first define a certain class of set functions (called "rudimentary functions"), and then define a hierarchy $\langle J_\alpha | \alpha \in OR \rangle$ (the *Jensen hierarchy*) such that each J_α is closed under the rudimentary functions, $L = \bigcup_{\alpha \in OR} J_\alpha$, and the two properties above hold for this hierarchy. In most cases, J_α will be a "constructibly inessential" extension of L_α, and in fact, if $\langle V_\alpha | \alpha \in OR \rangle$ denotes the familiar rank-hierarchy, the precise relationship between the J- and the L-hierarchies is easily seen to be $J_0 = L_0 = \emptyset$ and $L_{\omega+\alpha} = V_{\omega+\alpha} \cap J_{1+\alpha}$ for all α. Hence we have $J_\alpha = L_\alpha$ iff $\omega\alpha = \alpha$.

In §1 we give some basic definitions. In §2 we define the class of rudimentary functions and develop the elementary theory of this class. The reader may, if he wishes, safely skip all the proofs in this section without affecting the reading of the later parts. §3 is devoted to a very brief discussion of the concept of an admissible set. In §4 the Jensen hierarchy is defined and its elementary properties discussed. In §5 we investigate the fine structure of the Jensen hierarchy. A corresponding theory may also be developed for the L-hierarchy, the only difference being that some awkward complications arise because of the above mentioned defects in this definition.

§ 1. Preliminaries

We shall be concerned with first-order structures of the form $\mathbf{M} = \langle M, \in, A \rangle$, where M is a non-empty set and $A \subset M$. In general, we shall write $\langle M, A \rangle$ for $\langle M, \in, A \rangle$. The (first-order) language for such structures consists of the following:

(i) variables v_j, $j \in \omega$ (generally denoted by v, w, x, y, z, etc.) (Vbl.).
(ii) predicates $=, \in, A$.
(iii) bounded quantifiers $(\forall v_i \in v_j)$, $(\exists v_i \in v_j)$, $i, j \in \omega$, $i \neq j$.
(iv) unbounded quantifiers $(\forall v_i)$, $(\exists v_i)$, $i \in \omega$.
(v) connectives $\wedge, \vee, \neg, \rightarrow, \leftrightarrow$.

Finite strings of variables (or of elements of M) are denoted by $\boldsymbol{v}, \boldsymbol{x}$, etc. We write $\boldsymbol{a} \in X$ for $a_1 \in X \wedge \ldots \wedge a_n \in X$, where we have $\boldsymbol{a} = a_1, \ldots, a_n$. Similarly for $\exists \boldsymbol{v}$, $\forall \boldsymbol{v}$, etc.

The notions of primitive formula (PFml), formula (Fml), free variable, statement, and satisfaction are assumed known. A formula of this language is $\boldsymbol{\Sigma}_0$ (or $\boldsymbol{\Pi}_0$) if it contains no unbounded quantifiers. Let $n \geq 1$, and let Q_n denote $\forall$ if n is even and $\exists$ if n is odd. A formula is $\boldsymbol{\Sigma}_n(\boldsymbol{\Pi}_n)$ if it is of the form $\exists \boldsymbol{x}_1 \forall \boldsymbol{x}_2 \exists \boldsymbol{x}_3 \ldots Q_n \boldsymbol{x}_n \varphi (\forall \boldsymbol{x}_1 \exists \boldsymbol{x}_2 \forall \boldsymbol{x}_3 \ldots Q_{n+1} \boldsymbol{x}_n \varphi)$ where φ is $\boldsymbol{\Sigma}_0$. A formula in which the predicate A does not occur is called an $\in$-*formula.*

$\models_{\mathbf{M}}$ denotes the satisfaction relation for $\mathbf{M}$. Thus, $\models_{\mathbf{M}}$ is the set of all $\langle \varphi, \langle \boldsymbol{z} \rangle \rangle$ such that φ is a formula of the above language and $\boldsymbol{z} \in M$ and φ holds in $\mathbf{M}$ at the point $\langle \boldsymbol{z} \rangle$. We generally write $\models_{\mathbf{M}} \varphi[\boldsymbol{z}]$ for $\langle \varphi, \langle \boldsymbol{z} \rangle \rangle \in \models_{\mathbf{M}}$. $\models_{\mathbf{M}}^{\boldsymbol{\Sigma}_n}$ denotes the set of all $\langle \varphi, \langle \boldsymbol{z} \rangle \rangle \in \models_{\mathbf{M}}$ such that φ is $\boldsymbol{\Sigma}_n$.

Let $N \subset M$. A set $R \subset M$ is $\boldsymbol{\Sigma}_n^{\mathbf{M}}(N)$ $(\boldsymbol{\Pi}_n^{\mathbf{M}}(N))$ iff there is a $\boldsymbol{\Sigma}_n$ $(\boldsymbol{\Pi}_n)$ formula $\varphi(u, \boldsymbol{v})$ and elements $\boldsymbol{a} \in N$ such that for all $x \in M$, $R(x) \leftrightarrow \models_{\mathbf{M}} \varphi[x, \boldsymbol{a}]$. The set of all such R is also denoted by $\boldsymbol{\Sigma}_n^{\mathbf{M}}(N)$ $(\boldsymbol{\Pi}_n^{\mathbf{M}}(N))$.

Set $\boldsymbol{\Sigma}_\omega^{\mathbf{M}}(N) = \bigcup_{n \in \omega} \boldsymbol{\Sigma}_n^{\mathbf{M}}(N)$, $\boldsymbol{\Delta}_n^{\mathbf{M}}(N) = \boldsymbol{\Sigma}_n^{\mathbf{M}}(N) \cap \boldsymbol{\Pi}_n^{\mathbf{M}}(N)$.

Write $\boldsymbol{\Sigma}_n^{\mathbf{M}}$ for $\boldsymbol{\Sigma}_n^{\mathbf{M}}(\emptyset)$ and $\boldsymbol{\Sigma}_n(\mathbf{M})$ for $\boldsymbol{\Sigma}_n^{\mathbf{M}}(M)$. Similarly for $\boldsymbol{\Pi}$, $\boldsymbol{\Delta}$.

If φ is a formula, $\varphi^{\mathbf{M}}$ denotes the relation $\{x \in M \mid \models_{\mathbf{M}} \varphi[x]\}$. Similarly, and more generally, define $\varphi^{\mathbf{M}}[\boldsymbol{a}]$ for $\boldsymbol{a} \in \mathbf{M}$ as $\{x \in M \mid \models_{\mathbf{M}} \varphi[x, \boldsymbol{a}]\}$.

Let F be a class of structures of the form $\mathbf{M} = \langle M, A \rangle$. A relation R is *uniformly* $\Sigma_n(\mathbf{M})$ *for* $\mathbf{M} \in F$ iff there is a Σ_n formula $\varphi(u, \boldsymbol{v})$ and elements $\boldsymbol{a} \in \bigcap\{M \mid \mathbf{M} \in F\}$ such that whenever $\mathbf{M} \in F, R \cap M = \varphi^{\mathbf{M}}[\boldsymbol{a}]$.

§2. Rudimentary functions

A function $f: \mathrm{V}^n \to \mathrm{V}$ is *rudimentary* (rud) iff it is generated by the following schemata:

(i) $f(x_1, ..., x_n) = x_i, \quad 1 \leq i \leq n.$

(ii) $f(x_1, ..., x_n) = x_i - x_j, \quad 1 \leq i, j \leq n.$

(iii) $f(x_1, ..., x_n) = \{x_i, x_j\}, \quad 1 \leq i, j \leq n.$

(iv) $f(x_1, ..., x_n) = h(g_1(x_1, ..., x_n), ..., g_k(x_1, ..., x_n))$, where $g_1, ..., g_k, h$ are rud.

(v) $f(x_1, ..., x_n) = \bigcup_{y \in x_1} h(y, x_2, ..., x_n)$, where h is rud.

For example, the following functions are clearly rud:

$f(\boldsymbol{x}) = \bigcup x_i$

$f(\boldsymbol{x}) = x_i \cup x_j \; (= \bigcup\{x_i, x_j\})$

$f(\boldsymbol{x}) = \{\boldsymbol{x}\}$

$f(\boldsymbol{x}) = \langle \boldsymbol{x} \rangle = \{\{x_1\}, \{x_1, \langle x_2, ..., x_n \rangle\}\}$.

And if $f(y, \boldsymbol{x})$ is rud, so is $g(y, \boldsymbol{x}) = \langle f(z, \boldsymbol{x}) \mid z \in y \rangle \; (= \bigcup_{z \in y} \{\langle f(z, \boldsymbol{x}), z \rangle\})$.

We say that $R \subset \mathrm{V}^n$ is *rudimentary* (rud) iff there is a rud function $r: \mathrm{V}^n \to \mathrm{V}$ such that $R = \{\langle \boldsymbol{x} \rangle \mid r(\boldsymbol{x}) \neq \emptyset\}$. For example, $\notin$ is rud, since $y \notin x \leftrightarrow \{y\} - x \neq \emptyset$.

We list some basic properties of rudimentary functions and relations.

(1) If f, R are rud, so is

$$g(\boldsymbol{x}) = \begin{cases} f(\boldsymbol{x}), & \text{if } R(\boldsymbol{x}) \\ \emptyset, & \text{if } \neg R(\boldsymbol{x}). \end{cases}$$

Proof. By definition, there is a rud r such that $R(\boldsymbol{x}) \leftrightarrow r(\boldsymbol{x}) \neq \emptyset$. Then $g(\boldsymbol{x}) = \bigcup_{y \in r(\boldsymbol{x})} f(\boldsymbol{x})$.

(2) Let χ_R be the characteristic function of R. R is rud if χ_R is rud.

Proof. By (1).

(3) R is rud iff $\neg R$ is rud.

Proof. By (2), since $\chi_{\neg R}(\boldsymbol{x}) = 1 - \chi_R(\boldsymbol{x})$.

(4) Let $f_i : V^n \to V$ be rud, $i = 1, ..., m$. Let $R_i \subset V^n$ be rud and mutually disjoint, $i = 1, ..., m$, and such that $\bigcup_{i=1}^m R_i = V^n$. Define $f : V^n \to V$ by $f(\boldsymbol{x}) = f_i(\boldsymbol{x})$ iff $R_i(\boldsymbol{x})$. Then f is rud.

Proof. Set

$$\bar{f}_i(\boldsymbol{x}) = \begin{cases} f_i(\boldsymbol{x}), & \text{if } R_i(\boldsymbol{x}) \\ \emptyset, & \text{if } \neg R_i(\boldsymbol{x}), \end{cases} \qquad i = 1, ..., m.$$

By (1), $\bar{f}_i$ is rud. But $f(\boldsymbol{x}) = \bigcup_{i=1}^m \bar{f}_i(\boldsymbol{x})$.

(5) If $R(y, \boldsymbol{x})$ is rud, so is $f(y, \boldsymbol{x}) = y \cap \{z \mid R(z, \boldsymbol{x})\}$.

Proof. Set

$$h(y, \boldsymbol{x}) = \begin{cases} \{y\}, & \text{if } R(y, \boldsymbol{x}) \\ \emptyset, & \text{otherwise.} \end{cases}$$

Then h is rud. But $f(y, \boldsymbol{x}) = \bigcup_{z \in y} h(z, \boldsymbol{x})$.

(6) Suppose $R(y, \boldsymbol{x})$ is rud and $(\forall \boldsymbol{x})(\exists! y) R(y, \boldsymbol{x})$. Set

$$f(y, \boldsymbol{x}) = \begin{cases} \text{the unique } z \in y \text{ such that } R(z, \boldsymbol{x}), \text{ if such a } z \text{ exists.} \\ \emptyset, \text{ otherwise.} \end{cases}$$

Then f is rud.

Proof. $f(y, \boldsymbol{x}) = \bigcup [y \cap \{z \mid R(z, \boldsymbol{x})\}]$.

(7) If $R(y, \boldsymbol{x})$ is rud, so is $(\exists z \in y) R(z, \boldsymbol{x})$.

Proof. Take r rud so that $R(y, \boldsymbol{x}) \leftrightarrow r(y, \boldsymbol{x}) \neq \emptyset$. Then

$$(\exists z \in y) R(z, \boldsymbol{x}) \leftrightarrow \bigcup_{z \in y} r(z, \boldsymbol{x}) \neq \emptyset .$$

(8) If $R_i(\boldsymbol{x})$ are rud, $i = 1, \ldots, m$, then so are

$$\bigcup_{i=1}^{m} R_i , \qquad \bigcap_{i=1}^{m} R_i , \qquad \text{(Trivial)} .$$

(9) Let $(-)_0, (-)_1$, denote the inverse functions to $\langle -, - \rangle$. Then $(-)_0, (-)_1$ are rud. More generally, let $(-)_0^n, \ldots, (-)_{n-1}^n$ denote the inverse functions to $\langle x_1, \ldots, x_n \rangle$. Then $(-)_0^n, \ldots, (-)_{n-1}^n$ are rud.

Proof.

$$(x)_0 = \begin{cases} \text{the unique } z \in \bigcup x \text{ such that } (\exists u_1, u_2 \in \bigcup x)\,(x = \langle u_1, u_2 \rangle \wedge u_1 = z) \\ \emptyset, \quad \text{if no such } z \text{ exists.} \end{cases}$$

etc.

(10) The function

$$f(x,y) = x(y) = \begin{cases} \text{the unique } z \in \bigcup \bigcup x \text{ such that } \langle z, y \rangle \in x \\ \emptyset, \quad \text{if no such } z \text{ exists.} \end{cases}$$

is rud. (By definition.)

(11) dom and ran are rud.

Proof. $\operatorname{dom}(x) = \{z \in \bigcup\bigcup x \mid (\exists w \in \bigcup\bigcup x)(\langle w, z \rangle \in x)\}$
$\operatorname{ran}(x) = \{z \in \bigcup\bigcup x \mid (\exists w \in \bigcup\bigcup x)(\langle z, w \rangle \in x)\}$.

(12) $f(x,y) = x \times y = \bigcup_{u \in x} \bigcup_{v \in y} \{\langle u, v \rangle\}$ is rud.

(13) $f(x,y) = x \restriction y = x \cap (\operatorname{ran}(x) \times y)$ is rud.

(14) $f(x,y) = x''y = \operatorname{ran}(x \restriction y)$ is rud.

(15) $f(x) = x^{-1}$ is rud.

Proof. Set $h(z) = \langle (z)_1, (z)_0 \rangle$. Then h is rud. But clearly, $f(x) = x^{-1} = h''(x \cap (\mathrm{ran}(x) \times \mathrm{dom}(x)))$.

Recalling our preliminary discussion (§0), we observe that though rud functions increase rank, they only do so by a finite amount. More precisely, by induction on the rud definition * of a given rud function f, we see that there is a $p \in \omega$ such that for all $x_1, ..., x_n$, $\mathrm{rank}(f(x_1, ..., x_n)) < \max\{\mathrm{rank}(x_1), ..., \mathrm{rank}(x_n)\} + p$.

* *Note*: In future, we shall often refer to "the rud definition of f", or simply "the definition of f". We mean an arbitrary such definition, the actual choice being irrelevant, and hence assumed made once and for all.

We now prove that the rud functions do in fact encompass all of the "simply definable" functions we spoke about in §0. First, let us call a function $f: \mathrm{V}^n \to \mathrm{V}$ *simple* iff whenever $\varphi(z, \boldsymbol{y})$ is a $\boldsymbol{\Sigma}_0$ $\in$-formula, there is a $\boldsymbol{\Sigma}_0$ $\in$-formula ψ such that $\models \varphi(f(\boldsymbol{x}), \boldsymbol{y}) \leftrightarrow \psi(\boldsymbol{x}, \boldsymbol{y})$. A useful characterisation of this concept is given by the following:

Proposition. *A function* $f: \mathrm{V}^n \to \mathrm{V}$ *is simple iff*
 (i) *the predicate* $x \in f(\boldsymbol{y})$ *is* $\boldsymbol{\Sigma}_0^{\mathbf{V}}$; *and*
 (ii) *whenever* $A(x)$ *is* $\boldsymbol{\Sigma}_0^{\mathbf{V}}$, *so is* $(\forall x \in f(\boldsymbol{y}))A(x)$.

Proof. $(\to)$ By definition.
$(\leftarrow)$ Let f satisfy (i) and (ii), and let $\varphi(x, \boldsymbol{y})$ be a $\boldsymbol{\Sigma}_0$ $\in$-formula. An easy induction on the length of φ shows that $\varphi(f(\boldsymbol{x}), \boldsymbol{y})$ is equivalent to a $\boldsymbol{\Sigma}_0$ $\in$-formula; so f is simple.

Using this proposition, and easy induction on the definition of f yields:

Lemma 1. *If f is* rud *then f is simple.*

Now, since there are $\boldsymbol{\Sigma}_0^{\mathbf{V}}$ functions which increase rank by an infinite amount, it is clear that the converse to the above lemma is false. However, we do have:

Lemma 2. $R \subset \mathrm{V}^n$ *is* $\boldsymbol{\Sigma}_0^{\mathbf{V}}$ *iff R is* rud.

Proof. ($\rightarrow$) Let R be $\mathbf{\Sigma}_0^{\mathbf{V}}$. By (3), (7), and (8) above, an easy induction on the $\mathbf{\Sigma}_0^{\mathbf{V}}$ definition of R shows that R is rud.
($\leftarrow$) Let R be rud. Then χ_R is rud. So by Lemma 1, χ_R is simple. Using our above proposition, an easy induction on the rud definition of χ_R shows that χ_R, and hence R, is $\mathbf{\Sigma}_0^{\mathbf{V}}$.

We require some generalisations of these concepts.

Let $A \subset \mathrm{V}$. We say that a function f is *rud in A* iff f is generated by the schemata for rud functions and the function χ_A.

Let $p \in \mathrm{V}$. We say that a function f is rud in *parameter p* iff f is generated by the schemata for rud functions and the constant functions $h(\boldsymbol{x}) = p$.

Lemma 3. *If f is* rud *in $A \subset \mathrm{V}$, there are* rud *functions $g_1, ..., g_n$ such that f is expressible (in a uniform way with respect to the rud definition of f) as a composition of $g_1, ..., g_n$ and the function $h(x) = A \cap x$.*

Proof. By induction on the (rud) definition of f.

A set X is said to be *rud closed* if for all rud functions f, $f''X^n \subset X$.

A structure $\mathbf{M} = \langle M, A \rangle$ is said to be *rud closed* if for all functions f which are rud in A, $f''M^n \subset M$.

We say a structure $\mathbf{M} = \langle M, A \rangle$ is *amenable* if $u \in M \rightarrow A \cap u \in M$.

Lemma 4. *A structure $\mathbf{M} = \langle M, A \rangle$ is* rud *closed iff the set M is* rud *closed and $\mathbf{M}$ is amenable.*

Proof. By Lemma 3.

Lemma 5. *Let $A \subset \mathrm{V}$. If f is* rud *in A, then $f \restriction M^n$ is uniformly $\mathbf{\Sigma}_0(\langle M, A \cap M \rangle)$ for all transitive,* rud *closed $\mathbf{M} = \langle M, A \cap M \rangle$.*

Proof. By Lemmas 2 and 3.

The next result will be of considerable use to us later on.

Lemma 6. *Every* rud *function is a composition of some of the following* rud *functions*

$F_0(x,y) = \{x,y\}$
$F_1(x,y) = x - y$
$F_2(x,y) = x \times y$
$F_3(x,y) = \{\langle u,z,v\rangle \mid z \in x \wedge \langle u,v\rangle \in y\}$
$F_4(x,y) = \{\langle u,v,z\rangle \mid z \in x \wedge \langle u,v\rangle \in y\}$
$F_5(x,y) = \bigcup x$
$F_6(x,y) = \mathrm{dom}\,(x)$
$F_7(x,y) = \in \cap x^2$
$F_8(x,y) = \{x''\{z\} \mid z \in y\}$.

Proof. Let $\mathcal{C}$ denote the class of all functions obtainable from $F_0, ..., F_8$ by composition. We must show that f rud $\rightarrow f \in \mathcal{C}$.

For each $\in$-formula $\varphi(x_1, ..., x_n)$, set

$$t_\varphi(u) = \{\langle x_1, ..., x_n\rangle \mid x_1, ..., x_n \in u \wedge \models_{\langle u,\in\rangle} \varphi[x_1, ..., x_n]\}\,.$$

By induction on φ, we show that for all φ, $t_\varphi \in \mathcal{C}$. (The required result will then be proved using this fact.)

(a) $\varphi(\boldsymbol{x}) \equiv x_i \in x_j, \quad 1 \leq i < j \leq n\,.$

Write $F_x(y)$ for $F_3(x,y)$. Then $t_\varphi(u) = u^{i-1} \times F_u^{j-i}(F_4(\in \cap u^2, u^{n-j}))$ so $t_\varphi \in \mathcal{C}$.

(b) Let $\varphi_1(\boldsymbol{x}), ..., \varphi_p(\boldsymbol{x})$ be such that $t_{\varphi_1}, ..., t_{\varphi_p} \in \mathcal{C}$.

Let $\varphi(\boldsymbol{x})$ be any propositional combination of $\varphi_1, ..., \varphi_p$.

Since $x - y$, $x \cup y$ $(= \bigcup\{x,y\})$, $x \cap y$ $(= x - (x - y)) \in \mathcal{C}$, we clearly have $t_\varphi \in \mathcal{C}$.

(c) Let $\bar{\varphi}(y,\boldsymbol{x})$ be such that $t_{\bar{\varphi}} \in \mathcal{C}$. Let $\varphi(\boldsymbol{x}) \equiv \exists y \bar{\varphi}(y,\boldsymbol{x})$ or $\forall y \bar{\varphi}(y,\boldsymbol{x})$.

Clearly, $t_{\exists y \bar{\varphi}}(u) = \mathrm{dom}\,(t_{\bar{\varphi}}(u))$ and $t_{\forall y \bar{\varphi}}(u) = u^n - \mathrm{dom}\,(u^n - t_{\bar{\varphi}}(u))$. So in either case, $t_\varphi \in \mathcal{C}$.

(d) $\varphi(\boldsymbol{x}) \equiv x_i = x_j\,.$

Let $\theta(y,\boldsymbol{x}) \equiv y \in x_i \leftrightarrow y \in x_j$. By (a), (b), $t_\theta \in \mathcal{C}$. But look, $\models_{\langle u,\in\rangle} \varphi[\boldsymbol{x}]$ iff $(\forall y \in \mathbf{U}u)\,[\models_{\langle u \cup (\mathbf{U}u),\in\rangle} \theta[y,\boldsymbol{x}]]$.

Hence $t_\varphi(u) = u^n \cap t_{\forall y\theta}(u \cup (\mathbf{U}u))$, so $t_\varphi \in \mathcal{C}$, by (c).

(e) $\varphi(\boldsymbol{x}) \equiv x_i \in x_j , \qquad 1 \le j < i \le n .$

Let $\psi(y,z,\boldsymbol{x}) \equiv y \in z \wedge y = x_i \wedge z = x_j$. By (a), (b), (d), $t_\psi \in \mathcal{C}$. But $\varphi(\boldsymbol{x}) \leftrightarrow \exists y \exists z \psi(y,z,\boldsymbol{x})$, so by (c), $t_\varphi \in \mathcal{C}$.

Hence, for any $\in$-formula φ, $t_\varphi \in \mathcal{C}$.

If $f : \mathrm{V}^n \to \mathrm{V}$, define $f^* : \mathrm{V} \to \mathrm{V}$ by $f^*(u) = f''u^n$. Using our above result, we prove by induction on the rud definition of f, that f rud $\to f^* \in \mathcal{C}$. This easily implies the required result.

(a) $f(\boldsymbol{x}) = x_i$.

$f^*(u) = f''u^n = u = u - (u-u) \in \mathcal{C}$.

(b) $f(\boldsymbol{x}) = x_i - x_j$.

$f^*(u) = f''u^n = \{x-y \mid x,y \in u\}$. Let $\varphi(z,x,y) \equiv z \in x - y$. Let $F(u) = t_\varphi(u \cup (\mathbf{U}u)) \cap (\mathbf{U}u \times u^2) = \{\langle z,x,y\rangle \mid x,y \in u \wedge z \in x-y\}$. Then $f^*(u) = F_8(F(u),u^2) \in \mathcal{C}$, since $t_\varphi \in \mathcal{C}$.

(c) $f(\boldsymbol{x}) = \{x_i,x_j\}$.

$f^*(u) = f''u^n = \{\{x,y\} \mid x,y \in u\} = \mathbf{U}u^2 \in \mathcal{C}$.

(d) $f(\boldsymbol{x}) = h(g_1(\boldsymbol{x}), \ldots, g_k(\boldsymbol{x}))$.

Let $G(u) = \mathbf{U}_{i=1}^k g_i^*(u)$, $H(u) = h^*(\mathbf{U}_{i=1}^k g_i^*(u)) = h^*(G(u))$, and $K(u) = u^n \cup G(u) \cup H(u)$. By hypothesis, $G, H, K \in \mathcal{C}$. By Lemma 1, let $\varphi(y,\boldsymbol{x})$ be an $\in$-formula equivalent to the formula

$$\exists z_1 \ldots \exists z_k (z_1 = g_1(\boldsymbol{x}) \wedge \ldots \wedge z_k = g_k(\boldsymbol{x}) \wedge y = h(z_1, \ldots, z_k)) .$$

Clearly, $f^*(u) = F_8(([t_\varphi(K(u))] \cap [H(u) \times u^n]), u^n) \in \mathcal{C}$.

(e) $f(y,\boldsymbol{x}) = \bigcup_{z \in y} g(z,\boldsymbol{x})$.

Let $G(u) = \{\langle z,y,\boldsymbol{x}\rangle \mid (\exists v \in y)\,[z \in g(v,\boldsymbol{x})] \wedge \boldsymbol{x} \in u\}$. As above $G \in \mathcal{C}$. But $f^*(u) = F_8(G(u), u^{n+1}) \in \mathcal{C}$.

Hence f rud $\to f^* \in \mathcal{C}$, for all f.

Finally, let f be rud. We show that $f \in \mathcal{C}$.

Set $\bar{f}(\langle \boldsymbol{z}\rangle) = f(\boldsymbol{z})$, $\bar{f}(y) = \emptyset$ in all other cases. Thus $\bar{f}$ is rud. So by the above, $\bar{f}^* \in \mathcal{C}$. Let $P(\boldsymbol{x}) = \{\langle \boldsymbol{x}\rangle\}$. Thus $P \in \mathcal{C}$. But look, $f(\boldsymbol{x}) = \bigcup\bigcup\{\{f(\boldsymbol{x})\}\} = \bigcup\bigcup\{\bar{f}\{\langle \boldsymbol{x}\rangle\}\} = \bigcup\bigcup F_8(\bar{f}^*(P(\boldsymbol{x})), P(\boldsymbol{x})) \in \mathcal{C}$.

As an immediate corollary of Lemmas 3 and 6 we have:

Lemma 7. *Let $A \subset V$ and define F_9 by $F_9(x,y) = A \cap x$. Every function* rud *in A may be expressed as a composition of some of the* (rud *in A*) *functions $F_0, \ldots, F_9$.*

We shall make immediate use of Lemma 7 in investigating the logical complexity of the predicates $\models^{\Sigma_n}_{\mathbf{M}}$ for suitable $\mathbf{M}$. We assume, once and for all, that we have a fixed arithmetization of our language.

Lemma 8. $\models^{\Sigma_0}_{\mathbf{M}}$ *is uniformly $\Sigma_1^{\mathbf{M}}$ for transitive,* rud *closed* $\mathbf{M} = \langle M, A\rangle$.

Proof. Let $\mathcal{L}$ be the language consisting of:

(i) variables w_i, $i \in \omega$.

(ii) function symbols (binary) $f_0, \ldots, f_9$.

We shall assume we have a fixed arithmetization of $\mathcal{L}$. We also assume that the reader understands what is meant by a "term" of $\mathcal{L}$. Henceforth, let $\mathbf{M} = \langle M, A\rangle$ be arbitrary, transitive, and rud closed.

We first define precisely how $\mathcal{L}$ is to be interpreted in $\mathbf{M}$.

Let Q be the set of functions ρ mapping a finite subset of $\{w_i \mid i \in \omega\}$ into M. We may clearly assume Q is rud. Let C be the (rud) function which to each term τ of $\mathcal{L}$ assigns the set of all component terms of τ, including variables. Let $\mathrm{Vbl}_{\mathcal{L}}$ be the rud predicate defining the set $\{w_i \mid i \in \omega\}$. Let P be the predicate

$$P(u,g,v) \leftrightarrow [\mathrm{dom}(g) = u] \wedge (\forall x \in u)[[x \in \mathrm{Vbl} \to x \in \mathrm{dom}(v) \wedge g(x) = v(x)]$$
$$\wedge \bigwedge_{i=0}^{9} (\forall t_0, t_1 \in u)[x = f_i(t_0, t_1) \to g(x) = F_i(g(t_0), g(t_1))]] \, .$$

Thus P is rud in A.

We may now define the interpretation of a term τ of $\mathcal{L}$ at a "point" $\rho \in Q$ by

$$y = \tau^{\mathbf{M}}[\rho] \leftrightarrow \text{"}\tau \text{ is an } \mathcal{L}\text{-term"} \wedge \rho \in C \wedge \exists g[P(C(\tau), g, \rho) \wedge g(\tau) = y] \, .$$

Hence the function

$$f(\tau,\rho) = \begin{cases} \tau^{\mathbf{M}}[\rho], & \text{if } \tau \text{ is an } \mathcal{L}\text{-term and } \rho \in Q \\ \emptyset, & \text{otherwise} \end{cases}$$

is (uniformly) $\Sigma_1^{\mathbf{M}}$ (for transitive, rud closed $\mathbf{M}$). Since $\mathbf{M}$ is rud closed we can use the above result to define $\models_{\mathbf{M}}^{\Sigma_0}$ as an $\mathbf{M}$-predicate.

Let $\varphi \in \mathrm{Fml}^{\Sigma_0}$. By Lemma 2, $\varphi^{\mathbf{M}}$ is rud in A. Hence the function Γ defined by

$$\Gamma(\boldsymbol{x}) = \begin{cases} 1, & \text{if } \varphi^{\mathbf{M}}[\boldsymbol{x}] \\ 0, & \text{otherwise} \end{cases}$$

is rud in A. So, by Lemma 7, we may assume $\Gamma = \tau^{\mathbf{M}}$, where τ is a term of $\mathcal{L}$, under the above interpretation (i.e. with F_i interpreting f_i for each i). In fact, we may clearly pick a recursive function σ mapping Fml^{Σ_0} into the terms of $\mathcal{L}$ so that whenever $\varphi \in \mathrm{Fml}^{\Sigma_0}$, $\varphi^{\mathbf{M}}[\boldsymbol{x}] \leftrightarrow [\sigma(\varphi)]^{\mathbf{M}}[\boldsymbol{x}] = 1$. But by our above result, this implies that $\models_{\mathbf{M}}^{\Sigma_0}$ is (uniformly) $\Sigma_1^{\mathbf{M}}$ (for transitive, rud closed $\mathbf{M}$).

As an immediate consequence of this result, we have

Lemma 9. *Let $n \geq 1$. Then $\models_{\mathbf{M}}^{\Sigma_n}$ is uniformly $\Sigma_n^{\mathbf{M}}$ for transitive, rud closed $\mathbf{M} = \langle M, A \rangle$.*

We conclude this section with a few miscellaneous results of use later. The first two are technical, and will often be used without mention.

Lemma 10. *Let* $\mathbf{M} = \langle M, A \rangle$ *be* rud *closed. If* $R \subset M$ *is* $\Sigma_n(\mathbf{M})$, *there is a* $\Sigma_0(\mathbf{M})$ *relation* P *such that* $R(x) \leftrightarrow \exists x_1 \forall x_2 \exists x_3 \ldots Q_n x_n P(x, x_1, \ldots, x_n)$.

Proof. Suppose $R(x) \leftrightarrow \models_{\mathbf{M}} \exists \boldsymbol{v}_1 \forall \boldsymbol{v}_2 \exists \boldsymbol{v}_3 \ldots Q_n \boldsymbol{v}_n \varphi(v, \boldsymbol{v}_1, \ldots, \boldsymbol{v}_n)[x]$, where φ is a Σ_0-formula. Using the rud functions $\langle -, \ldots, - \rangle, (-)_0^m, \ldots, (-)_{m-1}^m$, we can easily obtain, via Lemma 1, a Σ_0-formula ψ such that $R(x) \leftrightarrow \models_{\mathbf{M}} \exists v_1 \forall v_2 \ldots Q_n v_n \psi(v, v_1, \ldots, v_n)[x]$. Then $R(x) \leftrightarrow \exists x_1 \forall x_2 \ldots Q_n x_n [\models_{\mathbf{M}} \psi[x, x_1, \ldots, x_n]]$, as required.

Lemma 11. *Let* $\mathbf{M} = \langle M, A \rangle$ *be* rud *closed. If* $R \subset M$ *is* $\Sigma_n(\mathbf{M})$, *there is a single element* $p \in M$ *such that* R *is* $\Sigma_n^{\mathbf{M}}(\{p\})$.

Proof. If R is $\Sigma_n^{\mathbf{M}}(\{p_1, \ldots, p_n\})$, then R is also $\Sigma_n^{\mathbf{M}}(\{\langle p_1, \ldots, p_n \rangle\})$.

Let $\mathbf{M} = \langle M, A \rangle$, $n \geq 0$. Write $X \prec_{\Sigma_n} \mathbf{M}$ iff $X \subset M$ and for every Σ_n formula φ and every $\boldsymbol{x} \in X$,

$$\models_{\langle X, A \cap X \rangle} \varphi[\boldsymbol{x}] \quad \text{iff} \quad \models_{\mathbf{M}} \varphi[\boldsymbol{x}] .$$

Clearly, if X, M are transitive and $X \subset M$, we always have $X \prec_{\Sigma_0} \mathbf{M}$. And for $n > 0$, we have $X \prec_{\Sigma_n} \mathbf{M}$ iff $X \subset M$ and for every $P \in \Sigma_n^{\mathbf{M}}(X), P \neq \emptyset \rightarrow P \cap X \neq \emptyset$.

Recall that if $\langle X, \in \rangle$ satisfies the axiom of extensionality, there is a unique isomorphism $\pi : \langle X, \in \rangle \cong \langle W, \in \rangle$, where W is a unique transitive set. Furthermore, if $Z \subset X$ is transitive, then $\pi \upharpoonright Z = \mathrm{id} \upharpoonright Z$. In fact, π is defined by $\in$-induction thus: $\pi(x) = \{\pi(y) \mid y \in x \cap X\}$ for each $x \in X$. The next result is of considerable importance.

Lemma 12. *Let* $\mathbf{M}$ *be transitive and* rud *closed. Let* $X \prec_{\Sigma_1} \mathbf{M}$. *Then* $\langle X, A \cap X \rangle$ *satisfies the axiom of extensionality and is* rud *closed. Let* $\pi : \langle X, A \cap X \rangle \cong \langle W, B \rangle$, *where* W *is transitive. Let* $f : M^n \rightarrow M$ *be* rud *in* A. *Then for all* $\boldsymbol{z} \in X$, $\pi(f(\boldsymbol{z})) = f(\pi(\boldsymbol{z}))$.

Proof. Since $\mathbf{M}$ is transitive, $\mathbf{M}$ satisfies the axiom of extensionality. Hence as $X \prec_{\Sigma_1} \mathbf{M}$, so does $\langle X, A \cap X \rangle$. Similarly, by Lemma 5, $\langle X, A \cap X \rangle$ is rud closed. Hence, in particular, $\boldsymbol{z} \in X \rightarrow f(\boldsymbol{z}) \in X$ for $f : M^n \rightarrow M$ rud in A. By induction on the (rud in A) definition of f, $\pi(f(\boldsymbol{z})) = f(\pi(\boldsymbol{z}))$ for each $\boldsymbol{z} \in X$.

§3. Admissible sets

Let $\mathbf{M} = \langle M, A\rangle$ be non-empty and transitive. We say $\mathbf{M}$ is *admissible* iff $\mathbf{M}$ is rud closed and satisfies the Σ_0-*Replacement Axiom*: for all Σ_0 formulas φ and all $\boldsymbol{a} \in M$, $\models_{\mathbf{M}} [\forall x \exists y \varphi \to \forall u \exists y (\forall x \in u)(\exists y \in v)\varphi][\boldsymbol{a}]$.

In case $A = \emptyset$ in the above, we call M an *admissible set.* More generally, $\mathbf{M}$ is Σ_n-*admissible* iff $\mathbf{M}$ is rud closed and satisfies the (analogous) Σ_n-*Replacement Axiom.* Likewise a Σ_n-*admissible set.* We prove below that $\mathbf{M}$ is admissible iff $\mathbf{M}$ is Σ_1-admissible. All our results extend trivially from admissibility to Σ_n-admissibility, with "Σ_n" everywhere replacing Σ_1, etc.

Roughly speaking, an admissible set (or structure) behaves like the universe as far as Σ_1 concepts are concerned. We give a few elementary results which set the tone for the rest of this exposition.

Convention: For the whole of this paper, we shall adopt the following abuse of notation. Suppose $\mathbf{M}$ is a structure, $\varphi(\boldsymbol{v})$ is a formula, and $\boldsymbol{x} \in M$. We shall write $\models_{\mathbf{M}} \varphi(\boldsymbol{x})$ rather than $\models_{\mathbf{M}} \varphi(\boldsymbol{v})[\boldsymbol{x}]$. Clearly, this is purely a notational convenience.

Firstly, we give the promised "stronger" form of the admissibility definition.

Lemma 13 (Σ_1-Replacement). *Let* $\mathbf{M}$ *be admissible, and let* φ *be a* Σ_1*-formula,* $\boldsymbol{a} \in M$. *Then*

$$\models_{\mathbf{M}} \forall x \exists y \varphi(x,y,\boldsymbol{a}) \to \forall u \exists v (\forall x \in u)(\exists y \in v)\varphi(x,y,\boldsymbol{a}).$$

Proof. Let ψ be a Σ_0-formula such that $\models_{\mathbf{M}} \varphi(x,y,\boldsymbol{a}) \leftrightarrow \exists z \psi(x,y,z,\boldsymbol{a})$. Then

$$\begin{aligned}\models_{\mathbf{M}} \forall x \exists y \varphi(x,y,\boldsymbol{a}) &\to \forall x \exists y \exists z \psi(x,y,z,\boldsymbol{a})\\ &\to \forall x \exists w \psi(x,(w)_0,(w)_1,\boldsymbol{a})\\ &\to \forall u \exists v (\forall x \in u)(\exists w \in v)\psi(x,(w)_0,(w)_1,\boldsymbol{a}),\\ &\qquad \text{by } \Sigma_0\text{-Replacement}\\ &\to \forall u \exists v (\forall x \in u)(\exists y \in v)\varphi(x,y,\boldsymbol{a}).\end{aligned}$$

Convention: The essentially superfluous role played by $\boldsymbol{a}$ in the above theorem leads us to extend our previous convention slightly by allowing formulas to contain members of **M** as parameters. Again, this is clearly an avoidable convenience.

Lemma 14. *Let* **M** *be admissible. If* $R(\boldsymbol{x},y)$ *is* $\boldsymbol{\Sigma}_1(\mathbf{M})$*, so is* $(\forall y \in z)R(\boldsymbol{x},y)$.

Proof. Let φ be a $\boldsymbol{\Sigma}_0$-formula with parameters from ("w.p.f.") M such that $R(\boldsymbol{x},y) \leftrightarrow \models_{\mathbf{M}} \exists w\varphi(\boldsymbol{x},y)$. Then

$$(\forall y \in z)R(\boldsymbol{x},y) \leftrightarrow \models_{\mathbf{M}} (\forall y \in z)\exists w\varphi(\boldsymbol{x},y)$$
$$\leftrightarrow \models_{\mathbf{M}} \forall y \exists w[(y \in z \wedge \varphi(\boldsymbol{x},y)) \vee (y \notin z)] \;.$$

So by $\boldsymbol{\Sigma}_0$-Replacement,

$$(\forall y \in z)R(\boldsymbol{x},y) \leftrightarrow \models_{\mathbf{M}} \exists v(\forall y \in z)(\exists w \in v)\varphi(\boldsymbol{x},y) \,,$$

which is $\boldsymbol{\Sigma}_1(\mathbf{M})$.

Lemma 15 ($\boldsymbol{\Delta}_1$-Comprehension). *Let* **M** *be admissible,* $P \in \boldsymbol{\Delta}_1(\mathbf{M})$. *Then* $u \in M \to P \cap u \in M$.

Proof. Let φ, ψ be $\boldsymbol{\Sigma}_0$-formulas w.p.f. M such that

$$P(z) \leftrightarrow \models_{\mathbf{M}} \forall x\varphi(x,z) \leftrightarrow \models_{\mathbf{M}} \exists y\psi(y,z) \,.$$

Then,

$$\models_{\mathbf{M}} \forall w_1 \exists w_2 [[w_1 \in u \wedge (\exists y(\psi \wedge w_2 = y) \vee \exists x(\neg\varphi \wedge w_2 = x))]$$
$$\vee\; [w_1 \notin u \wedge w_2 = \emptyset]] \,.$$

So by $\boldsymbol{\Sigma}_1$-Replacement there is $v \in M$ such that

$$\models_{\mathbf{M}} (\forall w_1 \in u)(\exists w_2 \in v)[\exists y(\psi \wedge w_2 = y) \vee \exists x(\neg\varphi \wedge w_2 = x)] \,.$$

So,

$$P \cap u = \{z \mid\, \models_{\mathbf{M}} \forall x\varphi(x,z)\} \cap u = \{z \mid\, \models_{\mathbf{M}} (\exists y \in v)\psi(y,z)\} \cap u \,.$$

But **M** is rud closed (so satisfies what might be called the Σ_0-Comprehension Axiom), and therefore we conclude that $P \cap u = \{z \in u \mid \models_{\mathbf{M}} (\exists y \in v)\psi(y,z)\} \in M$.

The next result has nothing specifically to do with admissibility, but is of considerable value. Let $f : \subset M \to M$ mean that $f : X \to M$ for some $X \subset M$.

Lemma 16. *Let* **M** *be arbitrary,* $f : \subset M \to M$ *be* $\mathbf{\Sigma}_1(\mathbf{M})$. *If* dom$(f)$ *is* $\mathbf{\Pi}_1(M)$, *then in fact f and* dom(f) *are* $\mathbf{\Delta}_1(\mathbf{M})$.

Proof.

(a) $$\underbrace{f(x) = y}_{\Sigma_1(\mathbf{M})} \leftrightarrow \underbrace{x \in \mathrm{dom}(f) \wedge \forall z (z \neq y \to f(x) \neq z)}_{\Pi_1(\mathbf{M})}$$

(b) $$\underbrace{x \in \mathrm{dom}(f)}_{\Pi_1(\mathbf{M})} \leftrightarrow \underbrace{\exists y (f(x) = y)}_{\Sigma_1(\mathbf{M})}.$$

It was necessary to state the above result explicitly because we shall frequently have to deal with functions which, though definable, are not total functions. A particular case of the above theorem would of course occur when dom$(f) \in M$, (whence dom(f) is $\Sigma_0(\mathbf{M})$).

As usual, we shall use the notation $f(x) \simeq g(x)$ for partial functions, with its usual meaning (i.e. $f(x)$ is defined iff $g(x)$ is defined, in which case $f(x) = g(x)$).

Lemma 17. *Let* **M** *be admissible,* $f : \subset M \to M$ *be* $\mathbf{\Sigma}_1(\mathbf{M})$. *If* $u \in M$ *and* $u \subset$ dom(f), *then* $f''u \in M$.

Proof. Since **M** is rud closed and $f''u = \mathrm{ran}(f \upharpoonright u)$, it suffices to prove that $f \upharpoonright u \in M$. Now, as $u \in M$, $f \upharpoonright u$ is $\mathbf{\Delta}_1(\mathbf{M})$ by Lemma 16. Let $\varphi(x,y)$ be a Σ_1-formula w.p.f. M such that $f(x) = y \leftrightarrow \models_{\mathbf{M}} \varphi(x,y)$.

Then $\models_{\mathbf{M}} \forall x \exists y [(x \in u \wedge \varphi(x,y)) \vee (x \notin u)]$, so by Σ_1-Replacement there is $v \in M$ such that $\models_{\mathbf{M}} (\forall x \in u)(\exists y \in v)\varphi(x,y)$. Hence $f \upharpoonright u \subset v \times u$. So, by Δ_1-Comprehension, $f \upharpoonright u = (f \upharpoonright u) \cap (v \times u) \in M$.

Theorem 18 (Recursion Theorem). *Let* **M** *be admissible. Let* $h : M^{n+1} \to M$ *be a* $\mathbf{\Sigma}_1(\mathbf{M})$ *function such that for all* $\boldsymbol{x} \in M$, $\{\langle z,y\rangle \mid z \in h(y,\boldsymbol{x})\}$ *is well-*

founded. Let $G = M^{n+2} \to M$ be $\Sigma_1(\mathbf{M})$. Then there is a unique $\Sigma_1(\mathbf{M})$ function F such that

(i) $\langle y, \boldsymbol{x} \rangle \in \mathrm{dom}(F) \leftrightarrow \{\langle z, \boldsymbol{x} \rangle \mid z \in h(y, \boldsymbol{x})\} \subset \mathrm{dom}(F)$

(ii) $F(y, \boldsymbol{x}) \simeq G(y, \boldsymbol{x}, \langle F(z, \boldsymbol{x}) \mid z \in h(y, \boldsymbol{x}) \rangle)$.

Proof. Let Φ be the predicate

$$\Phi(f, \boldsymbol{x}) \leftrightarrow \text{“}f \text{ is a function”} \wedge (\forall y \in \mathrm{dom}(f))(\forall z \in h(y, \boldsymbol{x}))(z \in \mathrm{dom}(f))$$

$$\wedge (\forall y \in \mathrm{dom}(f))[(f \restriction y) = G(y, \boldsymbol{x}, f \restriction h(y, \boldsymbol{x}))] .$$

By Lemma 16, h, G are $\Delta_1(\mathbf{M})$, so Φ is $\Delta_1(\mathbf{M})$.

Let φ be a Σ_1-formula w.p.f. M such that $\Phi(f, \boldsymbol{x}) \leftrightarrow \models_{\mathbf{M}} \varphi(f, \boldsymbol{x})$. Define a $\Sigma_1(\mathbf{M})$ predicate F by (using notation which will later be justified)

$$F(y, \boldsymbol{x}) = u \leftrightarrow \exists f [\Phi(f, \boldsymbol{x}) \wedge f(y) = u] .$$

We verify (i) for this F. Suppose first that $\langle y, \boldsymbol{x} \rangle \in \mathrm{dom}(F)$. Then, by definition, $\exists f [\Phi(f, \boldsymbol{x}) \wedge y \in \mathrm{dom}(f)]$. By definition of Φ, for such an f we must have $(\forall z \in h(y, \boldsymbol{x}))(z \in \mathrm{dom}(f))$. Hence $z \in h(y, \boldsymbol{x}) \to \langle z, \boldsymbol{x} \rangle \in \mathrm{dom}(F)$. Now suppose that $z \in h(y, \boldsymbol{x}) \to \langle z, \boldsymbol{x} \rangle \in \mathrm{dom}(F)$. Note that as M is transitive, $h(y, \boldsymbol{x}) \subset M$. By our supposition,

$$\models_{\mathbf{M}} \forall z \exists f [(z \in h(y, \boldsymbol{x}) \wedge z \in \mathrm{dom}(f) \wedge \varphi(f, \boldsymbol{x})) \vee (z \notin h(y, \boldsymbol{x}) \wedge f = \emptyset)] ,$$

so by Σ_1-Replacement,

$$\models_{\mathbf{M}} \exists v (\forall z \in h(y, \boldsymbol{x}))(\exists f \in v)[z \in \mathrm{dom}(f) \wedge \varphi(f, \boldsymbol{x})] .$$

Pick such a v. As Φ is $\Delta_1(\mathbf{M})$, by Δ_1-Comprehension we see that $w = v \cap \{f \mid \Phi(f, \boldsymbol{x})\} \in M$. Hence $\bigcup w \in M$. It is easily seen that $\Phi(\bigcup w, \boldsymbol{x})$. Noting that $h(y, \boldsymbol{x}) \subset \mathrm{dom}(\bigcup w)$, note that $\bigcup w \restriction h(y, \boldsymbol{x}) \in M$. Set $f = \bigcup w \restriction h(y, \boldsymbol{x}) \cup \{\langle G(y, \boldsymbol{x}, \bigcup w \restriction h(y, \boldsymbol{x})), y \rangle\}$. Clearly, $\Phi(f, \boldsymbol{x})$, so $\langle y, \boldsymbol{x} \rangle \in \mathrm{dom}(F)$. Hence (i) holds for this F.

We now show that F is a function and is unique. By (i), $\mathrm{dom}(F)$ is already uniquely determined, so for both of these it suffices to prove the following:

$$\Phi(f,\boldsymbol{x}) \wedge \Phi(f',\boldsymbol{x}) \wedge y \in \operatorname{dom}(f) \cap \operatorname{dom}(f') \to f(y) = f'(y) .$$

To this end, suppose not. Then $P = \{y \mid y \in \operatorname{dom}(f) \cap \operatorname{dom}(f') \wedge f(y) \neq f'(y)\} \neq \emptyset$. Let y_0 be an h-minimal element of P. Since $y_0 \in P, f(y_0) \neq f'(y_0)$. But $\Phi(f,\boldsymbol{x})$, $\Phi(f',\boldsymbol{x})$, so clearly $f(y_0) = f'(y_0)$ by the h-minimality of $y_0 \in P$. This contradiction suffices (and thus justifies our notation somewhat).

Finally, it is trivial to note that (ii) must hold, virtually by definition.

In view of the many set theoretic concepts defined by a recursion of the above type, it is clear that admissible sets play an important role in set theory.

Say **M** is *strongly admissible* iff **M** is non-empty, transitive, rud closed, and satisfies the *Strong* Σ_0*-Replacement Axiom*: for all Σ_0 formulas φ w.p.f. M, $\models_{\mathbf{M}} \forall u \exists v (\forall x \in u)[\exists y \varphi(x,y) \to (\exists y \in u)\varphi(x,y)]$. (Clearly, such an **M** will also satisfy the "Strong Σ_1-Replacement Axiom".)

Strongly admissible structures **M** are (for reasons to be indicated later) also called *non-projectible admissible structures*. The difference between admissibility and strong admissibility is closely connected with the difference between Σ_n predicates and Δ_n predicates, which is in turn closely connected with the difference between a function being partial and total. We shall have more to say on this matter later.

§4. The Jensen hierarchy

Let X be a set. The *rudimentary closure* of X is the smallest set $Y \supset X$ such that Y is rud closed.

Lemma 19. *If U is transitive, so is its* rud *closure.*

Proof. Let W be the rud closure of U. Since rud functions are closed under composition, we clearly have $W = \{f(\boldsymbol{x}) \mid \boldsymbol{x} \in U \wedge f \text{ is rud}\}$. An easy induction on the rud definition of any rud f shows that $\boldsymbol{x} \in U \to \mathrm{TC}(f(\boldsymbol{x})) \subset W$. Hence W is transitive. (TC denotes the transitive closure function.)

For U transitive, let $\operatorname{rud}(U)$ = the rud closure of $U \cup \{U\}$. Of crucial importance is:

Lemma 20. *Let U be transitive. Then $\mathcal{P}(U) \cap \mathrm{rud}(U) = \Sigma_\omega(U)$.*

Proof. Clearly, $\mathcal{P}(U) \cap \Sigma_0(U \cup \{U\}) = \Sigma_\omega(U)$, so it suffices to show that $\mathcal{P}(U) \cap \Sigma_0(U \cup \{U\}) = \mathcal{P}(U) \cap \mathrm{rud}(U)$. Let $X \in \mathcal{P}(U) \cap \Sigma_0(U \cup \{U\})$. Then, exactly as in the proof of Lemma 2, $X \in \mathrm{rud}(U)$ (by induction on the Σ_0 definition of X). Now let $X \in \mathcal{P}(U) \cap \mathrm{rud}(U)$. Then X is a $\Sigma_0(\mathrm{rud}(U))$ subset of U. By Lemma 1, we may in fact assume that X is $\Sigma_0^{\mathrm{rud}(U)}(U \cup \{U\})$. But $X \subset U \cup \{U\} \subset \mathrm{rud}(U)$ and $U \cup \{U\}$, $\mathrm{rud}(U)$ are transitive, so X is actually $\Sigma_0^{U \cup \{U\}}(U \cup \{U\}) = \Sigma_0(U \cup \{U\})$.

Also very relevant is:

Lemma 21. *There is a* rud *function* **S** *such that whenever U is transitive, $\mathbf{S}(U)$ is transitive, $U \cup \{U\} \subset \mathbf{S}(U)$ and $\bigcup_{n \in \omega} \mathbf{S}^n(U) = \mathrm{rud}(U)$.*

Proof. Set $\mathbf{S}(U) = (U \cup \{U\}) \cup (\bigcup_{i=0}^{8} F_i''(U \cup \{U\})^2)$. The result follows by Lemma 6.

Lemma 22. *There is a* rud *function* Wo *such that whenever r is a well-ordering of u,* Wo(r, u) *is an end-extension of r which well-orders $\mathbf{S}(u)$.*

Proof. Define i^u, j_1^u, j_2^u by:–

$i^u(x)$ = the least $i \leq 8$ such that $(\exists x_1, x_2 \in u)[F_i(x_1, x_2) = x]$

$j_1^u(x)$ = the r-least $x_1 \in u$ such that $(\exists x_2 \in u)[F_{i^u(x)}(x_1, x_2) = x]$

$j_2^u(x)$ = the r-least $x_2 \in u$ such that $F_{i^u(x)}(j_1^u(x), x_2) = x$.

Clearly, i^u, j_1^u, j_2^u are rud functions of u, x.

Define

$$\mathrm{Wo}(r,u) = \{\langle x,y\rangle \mid x,y \in u \wedge xry\} \cup \{\langle x,y\rangle \mid x \in u \wedge y \notin u\}$$

$$\cup \{\langle x,y\rangle \mid x \notin u \wedge y \notin u \wedge [i^u(x) < i^u(y) \vee i^u(x) =$$

$$i^u(y) \wedge [j_1^u(x)\, r\, j_1^u(y) \vee (j_1^u(x) = j_1^u(y) \wedge j_2^u(x)\, r\, j_2^u(y))]]\} .$$

The *Jensen hierarchy*, $\langle \mathrm{J}_\alpha \mid \alpha \in \mathrm{OR}\rangle$, is defined as follows:

$$\mathrm{J}_0 = \emptyset$$

$$\mathrm{J}_{\alpha+1} = \mathrm{rud}(\mathrm{J}_\alpha)$$

$$\mathrm{J}_\lambda = \bigcup_{\alpha<\lambda} \mathrm{J}_\alpha, \quad \text{if } \lim(\lambda).$$

Lemma 23. (i) *Each* J_α *is transitive.*

(ii) $\alpha \leq \beta \to J_\alpha \subset J_\beta$

(iii) $\mathrm{rank}(J_\alpha) = \mathrm{OR} \cap J_\alpha = \omega\alpha$.

Proof. (i) By Lemma 19.

(ii) Immediate.

(iii) By induction: $\mathrm{rank}(J_{\alpha+1}) = \mathrm{rank}(\mathrm{rud}(J_\alpha)) = \mathrm{rank}(J_\alpha) + \omega$ (by an earlier remark, this last step is easily verified).

To facilitate our handling of the hierarchy, we "stratify" the J_α's by defining an auxiliary hierarchy $\langle S_\alpha | \alpha \in \mathrm{OR}\rangle$ as follows:

$$S_0 = \emptyset$$

$$S_{\alpha+1} = \mathbf{S}(S_\alpha)$$

$$S_\lambda = \bigcup_{\alpha<\lambda} S_\alpha, \quad \text{if } \lim(\lambda).$$

Clearly, the J_α's are just the limit points of this sequence. In fact:

Lemma 24. (i) *Each* S_α *is transitive*

(ii) $\alpha \leq \beta \to S_\alpha \subset S_\beta$

(iii) $J_\alpha = \bigcup_{\nu<\omega\alpha} S_\nu = S_{\omega\alpha}$.

Proof. (i) By Lemma 21.

(ii) Immediate.

(iii) By induction: $J_{\alpha+1} = \mathrm{rud}(J_\alpha) = \bigcup_{n\in\omega} \mathbf{S}^n(J_\alpha) = \bigcup_{n\in\omega} \mathbf{S}^n(S_{\omega\alpha}) = \bigcup_{n\in\omega} S_{\omega\alpha+n} = S_{\omega\alpha+\omega} = S_{\omega(\alpha+1)}$.

Lemma 25. $\langle S_\nu | \nu < \omega\alpha\rangle$ *is uniformly* $\Sigma_1^{J_\alpha}$ *for all* α.

Proof. Set

$$\Phi(f) \equiv \text{"}f \text{ is a function"} \wedge \mathrm{dom}(f) \in \mathrm{OR} \wedge f(0) = \emptyset \wedge$$
$$(\forall \nu \in \mathrm{dom}(f))[(\mathrm{succ}(\nu) \to f(\nu) = \mathbf{S}(f(\nu-1))) \wedge$$
$$[\lim(\nu) \to f(\nu) = \bigcup_{\alpha\in\nu} f(\alpha)]] .$$

Clearly, Φ is uniformly $\Sigma_0^{J_\alpha}$. And by definition, $y = S_\nu \leftrightarrow \exists f(\Phi(f) \wedge y = f(\nu))$. Thus it suffices to show that for any α, $\nu < \omega\alpha$, the existential quantifier here can be restricted to J_α. In other words, we must show that whenever $\tau < \omega\alpha$, then $\langle S_\nu | \nu < \tau \rangle \in J_\alpha$. This is proved by induction on α. For $\alpha = 0$ it is trivial. For limit α the induction step is immediate. So assume $\alpha = \beta + 1$ and that $\tau < \omega\beta \to \langle S_\nu | \nu < \tau \rangle \in J_\beta$. Then, by our above remarks, it is clear that $\langle S_\nu | \nu < \omega\beta \rangle$ is $\Sigma_1^{J_\beta}$. So by Lemma 20, $\langle S_\nu | \nu < \omega\beta \rangle \in J_\alpha$. Thus for all $n < \omega$, $\langle S_\nu | \nu < \omega\beta + n \rangle = \langle S_\nu | \nu < \omega\beta \rangle \cup \{\langle \mathbf{S}^m(J_\beta), \omega\beta + m \rangle | m < n\} \in J_\alpha$, as J_α is rud closed.

Lemma 26. *$\langle J_\nu | \nu < \alpha \rangle$ is uniformly $\Sigma_1^{J_\alpha}$ for all α.*

Proof. By an easy induction, $\langle \omega\nu | \nu < \alpha \rangle$ is uniformly $\Sigma_1^{J_\alpha}$ for all α. Since $J_\nu = S_{\omega\nu}$, The result follows by Lemma 25.

Lemma 27. *There are well-orderings $<_\nu$ of the S_ν such that:*

(i) $\nu_1 < \nu_2 \to <_{\nu_1} \subset <_{\nu_2}$;

(ii) $<_{\nu+1}$ *is an end-extension of* $<_\nu$;

(iii) $\langle <_\nu | \nu < \omega\alpha \rangle$ *is uniformly* $\Sigma_1^{J_\alpha}$ *for all* α.

Proof. We use Lemma 22. Set $<_0 = \emptyset$, and by induction:

$$<_{\nu+1} = \mathrm{Wo}(<_\nu, S_\nu)$$

$$<_\lambda = \bigcup_{\nu\in\lambda} <_\nu, \quad \text{if } \lim(\lambda).$$

(i) and (ii) are immediate and (iii) is proved like Lemma 25.

Lemma 28. *There are well-orderings $<_{J_\alpha}$ of the J_α such that:*

(i) $\alpha_1 < \alpha_2 \to <_{J_{\alpha_1}} \subset <_{J_{\alpha_2}}$;

(ii) $<_{J_{\alpha+1}}$ *is an end-extension of* $<_{J_\alpha}$;

(iii) $\langle <_{J_\beta} | \beta < \alpha \rangle$ *is uniformly* $\Sigma_1^{J_\alpha}$;

(iv) $<_{J_\alpha}$ *is uniformly* $\Sigma_1^{J_\alpha}$;

(v) *the function* $\mathrm{pr}_\alpha(x) = \{z | z <_{J_\alpha} x\}$ *is uniformly* $\Sigma_1^{J_\alpha}$.
("pr" *stands for "predecessors" of course.*)

Proof. Set $<_{J_\alpha} = <_{\omega\alpha}$. (i)–(iii) are immediate by Lemma 27. For (iv), note simply that $x <_{J_\alpha} y \leftrightarrow \exists\nu(x <_\nu y)$. Finally, for (v), note that $y = \mathrm{pr}_\alpha(x) \leftrightarrow \exists\nu[x \in S_\nu \wedge y = \{z | z <_\nu x\}]$ (and that $<_\nu \in J_\alpha$), and use Lemma 27.

Lemmas 12 and 26 enable us to prove the following extremely powerful result (due in its original form to Gödel, the present version being Jensen's):

Theorem 29 (Condensation Lemma). *Let* $X \prec_{\Sigma_1} J_\alpha$. *Then for some* $\beta \leq \alpha$, $X \cong J_\beta$.

Proof. Let $X \prec_{\Sigma_1} J_\alpha$. Then by Lemma 12, let $\pi : X \cong W$, where W is transitive. We prove by induction on α that $W = J_\beta$ for $\beta = \pi''(X \cap \alpha)$.

Assume, therefore, that whenever $\nu < \alpha$ and $X^\nu \prec_{\Sigma_1} J_\nu$, the unique isomorphism π^ν of X^ν onto a transitive set W^ν yields $W^\nu = J_{\pi^{\nu}{}''(X^\nu \cap \nu)}$. Note that, as $\langle J_\nu | \nu < \alpha \rangle$ is $\Sigma_1^{J_\alpha}$, $\nu \in X \cap \alpha \leftrightarrow J_\nu \in X$.

Claim 1: For all $\nu \in X \cap \alpha$, $\pi(J_\nu) = J_{\pi(\nu)}$.

To see this, note first that for $\nu \in X \cap \alpha$, $X \cap J_\nu \prec_{\Sigma_1} J_\nu$. [For, let $A \in \Sigma_1^{J_\nu}(X \cap J_\nu)$. Since $J_\nu \in X$, $A \in \Sigma_1^{J_\alpha}(X)$. So, if $A \neq \emptyset$, then as $X \prec_{\Sigma_1} J_\alpha$, $A \cap X \neq \emptyset$. But $A \subset J_\nu$, so $A \cap (X \cap J_\nu) \neq \emptyset$.] Hence by induction hypothesis, $\pi' : X \cap J_\nu \cong J_{\pi'''(X \cap J_\nu \cap \nu)}$ for some unique π'. But look, J_α is an $\in$-end extension of J_ν, so π maps $X \cap J_\nu$ isomorphically onto a transitive set also. In other words, $\pi' = \pi \restriction X \cap J_\nu$, and $\pi'' X \cap J_\nu = J_{\pi''(X \cap \nu)}$. So, $\pi(J_\nu) = \pi''(X \cap J_\nu) = J_{\pi''(X \cap \nu)} = J_{\pi(\nu)}$, by the definition on π, as claimed.

For $\nu < \alpha$, define $\mathrm{rud}_X(J_\nu)$ = the rud closure of $X \cap (J_\nu \cup \{J_\nu\})$.

Claim 2: $X = \bigcup_{\nu \in X \cap \alpha} \mathrm{rud}_X(J_\nu)$.

To establish this claim, note that as $X \prec_{\Sigma_1} J_\alpha$, X is rud closed, so $\supset$ is obvious. For the converse, let $x \in X$. Then $x \in J_\alpha = \bigcup_{\nu < \alpha} \mathrm{rud}(J_\nu)$, so for some rud function f, $\models_{J_\alpha} (\exists \nu)(\exists p \in J_\nu)(x = f(p, J_\nu))$. But $X \prec_{\Sigma_1} J_\alpha$, so $(\exists \nu \in X \cap \alpha)(\exists p \in J_\nu \cap X)(x = f(p, J_\nu))$. In other words, $x \in \bigcup_{\nu \in X \cap \alpha} \mathrm{rud}_X(J_\nu)$. Hence Claim 2.

Claim 3: For $\nu \in X \cap \alpha$, $\pi'' \mathrm{rud}_X(J_\nu) = \mathrm{rud}(J_{\pi(\nu)})$.

To see this, let $\nu \in X \cap \alpha$. Suppose first that $x \in \mathrm{rud}_X(J_\nu)$. Then for some rud function f and some $p \in J_\nu \cap X$, $x = f(p, J_\nu)$. By Lemma 12 and Claim 1, $\pi(x) = f(\pi(p), J_{\pi(\nu)})$. But $p \in J_\nu \cap X$ so $\pi(p) \in J_{\pi(\nu)}$. Hence $\pi(x) \in \mathrm{rud}(J_{\pi(\nu)})$. This proves $\subset$. Conversely, suppose $y \in \mathrm{rud}(J_{\pi(\nu)})$. Then $y \in \mathrm{rud}(\pi(J_\nu))$, by Claim 1, so for some rud function f and some $p \in \pi(J_\nu)$, $y = f(p, \pi(J_\nu))$. Now, $\pi(J_\nu) = \pi''(J_\nu \cap X)$, so for some $q \in J_\nu \cap X$, $p = \pi(q)$ and we have $y = f(\pi(q), \pi(J_\nu)) = \pi(f(q, J_\nu)) \in \pi'' \mathrm{rud}_X(J_\nu)$. Hence $\supset$, and Claim 3 is proved.

By Claims 2 and 3, we have $W = \pi''X = \pi''(\bigcup_{\nu\in X\cap\alpha} \mathrm{rud}_X(J_\nu)) = \bigcup_{\nu\in X\cap\alpha} \pi''\,\mathrm{rud}_X(J_\nu) = \bigcup_{\nu\in X\cap\alpha} \mathrm{rud}(J_{\pi(\nu)}) = \bigcup_{\eta<\beta} \mathrm{rud}(J_\eta) = J_\beta$, where $\beta = \pi''(X\cap\alpha)$.

Note. It is easily seen that we may regard the following as part of the statement of Theorem 29: If $Y \subset X$ is transitive, then $\pi \restriction Y = \mathrm{id} \restriction Y$. And for $\nu \in X \cap \alpha$, $\pi(\nu) \leq \nu$, and for all $x \in X$, $\pi(x) \leq_{J_\alpha} x$.

By an argument well known to all set theorists, it is easily shown that $J = \bigcup_{\alpha\in OR} J_\alpha$ is a model of ZFC. (In fact, setting $<_J = \bigcup_{\alpha\in OR} <_{J_\alpha}$, $<_J$ is a J-definable well-ordering of the entire class J, so J satisfies the Axiom of Choice in a strong way.) Using the Condensation Lemma, an equally well-known argument shows that $J \models$ GCH. However, in the next section we will prove (and have already indicated this fact in our preamble) that J = L, so all that the above says is that we can use the Jensen hierarchy in place of the L-hierarchy in order to establish the classical results on the constructible universe.

§5. On the fine structure of the Jensen hierarchy

As mentioned in the introduction, a theory similar to the one following can be developed for the usual L-hierarchy, if desired.

Central in our discussion will be the concept of a "uniformising function" for a relation, which is a sort of "choice function" for a given relation. Specifically, a function r is said to *uniformise* a relation R iff $\mathrm{dom}(r) = \mathrm{dom}(R)$ and for all $\boldsymbol{x}$, $\exists y R(y, \boldsymbol{x}) \leftrightarrow R(r(\boldsymbol{x}), \boldsymbol{x})$.

Let $\mathbf{M} = \langle M, A\rangle$, $n \geq 1$. We say $\mathbf{M}$ is Σ_n*-uniformisable* iff every $\Sigma_n(\mathbf{M})$ relation on M is uniformised by a $\Sigma_n(\mathbf{M})$ function. A few moments reflection will reveal that Σ_n-uniformisability is a very strong condition to demand of an arbitrary structure $\mathbf{M}$, since in the more obvious cases, the definition of a uniformising function for a given relation would appear to increase the logical complexity by one or more quantifier switches. However, it will turn out that for all α, all $n \geq 1$, J_α *is* Σ_n-uniformisable. For $n = 1$, this will be easy to prove, but for $n > 1$, the corresponding argument will only work when J_α is Σ_{n-1}-admissible, so a more indirect approach will be necessary. We shall outline the approach required after we dispose of some of the more easy results. First, Σ_1-uniformisability. The $\Sigma_1(J_\alpha)$ well-ordering of each J_α gives us this with little effort. In fact, we have a much stronger result, of importance in applications of Σ_1-uniformisability.

Let F be a class of structures $\mathbf{M} = \langle M, A\rangle$, $n \geq 1$. Say F is *uniformly Σ_n-uniformisable* if, whenever φ is a Σ_n-formula w.p.f. $\bigcap\{M \mid \mathbf{M} \in F\}$ such that $\varphi^{\mathbf{M}}$ is a relation on M for each $\mathbf{M} \in F$, there is a Σ_n-formula ψ (w.p.f. $\bigcap\{M \mid \mathbf{M} \in F\}$) such that for each $\mathbf{M} \in F$, $\psi^{\mathbf{M}}$ is a function uniformising $\varphi^{\mathbf{M}}$.

Theorem 30. $\langle \mathrm{J}_\alpha, A\rangle$ *is Σ_1-uniformisable. In fact, the class of all* $\langle \mathrm{J}_\alpha, A\rangle$ *is uniformly Σ_1-uniformisable.*

Proof. Let φ be a Σ_1-formula w.p.f. J_α such that $[\varphi(y,\boldsymbol{x})]^{\langle \mathrm{J}_\alpha, A\rangle}$ is a Σ_1 relation on J_α. By contraction of quantifiers, we can, in a uniform way, find a Σ_0 formula ψ (w.p.f. J_α) such that $\models_{\langle \mathrm{J}_\alpha, A\rangle} \varphi(y,\boldsymbol{x}) \leftrightarrow \exists z\, \psi(z,y,\boldsymbol{x})$. Define g by: $g(\boldsymbol{x}) \simeq$ the $<_{\mathrm{J}}$-least w such that $\models_{\langle \mathrm{J}_\alpha, A\rangle} \psi((w)_0,(w)_1,\boldsymbol{x})$. Then g is (uniformly) $\Sigma_1(\langle \mathrm{J}_\alpha, A\rangle)$, since

$$w = g(\boldsymbol{x}) \leftrightarrow \models_{\langle \mathrm{J}_\alpha, A\rangle} \psi((w)_0,(w)_1,\boldsymbol{x})$$

$$\wedge\ \exists t\, [t = \mathrm{pr}_\alpha(w) \wedge (\forall w' \in t) \neg \psi((w')_0,(w')_1,\boldsymbol{x})]\ .$$

Set $r(\boldsymbol{x}) \simeq (g(\boldsymbol{x}))_1$. Then r is (uniformly) $\Sigma_1(\langle \mathrm{J}_\alpha, A\rangle)$ and clearly uniformises $[\varphi(y,\boldsymbol{x})]^{\langle \mathrm{J}_\alpha, A\rangle}$.

Remark. We call the above construction the *canonical* Σ_1-uniformisation procedure. Observe that if $R(y,\boldsymbol{x})$ is a $\Sigma_1(\langle \mathrm{J}_\alpha, A\rangle)$ predicate, then the canonical Σ_1-uniformisation of R is a function whose $\Sigma_1(\langle \mathrm{J}_\alpha, A\rangle)$ definition involves *only those parameters which occur in the definition of R.*

Let us take a little time off to examine the above construction more closely. Suppose $R(y,\boldsymbol{x})$ is a given Σ_1 relation, say $R(y,\boldsymbol{x}) \leftrightarrow \exists z\, P(z,y,\boldsymbol{x})$, where P is Σ_0. To obtain the Σ_1 uniformisation of R, we first obtain a Σ_1 uniformisation of the Σ_0 relation $\{\langle w,\boldsymbol{x}\rangle \mid P((w)_0,(w)_1,\boldsymbol{x})\}$, and then simply pick out the requisite component of the result as our required function. And since $<_{\mathrm{J}}$ is a Σ_1 well-order of J_α the result is also Σ_1. However, returning now to the notation of Theorem 30, we see that, if we try to extend this procedure to the case $n > 1$, we cannot conclude that the function g is Σ_n, the problem being the last conjunct in the explicit definition of g. Let $\Psi(w,\boldsymbol{x})$

denote the predicate $[\neg\psi((w)_0,(w)_1,\boldsymbol{x})]^{\langle J_\alpha,A\rangle}$. For $n=1$, there was no problem, since if $\Psi(w,\boldsymbol{x})$ is $\boldsymbol{\Sigma}_0$, so is $(\forall w\in t)\Psi(w,\boldsymbol{x})$. However, for $n>1$, $\Psi(w,\boldsymbol{x})$ is $\boldsymbol{\Sigma}_{n-1}$, and we can only conclude that $(\forall w\in t)\Psi(w,\boldsymbol{x})$ is $\boldsymbol{\Sigma}_{n-1}$ if $\langle J_\alpha, A\rangle$ is $\boldsymbol{\Sigma}_{n-1}$-admissible. Otherwise it is merely $\boldsymbol{\Pi}_n$ of course, and so the resulting uniformisation of the original $\boldsymbol{\Sigma}_n$ relation turns out to be $\boldsymbol{\Sigma}_{n+1}$. So, in order to establish the general Σ_n-uniformisation lemma, it is not altogether unreasonable to try and "reduce" all $\boldsymbol{\Sigma}_n$ relations on an arbitrary J_α to $\boldsymbol{\Sigma}_n$ relations on some $\boldsymbol{\Sigma}_{n-1}$-admissible J_β, for which we have a $\boldsymbol{\Sigma}_n$-uniformisation procedure. In practice, it will turn out that this hint is slightly off target, but in its general tone it is worth bearing in mind.

Closely connected with Σ_n-uniformisability is the notion of a "Σ_n Skolem function". Let $\mathbf{M}=\langle M,A\rangle$ be transitive and rud closed. By a Σ_n *Skolem function for* $\mathbf{M}$ we mean a $\Sigma_n(\mathbf{M})$ function h with $\mathrm{dom}(h)\subset\omega\times M$, such that for some $p\in M$, h is $\Sigma_n^{\mathbf{M}}(\{p\})$, and whenever $P\in\Sigma_n^{\mathbf{M}}(\{x,p\})$ for some $x\in M$, then $\exists yP(y)\rightarrow(\exists i\in\omega)P(h(i,x))$. (With h,p as above, we say that p is a *good* parameter for h.) Note that $\boldsymbol{\Sigma}_n$ Skolem functions need not be (and in general are not) total! As far as existence of $\boldsymbol{\Sigma}_n$ Skolem functions is concerned, we can get away with slightly less than might first appear. In fact:

Lemma 31. *Let* $\mathbf{M}=\langle M,A\rangle$ *be transitive and* rud *closed. Let h be a* $\boldsymbol{\Sigma}_n^{\mathbf{M}}(\{p\})$ *function with* $\mathrm{dom}(h)\subset\omega\times M$. *Suppose that whenever* $P\in\Sigma_n^{\mathbf{M}}(\{x\})$ *for some* $x\in M$, *then* $\exists yP(y)\rightarrow(\exists i\in\omega)P(h(i,x))$. *Then* $\mathbf{M}$ *has a* $\boldsymbol{\Sigma}_n$ *Skolem function.*

Proof. Set $\tilde{h}(i,x)\simeq h(i,\langle x,p\rangle)$. It is easily seen that $\tilde{h}$ is a $\boldsymbol{\Sigma}_n$ Skolem function for $\mathbf{M}$.

Note that in the above, if h is actually $\boldsymbol{\Sigma}_n^{\mathbf{M}}$, then $\tilde{h}=h$. This is used in establishing the following result:

Lemma 32. *If* $\langle J_\alpha, A\rangle$ *is amenable, then it has a* $\boldsymbol{\Sigma}_1$ *Skolem function. In fact, there is a* $\boldsymbol{\Sigma}_1$ *Skolem function* $h_{\alpha,A}$ *for* $\langle J_\alpha, A\rangle$ *which is uniformly* $\boldsymbol{\Sigma}_1^{\langle J_\alpha,A\rangle}$ *for all amenable* $\langle J_\alpha, A\rangle$.

Proof. Let $\langle\varphi_i\mid i<\omega\rangle$ be a recursive enumeration of $\mathrm{Fml}^{\boldsymbol{\Sigma}_1}$. Let $\langle J_\alpha, A\rangle$ be amenable. By Lemma 9, $\models^{\boldsymbol{\Sigma}_1}_{\langle J_\alpha,A\rangle}$ is (uniformly) $\boldsymbol{\Sigma}_1^{\langle J_\alpha,A\rangle}$. Let $h=h_{\alpha,A}$ be the canonical $\boldsymbol{\Sigma}_1$-uniformisation of the $\boldsymbol{\Sigma}_1^{\langle J_\alpha,A\rangle}$ relation $\{\langle y,i,x\rangle\mid\models^{\boldsymbol{\Sigma}_1}_{\langle J_\alpha,A\rangle}\varphi_i[y,x]\}$.

(By Lemma 30 and the ensuing remark, h is thus uniformly $\Sigma_1^{\langle J_\alpha, A\rangle}$ for amenable $\langle J_\alpha, A\rangle$.) By the remark following Lemma 31, it is clear that h is a Σ_1 Skolem function for $\langle J_\alpha, A\rangle$.

We refer to $h_{\alpha,A}$ as the *canonical* Σ_1 *Skolem function for* (amenable) $\langle J_\alpha, A\rangle$.

By a similar argument, we have:

Lemma 33. *If $\langle J_\alpha, A\rangle$ is amenable and Σ_n-uniformisable, it has a Σ_n Skolem function.*

The following lemmas indicate our reason for using the word "Skolem" here.

Lemma 34. *Let* $\mathbf{M}$ *be transitive and* rud *closed, and let h be a Σ_n Skolem function for* $\mathbf{M}$. *Then whenever $x \in M$, $x \in h''(\omega \times \{x\}) \prec_{\Sigma_n} \mathbf{M}$.*

Proof. Set $X = h''(\omega \times \{x\})$. Clearly, $x \in X$. Let $P \in \Sigma_n^{\mathbf{M}}(X)$, $P \neq \emptyset$. We must show that $P \cap X \neq \emptyset$. Let p be a good parameter for h, and pick $y_1, ..., y_m \in X$ with $P \in \Sigma_n^{\mathbf{M}}(\{y_1, ..., y_m\})$. By definition of X, there are $j_1, ..., j_m \in \omega$ such that $y_1 = h(j_1, x), ..., y_m = h(j_m, x)$. Since h is $\Sigma_n^{\mathbf{M}}(\{p\})$, it follows that $P \in \Sigma_n^{\mathbf{M}}(\{p, x\})$. Hence, $P \neq \emptyset \to \exists y P(y) \to (\exists i \in \omega) P(h(i, x)) \to (\exists y \in X) P(y)$.

Lemma 35. *Let* $\mathbf{M}$ *be a transitive and* rud *closed, and let h be a Σ_n Skolem function for* $\mathbf{M}$. *If $X \subset M$ is closed under ordered pairs, then $X \subset h''(\omega \times X) \prec_{\Sigma_n} \mathbf{M}$.*

Proof. Set $Y = h''(\omega \times X)$. By Lemma 34, $X \subset Y$. Let $P \in \Sigma_n^{\mathbf{M}}(Y)$, $P \neq \emptyset$. We must show that $P \cap Y \neq \emptyset$. Let p be a good parameter for h, and pick $y_1, ..., y_m \in Y$ with $P \in \Sigma_n^{\mathbf{M}}(\{y_1, ..., y_m\})$. Pick $j_1, ..., j_m \in \omega$ and $x_1, ..., x_m \in X$ such that $y_1 = h(j_1, x_1), ..., y_m = h(j_m, x_m)$. Let $x = \langle x_1, ..., x_m\rangle$. By assumption, $x \in X$. But clearly, as h is $\Sigma_n^{\mathbf{M}}(\{p\})$, P is then $\Sigma_n^{\mathbf{M}}(\{p, x\})$, so $P \neq \emptyset \to \exists y P(y) \to (\exists i \in \omega) P(h(i, x)) \to (\exists y \in Y) P(y)$.

Corollary 36. *Let* $\mathbf{M}$, *h be as above. Let $X \subset M$ and suppose $h''(\omega \times X)$ is closed under ordered pairs. Then $X \subset h''(\omega \times X) \prec_{\Sigma_n} \mathbf{M}$.*

Proof. Let $Y = h''(\omega \times X)$. Clearly, $Y = h''(\omega \times Y)$, so the result follows by the lemma.

Lemma 37 (Gödel). *There is a bijection* $\Phi : \mathrm{OR}^2 \leftrightarrow \mathrm{OR}$ *such that* $\Phi(\alpha,\beta) \geq \alpha, \beta$ *for all* α, β, *and* $\Phi^{-1} \restriction \omega\alpha$ *is uniformly* $\boldsymbol{\Sigma}_1^{\mathrm{J}_\alpha}$ *for all* α.

Proof. Define a well-order $<^*$ of OR^2 by

$$(\alpha,\beta) <^* (\gamma,\delta) \leftrightarrow [\max(\alpha,\beta) < \max(\gamma,\delta)] \vee [\max(\alpha,\beta) = \max(\gamma,\delta) \wedge (\alpha < \gamma \vee (\alpha = \gamma \wedge \beta < \delta))] .$$

Let $\Phi : <^* \cong \mathrm{OR}$. By induction on α, $\Phi^{-1} \restriction \omega\alpha$ is $\boldsymbol{\Sigma}_1^{\mathrm{J}_\alpha}$ (uniformly).

Lemma 38. *There is a* $\boldsymbol{\Sigma}_1(\mathrm{J}_\alpha)$ *map of* $\omega\alpha$ *onto* $(\omega\alpha)^2$ *for all* α.

Proof. Let $Q = \{\alpha \mid \Phi(0,\alpha) = \alpha\}$. Then Q is closed and unbounded in OR. In fact, $Q = \{\alpha \mid \Phi : \alpha^2 \leftrightarrow \alpha\}$, so $\omega\alpha \in Q \rightarrow \omega\alpha = \alpha$. We prove the lemma by induction on α. Assume it is true for all $\nu < \alpha$.

Case 1: $\omega\alpha \in Q$. Then $\Phi^{-1} \restriction \alpha$ suffices.

Case 2: $\alpha = \beta + 1$. If $\beta = 0$, then $\omega\alpha = \omega \in Q$, so we are done by Case 1. Hence we may assume $\beta \geq 1$. Then clearly, there is a $\boldsymbol{\Sigma}_1(\mathrm{J}_\alpha)$ map $j : \omega\alpha \leftrightarrow \omega\beta$. By hypothesis, there is a $\boldsymbol{\Sigma}_1(\mathrm{J}_\beta)$ map of $\omega\beta$ onto $(\omega\beta)^2$, so there is certainly a $\boldsymbol{\Sigma}_1(\mathrm{J}_\beta)$ map g of $(\omega\beta)^2$ one-one into $\omega\beta$. Then $g \in \mathrm{rud}(\mathrm{J}_\beta) = \mathrm{J}_\alpha$, so for $\nu, \gamma \in \omega\alpha$, define

$$f(\langle \nu, \tau \rangle) = g(\langle j(\nu), j(\tau) \rangle) .$$

Then f is $\boldsymbol{\Sigma}_1(\mathrm{J}_\alpha)$ and f maps $(\omega\alpha)^2$ one-one into $\omega\beta$. Now $\mathrm{ran}(f) = \mathrm{ran}(g) \in \mathrm{J}_\alpha$, so if we define h by (for $\nu \in \omega\alpha$)

$$h(\nu) = \begin{cases} f^{-1}(\nu), & \text{if } \nu \in \mathrm{ran}(f) \\ \langle 0, 0 \rangle, & \text{otherwise} \end{cases}$$

we see that h is $\boldsymbol{\Sigma}_1(\mathrm{J}_\alpha)$ and $h : \omega\alpha \xrightarrow{\text{onto}} (\omega\alpha)^2$.

Case 3: $\lim(\alpha) \wedge \omega\alpha \notin Q$. In this case let $\langle \nu, \tau \rangle = \Phi^{-1}(\omega\alpha)$.

Thus $\nu, \tau < \omega\alpha$. Set $c = \{z \mid z <^* \langle \nu, \tau \rangle\}$ $(\in J_\alpha)$. Thus $\Phi \restriction c$ maps c one-one onto $\omega\alpha$, and is $\Sigma_1(J_\alpha)$. Pick $\gamma < \alpha$ with $\nu, \tau < \omega\gamma$. Then $\Phi^{-1} \restriction \omega\alpha$ is a $\Sigma_1(J_\alpha)$ map of $\omega\alpha$ one-one into $\omega\gamma$. And by assumption, there is a map $g \in J_\alpha$ mapping $(\omega\gamma)^2$ one-one into $\omega\gamma$. So, setting $f(\langle \iota, \theta \rangle) = g(\langle g(\Phi^{-1}(\iota)), g(\Phi^{-1}(\theta)) \rangle)$, $\iota, \theta < \omega\alpha$, we see that f is a $\Sigma_1(J_\alpha)$ map of $(\omega\alpha)^2$ one-one into $d = g''(g''c)^2$. But $d \in J_\alpha$, so we can define a $\Sigma_1(J_\alpha)$ map h on $\omega\alpha$ by

$$h(\theta) = \begin{cases} f^{-1}(\theta), & \text{if } \theta \in d \\ \langle 0, 0 \rangle, & \text{otherwise.} \end{cases}$$

Clearly, $h = \omega\alpha \xrightarrow{\text{onto}} (\omega\alpha)^2$.

The lemma is proved.

Using this lemma, we may now establish the following important result:

Theorem 39. *There is a $\Sigma_1(J_\alpha)$ map of $\omega\alpha$ onto J_α for all α.*

Proof. Let $f : \omega\alpha \xrightarrow{\text{onto}} (\omega\alpha)^2$ be $\Sigma_1^{J_\alpha}(\{p\})$, where $p \in J_\alpha$ is the $<_J$-least element of J_α for which such an f exists. Define f^0, f^1 by demanding that $f(\nu) = \langle f^0(\nu), f^1(\nu) \rangle$ for all $\nu \in \omega\alpha$. By induction, define $f_n : \omega\alpha \xrightarrow{\text{onto}} (\omega\alpha)^n$ thus: $f_0 = \text{id} \restriction \omega\alpha$; $f_{n+1}(\nu) = \langle f^0(\nu), f_n \cdot f^1(\nu) \rangle$. Hence each f_n is $\Sigma_1^{J_\alpha}(\{p\})$. Let $h = h_\alpha$, the canonical Σ_1 Skolem function for J_α. Set $X = h''(\omega \times (\omega\alpha \times \{p\}))$.

Claim 1. *X is closed under ordered pairs.*

To see this, let $y_1, y_2 \in X$, say $y_1 = h(j_1, \langle \nu_1, p \rangle)$, $y_2 = h(j_2, \langle \nu_2, p \rangle)$. Let $\langle \nu_1, \nu_2 \rangle = f_2(\tau)$. Then $\{\langle y_1, y_2 \rangle\}$ is a $\Sigma_1^{J_\alpha}(\{\tau, p\})$ predicate, so by definition of h, $\langle y_1, y_2 \rangle \in X$, as claimed.

So by Corollary 36, $X \prec_{\Sigma_1} J_\alpha$. By the condensation lemma, let $\pi : X \cong J_\beta$, $\beta \leq \alpha$. Since $\omega\alpha \subset X$, we clearly have $\beta = \alpha$ here.

Claim 2. *For all $i \in \omega$, $x \in X$, $\pi(h(i, x)) \simeq h(i, \pi(x))$.*

To see this, observe first that as h is $\Sigma_1^{J_\alpha}$, there is a rud function H such that $y = h(i,x) \leftrightarrow (\exists t \in J_\alpha)[H(t,i,x,y) = 1]$. Now let $i \in \omega$, $x \in X$. Since $X \prec_{\Sigma_1} J_\alpha$, $y = h(i,x) \in X$ (if defined). Thus, by the above, since $x,y \in X \prec_{\Sigma_1} J_\alpha$, $(\exists t \in X)[H(t,i,x,y) = 1]$. By Lemma 12, therefore, $(\exists t \in X)[H(\pi(t), i, \pi(x), \pi(y)) = 1]$. Since $\pi''X = J_\alpha$, this can be rewritten as $(\exists t \in J_\alpha)[H(t, i, \pi(x), \pi(y)) = 1]$. Thus $\pi(y) = h(i, \pi(x))$, as claimed.

Now, $f \subset (\omega\alpha)^3$, so as $\pi \restriction \omega\alpha = \mathrm{id} \restriction \omega\alpha$, $\pi''f = f$. And by isomorphism, $\pi''f$ is $\Sigma_1^{J_\alpha}(\{\pi(p)\})$. So as $\pi(p) \leq_J p$, the choice of p shows that $\pi(p) = p$. So, by Claim 2, if $i \in \omega$, $\nu \in \omega\alpha$, $\pi(h(i, \langle \nu, p \rangle)) \simeq h(i, \langle \nu, p\rangle)$, which is to say $\pi \restriction X = \mathrm{id} \restriction X$. Thus $X = J_\alpha$.

Now define $\tilde{h} : (\omega\alpha)^3 \to J_\alpha$ by setting

$$\tilde{h}(i,\nu,\tau) = \begin{cases} y, & \text{if } (\exists t \in S_\tau)[H(t,i,\langle \nu, p\rangle, y) = 1] \\ \emptyset, & \text{otherwise.} \end{cases}$$

Then $\tilde{h}$ is $\Sigma_1(J_\alpha)$, and clearly $\tilde{h}''(\omega\alpha)^3 = h''(\omega \times (\omega\alpha \times \{p\})) = X = J_\alpha$. Therefore, $\tilde{h} \cdot f_3$ is as required by the theorem.

Observe that in Lemmas 38, 39, the maps constructed generally have parameters in their definitions. Note also that, being total, these maps are in fact $\Delta_1(J_\alpha)$.

Recalling the results of §3, we now investigate those ordinals α for which J_α is an admissible set.

Let us call an ordinal α *admissible* iff $\alpha = \omega\beta$ and J_β is an admissible set.

Theorem 40. *$\omega\alpha$ is admissible iff there is no $\Sigma_1(J_\alpha)$ map of any $\gamma < \omega\alpha$ cofinally into $\omega\alpha$. (Note that such a map, having domain $\gamma \in J_\alpha$, would in fact be $\Delta_1(J_\alpha)$.)*

Proof. ($\rightarrow$). Let $\gamma < \omega\alpha$ and suppose $f : \gamma \to \omega\alpha$ is $\Sigma_1(J_\alpha)$. Then $(\forall \xi \in \gamma)(\exists \zeta \in \omega\alpha)(f(\xi) = \zeta)$. If J_α is admissible, then by Σ_1-Replacement, $(\exists \eta \in \omega\alpha)(\forall \xi \in \gamma)(\exists \zeta \in \eta)(f(\xi) = \zeta)$, so f is not cofinal in $\omega\alpha$.

($\leftarrow$). Assume $\omega\alpha$ is not admissible. If $\alpha = \beta + 1$, then the $\Sigma_1(J_\alpha)$ map $\{\langle \omega\beta + n, n\rangle \mid n \in \omega\}$ maps ω cofinally into $\omega\alpha$, so we are done. Assume then that $\lim(\alpha)$. Since J_α is not admissible, there must be a $\Sigma_1(J_\alpha)$ relation R and a $u \in J_\alpha$ such that $(\forall x \in u)(\exists y) R(x,y)$ but for all $z \in J_\alpha$,

$\neg(\forall x \in u)(\exists y \in z)R(x,y)$. Take $\gamma < \alpha$ with $u \in J_\gamma$. By Theorem 39, let f be a $\Sigma_1(J_\gamma)$ map of $\omega\gamma$ onto J_γ. Thus $f \in J_\alpha$, and $u \subset f''\omega\gamma$. Define $g : \omega\gamma \to \omega\alpha$ by

$$g(\nu) = \begin{cases} \text{the least } \tau \text{ such that } (\exists y \in S_\tau)R(f(\nu), y), & \text{if } f(\nu) \in u \\ 0, \text{ if } f(\nu) \notin u. \end{cases}$$

Then g is a $\Sigma_1(J_\alpha)$ map of $\omega\gamma$ cofinally into $\omega\alpha$.

Recalling our discussion at the end of §3, let us call an ordinal α *strongly admissible* (or *non-projectible admissible*) iff $\alpha = \omega\beta$ and J_β is strongly admissible. Imitating the proof of Theorem 40, we have:

Theorem 41. *$\omega\alpha$ is strongly admissible iff there is no $\Sigma_1(J_\alpha)$ map of a bounded subset of $\omega\alpha$ cofinally into $\omega\alpha$.*

The above two results illustrate our earlier remark concerning the difference between a function being partial and being total, and the corresponding difference between a predicate being Σ_n and being Δ_n. The next two results, which strengthen the last two, and are also due to Kripke and Platek, also highlight this distinction.

Theorem 42. *The following are equivalent*:

(i) *$\omega\alpha$ is admissible.*

(ii) *$\langle J_\alpha, A\rangle$ is amenable for all $A \in \Delta_1(J_\alpha)$.*

(iii) *There is no $\Sigma_1(J_\alpha)$ function mapping a $\gamma < \omega\alpha$ onto J_α.*
(Of course, any such function would in fact be $\Delta_1(J_\alpha)$.)

Proof. (i) → (ii). By Lemma 15 (Δ_1-Comprehension).
(ii) → (iii). Assume (ii) ∧ ¬(iii). Let $\gamma < \omega\alpha$, and let $f : \gamma \xrightarrow{\text{onto}} J_\alpha$ be $\Sigma_1(J_\alpha)$. Then f is $\Delta_1(J_\alpha)$, so $d = \{\nu \mid \nu \notin f(\nu)\}$ is $\Delta_1(J_\alpha)$. Thus by (ii), $d = d \cap \gamma \in J_\alpha$. So, $d = f(\nu)$ for some $\nu < \gamma$, so $\nu \in f(\nu) \leftrightarrow \nu \in d \leftrightarrow \nu \notin f(\nu)$, a contradiction.
(iii) → (i). Assume (iii) ∧ ¬(i). If $\alpha = \beta + 1$, we can easily construct a $\Sigma_1(J_\alpha)$ map of $\omega\beta$ onto $\omega\alpha$, so Theorem 39 yields the required contradiction. Assume $\lim(\alpha)$. By Theorem 40, there must be a $\tau < \omega\alpha$ and a $\Sigma_1(J_\alpha)$ map f of τ cofinally into $\omega\alpha$. Let f be $\Sigma_1^{J_\alpha}(\{p\})$. Pick $\gamma < \alpha$ with $\tau, p \in J_\gamma$. Let $h = h_\alpha$ be the canonical Σ_1 Skolem function for J_α. Set $X = h''(\omega \times J_\gamma)$. As J_γ is closed

under ordered pairs, Lemma 35 tells us that $X \prec_{\Sigma_1} J_\alpha$. Let $\pi : X \cong J_\beta$. Thus $\pi \restriction J_\gamma = \mathrm{id} \restriction J_\gamma$. By an argument as in Theorem 39, $\pi \restriction X = \mathrm{id} \restriction X$, so $X = J_\beta$. Now, f is $\Sigma_1^{J_\alpha}(\{p\})$ and $p \in X \prec_{\Sigma_1} J_\alpha$, so X is closed under f. But $\tau \subset X$ and so $f''\tau \subset X$, which means, since $f''\tau$ is cofinal in $\omega\alpha$ and $X = J_\beta$ is transitive, that $\omega\alpha \subset J_\beta$. Thus $\beta = \alpha$, and $X = J_\alpha$. Define a $\Sigma_1(J_\alpha)$ map $\tilde{h} : \omega \times \tau \times J_\gamma \to J_\alpha$ as follows. Let H be a $\Sigma_0^{J_\alpha}$ relation such that $y = h(i,x) \leftrightarrow (\exists t \in J_\alpha) H(t,i,x,y)$. Set

$$\tilde{h}(i,\nu,x) = \begin{cases} y, & \text{if } (\exists t \in S_{f(\nu)}) H(t,i,x,y) \\ \emptyset, & \text{otherwise.} \end{cases}$$

Then $\tilde{h}$ is total on $\omega \times \tau \times J_\gamma$, and $\tilde{h}''(\omega \times \tau \times \{x\}) = h''(\omega \times \{x\})$ for any x, as $f''\tau$ is cofinal in $\omega\alpha$. Hence $\tilde{h}''(\omega \times \tau \times J_\gamma) = X = J_\alpha$. By Theorem 39 there is $g \in J_\alpha$, $g : \omega\gamma \xrightarrow{\text{onto}} \omega \times \tau \times J_\gamma$. Then $\tilde{h} \cdot g$ is a $\Sigma_1(J_\alpha)$ map of $\omega\gamma$ onto J_α, contrary to (iii).

Theorem 43. *The following are equivalent*:

(i) *$\omega\alpha$ is strongly admissible.*

(ii) *$\langle J_\alpha, A\rangle$ is amenable for all $A \in \Sigma_1(J_\alpha)$.*

(iii) *There is no $\Sigma_1(J_\alpha)$ function mapping a bounded subset of $\omega\alpha$ onto J_α.*

Proof. (i) → (ii) → (iii). Similar to the above.
(iii) → (i). Assume (iii) $\cap \neg$(i) and proceed much as before. So, we assume $\lim(\alpha)$, f is (by Theorem 41) a $\Sigma_1(J_\alpha)$ map of some $a \subset \tau < \omega\alpha$ cofinally into $\omega\alpha$, $f \in \Sigma_1^{J_\alpha}(\{p\})$, and $\tau < \omega\gamma$, $p \in J_\gamma$, $\gamma < \alpha$. As before, if $h = h_\alpha$ and $X = h''(\omega \times J_\gamma)$, then $X = J_\alpha$. Now, since we do not need to bother about functions being total, we can easily contradict (iii). By Theorem 39, let $g \in J_\alpha$, $g : \omega\gamma \xrightarrow{\text{onto}} \omega \times J_\gamma$. Set $\bar{f}(\nu) \simeq h(g(\nu))$. Then $\bar{f}$ is a $\Sigma_1(J_\alpha)$ map of a subset of $\omega\gamma$ onto J_α.

Note that an immediate corollary of Theorem 42 is:

Theorem 44. *If κ is a cardinal, then κ is an admissible ordinal.*

Using admissibility theory, we can give a quick proof that $L = \bigcup_{\alpha \in OR} J_\alpha$.

Theorem 45. *If $\omega\alpha$ is admissible, then $J_\alpha = L_{\omega\alpha}$.*

Proof. If $\alpha = 1$, then $J_1 = L_\omega$ = the hereditarily finite sets. Assume $\alpha > 1$. Thus $\omega \in J_\alpha$. Since J_α is admissible, the recursion theorem tells us that $\mathrm{rud}(x) = \bigcup_{n<\omega} S^n(x)$ is $\Sigma_1(J_\alpha)$. But if u is transitive, then $\Sigma_\omega(u) = \mathcal{P}(u) \cap \mathrm{rud}(u)$. Hence the map $L_\gamma \mapsto \Sigma_\omega(L_\gamma) = L_{\gamma+1}$ is $\Sigma_1(J_\alpha)(\gamma < \omega\alpha)$. So, by the recursion theorem again, we see that $\langle L_\nu | \nu < \omega\alpha \rangle$ is $\ _1(J_\alpha)$. Hence $L_{\omega\alpha} = \bigcup_{\nu<\omega\alpha} L_\nu \subset J_\alpha$. For the converse inclusion, it suffices to show that $L_{\omega\alpha}$ is admissible. (For then, by the recursion theorem, $\langle S_\nu | \nu < \omega\alpha \rangle$ is $\Sigma_1(L_{\omega\alpha})$, so $J_\alpha = \bigcup_{\nu<\omega\alpha} S_\nu \subset L_{\omega\alpha}$.) Let R be $\Sigma_0(L_{\omega\alpha})$, $x \in L_{\omega\alpha}$, and assume $(\forall z \in x)\, \exists y R(y,z)$. Since $\langle L_\nu | \nu < \omega\alpha \rangle$ is $\Sigma_1(J_\alpha)$, we may define a $\Sigma_1(J_\alpha)$ predicate R' by $R'(\nu, z) \leftrightarrow z \in x \wedge (\exists y \in L_\nu) R(y,z)$. Since J_α is admissible, there is $\tau < \omega\alpha$ with $(\forall z \in x)(\exists \nu < \tau) R'(\nu, z)$. Hence $(\forall z \in x)(\exists y \in L_\tau) R(y,z)$. So as $L_\tau \in L_{\omega\alpha}$, $L_{\omega\alpha}$ satisfies the Σ_0-Replacement axiom. Since $\lim(\omega\alpha)$, it follows easily that $L_{\omega\alpha}$ is admissible.

Let $\alpha, n \geq 0$. The Σ_n-*projectum* of α, ρ^n_α, is the largest $\rho \leq \alpha$ such that $\langle J_\rho, A \rangle$ is amenable for all $A \in \Sigma_n(J_\alpha) \cap \mathcal{P}(J_\rho)$.

Roughly speaking, our reason for introducing the Σ_n projectum is this. We have seen that, for example, we can reasonably handle $\Sigma_n(J_\alpha)$ predicates when J_α is Σ_n-admissible. This is because Σ_n-admissibility is a sort of "hardness" condition on J_α for Σ_n predicates. For, if we take an arbitrary J_α, it may be "soft" for $\Sigma_n(J_\alpha)$ predicates; we may, for instance, find $\Sigma_n(J_\alpha)$ subsets of members of J_α which are not themselves members of J_α, or even $\Sigma_n(J_\alpha)$ functions which *project* a subset of a member of J_α onto all of J_α. But if J_α is Σ_n-admissible, none of these situations can arise. Thus, we try to isolate that part of J_α which *is* "hard" for $\Sigma_n(J_\alpha)$ predicates, a sort of "Σ_n-admissible core" of J_α. One natural way of formalising these ideas is provided by the Σ_n projectum. Clearly, $J_{\rho^n_\alpha}$ is a reasonable interpretation of the notion of a "Σ_n-hard core" of J_α. We shall eventually give two characterisations of the Σ_n-projectum which make it appear even more reasonable – if not inevitable. One of these is that ρ^n_α is the smallest $\rho \leq \alpha$ for which there is a $\Sigma_n(J_\alpha)$ map of a subset of $\omega\rho$ onto J_α. Then, since we clearly have, for $\omega\alpha$ admissible, that $\omega\alpha$ is strongly admissible iff $\rho^1_\alpha = \alpha$, we obtain some justification for our alternative name of "non-projectible admissible" for strong admissibility.

It is convenient, at this point, for us to define an obvious generalisation of the notion of the Σ_n-projectum of an ordinal.

Let $\langle J_\alpha, A \rangle$ be amenable. The Σ_n-*projectum* of $\langle J_\alpha, A \rangle$, $\rho^n_{\alpha,A}$, is the largest

$\rho \leq \alpha$ such that $\langle J_\rho, B\rangle$ is amenable for all $B \in \Sigma_n(\langle J_\alpha, A\rangle)$.

Note that by Theorem 43, $\omega\rho^n_{\alpha,A}$ is always strongly admissible.

We shall make strong use of the Σ_n-projectum in proving that every J_α is Σ_n-uniformisable, all $n \geq 1$. Since most of the following lemmas are directed towards this goal, it is worth indicating briefly our strategy.

We already know that $\langle J_\delta, A\rangle$ is Σ_1-uniformisable for all $\langle J_\delta, A\rangle$. What we shall do is attempt to "reduce" $\Sigma_n(J_\alpha)$ predicates to $\Sigma_1(\langle J_{\rho^n_\alpha}, A\rangle)$ predicates for some $A \subset J_{\rho^n_\alpha}$ which is itself $\Sigma_n(J_\alpha)$. To carry out this reduction, we need to have at our disposal a $\Sigma_n(J_\alpha)$ map of a subset (at least) of $J_{\rho^n_\alpha}$ onto J_α. Thus, what we shall do is to simultaneously prove, by induction on n, α, the following two propositions:

(P 1) J_α is Σ_{n+1}-uniformisable

(P 2) There is a $\Sigma_n(J_\alpha)$ map of a subset of $\omega\rho^n_\alpha$ onto J_α.

The proof of (P 1) goes roughly as follows. Let R be a $\Sigma_{n+1}(J_\alpha)$ predicate on J_α. Let $f : \subset \omega\rho^n_\alpha \xrightarrow{\text{onto}} J_\alpha$ be $\Sigma_n(J_\alpha)$. Now, f^{-1} is a $\Sigma_n(J_\alpha)$ relation, so by assuming Σ_n-uniformisability, f^{-1} can be "shrunk" to a $\Sigma_n(J_\alpha)$ map of J_α into $\omega\rho^n_\alpha$. This reduces R to a $\Sigma_{n+1}(J_\alpha)$ predicate R' on $J_{\rho^n_\alpha}$. Now find a $\Sigma_n(J_\alpha)$ predicate $A \subset J_{\rho^n_\alpha}$ such that R' is in fact $\Sigma_1(\langle J_{\rho^n_\alpha}, A\rangle)$. Uniformise R' by a $\Sigma_1(\langle J_{\rho^n_\alpha}, A\rangle)$ function, and then reverse the procedure to recover a $\Sigma_{n+1}(J_\alpha)$ uniformising function for R. There is one doubtful point in the above outline. Can we in fact find a set A as required. That we can has to be proved as we proceed, so we shall in fact simultaneously prove *three* propositions, (P 1), (P 2), and a proposition (P 3) to be formulated precisely later.

Lemma 46. *Let $n \geq 1$, and assume J_α is Σ_n-uniformisable. Let $\gamma \leq \alpha$ be the least ordinal such that $\mathcal{P}(\omega\gamma) \cap \Sigma_n(J_\alpha) \not\subset J_\alpha$. Then there is a $\Sigma_n(J_\alpha)$ map of a subset of $\omega\gamma$ onto J_α.*

Proof. By Lemma 33, J_α has a Σ_n Skolem function, h. Let h be $\Sigma_n^{J_\alpha}(\{p\})$. We may assume p is the $<_J$-least element of J_α for which such an h exists. Let $a \subset \omega\gamma$, $a \in \Sigma_n(J_\alpha)$, $a \notin J_\alpha$. Let q be the $<_J$-least element of J_α such that $a \in \Sigma_n^{J_\alpha}(\{q\})$. Define $\tilde{h}$ by $\tilde{h}(i,x) \simeq h(i, \langle x, p, q\rangle)$. It is easily seen that $\tilde{h}$ is a Σ_n Skolem function for J_α and that $\langle p,q\rangle$ is a good parameter for $\tilde{h}$. Set $X = \tilde{h}''(\omega \times J_\gamma)$. Now, there is a $\Sigma_1(J_\gamma)$ map $g : \omega\gamma \xrightarrow{\text{onto}} J_\gamma$, so $\tilde{h} \cdot g$ is a $\Sigma_n(J_\alpha)$ map of a subset of $\omega\gamma$ onto X. Hence it suffices to show that $X = J_\alpha$.

Clearly, $X \prec_{\Sigma_n} J_\alpha$. Let $\pi : X \cong J_\beta, \beta \leq \alpha$. Then $\pi \upharpoonright J_\gamma = \mathrm{id} \upharpoonright J_\gamma$, so in particular, $\pi''a = a$. Also $\pi''a$ is $\Sigma_n^{J_\beta}(\{\pi(q)\})$. But look, this implies that $a = \pi''a \in J_{\beta+1}$. Hence we must have $\beta = \alpha$ (and here we have used our hypothesis that $\mathcal{P}(\omega\gamma) \cap \Sigma_n(J_\alpha) \not\subset J_\alpha$!). Thus, in particular, $a = \pi''a$ is $\Sigma_n^{J_\alpha}(\{\pi(q)\})$, so by the choice of q, we see that $\pi(q) = q$. Again, it is easy to see that $h' = \pi \cdot h \cdot \pi^{-1}$ is a $\Sigma_n^{J_\alpha}(\{\pi(p)\})$ Σ_n Skolem function for J_α, so by choice of p, $\pi(p) = p$. But then h, h' are both defined by the same Σ_n formula (with parameter p) in J_α, so $h = h'$. It follows immediately that $\pi \cdot \tilde{h} \cdot \pi^{-1} = \tilde{h}$, of course. So for $i \in \omega$, $x \in J_\gamma$, $\pi \cdot \tilde{h}(i,x) \simeq \tilde{h} \cdot \pi(i,x) \simeq \tilde{h}(i,x)$. Thus $\pi \upharpoonright X = \mathrm{id} \upharpoonright X$, and $X = J_\alpha$.

Lemma 46 plays a direct part in the proof of (P 1)–(P 3). The next lemma, however, is only used during the proof of the lemma which follows it, and may, on first sight, appear somewhat uninspiring.

Lemma 47. *Let $\langle J_\alpha, A\rangle$ be amenable, $\rho = \rho^1_{\alpha,A}$. If $B \subset J_\rho$ is $\Sigma_1(\langle J_\alpha, A\rangle)$, then $\Sigma_1(\langle J_\rho, B\rangle) \subset \Sigma_2(\langle J_\alpha, A\rangle)$.*

Proof. *Case 1.* There is a $\Sigma_1(\langle J_\alpha, A\rangle)$ map of some $\gamma < \omega\rho$ cofinally into $\omega\alpha$.

Let g be such a map, and let $\bar{B}$ be $\Sigma_0(\langle J_\alpha, A\rangle)$ such that $B(x) \leftrightarrow \exists z \bar{B}(z,x)$ for each $x \in J_\rho$. Define B' by $B'(\langle \nu, x\rangle) \leftrightarrow (\exists z \in S_{g(\nu)}) \bar{B}(z,x)$, for $\nu \in \gamma$, $x \in J_\rho$. Thus B' is $\Delta_1(\langle J_\alpha, A\rangle)$. And since $B(x) \leftrightarrow (\exists \nu \in \gamma) B'(\langle \nu, x\rangle)$, $\Sigma_1(\langle J_\rho, B\rangle) \subset \Sigma_1(\langle J_\rho, B'\rangle)$. Thus, we need only prove that $\Sigma_1(\langle J_\rho, B'\rangle \subset \Sigma_2(\langle J_\alpha, A\rangle)$. It clearly suffices to prove that $\Sigma_0(\langle J_\rho, B'\rangle) \subset \Sigma_2(\langle J_\alpha, A\rangle)$. Let R be $\Sigma_0(\langle J_\rho, B'\rangle)$. Thus R is rud in B' and some parameter $p \in J_\rho$. By choice of ρ, $\langle J_\rho, B'\rangle$ is amenable, so by Lemmas 3 and 4, there is a $\Sigma_0(J_\rho)$ predicate P and functions $f_1, \ldots, f_{m+k}$, rud in parameter p, such that $R(\boldsymbol{x}) \leftrightarrow P(\boldsymbol{x}, f_1(\boldsymbol{x}), \ldots, f_m(\boldsymbol{x}), B' \cap f_{m+1}(\boldsymbol{x}), \ldots, B' \cap f_{m+k}(\boldsymbol{x}))$. Hence $R(\boldsymbol{x}) \leftrightarrow \exists y_1, \ldots, \exists y_k [y_1 = B' \cap f_{m+1}(\boldsymbol{x}) \wedge \ldots \wedge y_k = B' \cap f_{m+k}(\boldsymbol{x}) \wedge P(\boldsymbol{x}, f_1(\boldsymbol{x}), \ldots, f_m(\boldsymbol{x}), y_1, \ldots, y_k)]$. Now P is certainly $\Sigma_0(J_\alpha)$, and $f_1, \ldots, f_{m+k}$ are rud in parameter p, so it suffices to show that the function $b(u) = B' \cap u$ is $\Sigma_2(\langle J_\alpha, A\rangle)$. It is in fact $\Pi_1(\langle J_\alpha, A\rangle)$, because: $y = b(u) \leftrightarrow \forall x [x \in y \leftrightarrow x \in u \wedge B'(x)]$, and B' is $\Delta_1(\langle J_\alpha, A\rangle)$.

Case 2. Otherwise.

As before, we must show that $\Sigma_0(\langle J_\rho, B\rangle) \subset \Sigma_2(\langle J_\alpha, A\rangle)$. Again as before, this reduces, by the amenability of $\langle J_\rho, B\rangle$, to proving that the function $b(u) = B \cap u$ is $\Sigma_2(\langle J_\alpha, A\rangle)$ on J_ρ. Now, we clearly have

$$y = b(u) \leftrightarrow (\forall x \in y)(x \in u \wedge B(x)) \wedge (\forall x \in u)(B(x) \to x \in y).$$

Now, the second conjunct here is $\mathbf{\Pi}_1(\langle J_\alpha, A\rangle)$. We show that the first conjunct is $\mathbf{\Sigma}_1(\langle J_\alpha, A\rangle)$, which is sufficient. It reduces to showing that $(\forall x \in y) B(x)$ is $\mathbf{\Sigma}_1(\langle J_\alpha, A\rangle)$. But look, we know that Case 1 fails to hold, so this is proved just as in Lemma 14.

The next lemma is the key step involved in proving, by induction, the as yet unformulated (P 3).

Lemma 48. *Let $\langle J_\alpha, A\rangle$ be amenable, $\rho = \rho^1_{\alpha,A}$. Suppose there is a $\mathbf{\Sigma}_1(\langle J_\alpha, A\rangle)$ map of a subset of $\omega\rho$ onto J_α. Then there is a $B \subset J_\rho$, $B \in \mathbf{\Sigma}_1(\langle J_\alpha, A\rangle)$, such that $\mathbf{\Sigma}_n(\langle J_\rho, B\rangle) = \mathcal{P}(J_\rho) \cap \mathbf{\Sigma}_{n+1}(\langle J_\alpha, A\rangle)$ for all $n \geq 1$.*

Proof. Let $u \subset \omega\rho$, and let $f : u \xrightarrow{\text{onto}} J_\alpha$ be $\mathbf{\Sigma}_1(\langle J_\alpha, A\rangle)$. Pick $p \in J_\alpha$ such that f is $\Sigma_1^{\langle J_\alpha, A\rangle}(\{p\})$. Let $\langle \varphi_i \mid i < \omega\rangle$ be a recursive enumeration of Fml^{Σ_1}. Set

$$B = \{\langle i, x\rangle \mid i \in \omega \wedge x \in J_\rho \wedge \models^{\Sigma_1}_{\langle J_\alpha, A\rangle} \varphi_i [x, p]\}.$$

Now, $\langle J_\alpha, A\rangle$ is amenable, and hence rud closed, so by Lemma 9, $B \in \mathbf{\Sigma}_1(\langle J_\alpha, A\rangle)$. And of course $B \subset J_\rho$.

Commencing with Lemma 47, an easy induction shows that for all $n \geq 1$, $\mathbf{\Sigma}_n(\langle J_\rho, B\rangle) \subset \mathbf{\Sigma}_{n+1}(\langle J_\alpha, A\rangle)$.

For the converse, let $R(\boldsymbol{x})$ be a $\mathbf{\Sigma}_{n+1}(\langle J_\alpha, A\rangle)$ relation on J_ρ, $n \geq 1$. Assume, for the sake of argument, that n is even. Let P be a $\mathbf{\Sigma}_1(\langle J_\alpha, A\rangle)$ relation such that, for $\boldsymbol{x} \in J_\rho$, $R(\boldsymbol{x}) \leftrightarrow \exists y_1 \forall y_2 \ldots \forall y_n P(\boldsymbol{y}, \boldsymbol{x})$. Define $\widetilde{P}$ by $\widetilde{P}(\boldsymbol{z}, \boldsymbol{x}) \leftrightarrow [\boldsymbol{z}, \boldsymbol{x} \in J_\rho \wedge P(f(\boldsymbol{z}), \boldsymbol{x})]$. By choice of f, any $x \in J_\alpha$ is $\Sigma_1^{\langle J_\alpha, A\rangle}(\{p, \nu\})$ for some $\nu < \omega\rho$, so by definition of B, $\widetilde{P}$ is rud in B and some parameter $\nu < \omega\rho$. In particular, $\widetilde{P}$ is $\Delta_1(\langle J_\rho, B\rangle)$. Again, $D = \mathrm{dom}(f)$ is rud in B and some parameter $\tau < \omega\rho$, so D is also $\Delta_1(\langle J_\rho, B\rangle)$. But for $\boldsymbol{x} \in J_\rho$, $R(\boldsymbol{x}) \leftrightarrow (\exists z_1 \in D)(\forall z_2 \in D) \ldots (\forall z_n \in D) \widetilde{P}(\boldsymbol{z}, \boldsymbol{x})$, which is thus $\mathbf{\Sigma}_n(\langle J_\rho, B\rangle)$.

We are now ready to formulate (P 3) and prove our promised uniformisation theorem.

Let $\alpha, n \geq 0$. A $\mathbf{\Sigma}_n$ *master code* for J_α is a set $A \subset J_{\rho^n_\alpha}$, $A \in \mathbf{\Sigma}_n(J_\alpha)$, such that whenever $m \geq 1$, $\mathbf{\Sigma}_m(\langle J_{\rho^n_\alpha}, A\rangle) = \mathcal{P}(J_{\rho^n_\alpha}) \cap \mathbf{\Sigma}_{n+m}(J_\alpha)$.

Theorem 49. *Let $\alpha, n \geq 0$. Then:*

(P 1) J_α *is* Σ_{n+1}*-uniformisable.*

(P 2) *There is a* $\Sigma_n(J_\alpha)$ *map of a subset of* $\omega\rho_\alpha^n$ *onto* J_α. *(Unless $n = 0$, when it is $\Sigma_1(J_\alpha)$.)*

(P 3) J_α *has a* Σ_n *master code.*

Proof. We prove the theorem (for all n) by induction on α. For $\alpha = 0$, it is trivial. So assume $\alpha > 0$ and that (P 1)–(P 3) hold (for all n) for all $\beta < \alpha$. We prove (P 1)–(P 3) at α by induction on n.

Case 1: $n = 0$. (P 1) is already proved (Theorem 30).

(P 2) $\rho_\alpha^0 = \alpha$, so (P 2) is already proved (Theorem 39)

(P 3) since $\rho_\alpha^0 = \alpha$, $A = \emptyset$ is a Σ_0 master code for J_α.

Case 2: $n = m + 1, m \geq 0$. Let $\rho = \rho_\alpha^n$ for convenience.

We first prove that ρ is the least ordinal such that some $\Sigma_n(J_\alpha)$ function maps a subset of $\omega\rho$ onto J_α. To this end, let δ be the least such ordinal. Suppose first that $\delta < \rho$. Then $B = \{\xi \in \omega\delta \mid \xi \notin f(\xi)\}$ is a $\Sigma_n(J_\alpha)$ subset of J_ρ, so by definition of ρ, $\langle J_\rho, B\rangle$ is amenable. Thus, as $\delta < \rho$, $B = B \cap \omega\delta \in J_\rho \subset J_\alpha$. So $B = f(\xi)$ for some $\xi \in \omega\delta$, whence $\xi \in f(\xi) \leftrightarrow \xi \in B \leftrightarrow \xi \notin f(\xi)$, which is absurd. Hence $\rho \leq \delta$. Suppose $\rho < \delta$. By definition of ρ, this means that for some $\Sigma_n(J_\alpha)$ set $B \subset J_\delta$, $\langle J_\delta, B\rangle$ is not amenable. Since $\langle J_1, B\rangle$ must be amenable, $\delta > 1$. If $\delta = \gamma + 1$, then since there is a $\Sigma_1(J_\alpha)$ map of $\omega\gamma$ onto $\omega\delta$, there is a $\Sigma_n(J_\alpha)$ map of a subset of $\omega\gamma$ onto J_α, contrary to the choice of δ. Hence $\lim(\delta)$. It follows, since $\langle J_\delta, B\rangle$ is not amenable, that there is $\tau < \delta$ with $B \cap J_\tau \notin J_\delta$. By induction hypothesis, J_α is Σ_n-uniformisable. So as $\tau < \delta$, Lemma 46 implies that $\mathcal{P}(\omega\tau) \cap \Sigma_n(J_\alpha) \subset J_\alpha$. But there is $h \in J_\alpha$, $h : \omega\tau \xrightarrow{\text{onto}} J_\tau$, so this implies $\mathcal{P}(J_\tau) \cap \Sigma_n(J_\alpha) \subset J_\alpha$. In particular, $B \cap J_\tau \in J_\alpha$. Hence for some $\beta < \alpha$, $B \cap J_\tau$ is J_β-definable. Let β be the least such, and let r be least such that $B \cap J_\tau$ is $\Sigma_r(J_\beta)$. By definition, $\langle J_{\rho_\beta^r}, B \cap J_\tau\rangle$ is amenable, so if $\tau < \rho_\beta^r$, then $B \cap J_\tau = (B \cap J_\tau) \cap J_\tau \in J_{\rho_\beta^r} \subset J_\beta$, contrary to the choice of β. Hence $\tau \geq \rho_\beta^r$. By induction hypothesis, there is a $\Sigma_r(J_\beta)$ map g from a subset of $\omega\rho_\beta^r$ onto J_β. And since $B \cap J_\tau \in J_{\beta+1}$ and $B \cap J_\tau \notin J_\delta$, $\beta + 1 > \delta$, or $\beta \geq \delta$. Hence there is a $\Sigma_r(J_\beta)$ map g' from a subset of $\omega\rho_\beta^r$ onto $\omega\delta$. Then $f \cdot g'$ is a $\Sigma_n(J_\alpha)$ map of a subset of $\omega\rho_\beta^r$ onto J_α. But we have established that $\rho_\beta^r \leq \tau < \delta$, so this contradicts the choice of δ. Hence $\delta = \rho$.

(P 2) follows immediately from the above result of course.

We turn now to (P 3). By induction hypothesis, let A be a Σ_m master code for J_α. Set $\eta = \rho_\alpha^m$ for convenience. By the above, let f be a $\Sigma_n(J_\alpha)$ map of a subset of $\omega\rho$ onto J_α. By choice of A, $f' = f \restriction (f^{-1}{}''J_\eta)$ is a $\Sigma_1(\langle J_\eta, A\rangle)$ map of a subset of $\omega\rho$ onto J_η. By choice of A, it is clear that $\rho = \rho_\alpha^n = \rho_{\eta,A}^1$. Finally, of course, $\langle J_\eta, A\rangle$ is amenable. So, we may apply Lemma 48 to $\langle J_\eta, A\rangle$ to obtain a $\Sigma_1(\langle J_\eta, A\rangle)$ set $B \subset J_\rho$ such that $\Sigma_r(\langle J_\rho, B\rangle) = \mathcal{P}(J_\rho) \cap \Sigma_{r+1}(\langle J_\eta, A\rangle)$ for all $r \geq 1$. By choice of A, $B \in \Sigma_n(J_\alpha)$ and $\Sigma_r(\langle J_\rho, B\rangle) = \mathcal{P}(J_\rho) \cap \Sigma_{n+r}(J_\alpha)$ for all $r \geq 1$. Hence B is a Σ_n master code for J_α.

Finally we prove (P 1). Let B be, as above, a Σ_n master code for J_α. Let $R(y, \underset{\sim}{x})$ be a $\Sigma_{n+1}(J_\alpha)$ relation on J_α. Define, with f as above, a relation $\widetilde{R}$ on J_ρ by $\widetilde{R}(y, \underset{\sim}{x}) \leftrightarrow [y, \underset{\sim}{x} \in J_\rho \wedge R(f(y), f(\underset{\sim}{x}))]$. Then $\widetilde{R}$ is $\Sigma_{n+1}(J_\alpha)$, and hence $\Sigma_1(\langle J_\rho, B\rangle)$. Let $\widetilde{r}$ be a $\Sigma_1(\langle J_\rho, B\rangle)$ function uniformising $\widetilde{R}$. Since f is $\Sigma_n(J_\alpha)$, so is f^{-1}. But J_α is Σ_n-uniformisable, by induction hypothesis, so we can let $\widetilde{f}$ be a $\Sigma_n(J_\alpha)$ function uniformising f^{-1}. Set $r = f \circ \widetilde{r} \circ f^{-1}$. It is clear that r is a $\Sigma_{n+1}(J_\alpha)$ function which uniformises R. The proof is complete.

The above results give us two (intuitive) equivalent formulations of the Σ_n-projectum:

Theorem 50. *Let $\alpha, n \geq 0$. Let δ be the least ordinal such that some $\Sigma_n(J_\alpha)$ function maps a subset of $\omega\delta$ onto J_α. Let γ be the least ordinal such that $\mathcal{P}(\omega\gamma) \cap \Sigma_n(J_\alpha) \not\subset J_\alpha$. Then $\delta = \gamma = \rho_\alpha^n$.*

Proof. That $\delta = \rho_\alpha^n$ was actually proved during the proof of Theorem 49. Since we now *know* that J_α is Σ_n-uniformisable, Lemma 46 tells us that $\delta \leq \gamma$. Assume $\delta < \gamma$. Now by definition, let $u \subset \omega\delta$, and let $f: u \xrightarrow{\text{onto}} J_\alpha$ be $\Sigma_n(J_\alpha)$. Let $Z = \{\xi \mid \xi \notin f(\xi)\}$. Then $Z \subset \omega\delta$ and $Z \in \Sigma_n(J_\alpha)$, so by definition of γ, $Z \in J_\alpha$. Thus $Z = f(\xi)$ for some ξ, so $\xi \in f(\xi) \leftrightarrow \xi \notin f(\xi)$, which is absurd. Hence $\delta = \gamma$.

There is, of course, a concept which, for Δ_n predicates, plays the role that the Σ_n projectum plays for Σ_n predicates. And, as might be expected, there is a corresponding "total function" or Δ_n equivalent of Theorem 50 for this concept.

Let $\alpha, n \geq 0$. The *Δ_n-projectum* of α (sometimes called the *weak Σ_n-projectum*), η_α^n, is the largest $\eta \leq \alpha$ such that $\langle J_\eta, A\rangle$ is amenable for all $\Delta_n(J_\alpha)$ sets $A \subset J_\eta$.

Thus the $\mathbf{\Delta}_n$-projectum of α represents the "hard core" of J_α with regards to $\mathbf{\Delta}_n$ predicates on J_α. Clearly, $\eta^n_\alpha \geq \rho^n_\alpha$. We do not, however, necessarily have equality here. For example, let α be the first admissible ordinal $> \omega$. Then it is easily seen that $\eta^1_\alpha = \alpha$, whereas $\rho^1_\alpha = \omega$.

Corresponding to Lemma 46, we have:

Lemma 51. *Let $n \geq 1$, and let γ be the least ordinal such that $\mathcal{P}(\omega\gamma) \cap \mathbf{\Delta}_n(J_\alpha) \not\subset J_\alpha$. Then there is a $\mathbf{\Sigma}_n(J_\alpha)$ (and hence $\mathbf{\Delta}_n(J_\alpha)$) map of $\omega\gamma$ onto J_α.*

Proof. Let $n = m + 1, n \geq 0$. Since $\mathbf{\Sigma}_m(J_\alpha) \subset \mathbf{\Delta}_n(J_\alpha) \subset \mathbf{\Sigma}_n(J_\alpha)$, Theorem 50 implies that $\rho^n_\alpha \leq \gamma \leq \rho^m_\alpha$. Theorem 50 also implies that there is a $\mathbf{\Sigma}_m(J_\alpha)$ map of a subset of $\omega\rho^m_\alpha$ onto J_α. So, we can clearly define a $\mathbf{\Sigma}_n(J_\alpha)$ map of $\omega\rho^m_\alpha$ itself onto J_α. This reduces our problem to showing that there is a $\mathbf{\Sigma}_n(J_\alpha)$ map of $\omega\gamma$ onto $\omega\rho^m_\alpha$. As a first step, we have the:

Claim. *There is a $\mathbf{\Sigma}_n(J_\alpha)$ map g from $\omega\gamma$ cofinally into $\omega\rho^m_\alpha$.*

Let A be a $\mathbf{\Sigma}_m$ master code for J_α. By hypothesis, let $b \subset \omega\gamma$, $b \in \mathbf{\Delta}_n(J_\alpha)$, $b \notin J_\alpha$. By choice of A, b is $\mathbf{\Delta}_1(\langle J_{\rho^m_\alpha}, A\rangle)$. Suppose b is in fact defined by:

$$\nu \in b \leftrightarrow \exists y B_0(y,\nu)\,, \qquad \nu \notin b \leftrightarrow \exists y B_1(y,\nu)\,,$$

where B_0, B_1 are $\mathbf{\Sigma}_0(\langle J_{\rho^m_\alpha}, A\rangle)$. Then

$$(\forall \nu \in \omega\gamma)\, \exists y\, [B_0(y,\nu) \vee B_1(y,\nu)]\,.$$

But $\langle J_{\rho^m_\alpha}, A\rangle$ is amenable, and hence rud closed, so as $b \notin J_{\rho^m_\alpha}$, there can be no $\tau < \omega\rho^m_\alpha$ such that

$$(\forall \nu \in \omega\gamma)(\exists y \in S_\tau)[B_0(y,\nu) \vee B_1(y,\nu)]\,.$$

Define $g : \omega\gamma \to \omega\rho^m_\alpha$ by

$$g(\nu) = \text{the least } \tau \text{ such that } (\exists y \in S_\tau)[B_0(y,\nu) \vee B_1(y,\nu)]\,.$$

Clearly, g is $\mathbf{\Sigma}_n(J_\alpha)$ and cofinal in $\omega\rho^m_\alpha$, proving the claim.

We now prove the lemma. Since $\rho_\alpha^n \leq \gamma$, there must be a $\Sigma_n(J_\alpha)$ map f from a subset of $\omega\gamma$ onto $\omega\rho_\alpha^m$ ($\leq \omega\alpha$), and of course such an f will then be $\Sigma_1(\langle J_{\rho_\alpha^m}, A\rangle)$. Define $\bar{f} : (\omega\gamma)^2 \xrightarrow{\text{onto}} \omega\rho_\alpha^m$ as follows. Let f be given by $f(\nu) = \tau \leftrightarrow \exists y F(y, \tau, \nu)$, where F is $\Sigma_0(\langle J_{\rho_\alpha^m}, A\rangle)$. Set

$$\bar{f}(\nu, \tau) = \begin{cases} \theta, & \text{if } (\exists y \in S_{y(\tau)}) F(y, \theta, \nu) \\ 0, & \text{otherwise.} \end{cases}$$

Then $\bar{f}$ is $\Sigma_1(\langle J_{\rho_\alpha^m}, A\rangle)$, and hence $\Sigma_n(J_\alpha)$. And $\bar{f}$ clearly maps $(\omega\gamma)^2$ onto $\omega\rho_\alpha^m$, as g cofinal in $\omega\rho_\alpha^m$.

Since we have (by Lemma 38) a $\Sigma_1(J_\alpha)$ map of $\omega\gamma$ onto $(\omega\gamma)^2$, the lemma follows.

Corresponding to Theorem 50, we have:

Theorem 52. *Let $\alpha, n > 0$. Let δ be the least ordinal such that some $\Sigma_n(J_\alpha)$ (and hence $\Delta_n(J_\alpha)$) function maps $\omega\delta$ onto J_α. Let γ be the least ordinal such that $\mathcal{P}(\omega\gamma) \cap \Delta_n(J_\alpha) \not\subset J_\alpha$. Then $\delta = \gamma = \eta_\alpha^n$.*

Proof. Suppose $\gamma < \eta_\alpha^n$. Let $B \subset \omega\gamma$, $B \in \Delta_n(J_\alpha)$, $B \notin J_\alpha$. Then $\omega\gamma \cap B = B \notin J_{\eta_\alpha^n}$, contrary to $\langle J_{\eta_\alpha^n}, B\rangle$ being amenable.

Suppose now that $\eta_\alpha^n < \gamma$. Then there is $A \subset J_\gamma$, $A \in \Delta_n(J_\alpha)$, such that $\langle J_\alpha, A\rangle$ is not amenable. In particular, $\gamma > 1$. Suppose $\gamma = \xi + 1$. There is then a $\Sigma_1(J_\gamma)$ map of $\omega\xi$ onto $\omega\gamma$, so by Lemma 51 there is a $\Sigma_n(J_\alpha)$ map, f, of $\omega\xi$ onto J_α. Then $Z = \{\iota \in \omega\xi \mid \iota \notin f(\iota)\}$ is a $\Delta_n(J_\alpha)$ subset of $\omega\xi$. Clearly, $Z \notin J_\alpha$, so this contradicts the choice of γ. Hence $\lim(\gamma)$. Thus as $\langle J_\gamma, A\rangle$ is not amenable, there must be $\tau < \gamma$ with $A \cap J_\tau \notin J_\gamma$. But by choice of γ, $\tau < \gamma \rightarrow A \cap J_\tau \in J_\alpha$, so for some $\theta < \alpha$, $A \cap J_\tau$ *is* J_θ-definable. Let θ be the least such. Then $A \cap J_\tau \in \mathcal{P}(J_\tau) \cap \Delta_m(J_\theta)$ for some $m \in \omega$, and $A \cap J_\tau \notin J_\theta$. Thus by Lemma 51 there is a $\Sigma_m(J_\theta)$ map f of $\omega\tau$ onto J_θ. (Actually the hypotheses of Lemma 51 require that we have a $\Delta_m(J_\theta)$ subset of $\omega\tau$ not in J_θ, whereas we have only exhibited a subset of J_τ with these properties. However, since there is available a $\Sigma_1(J_\tau)$ map of $\omega\tau$ onto J_τ, this point causes no problem.) Since $\theta < \alpha$, $f \in J_\alpha$. But $A \cap J_\tau \notin J_\gamma$, and $A \cap J_\tau \in J_{\theta+1}$, so $\theta \geq \gamma$, and there is thus a map $f' \in J_\alpha$ of $\omega\tau$ onto $\omega\gamma$. By Lemma 51, again, this gives us a $\Sigma_n(J_\alpha)$ map k of $\omega\tau$ onto J_α. Then, clearly, $K =$

$\{\iota \mid \iota \notin K(\iota)\}$ is a $\mathbf{\Delta}_n(\mathrm{J}_\alpha)$ subset of $\omega\tau$ not lying in J_α, contrary to $\tau < \gamma$.

Hence $\gamma = \eta_\alpha^n$. Now, by Lemma 51, we have $\delta \leq \gamma$. Suppose $\delta < \gamma$. Let $f : \omega\delta \xrightarrow{\text{onto}} \mathrm{J}_\alpha$ be $\Sigma_n(\mathrm{J}_\alpha)$. Let $Z = \{\nu \mid \nu \notin f(\nu)\}$. Then $Z \in \mathcal{P}(\omega\delta) \cap \mathbf{\Delta}_n(\mathrm{J}_\alpha) - \mathrm{J}_\alpha$. But this contradicts the choice of γ. Hence $\delta = \gamma$.

Remark. Lemmas 46 and 51 can be regarded as much sharper versions of the following, much earlier theorem of Putman:

Suppose $\mathcal{P}(\gamma) \cap \mathrm{L}_{\alpha+1} \not\subset \mathrm{L}_\alpha$. Then $\mathrm{L}_{\alpha+1}$ contains a well-ordering of γ of order type α. (For $\gamma \geq \omega$.)

Putman actually proved this result for the case $\rho = \omega$, but his proof works in the general case.

The methods described above have, of course, many uses. We give just one, very general, example, showing that (in certain circumstances) it is possible to carry out Löwenheim–Skolem arguments for non-regular ordinals α which can generally only be done when α is actually a regular cardinal.

More precisely, the following theorem, is well-known:

Theorem 53. *Let κ be a regular cardinal. Let $\gamma < \kappa \leq \omega\beta$, and suppose that $Y \subset \mathrm{J}_\beta$, $|Y| < \kappa$. Then there is $X \prec \mathrm{J}_\beta$ such that $Y \cup \gamma \subset X$ and $\kappa \cap X \in \kappa$.*

To prove this, one simply forms an ω-chain $X_0 \prec X_1 \prec \ldots \prec X_n \prec \ldots \prec \mathrm{J}_\beta$ of elementary submodels of J_β, taking X_0 as the Skolem hull of $Y \cup \gamma$ in J_β, and X_{n+1} as the Skolem hull of $X_n \cup \sup(\kappa \cap X_n)$ in J_β, and then $X = \bigcup_{n<\omega} X_n$ is the required submodel of J_β. By construction, $\kappa \cap X$ is transitive, and hence an ordinal, and *since κ is regular*, $|X| < \kappa$, so $\kappa \cap X \in \kappa$. It should be observed that κ being regular is a necessary condition for the above procedure to work (in general). However, providing we can, in some way, ensure that for each n, $\sup(\kappa \cap X_n) < \kappa$, then we can, of course, get by with just $\mathrm{cf}(\kappa) > \omega$. The theorem below shows that, in certain cases we can do just this, providing we relax our demands somewhat.

Let $n \geq 1$, $\alpha \leq \omega\beta$. We say that α is *$\mathbf{\Sigma}_n$-regular at* β iff there is no $\mathbf{\Sigma}_n(\mathrm{J}_\beta)$ map of a bounded subset of α cofinally into α.

For example, by Theorem 43, $\omega\alpha$ is strongly admissible iff $\omega\alpha$ is $\mathbf{\Sigma}_1$-regular at α.

Theorem 54. *Let $n \geq 1$, $\omega\beta \geq \alpha \geq 1$. Suppose α is Σ_n-regular at β. Let $Y \subset J_\beta$, $\omega \leq |Y| < \mathrm{cf}(\alpha)$, and let $\gamma < \alpha$. Then there is an $X \prec_{\Sigma_n} J_\beta$ such that $Y \cup \gamma \subset X$ and $\alpha \cap X \in \alpha$.*

Proof. Since $\mathrm{cf}(\alpha) > \omega$, it clearly suffices to prove that, under the stated hypotheses, there is $X \prec_{\Sigma_n} J_\beta$ such that $Y \cup \gamma \subset X$ and $\sup(\alpha \cap X) < \alpha$.

Let h be a Σ_n Skolem function for J_β (by Theorem 49 and Lemma 33). Since $\omega \leq |Y| < \mathrm{cf}(\alpha)$, we may, without loss of generality, assume that Y is closed under ordered pairs. Furthermore, let Φ be the function defined in Lemma 37. Since α is Σ_n-regular at β, α is certainly strongly admissible. Hence, by Lemma 37, $\{\xi \in \alpha \mid \Phi''\xi^2 \subset \xi\}$ is unbounded in α. It follows that we may also, without loss of generality, assume that $\Phi''\gamma^2 \subset \gamma$. Recall that $\Phi \restriction \gamma^2$ is $\Sigma_1^{J_\beta}$.

Let $X = h''(\omega \times (Y \times \gamma))$. Then we claim that X is closed under ordered pairs. To see this, let $x_1, x_2 \in X$, say $x_1 = h(i_1, \langle y_1, \nu_1 \rangle)$, $x_2 = h(i_2, \langle y_2, \nu_2 \rangle)$. Let $y = \langle y_1, y_2 \rangle \in Y$ and $\nu = \Phi(\langle \nu_1, \nu_2 \rangle) \in \gamma$. Clearly $\{\langle x_1, x_2 \rangle\}$ is $\Sigma_1^{J_\beta}(\{p, \langle y, \nu \rangle\})$, where p is a good parameter for h. Thus for some $i \in \omega$, $\langle x_1, x_2 \rangle = h(i, \langle y, \nu \rangle) \in X$, as required. So, by Corollary 36, $X \prec_{\Sigma_n} J_\beta$. And of course, we clearly have $Y \cup \gamma \subset X$. We show that $\sup(\alpha \cap X) < \alpha$.

For $y \in Y$, $i \in \omega$, define $h_{i,y} : \subset \gamma \to \alpha$ by $h_{i,y}(\nu) \simeq h(i, \langle y, \nu \rangle)$. Thus $h_{i,y}$ is $\Sigma_n(J_\beta)$, and so as α is Σ_n-regular at β, $\sup(h_{i,y}''\gamma) \simeq \gamma(i,y) < \alpha$. Since $|Y| < \mathrm{cf}(\alpha)$, it follows that $\sup_{y \in Y} \gamma(i,y) \simeq \gamma(i) < \alpha$. Since $\mathrm{cf}(\alpha) > \omega$, we conclude finally that $\sup_{i \in \omega} \gamma(i) < \alpha$. But clearly, $\sup_{i \in \omega} \gamma(i) = \sup(\alpha \cap X)$, so we are done.

The above lemma may be used to prove that if $V = L$ and κ is a regular uncountable, non-weakly compact cardinal, then there is a Souslin κ-tree. (Jensen.)

J.E.Fenstad, P.G.Hinman (eds.), Generalized Recursion Theory

DEGREE THEORY ON ADMISSIBLE ORDINALS [1]

Stephen G. SIMPSON

Yale University and The University of California, Berkeley

0. Introduction

The study of recursive functions on the ordinal numbers was initiated by Takeuti in the late 1950's. Takeuti's concept was generalized by several authors to that of α-recursive functions on admissible initial segments α of the ordinals. An intensive study of the generalized concept was begun by Sacks in 1964 [11].

Sacks' unstated program was to suitably generalize all the theorems of ordinary recursion theory to α-recursion theory. A typical theorem of ordinary recursion theory is the Friedberg–Muchnik solution to Post's problem: there are two recursively enumerable subsets of ω such that neither is recursive in the other. A recent paper of Sacks and Simpson [13] generalizes the Friedberg–Muchnik theorem as follows: for any admissible ordinal α, there are two α-recursively enumerable subsets of α such that neither is α-recursive in the other. (The Sacks–Simpson proof is of some methodological interest in that it involves a rather delicate downward Löwenheim–Skolem argument within L_α, the αth constructible level.)

Our object in the present paper is to survey the recent work in α-degree theory for arbitrary admissible ordinals α. (An α-degree is an equivalence class under the relation: A is α-recursive in B and B is α-recursive in A.) The flavor of α-degree theory can be savored only in the proofs; we therefore give detailed proofs of a few theorems. The reader must understand that in this survey

[1] The preparation of this paper was partially supported by NSF Contracts GP-34088X and GP-24352. The main result of Section 4, due to Sacks and the author, was presented by the author in an invited address to the 1971 Summer School in Mathematical Logic at Cambridge, England, sponsored in part by the North Atlantic Treaty Organization.

paper we are leaving many important matters out of account, e.g. (in decreasing order of importance):

1. reasons for generalizing recursion theory [2];
2. the lattice of α-r.e. sets [3];
3. alternative notions of relative α-recursiveness [4];
4. non-regular and non-hyperregular α-r.e. sets [5].

A detailed outline of the paper follows.

In Section 1 we review the basic notions of α-recursion theory such as α-degree, α-r.e. sets, and the α-jump of a set. We also discuss two important auxiliary notions, regularity and hyperregularity, which have no counterpart in ordinary recursion theory.

In Section 2 we prove the following generalization of Friedberg's Completeness Theorem: an α-degree $\boldsymbol{b} \geq \boldsymbol{0}'$ is regular if and only if it is the jump of a regular, hyperregular α-degree. We then give a necessary and sufficient condition on α that there exist an α-degree $\boldsymbol{d}$ such that every $\boldsymbol{b} \geq \boldsymbol{d}$ is regular. We show in particular that $\boldsymbol{d}$ exists if α is countable.

In Section 4 we reprove the main theorem of Sacks–Simpson [13] in the following strong form: for any admissible ordinal α there are α-r.e. degrees $\boldsymbol{a}, \boldsymbol{b}$ such that $\boldsymbol{a} \not\leq \boldsymbol{b}$, $\boldsymbol{b} \not\leq \boldsymbol{a}$, $\boldsymbol{a} \cup \boldsymbol{b}$ is regular and hyperregular, and $(\boldsymbol{a} \cup \boldsymbol{b})' = \boldsymbol{0}'$.

In Section 5 we survey recent work of Manny Lerman, Richard Shore, and others. Their work is further evidence that most if not all theorems of ordinary degree theory can be generalized (not straightforwardly!) to α-degree theory.

In Section 6 we discuss briefly the connection between α-recursion theory and Jensen's "fine structure" theory.

1. Fundamental notions

We employ von Neumann's notion of ordinality; thus an ordinal is identified with the set of smaller ordinals. Our set theoretical notation is standard.

[2] See Kreisel [2] especially Section 3.

[3] See Machtey [8]; Chapter 1 of Simpson [17]; Lerman–Simpson [7]; and Lerman [4].

[4] See pp. 157–158 of Kreisel [2] and pp. 39–44, 92–94 of Simpson [17].

[5] See Chapters 2 and 3 of Simpson [17].

In particular $\bigcup x$ (union of x), $x \cap y$ (intersection of x and y), $x \times y$ (Cartesian product of x and y), $x \restriction y$ (x restricted to y), $x''y$ (the range of $x \restriction y$), $x - y$ (set theoretical difference of x and y), $x \subseteq y$ (x is a subset of y), $x \in y$ (x is an element of y), $\emptyset$ (empty set), $\langle x,y \rangle$ (ordered pair), dom(x), rng(x) (domain and range of x), $x : y \to z$ (x is a function from y into z) have their usual meanings.

Throughout this paper α is a fixed but arbitrary admissible ordinal. Lower case Greek letters denote ordinals less than α, i.e. elements of α, except for β and λ which may denote limit ordinals less than or equal to α. Upper case Roman letters denote subsets of α. A set $X \subseteq \beta$ is *unbounded in* β if $\bigcup X = \beta$; otherwise it is *bounded below* β. We sometimes write *unbounded* for unbounded in α and *bounded* for bounded below α.

A partial function from α into α (i.e. a function whose domain and range are subsets of α) is *α-recursive* if its graph can be enumerated via an equation calculus resembling Kleene's but allowing infinitary bounded quantifications such as $(\exists x < \delta)$ where $\delta < \alpha$. (For details see [12].) A subset of α is *α-recursive* if its characteristic function is α-recursive. A subset of α is *α-recursively enumerable* (abbreviated α-r.e.) if it is the domain or range of an α-recursive partial function. Every α-r.e. set is the range of a one-one α-recursive function whose domain is an ordinal less than or equal to α.

A subset of α is *α-finite* if it is α-recursive and bounded. A basic [6] principle of α-recursion theory is:

$$(*)\quad \begin{cases} \text{if } f \text{ is an } \alpha\text{-recursive partial function} \\ \text{and } K \subseteq \text{dom}(f) \text{ is } \alpha\text{-finite then } f''K \\ \text{is } \alpha\text{-finite.} \end{cases}$$

In the usual way one defines an α-recursive function $k : \alpha \times \alpha \to \alpha$ such that: (i) $k(\gamma,\eta) = 0$ implies $\gamma < \eta$; and (ii) $\{\gamma \mid k(\gamma,\eta) = 0\}$ ranges over the α-finite sets as η ranges over α. We fix the notation $K_\eta = \{\gamma \mid k(\gamma,\eta) = 0\}$ and call η a *canonical index* for K if $K = K_\eta$.

The *projectum* of α, denoted α^*, is the least β such that there is a one-one α-recursive function $f : \alpha \to \beta$. It is easy to see that α^* is equal to the least β

[6] For any limit ordinal α the notion of α-recursive partial function can be defined. One can then define α-finiteness as above and prove: α is admissible if and only if principle $(*)$ holds. Note that if f is α-recursive and $\gamma < \alpha$ then $\lambda\xi f(\langle \xi, \gamma \rangle)$ is α-recursive.

such that not every α-r.e. subset of β is α-finite. If K is an α-finite set, the *α-cardinality* of K is the least β such that there is an α-finite one-one correspondence between K and β. An ordinal less than α is an *α-cardinal* if it is equal to its own α-cardinality. An α-cardinal β is *regular* if every α-finite subset of β of order type less than β is bounded below β. An α-cardinal is *singular* if it is not regular. It is easy to see that if $\alpha^* < \alpha$ then α^* is the largest α-cardinal.

As in ordinary recursion theory one defines an α-recursive function $r : \alpha \times \alpha^* \to \alpha$ such that: (i) $K_{r(\sigma,\epsilon)} \subseteq K_{r(\tau,\epsilon)} \subseteq \tau$ whenever $\sigma < \tau < \alpha$; (ii) $\bigcup\{K_{r(\sigma,\epsilon)} | \sigma < \alpha\}$ ranges over the α-r.e. sets as ϵ ranges over α^*. We fix the notations $R^\sigma_\epsilon = K_{r(\sigma,\epsilon)}$ and $R_\epsilon = \bigcup\{R^\sigma_\epsilon | \sigma < \alpha\}$. We call $\epsilon < \alpha^*$ an *index* for R if $R = R_\epsilon$. As in ordinary recursion theory, one can α-effectively compute an index for an α-finite set K from a canonical index for K but not conversely.

We now discuss relative α-recursiveness. For $A, B \subseteq \alpha$ we say A is *α-many-one reducible* to B (abbreviated $A \leq_{m\alpha} B$) if there is an α-recursive function $f : \alpha \to \alpha$ such that for all γ

$$\gamma \in A \leftrightarrow f(\gamma) \in B \ .$$

A is *α-recursive in B* (abbreviated $A \leq_\alpha B$) if there is an index $\epsilon < \alpha^*$ such that for all γ, δ

$$[K_\gamma \subseteq A \ \& \ K_\delta \cap A = \emptyset] \leftrightarrow$$

$$(\exists\xi)(\exists\eta)[\langle\gamma,\delta,\xi,\eta\rangle \in R_\epsilon \ \& \ K_\xi \subseteq B \ \& \ K_\eta \cap B = \emptyset] \ .$$

Clearly $A \leq_{m\alpha} B$ implies $A \leq_\alpha B$ but not conversely. The *α-jump* of B is defined by

$$B' = \{\epsilon < \alpha^* | (\exists\xi)(\exists\eta)[\langle\xi,\eta\rangle \in R_\epsilon \ \& \ K_\xi \subseteq B \ \& \ K_\eta \cap B = \emptyset]\} \ .$$

If f is a partial function from α into α, we say f is *weakly α-recursive in B* (abbreviated $f \leq_{w\alpha} B$) if there is an index $\epsilon < \alpha^*$ such that for all γ, δ

$$f(\gamma) = \delta \leftrightarrow (\exists\xi)(\exists\eta)[\langle\gamma,\delta,\xi,\eta\rangle \in R_\epsilon \ \& \ K_\xi \subseteq B \ \& \ K_\eta \cap B = \emptyset] \ .$$

For such an ϵ we write $f = [\epsilon]^B$ and call ϵ an *index for f from B*. We say A is *α-recursively enumerable in B* if A is the domain or range of a partial function weakly α-recursive in B.

If $A \subseteq \alpha$ we denote by c_A the characteristic function of A. Caution: $A \leq_\alpha B$ implies $c_A \leq_{w\alpha} B$ but not conversely. However, the following facts are easily verified.

1.1 $\leq_\alpha$ is transitive, and $f \leq_{w\alpha} A \leq_\alpha B$ implies $f \leq_{w\alpha} B$.
1.2 $B \leq_{m\alpha} B'$ but not $B' \leq_\alpha B$.
1.3 $A \leq_\alpha B$ if and only if $A' \leq_{m\alpha} B'$.
1.4 A is α-r.e. in B if and only if $A \leq_{m\alpha} B'$.

Two sets A and B are said to have the same *α-degree* if $A \equiv_\alpha B$ i.e. $A \leq_\alpha B$ and $B \leq_\alpha A$. Lower case boldface letters $\boldsymbol{a}, \boldsymbol{b}$, ... are used to denote α-degrees. If $\boldsymbol{a}, \boldsymbol{b}$ are the α-degrees of A, B respectively then we write: $\boldsymbol{a} \leq \boldsymbol{b}$ for $A \leq_\alpha B$; $\boldsymbol{a} \cup \boldsymbol{b}$ for the α-degree of $A \oplus B$ where

$$A \oplus B = \{2\gamma \mid \gamma \in A\} \cup \{2\gamma + 1 \mid \gamma \in B\};$$

$\boldsymbol{0}$ for the α-degree of $0 = \emptyset$; $\boldsymbol{a}'$ or jump $(\boldsymbol{a})$ for the α-degree of A'. Thus the α-degrees are an upper semilattice under $\leq, \cup$; moreover the *jump operator* is well defined, and increasing on α-degrees.

We now discuss two important auxiliary notions due to Sacks.

Definition. $A \subseteq \alpha$ is *α-regular* if $A \cap \beta$ is α-finite for all $\beta < \alpha$.

Definition. $A \subseteq \alpha$ is *α-hyperregular* if $f''\beta$ is bounded whenever $\beta < \alpha$, $f: \beta \to \alpha$, $f \leq_{w\alpha} A$.

Definition. An α-degree is *regular* (resp. *hyperregular*) if it contains an α-regular (resp. α-hyperregular) set.

Clearly every subset of ω is ω-regular and ω-hyperregular. The existence of non-α-hyperregular sets is what makes α-recursion theory more intricate than ordinary recursion theory. For instance, if $\alpha^* < \alpha$ then not even every α-r.e. set is α-regular (and conversely). This means that the difficulties associated with non-α-regularity can arise even in arguments concerning α-r.e. sets. Such difficulties are alleviated somewhat by the following theorem.

1.5 (Sacks [11]). *To every α-r.e. set A there is an α-regular α-r.e. set B such that $B \equiv_\alpha A$.*

A short proof of 1.5 can be found in Chapter 2 of [17].

1.6 (Sacks [11]). *For* $A \subseteq \alpha$ *the following are equivalent* [7]:
(i) A *is* α*-regular and* α*-hyperregular.*
(ii) $f''K$ *is* α*-finite whenever* $f \leq_{w\alpha} A$, $K \subseteq \mathrm{dom}(f)$, K α*-finite.*

Proof. That (ii) implies (i) is immediate from the definitions. To prove (ii) assuming (i), suppose $f \leq_{w\alpha} A$, $K \subseteq \mathrm{dom}(f)$, K α-finite. Let $\epsilon < \alpha^*$ be an index for f from A. Let $\beta < \alpha$ be the α-cardinality of K, and let i be an α-finite function from β onto K. For each $\gamma < \beta$ let $g(\gamma)$ be the least σ such that $K_\xi \subseteq A$ and $K_\eta \cap A = \emptyset$ for some δ, ξ, η such that $\langle i(\gamma), \delta, \xi, \eta \rangle \in R^\sigma_\epsilon$. Then $g \leq_{w\alpha} A$ since A is α-regular. Hence $g''\beta$ is bounded since A is α-hyperregular. Let $\tau < \alpha$ be such that $g''\beta \subseteq \tau$. Let $H = A \cap \tau$. Then H is α-finite since A is α-regular. Note that $\langle i(\gamma), \delta, \xi, \eta \rangle \in R^\tau_\epsilon$ implies $\delta, \xi, \eta < \tau$ which implies $K_\xi \cup K_\eta \subseteq \tau$. Thus $\delta \in f''K$ if and only if

$$(\exists \gamma, \xi, \eta)[\langle i(\gamma), \delta, \xi, \eta \rangle \in R^\tau_\epsilon \;\&\; K_\xi \subseteq H \;\&\; K_\eta \cap H = \emptyset] .$$

Thus $f''K$ is α-finite. ■

The end of a proof is indicated by ■.

We close this section with some useful technical lemmas concerning $\mathbf{\Sigma}_2$ functions. Following Rogers [10: pp. 301–307] we define the α*-arithmetical* hierarchy. Thus a relation on α is $\mathbf{\Sigma}_0$ if it is α-recursive, $\mathbf{\Pi}_n$ if its complement is $\mathbf{\Sigma}_n$, $\mathbf{\Sigma}_{n+1}$ if it is the projection of a $\mathbf{\Pi}_n$ relation, and $\mathbf{\Delta}_n$ if it is both $\mathbf{\Sigma}_n$ and $\mathbf{\Pi}_n$. In particular a relation is $\mathbf{\Sigma}_1$ if and only if it is α-r.e. A partial function f is said to be $\mathbf{\Sigma}_n$ if its graph is $\mathbf{\Sigma}_n$. In particular f is $\mathbf{\Sigma}_1$ if and only if it is α-recursive. Caution: the bounded quantifier manipulations described in Rogers [10: p. 311] do not generalize to α-recursion theory except in very special circumstances. However, as in ordinary recursion theory, one has:

1.7. Lemma. *Let* f *be a partial function from* α *into* α. *Then the following assertions are pairwise equivalent*:

[7] It can be shown that $A \subseteq \alpha$ is α-regular and α-hyperregular if and only if the structure $\langle L_\alpha, \in, A \rangle$ is admissible.

(i) *f is* Σ_2;
(ii) *f is weakly α-recursive in* $\boldsymbol{0}'$;
(iii) *there is an α-recursive function* $f' : \alpha \times \alpha \to \alpha$ *such that for all* γ

$$f(\gamma) \simeq \lim_\sigma f'(\sigma, \gamma) .$$

In (iii) the limit is taken in the discrete topology as σ approaches α. As usual, $x \simeq y$ means that x is defined iff y is defined in which case $x = y$. Thus in (iii) $f(\gamma) = \delta$ iff $(\exists\sigma)(\forall\tau \geq \sigma) f'(\tau, \gamma) = \delta$.

For each limit ordinal $\beta \leq \alpha$ we define $\boldsymbol{\Delta}_2 - \mathrm{cf}(\beta)$ to be the least λ such that there is a $\boldsymbol{\Delta}_2$ function with domain λ and range unbounded in β.

1.8 Lemma. $\boldsymbol{\Delta}_2 - \mathrm{cf}(\alpha) = \boldsymbol{\Delta}_2 - \mathrm{cf}(\alpha^*)$.

Proof. Let $f : \alpha \to \alpha^*$ be a one-one α-recursive function. Note that $\gamma \cap \mathrm{rng}(f)$ is α-finite for each $\gamma < \alpha^*$ since $\mathrm{rng}(f)$ is α-r.e. Hence $\{\sigma \mid f(\sigma) < \gamma\}$ is bounded for each $\gamma < \alpha^*$. Hence $f''X$ is unbounded in α^* whenever X is unbounded in α.

Suppose $\lambda \leq \alpha$ and $g : \lambda \to \alpha$ is $\boldsymbol{\Delta}_2$ and $\mathrm{rng}(g)$ is unbounded in α. Then $fg : \lambda \to \alpha^*$ is $\boldsymbol{\Delta}_2$ and by the previous paragraph $\mathrm{rng}(f)$ is unbounded in α^*. So $\boldsymbol{\Delta}_2 - \mathrm{cf}(\alpha) \geq \boldsymbol{\Delta}_2 - \mathrm{cf}(\alpha^*)$.

For the converse suppose $h : \lambda \to \alpha^*$ is $\boldsymbol{\Delta}_2$ and $\mathrm{rng}(h)$ unbounded in α^*. Define $k : \lambda \to \alpha$ by

$$k(\nu) = \bigcup \{\sigma \mid f(\sigma) < h(\nu)\} .$$

Using 1.7 it is easy to see that k is $\boldsymbol{\Delta}_2$. Furthermore $\mathrm{rng}(k)$ is obviously unbounded in α. So $\boldsymbol{\Delta}_2 - \mathrm{cf}(\alpha) \leq \boldsymbol{\Delta}_2 - \mathrm{cf}(\alpha^*)$. ∎

If $\langle I_\mu \mid \mu < \nu \rangle$ is an α-finite sequence of (canonical indices for) α-finite sets, then of course $\bigcup \{I_\mu \mid \mu < \nu\}$ is α-finite. The following trivial lemma provides another useful sufficient condition for the union of a sequence of α-finite sets to be α-finite.

1.9 Lemma. *Suppose* $\nu < \boldsymbol{\Delta}_2 - \mathrm{cf}(\alpha)$ *and let* $\langle I_\mu \mid \mu < \nu \rangle$ *be a simultaneously α-r.e. sequence of α-finite sets. Then* $\bigcup \{I_\mu \mid \mu < \nu\}$ *is α-finite.*

Corollary. *Put* $\lambda = \mathbf{\Delta}_2 - \mathrm{cf}(\alpha)$ *and suppose* $f : \lambda \to \alpha$ *is* $\mathbf{\Delta}_2$. *Then*

(i) $\langle f(\mu) \mid \mu < \nu \rangle$ *is α-finite for each* $\nu < \lambda$;

(ii) *the sequence* $\langle\langle f(\mu) \mid \mu < \nu \rangle \mid \nu < \lambda \rangle$ *is* $\mathbf{\Delta}_2$.

Proof. Let $f : \alpha \times \lambda \to \alpha$ be an α-recursive function such that $f(\nu) = \lim_\sigma f(\sigma, \nu)$ for all $\nu < \lambda$. Such an f exists by 1.7. Let us say that $f(\nu)$ *changes value at stage* σ if $(\forall \sigma' < \sigma)(\exists \tau)(\sigma' \leq \tau < \sigma \; \& \; f(\tau, \nu) \neq f(\sigma, \nu))$. Let I_ν be the set of all σ such that $f(\nu)$ changes value at stage σ. Then each I_ν is α-finite, and the sequence $\langle I_\nu | \nu < \lambda \rangle$ is simultaneously α-recursive. Conclusions (i) and (ii) are now immediate from 1.9. ∎

2. The α-jump operator [8]

Friedberg determined the range of the jump operator in ordinary degree theory by proving the following theorem: for every ω-degree $\boldsymbol{b} \geq \boldsymbol{0}'$ there is an ω-degree $\boldsymbol{a}$ such that $\boldsymbol{a}' = \boldsymbol{a} \cup \boldsymbol{0}' = \boldsymbol{b}$. (For a proof see Rogers [10: p. 265].) We now generalize Friedberg's theorem and its proof to α-degree theory as follows:

2.1 Theorem. *Let* $\boldsymbol{b}$ *be an α-degree* $\geq \boldsymbol{0}'$. *Then* $\boldsymbol{b}$ *is regular if and only if there is a regular, hyperregular α-degree* $\boldsymbol{a}$ *such that*

$$\boldsymbol{a}' = \boldsymbol{a} \cup \boldsymbol{0}' = \boldsymbol{b} \, .$$

Proof. Theorem 1.5 implies that the α-degree $\boldsymbol{0}'$ is regular. (In fact, it can be shown that $\boldsymbol{a}'$ is regular whenever $\boldsymbol{a}$ is regular and hyperregular.) The "if" part of 2.1 follows immediately.

For the "only if" part we shall employ a forcing construction. This in itself is not surprising since Friedberg's original proof may be viewed as a forcing argument. However, unlike Friedberg or Cohen, we shall do our forcing over a possibly uncountable ground model, L_α.

A *condition* is an α-finite sequence of 0's and 1's. We use $p, q, \ldots$ as variables ranging over conditions. Thus p is an α-finite function from an ordinal $\mathrm{lh}(p)$,

[8] The main result of Section 2 was first proved in August 1971. It was presented in an invited address to this Symposium.

the *length* of p, into $\{0, 1\}$. We say q *extends* p if $p \subseteq q$. Let Cond be the set of conditions. A set $D \subseteq$ Cond is *dense* if every $p \in$ Cond is extended by some $q \in D$.

For $X \subseteq \alpha$ we denote by c_X the characteristic function of X. Conditions are thought of as α-finite approximations to c_X. If D is a set of conditions, we say X *meets* D if $c_X \upharpoonright \gamma \in D$ for some $\gamma < \alpha$.

Our notion of *forcing* is defined as follows [9]:

(i) $p \Vdash [\epsilon](\gamma) = \delta$
iff $(\exists \xi)(\exists \eta)[\langle \gamma, \delta, \xi, \eta \rangle \in R_\epsilon^{\mathrm{lh}(p)}\ \&\ p''K_\xi \subseteq \{1\}\ \&\ p''K_\eta \subseteq \{0\}]$;
(ii) p *decides* $[\epsilon](\gamma)$
iff $p \Vdash [\epsilon](\gamma) = \delta$ for some δ;
(iii) $p \Vdash ([\epsilon](\gamma)$ is undefined)
iff no $q \supseteq p$ decides $[\epsilon](\gamma)$;
(iv) p *determines* $\langle [\epsilon](\gamma) \mid \gamma < \beta \rangle$
iff either p decides $[\epsilon](\gamma)$ for all $\gamma < \beta$ or $p \Vdash ([\epsilon](\gamma_0)$ is undefined) where γ_0 is the least γ such that p does not decide $[\epsilon](\gamma)$.

Note that "p decides $[\epsilon](\gamma)$" is α-recursive as a relation of p, ϵ, γ. We now define certain important sets of conditions. The definition splits into cases depending on the nature of α.

Definition. *Case I.* α^* is a regular α-cardinal. Then for each $\beta < \alpha^*$ put

$$D_\beta = \{p \in \mathrm{Cond} \mid p \text{ determines } \langle [\epsilon](\gamma) \mid \gamma < \alpha^* \rangle \text{ for each } \epsilon < \beta\}.$$

Case II. Otherwise, i.e. $\alpha^* = \alpha$ or α^* is a singular α-cardinal. Then for each $\beta < \alpha^*$ put

$$D_\beta = \{p \in \mathrm{Cond} \mid p \text{ determines } \langle [\epsilon](\gamma) \mid \gamma < \beta \rangle \text{ for each } \epsilon < \beta\}.$$

2.2 Lemma. *For each* $\beta < \alpha^*$, D_β *is dense.*

Proof. The proof splits into cases following the definition of D_β.

[9] A more conventional way of writing $c_B \upharpoonright \sigma \Vdash [\epsilon](\gamma) = \delta$ is: $[\epsilon]_\sigma^B(\gamma) = \delta$.

Case I. Define an α-recursive function

$$m : \text{Cond} \times \alpha^* \to \alpha^* + 1$$

by $m(p, \epsilon)$ = the least $\gamma < \alpha^*$ such that p does not decide $[\epsilon](\gamma)$; $m(p, \epsilon) = \alpha^*$ if no such γ exists. Fix $\beta < \alpha^*$ and define an α-recursive partial function θ from Cond into Cond by $\theta(p) \simeq$ the least q (in the canonical α-recursive well ordering of Cond) such that $p \subseteq q$ and $m(p, \epsilon) < m(q, \epsilon)$ for some $\epsilon < \beta$. Then, given a condition p, define an α-recursive increasing sequence of conditions by $q_0 = p$; $q_\eta = \bigcup \{q_\xi \mid \xi < \eta\}$ if η is a limit ordinal; $q_{\xi+1} = \theta(q_\xi)$ if $\theta(q_\xi)$ is defined; ξ_0 = the least ξ such that $\theta(q_\xi)$ is undefined.

We claim that $\xi_0 < \alpha$. Suppose to the contrary that $\xi_0 = \alpha$. For each $\epsilon < \beta$ define

$$I_\epsilon = \{\xi \mid m(q_\xi, \epsilon) < m(q_{\xi+1}, \epsilon)\} .$$

Then the I_ϵ's are simultaneously α-recursive and $\alpha = \bigcup \{I_\epsilon \mid \epsilon < \beta\}$. Furthermore $m(q_0, \epsilon) \leq \ldots \leq m(q_\xi, \epsilon) \leq m(q_{\xi+1}) \leq \ldots \leq \alpha^*$ so each I_ϵ has order type less than or equal to α^*. Let H be the set of $\epsilon < \beta$ such that I_ϵ has order type α^*. Then H is an α-r.e. subset of $\beta < \alpha^*$. Hence H is α-finite. It follows at once that $\bigcup \{I_\epsilon \mid \epsilon \in H\}$ is α-finite. Let $\gamma < \alpha$ be an upper bound for $\bigcup \{I_\epsilon \mid \epsilon \in H\}$. Let $J_\epsilon = I_\epsilon \cap \{\delta \mid \gamma \leq \delta < \gamma + \alpha^*\}$. Then $\langle J_\epsilon \mid \epsilon < \beta \rangle$ is an α-finite sequence of α-finite sets each having order type less than α^*. Furthermore $\bigcup \{J_\epsilon \mid \epsilon < \beta\} = \{\delta \mid \gamma \leq \delta < \gamma + \alpha^*\}$. This contradicts the Case assumption that α^* is a regular α-cardinal. The claim is proved.

Thus q_{ξ_0} is a condition extending p. For each $\epsilon < \beta$ it is clear from the construction that q_{ξ_0} decides $[\epsilon](\gamma)$ for each $\gamma < m(q_{\xi_0}, \epsilon)$; furthermore if $m(q_{\xi_0}, \epsilon) < \alpha^*$ then $q_{\xi_0} \Vdash ([\epsilon](m(q_{\xi_0}, \epsilon))$ undefined). Thus $q_{\xi_0} \in D_\beta$. Since p is arbitrary, D_β is dense.

Case II. Fix $\beta < \alpha^*$ and define an α-recursive function

$$m : \text{Cond} \times \alpha^* \to \beta + 1$$

by $m(p, \epsilon)$ = the least $\gamma < \beta$ such that p does not decide $[\epsilon](\gamma)$; $m(p, \epsilon) = \beta$ if no such γ exists. Now define θ and the sequence $\langle q_\xi \mid \xi \leq \xi_0 \rangle$ as in Case I.

Again we claim $\xi_0 < \alpha$. Suppose to the contrary that $\xi_0 = \alpha$. Define I_ϵ, $\epsilon < \beta$, as in Case I. Again the I_ϵ's are simultaneously α-recursive and

$\alpha = \bigcup \{I_\epsilon | \epsilon < \beta\}$. Furthermore $m(q_0, \epsilon) \leq \ldots \leq m(q_\xi, \epsilon) \leq m(q_{\xi+1}, \epsilon) \leq \ldots \leq \beta$ so each I_ϵ has order type $\leq \beta$. Hence there is an α-recursive partial function i from a subset of $\beta \times \beta$ onto α (namely $i(\xi, \epsilon) \simeq$ the ξth member of I_ϵ). This is impossible since $\beta < \alpha^*$.

Thus q_{ξ_0} is a condition extending p. As in Case I we note that $q_{\xi_0} \in D_\beta$ hence D_β is dense. ∎

Remark. From the above proof for Case I we can extract the following combinatorial lemma:

Assume that $\beta < \alpha^*$ and that α^* is a regular α-cardinal. Let $\langle I_\nu | \nu < \beta \rangle$ be a simultaneously α-recursive sequence of α-finite sets each of order type less than or equal to α^*. Then $\bigcup\{I_\nu | \nu < \beta\}$ is α-finite.

One can extract a similar but weaker combinatorial lemma from the proof for Case II.

Corollary. *There is a Δ_2 function $q : \alpha^* \times \text{Cond} \to \text{Cond}$ such that for all $p \in \text{Cond}$ and $\beta < \alpha^*$,*

$$p \subseteq q(\beta, p) \in D_\beta .$$

Proof. Define $q(\beta, p) = q_{\xi_0}$ where q_{ξ_0} is as in the proof of Lemma 2.2. Using Lemma 1.7 it is easy to see that this q is a Δ_2 function. ∎

A set $A \subseteq \alpha$ is said to be *generic* if A meets D_β for each $\beta < \alpha^*$. The following Lemma and its proof show that our notion of genericity is not so special as it appears. (It can be shown that $A \subseteq \alpha$ is generic if and only if $\langle L_\alpha, A \rangle$ is admissible and every Σ_2 sentence true in $\langle L_\alpha, A \rangle$ is forced by some condition $p \subseteq c_A$. However, this remark does not simplify the proof of Theorem 2.1.)

2.3 Lemma. *Let $A \subseteq \alpha$ be a generic set. Then A is α-regular and α-hyperregular, and*

$$A' \equiv_\alpha A \oplus 0' .$$

Proof. For each $\gamma < \alpha$ one can find $\epsilon < \alpha^*$ such that p decides $[\epsilon](0)$ if and only if $\text{lh}(p) \geq \gamma$. Hence for each $\gamma < \alpha$ there is $\beta < \alpha^*$ such that $\text{lh}(p) \geq \gamma$ for all $p \in D_\beta$. It follows that A is α-regular.

Suppose A is not α-hyperregular. Let κ be the least ordinal such that for some $\epsilon < \alpha^*$, $\kappa \subseteq \mathrm{dom}([\epsilon]^A)$ and $\{[\epsilon]^A(\gamma) \mid \gamma < \kappa\}$ is unbounded. Obviously $\kappa < \alpha$ since A is not α-hyperregular. It is also clear that κ is a limit ordinal.

We claim that κ is a regular α-cardinal. Suppose not and let $\langle \kappa_\mu \mid \mu < \pi\rangle$ be an α-finite sequence such that $\pi < \kappa = \bigcup\{\kappa_\mu \mid \mu < \pi\}$. For each $\mu < \pi$ let $h(\mu)$ be the least σ such that for all $\gamma < \kappa_\mu$ there exists $\langle \xi, \eta, \gamma, \delta\rangle \in R^\sigma_\epsilon$ with $K_\xi \subseteq A$ and $K_\eta \cap A = \emptyset$. Clearly $h(\mu)$ is less than α since $\kappa_\mu < \kappa$. Also $h \leq_{w\alpha} A$ since A is α-regular. Hence $\{h(\mu) \mid \mu < \pi\}$ is bounded since $\pi < \kappa$. It follows that $\{[\epsilon]^A(\gamma) \mid \gamma < \kappa\}$ is bounded, a contradiction. The claim is proved.

We now split into Cases as in the definition of the D_β's. In Case I we have $\kappa \leq \alpha^*$ and we set $\beta = \epsilon + 1$. In Case II we have $\kappa < \alpha^*$ and we set $\beta = \max\{\kappa, \epsilon + 1\}$. In either case A is generic so let $p \in D_\beta$ be such that $p \subseteq c_A$. Then for each $\gamma < \kappa$ it is clear that $[\epsilon]^A(\gamma)$ is the unique δ such that $p \Vdash [\epsilon](\gamma) = \delta$. Hence $\{[\epsilon]^A(\gamma) \mid \gamma < \kappa\}$ is α-finite, a contradiction. So A is α-hyperregular.

Obviously $A \oplus 0' \leq_\alpha A'$. That $A' \leq_\alpha A \oplus 0'$ follows from the existence of α-recursive functions $f, g : \alpha \to \alpha^*$ such that for all γ, δ

$$K_\gamma \subseteq A' \leftrightarrow (\exists p \subseteq c_A)[p \Vdash [f(\gamma)](0) = 0]$$

and

$$K_\delta \cap A' = \emptyset \leftrightarrow (\exists p \subseteq c_A)[p \Vdash ([g(\delta)](0) \text{ undefined})] .$$

The existence of f is a consequence (via 1.6) of the fact that A is α-regular and α-hyperregular. The existence of g follows directly from the genericity of A. ■

We are now ready to complete the proof of Theorem 2.1. Let B be an α-regular set. Thus, for each $\gamma < \alpha$, $c_B \restriction \gamma$ is a condition. If p and q are conditions we write

$$p * q = p \cup \{\langle q(\xi), \mathrm{lh}(p) + \xi\rangle \mid \xi < \mathrm{lh}(q)\}$$

i.e. $p * q$ is the concatenation of p and q. Let $\lambda = \mathbf{\Delta}_2 - \mathrm{cf}(\alpha)$. Recall Lemmas 1.7, 1.8, and 1.9 which state several useful properties of λ. Let F (resp. G) be a $\mathbf{\Delta}_2$ function from λ onto an unbounded subset of α (resp. α^*). Define

$$p_\nu = \bigcup\{p_\mu \mid \mu < \nu\} \quad \text{if} \quad \nu = \bigcup \nu ;$$

$$p_{2\nu+1} = p_{2\nu} * c_B \restriction F(\nu) ;$$

$$p_{2\nu+2} = q(G(\nu), p_{2\nu+1})$$

where q is the $\boldsymbol{\Delta}_2$ function mentioned in the Corollary to Lemma 2.2.

For each $\nu < \lambda$, $\{F(\mu) \mid \mu < \nu\}$ is bounded below α. Hence only α-finitely much of B is used in the definition of $\langle p_\mu \mid \mu < 2\nu \rangle$. Hence, by 1.7 and 1.9, $\langle p_\mu \mid \mu < 2\nu \rangle$ is α-finite. Thus $\langle p_\nu \mid \nu < \lambda \rangle$ is an increasing sequence of conditions. Moreover $\bigcup \{\mathrm{lh}(p_\nu) \mid \nu < \lambda\} = \alpha$ since $F(\nu) \leq \mathrm{lh}(p_{2\nu+1})$. Hence there is a unique set $A \subseteq \alpha$ such that $c_A = \bigcup \{p_\nu \mid \nu < \lambda\}$. For each $\nu < \lambda$ we have $p_{2\nu+2} \in D_{G(\nu)}$; hence A is generic. It follows by Lemma 2.3 that A is α-regular and α-hyperregular, and that

$$A' \equiv_\alpha A \oplus 0' .$$

It remains to show that

$$A \oplus 0' \equiv_\alpha B \oplus 0' .$$

This is an immediate consequence of the following claim. For each $\nu < \lambda$ let $P(\nu)$ be (a canonical index for) $\langle p_\mu \mid \mu < 2\nu \rangle$.

Claim: P is weakly α-recursive in $B \oplus 0'$ and in $A \oplus 0'$.

We shall give an informal, "Church's thesis" type of argument for the claim.

To compute $P(\nu)$ from $B \oplus 0'$. First compute $\gamma = \bigcup \{F(\mu) \mid \mu < \nu\}$. By Lemmas 1.7 and 1.9 this computation uses only an α-finite amount of information about $0'$. Then use $0'$ and the α-finite sequence $c_B \restriction \gamma$ to compute $p_0, p_1, \ldots, p_\mu, \ldots$ $(\mu < 2\nu)$ in order. Again by 1.7 and 1.9 this uses only α-finitely much information about $0'$.

To compute $P(\nu)$ from $A \oplus 0'$. First, compute $\langle F(\mu) \mid \mu < \nu \rangle$ and $\langle G(\mu) \mid \mu < \nu \rangle$ as above. Next, start generating $\boldsymbol{\Pi}_\nu$, the set of all (canonical indices for) α-finite sequences of conditions $\langle p'_\mu \mid \mu < 2\nu \rangle$ such that $\bigcup \{p'_\mu \mid \mu < 2\nu\} \subseteq c_A$ and

$$\mathrm{lh}(p'_{2\mu+1}) = \mathrm{lh}(p'_{2\mu}) + F(\mu)$$

for all $\mu < \nu$. Clearly $\mathbf{\Pi}_\nu$ is α-recursive in A. As you generate $\mathbf{\Pi}_\nu$, examine simultaneously all the members of $\mathbf{\Pi}_\nu$ trying to find one with the property that

$$p'_{2\mu+2} = q(G(\mu), p'_{2\mu+1})$$

for all $\mu < \nu$. This involves computing pieces of the $\mathbf{\Delta}_2$ function q. Do this α-recursively in $0'$, taking care to dovetail the examination of the various members of $\mathbf{\Pi}_\nu$ so that you don't spend too much time on any α-finite subset of $\mathbf{\Pi}_\nu$. (By Lemma 1.9 the examination of any single member of $\mathbf{\Pi}_\nu$ will eventually terminate having used only α-finitely such information about $0'$; however, there is no guarantee that the examination of an α-finite subset of $\mathbf{\Pi}_\nu$ will so terminate.) Eventually you find a sequence $\langle p'_\mu \mid \mu < 2\nu \rangle$ in $\mathbf{\Pi}_\nu$ with the desired property. Then $P(\nu) = \langle p'_\mu \mid \mu < 2\nu \rangle$. More importantly, you have used only α-finitely such information about $A \oplus 0'$.

We have informally described computation procedures showing that $P \leq_{w\alpha} B \oplus 0'$ and $P \leq_{w\alpha} A \oplus 0'$. The proof of Theorem 2.1 is complete. ■

One can use the same method to prove the following generalization of a theorem of Spector (see Rogers [10: p. 267]):

2.4 Theorem. *There exist regular, hyperregular α-degrees $\boldsymbol{a}$, $\boldsymbol{b}$ such that $\boldsymbol{a} \cup \boldsymbol{b} = \boldsymbol{a}' = \boldsymbol{b}' = \boldsymbol{0}'$.*

3. Upper segments of regular α-degrees

An interesting corollary of Friedberg's jump theorem is this: there is an ω-degree $\boldsymbol{d}$ such that every ω-degree $\geq \boldsymbol{d}$ is the jump of some ω-degree (namely $\boldsymbol{d} = \boldsymbol{0}'$). We now ask: for which admissible α does there exist an α-degree $\boldsymbol{d}$ such that every α-degree $\geq \boldsymbol{d}$ is the jump of some regular, hyperregular α-degree? By Theorem 2.1 this is equivalent to asking: for which admissible α does there exist an α-degree $\boldsymbol{d}$ such that every α-degree $\geq \boldsymbol{d}$ is regular?

We write $|\alpha|$ for the (von Neumann) *cardinality* of α i.e. the least β such that there exists a one-one correspondence between α and β. We write $\mathrm{cf}(\alpha)$

for the *cofinality* of α, i.e. the least λ such that there is a function from λ onto an unbounded subset of α.

3.1 Theorem. *Assume the Generalized Continuum Hypothesis. Let α be an admissible ordinal such that $|\alpha|$ is a regular cardinal. Then the following assertions about α are pairwise equivalent.*

(i) *There is an α-degree $\mathbf{d}$ such that every α-degree $\geq \mathbf{d}$ is regular.*

(ii) *To every α-degree $\mathbf{a}$ there is a regular α-degree $\mathbf{b} \geq \mathbf{a}$.*

(iii) *Every $X \subseteq \alpha$ of cardinality less than $|\alpha|$ is α-finite.*

Before proving 3.1 we give two lemmas in which $|\alpha|$ is not assumed to be regular.

3.2 Lemma. *Let α be an admissible ordinal such that Assertion* 3.1 (ii) *holds. Then every $X \subseteq \alpha$ of cardinality less than* $\mathrm{cf}(\alpha)$ *is α-finite.*

Proof. Let $X \subseteq \alpha$ have cardinality $< \mathrm{cf}(\alpha)$. By 3.1 (ii) there is an α-regular set B such that $c_X \leq_{w\alpha} B$. Let $\epsilon < \alpha^*$ be such that $c_X = [\epsilon]^B$. For each $\gamma \in X$ let $h(\gamma)$ be the least σ such that [9]

$$c_B \upharpoonright \sigma \Vdash [\epsilon](\gamma) = 1 .$$

Since $|X| < \mathrm{cf}(\alpha)$, $\{h(\gamma) | \gamma \in X\}$ must be bounded below α, say by τ. Then for all γ we must have $\gamma \in X$ iff $c_B \upharpoonright \tau \Vdash [\epsilon](\gamma) = 1$. Hence X is α-finite. ∎

3.3 Lemma. *Assume the Generalized Continuum Hypothesis. Let α be an admissible ordinal such that Assertion* 3.1 (ii) *holds. Then* $\mathrm{cf}(\alpha) = \mathrm{cf}(|\alpha|)$.

Proof. The hypothesis 3.1 (ii) clearly implies that the set of regular α-degrees has cardinality $2^{|\alpha|}$. On the other hand, the set of regular α-degrees clearly has cardinality at most $|\alpha|^{\mathrm{cf}(\alpha)}$. If $\mathrm{cf}(\alpha) < \mathrm{cf}(|\alpha|)$ then by the GCH we would have $2^{|\alpha|} \leq |\alpha|^{\mathrm{cf}(\alpha)} = |\alpha|$ contradicting Cantor's theorem of cardinal arithmetic. So $\mathrm{cf}(\alpha) \geq \mathrm{cf}(|\alpha|)$.

For the other direction recall König's theorem of cardinal arithmetic which

[9] A more conventional way of writing $c_B \upharpoonright \sigma \Vdash [\epsilon](\gamma) = \delta$ is: $[\epsilon]^B_\sigma(\gamma) = \delta$.

says $|\alpha|^{\mathrm{cf}(|\alpha|)} > |\alpha|$. Hence there is a non-$\alpha$-finite set $X \subseteq |\alpha|$ of cardinality $\mathrm{cf}(|\alpha|)$. By Lemma 3.2 X cannot have cardinality less than $\mathrm{cf}(\alpha)$. So $\mathrm{cf}(|\alpha|) \geq \mathrm{cf}(\alpha)$. ■

Proof of 3.1. The implication (i) ⇒ (ii) is trivial since any two α-degrees have an upper bound. The implication (ii) ⇒ (iii) is a special case of Lemma 3.2 in view of Lemma 3.3.

To prove (iii) ⇒ (i) first note that (iii) implies $\mathrm{cf}(\alpha) = |\alpha|$. Let $\langle \gamma_\nu | \nu < |\alpha| \rangle$ be an increasing sequence of ordinals such that $\alpha = \bigcup \{\gamma_\nu | \nu < |\alpha|\}$. Let g be a function from $|\alpha|$ onto α. Define

$$D = \{\langle g(\nu), \gamma_\nu \rangle | \nu < |\alpha|\}$$

and let $\boldsymbol{d}$ be the α-degree of D. Given $B \subseteq \alpha$ define

$$B^* = \{2\gamma_\nu | \nu < |\alpha| \ \& \ K_{g(\nu)} \subseteq B\}$$

$$\cup \{2\gamma_\nu + 1 | \nu < |\alpha| \ \& \ K_{g(\nu)} \cap B = \emptyset\} .$$

Then D and B^* are α-regular in view of (iii). Furthermore it is easy to see that

$$B \oplus D \equiv_\alpha B^* \oplus D .$$

Thus every α-degree $\geq \boldsymbol{d}$ is regular. ■

Corollary. *Let α be a countable admissible ordinal. Then there is an α-degree $\boldsymbol{d}$ such that every α-degree $\geq \boldsymbol{d}$ is regular.*

Remarks. 1. It is not hard to construct (in ZFC) an admissible ordinal α such that $|\alpha| = \mathrm{cf}(\alpha) = \omega_1$ but not every constructibly countable $X \subseteq \alpha$ is α-finite. Such an α falsifies some attractive conjectures.

2. If $|\alpha|$ is not regular then we conjecture that the equivalence 3.1 (i) ⟺ 3.1 (ii) still holds. We can prove in ZF + V = L [10] that 3.1 (ii) is equivalent to

[10] The set-theoretical hypothesis V = L can be expressed in α-recursion theoretic language as follows: for every infinite cardinal κ, every subset of κ bounded below κ is κ-finite.

the assertion that there is a function $f : |\alpha| \overset{\text{onto}}{\rightarrow} \alpha$ such that $f \restriction \gamma$ is α-finite for each $\gamma < |\alpha|$.

4. Incomparable α-r.e. degrees

In this section we generalize to α-recursion theory a strong form of the Friedberg–Muchnik theorem.

4.1 Theorem. *Let α be an admissible ordinal. Then there are α-r.e. degrees $\boldsymbol{a}, \boldsymbol{b}$ such that*

(i) $\boldsymbol{a} \not\leq \boldsymbol{b}$ *and* $\boldsymbol{b} \not\leq \boldsymbol{a}$;

(ii) $\boldsymbol{a} \cup \boldsymbol{b}$ *is hyperregular*;

(iii) $(\boldsymbol{a} \cup \boldsymbol{b})' = \boldsymbol{0}'$.

(*By Theorem* 1.5 $\boldsymbol{a} \cup \boldsymbol{b}$ *is necessarily regular*.)

In the proof we shall define α-recursive functions $A^\sigma, B^\sigma, f(\sigma, \epsilon)$, and $g(\sigma, \epsilon)$ $(\epsilon < \alpha^*)$. Here $\langle A^\sigma | \sigma < \alpha \rangle$ and $\langle B^\sigma | \sigma < \alpha \rangle$ will be nondecreasing sequences of α-finite sets. We shall then set $A = \bigcup \{A^\sigma | \sigma < \alpha\}$ and $B = \bigcup \{B^\sigma | \sigma < \alpha\}$. Thus A and B will be α-r.e. For each $\epsilon < \alpha^*$ $f(\sigma, \epsilon)$ and $g(\sigma, \epsilon)$ will be nondecreasing as functions of σ. It will turn out that $f(\epsilon) = \lim_\sigma f(\sigma, \epsilon)$ and $g(\epsilon) = \lim_\sigma g(\sigma, \epsilon)$ are α-finite. Thus by 1.7 $f, g : \alpha^* \to \alpha$ will be $\boldsymbol{\Delta}_2$ functions. The construction will be designed to insure that

$$\langle f(\epsilon), \epsilon \rangle \in A \leftrightarrow \exists \eta [\langle f(\epsilon), \epsilon, \eta \rangle \in R_\epsilon \;\&\; K_\eta \cap B = \emptyset]$$

and

$$\langle g(\epsilon), \epsilon \rangle \in B \leftrightarrow \exists \eta [\langle g(\epsilon), \epsilon, \eta \rangle \in R_\epsilon \;\&\; K_\eta \cap A = \emptyset]$$

for each $\epsilon < \alpha^*$. From this and the α-recursive enumerability of A and B it will follow that $A \not\leq_\alpha B$ and $B \not\leq_\alpha A$.

At the same time we shall define a certain auxiliary α-recursive function $p(\sigma, \epsilon)$ whose purpose will be to preserve certain computations so as to make $A \oplus B$ α-regular and α-hyperregular and $(A \oplus B)' \equiv_\alpha 0'$. The definition of $p(\sigma, \epsilon)$ will closely parallel the key definitions in Section 2 whose purpose was likewise to make a certain set α-regular and α-hyperregular. (Thus for instance there will be a split into Cases I and II exactly as in Section 2.)

Remark. The construction and proof below are arranged so that the reader not interested in conclusions 4.1 (ii) and 4.1 (iii) can skip certain parts. Specifically, such a reader should (a) ignore the functions $y(\sigma,\gamma,\epsilon)$, $m(\sigma,\epsilon)$, and $p(\sigma,\epsilon)$; (b) replace the below definitions of $f(<\sigma,\epsilon)$ and $g(<\sigma,\epsilon)$ by

$$f(<\sigma,\epsilon) = \bigcup\{f(\tau,\epsilon) \mid \tau < \sigma\}$$

and

$$g(<\sigma,\epsilon) = \bigcup\{g(\tau,\epsilon) \mid \tau < \sigma\}\,;$$

(c) regard Lemma 4.6 as an immediate consequence of Lemma 4.2; (d) disregard entirely Lemmas 4.3, 4.4, and 4.5.

Before describing the construction we must define some important functions and parameters which do not depend on the construction.

(1) Let λ be the Δ_2 cofinality of α. By Lemma 1.8 λ is equal to the Δ_2 cofinality of α^* so let $G : \lambda \to \alpha^*$ be a Δ_2 function whose range is unbounded in α^*. For each $\nu < \lambda$ define $H(\nu) = \bigcup\{G(\mu) \mid \mu < \nu\}$. Thus we have

$$0 = H(0) \leq \ldots \leq H(\nu) \leq H(\nu+1) \leq \ldots < \alpha^*$$

and each $\epsilon < \alpha^*$ satisfies $H(\nu) \leq \epsilon < H(\nu+1)$ for a unique $\nu < \lambda$.

Remark. Thus H partitions α^* into λ "blocks" each of size less than α^*. The use of H was suggested by R. Shore and has the effect of making the present proof somewhat simpler than that of Sacks–Simpson [13]. Shore has also put this "blocking" idea to good use in his own work (see Section 5).

(2) Let $G : \alpha \times \lambda \to \alpha^*$ be an α-recursive function such that $G(\nu) = \lim_\sigma G(\sigma,\nu)$ for all $\nu < \lambda$. (Such a G exists by Lemma 1.7.) Define $H(\sigma,\nu) = \bigcup\{G(\sigma,\mu) \mid \mu < \nu\}$. Then $H(\sigma,\nu)$ is α-recursive and $H(\nu) = \lim_\sigma H(\sigma,\nu)$; in fact, by Lemma 1.9, we have

$$(\forall\nu < \lambda)(\exists\sigma)(\forall\tau \geq \sigma)(\forall\mu \leq \nu)\, H(\tau,\mu) = H(\mu)\,.$$

Furthermore, for each $\sigma < \alpha$ we have

$$0 = H(\sigma,0) \leq \ldots \leq H(\sigma,\nu) \leq H(\sigma,\nu+1) \leq \ldots < \alpha^*\,.$$

(3) Let $L:\alpha \to \lambda$ be an α-recursive function such that $L^{-1}(\{\nu\})$ is unbounded in α for each $\nu < \lambda$.

We can now present the construction. As a preliminary to stage σ of the construction we set

$$A^{<\sigma} = \bigcup\{A^{\tau} \mid \tau < \sigma\} ;$$

$$B^{<\sigma} = \bigcup\{B^{\tau} \mid \tau < \sigma\} ;$$

$$C^{<\sigma} = A^{<\sigma} \oplus B^{<\sigma} ;$$

$$y(\sigma,\gamma,\epsilon) = (\mu\tau)_{\tau<\sigma}(\exists\eta)(\exists\delta)[\langle\gamma,\delta,\eta\rangle \in R^{\tau}_{\epsilon} \ \&\ K_{\eta} \cap C^{<\sigma} = \emptyset] ;$$

$$m(\sigma,\epsilon) = \begin{cases} (\mu\gamma)_{\gamma<\alpha^*}\, y(\sigma,\gamma,\epsilon) = \sigma & \text{if } \alpha^* \text{ is a regular } \alpha\text{-cardinal,} \\ (\mu\gamma)_{\gamma<\epsilon}\, y(\sigma,\gamma,\epsilon) = \sigma & \text{otherwise ;} \end{cases}$$

$$p(\sigma,\epsilon) = \bigcup\{y(\sigma,\gamma,\epsilon') \mid \epsilon' < \epsilon \ \&\ \gamma < m(\sigma,\epsilon')\} ;$$

$$f(<\sigma,\epsilon) = \begin{cases} \varphi & \text{if } \langle\varphi,\epsilon\rangle \in A^{<\sigma}, \\ \max\{\varphi, p(\sigma,\epsilon)\} & \text{otherwise} \end{cases}$$

where

$$\varphi = \bigcup\{f(\tau,\epsilon) \mid \tau < \sigma\} ;$$

$$g(<\sigma,\epsilon) = \begin{cases} \psi & \text{if } \langle\psi,\epsilon\rangle \in B^{<\sigma} \\ \max\{\psi, p(\sigma,\epsilon)\} & \text{otherwise} \end{cases}$$

where

$$\psi = \bigcup\{g(\tau,\epsilon) \mid \tau < \sigma\} .$$

Note: In the definitions of $y(\sigma,\gamma,\epsilon)$ and $m(\sigma,\epsilon)$ the usual convention concerning the *bounded* μ*-operator* is observed. Namely, $(\mu x)_{x<y}$ ... is defined

to be the least x less than y such that ... if such an x exists and y otherwise.

Stage σ of the construction is now described as follows. Put $\nu = L(\sigma)$ where L is as in (3) above. Define

$$A^\sigma = A^{<\sigma} \cup \{\langle f(<\sigma,\epsilon),\epsilon\rangle \mid H(\sigma,\nu) \leq \epsilon < H(\sigma,\nu+1)\ \&\ (4)\}$$

where

$$(4) \qquad \exists\eta[\langle f(<\sigma,\epsilon),\epsilon,\eta\rangle \in R^\sigma_\epsilon\ \&\ K_\eta \cap B^{<\sigma} = \emptyset]\ .$$

Define

$$g(\sigma,\epsilon') = \begin{cases} \sigma+1 & \text{if } \epsilon' \geq H(\sigma,\nu)\ \&\ A^\sigma \neq A^{<\sigma}, \\ \sigma+1 & \text{if } \epsilon' \geq H(\sigma,\mu) \text{ for some } \mu \text{ such that} \\ & \quad H(\mu) \text{ changes value at stage } \sigma, \\ g(<\sigma,\epsilon') & \text{otherwise.} \end{cases}$$

Define

$$B^\sigma = B^{<\sigma} \cup \{\langle g(\sigma,\epsilon),\epsilon\rangle \mid H(\sigma,\nu) \leq \epsilon < H(\sigma,\nu+1)\ \&\ (5)\}$$

where

$$(5) \qquad \exists\eta[\langle g(\sigma,\epsilon),\epsilon,\eta\rangle \in R^\sigma_\epsilon\ \&\ K_\eta \cap A^\sigma = \emptyset]\ .$$

Define

$$f(\sigma,\epsilon') = \begin{cases} \sigma+1 & \text{if } \epsilon' \geq H(\sigma,\nu+1)\ \&\ B^\sigma \neq B^{<\sigma}, \\ \sigma+1 & \text{if } \epsilon' \geq H(\sigma,\mu) \text{ for some } \mu \text{ such that} \\ & \quad H(\mu) \text{ changes value at stage } \sigma, \\ f(<\sigma,\epsilon') & \text{otherwise.} \end{cases}$$

This concludes stage σ and completes our description of the construction.

We now prove a series of lemmas leading to the conclusion that the construction works. For each $\nu < \lambda$ put

$$I_\nu = \{\sigma \mid (\exists\epsilon < H(\sigma,\nu+1))(f(\sigma,\epsilon) \neq f(<\sigma,\epsilon) \text{ or } g(\sigma,\epsilon) \neq g(<\sigma,\epsilon))\}\ .$$

Obviously the I_ν's are simultaneously α-recursive. Our first lemma corresponds to the Friedberg–Muchnik observation that in their construction each requirement is injured only finitely often (see Rogers [10: p. 166]).

4.2 Lemma. *Each I_ν is α-finite.*

Proof. By induction on ν. The induction hypothesis tells us that each I_μ, $\mu < \nu$, is α-finite. Since $\nu < \lambda = \Delta_2 - \mathrm{cf}(\alpha)$ it follows by Lemma 1.9 that $\bigcup\{I_\mu \mid \mu < \nu\}$ is α-finite. Let σ_0 be such that $\{I_\mu \mid \mu < \nu\} \subseteq \sigma_0$. Let $\sigma_1 \geq \sigma_0$ be such that no $H(\mu)$, $\mu \leq \nu + 1$, changes value at any stage $\sigma \geq \sigma_1$. For each $\epsilon < H(\nu)$ put

$$\varphi(\epsilon) \simeq \text{the least } \sigma \geq \sigma_1 \text{ such that } \langle g(\sigma, \epsilon), \epsilon \rangle \in B^\sigma .$$

We claim: (6) if $\epsilon < H(\nu)$ and $\varphi(\epsilon)$ is defined then $g(\sigma, \epsilon) = g(\varphi(\epsilon), \epsilon)$ for all $\sigma > \varphi(\epsilon)$.

Claim (6) is proved by induction on $\sigma > \varphi(\epsilon)$. The induction hypothesis tells us that $g(<\sigma, \epsilon) = g(\varphi(\epsilon), \epsilon)$ since $\langle g(\varphi(\epsilon), \epsilon), \epsilon \rangle \in B^{<\sigma}$. But $g(\sigma, \epsilon) = g(<\sigma, \epsilon)$ since $H(\nu) = H(\sigma, \nu)$ and $\sigma \geq \sigma_0$.

Our φ is an α-recursive partial function and $\mathrm{dom}(\varphi) \subseteq H(\nu) < \alpha^*$. Hence φ is α-finite. Let $\sigma_2 \geq \sigma_1$ be such that $\mathrm{rng}(\varphi) \subseteq \sigma_2$.

We now claim: (7) for each $\epsilon < H(\sigma, \nu+1)$ and $\sigma \geq \sigma_2$, $f(\sigma, \epsilon) = f(<\sigma, \epsilon)$.

To prove claim (7), suppose $\sigma \geq \sigma_2$ and $\epsilon' < H(\sigma, \nu+1)$ and $f(\sigma, \epsilon') \neq f(<\sigma, \epsilon')$. Since $\sigma \geq \sigma_1$ we must have $\epsilon' \geq H(\sigma, L(\sigma)+1)$ and $B^\sigma \neq B^{<\sigma}$. Hence $L(\sigma) \leq \nu$ and there must be an ϵ such that $H(\sigma, L(\sigma)) \leq \epsilon < H(\sigma, L(\sigma)+1)$ and $\langle g(\sigma, \epsilon), \epsilon \rangle \in B^\sigma - B^{<\sigma}$. Then $\epsilon < H(\sigma, \nu) = H(\nu)$. Hence $\varphi(\epsilon)$ is defined and $\varphi(\epsilon) \leq \sigma$. But then $g(\sigma, \epsilon) = g(\varphi(\epsilon), \epsilon)$ by claim (6) so in fact $\varphi(\epsilon) = \sigma$. Hence $\sigma < \sigma_2$ a contradiction.

Now for each $\epsilon < H(\nu + 1)$ put

$$\psi(\epsilon) \simeq \text{the least } \sigma \geq \sigma_2 \text{ such that } \langle f(<\sigma, \epsilon), \epsilon \rangle \in A^\sigma .$$

Again ψ is an α-finite function, and we let $\sigma_3 \geq \sigma_2$ be such that $\mathrm{rng}(\psi) \subseteq \sigma_3$. Again we make two claims: (6′) if $\epsilon < H(\nu+1)$ and $\psi(\epsilon)$ is defined then $f(\sigma, \epsilon) = f(<\psi(\epsilon), \epsilon)$ for all $\sigma \geq \psi(\epsilon)$; (7′) for each $\epsilon < H(\sigma, \nu+1)$ and $\sigma \geq \sigma_3$, $g(\sigma, \epsilon) = g(<\sigma, \epsilon)$. The proofs of (6′) and (7′) are similar to the proofs of (6) and (7).

It follows from (7) and (7′) that $I_\nu \subseteq \sigma_3$ hence I_ν is α-finite. ∎

4.3 Lemma. *Let ν and σ_3 be as in the proof of Lemma 4.2. Suppose $\epsilon < H(\nu+1)$ and $\gamma < m(\sigma,\epsilon)$, $\sigma \geq \sigma_3$. Then $y(\sigma,\gamma,\epsilon) = y(\tau,\gamma,\epsilon)$ and $\gamma < m(\tau,\epsilon)$ for all $\tau \geq \sigma$.*

Proof. Let us write $C^\tau = A^\tau \oplus B^\tau$. It suffices to show that $C^\tau \cap y(\sigma,\gamma,\epsilon) = C^{<\sigma} \cap y(\sigma,\gamma,\epsilon)$ for all $\tau \geq \sigma$. Suppose not and consider the least $\tau \geq \sigma$ such that this fails. We have $y(\tau,\gamma,\epsilon) = y(\sigma,\gamma,\epsilon)$ hence $y(\sigma,\gamma,\epsilon) \leq p(\tau,\epsilon'') \leq \min\{f(<\tau,\epsilon''), g(<\tau,\epsilon'')\}$ for all $\epsilon'' > \epsilon$. Hence there must be an $\epsilon' \leq \epsilon$ such that $\langle f(<\tau,\epsilon'),\epsilon'\rangle \in A^\tau - A^{<\tau}$ or $\langle g(\tau,\epsilon'),\epsilon'\rangle \in B^\tau - B^{<\tau}$. This is impossible since $\tau \geq \sigma_3$. ■

4.4 Lemma. *For each $\epsilon < \alpha^*$*

$$\{y(\sigma,\gamma,\epsilon') \mid \tau < \alpha \ \&\ \gamma < m(\sigma,\epsilon') \ \&\ \epsilon' < \epsilon\}$$

is α-finite.

Proof. Let ν and σ_3 be as in Lemma 4.3 where $\epsilon < H(\nu+1)$. Thus, for each $\epsilon' < \epsilon$, $m(\sigma,\epsilon')$ is nondecreasing as a function of $\sigma \geq \sigma_3$. Let $m(\epsilon') = \lim_\sigma m(\sigma,\epsilon')$. Note that $m(\epsilon') \leq \alpha^*$. We claim that $\langle m(\epsilon') \mid \epsilon' < \epsilon\rangle$ is α-finite. The proof of this claim splits into cases exactly as did the proof of Lemma 2.2. (If α^* is a regular α-cardinal then we can directly employ the combinatorial lemma mentioned in the Remark following the proof of 2.2.) In either case we see that there is a stage $\sigma(\epsilon) \geq \sigma_3$ such that $m(\sigma,\epsilon') = m(\epsilon')$ for all $\epsilon' < \epsilon$, $\sigma \geq \sigma(\epsilon)$. From this and Lemma 4.3, Lemma 4.4 is immediate. ■

Put $C = A \oplus B$.

4.5 Lemma. *C is α-regular and α-hyperregular and $C' \equiv_\alpha 0'$.*

Proof. Suppose C is not α-regular and α-hyperregular. Let κ be the least ordinal such that $f''\kappa$ is not α-finite.for some $f \leq_{w\alpha} C$, $\kappa \subseteq \operatorname{dom}(f)$. As in the proof of Lemma 2.3 we can show that κ is a regular α-cardinal. In particular $\kappa \leq \alpha^*$. Note that $\{\xi \mid K_\xi \subseteq C\}$ is α-r.e. since C is. Hence there is $\epsilon < \alpha^*$ such that for all γ, δ

$$f(\gamma) = \delta \leftrightarrow \exists\eta\,[\langle\gamma,\delta,\eta\rangle \in R_\epsilon \ \&\ K_\eta \cap C = \emptyset]\,.$$

Thus $y(\gamma, \epsilon) = \lim_\sigma y(\sigma, \gamma, \epsilon)$ exists for all $\gamma < \kappa$. Furthermore

$$m(\epsilon) = \lim_\sigma m(\sigma, \epsilon) = \begin{cases} (\mu\gamma)_{\gamma<\alpha^*}(y(\gamma, \epsilon) \text{ does not exist}) \\ \qquad\qquad \text{if } \alpha^* \text{ is a regular } \alpha\text{-cardinal;} \\ (\mu\gamma)_{\gamma<\epsilon}(y(\gamma, \epsilon) \text{ does not exist}) \\ \qquad\qquad \text{otherwise.} \end{cases}$$

The proof now splits into cases. If α^* is a regular α-cardinal then $\kappa \leq m(\epsilon)$ obviously. If α^* is not a regular α-cardinal, then $\kappa < \alpha^*$ so we may safely assume $\kappa \leq \epsilon$ hence again $\kappa \leq m(\epsilon)$. In either case Lemma 4.4 shows that $\{y(\gamma, \epsilon) \mid \gamma < \kappa\}$ is α-finite, a contradiction. So C is α-regular and α-hyper-regular.

It follows that there are α-recursive functions $h, k : \alpha \to \alpha^*$ such that for all γ, δ

$$K_\gamma \subseteq C' \leftrightarrow h(\gamma) \in C'$$

and

$$K_\delta \cap C' = \phi \leftrightarrow k(\gamma) \notin C'.$$

By Lemma 4.4 there is an α-recursive function $t : \alpha^* \to \alpha^*$ such that for all $\epsilon < \alpha^*$

$$\epsilon \in C' \leftrightarrow (\exists\sigma)(\forall\tau \geq \sigma)(y(\sigma, 0, t(\epsilon)) = y(\tau, 0, t(\epsilon)))$$

$$\leftrightarrow (\forall\sigma)(\exists\tau \geq \sigma)(y(\tau, 0, t(\epsilon)) < \tau).$$

From the existence of h, k, and t it follows that $C' \leq_\alpha 0'$. ■

4.6 Lemma. *For each $\epsilon < \alpha^*$ there exists σ such that $f(\sigma, \epsilon) = f(\tau, \epsilon)$ and $g(\sigma, \epsilon) = g(\tau, \epsilon)$ for all $\tau \geq \sigma$.*

Proof. Immediate from Lemmas 4.2 and 4.4. ■

Thus $f(\epsilon) = \lim_\sigma f(\sigma, \epsilon)$ and $g(\epsilon) = \lim_\sigma g(\sigma, \epsilon)$ are α-finite. Define $A = \bigcup\{A^\sigma \mid \sigma < \alpha\}$ and $B = \bigcup\{B^\sigma \mid \sigma < \alpha\}$. It is perhaps amusing to note that clauses (4) and (5) of the construction have played no role in the proofs of Lemmas 4.2–4.6.

4.7 Lemma. *For each* $\epsilon < \alpha^*$,

$$\langle f(\epsilon), \epsilon\rangle \in A \leftrightarrow \exists\eta\,[\langle f(\epsilon), \epsilon, \eta\rangle \in R_\epsilon \,\&\, K_\eta \cap B = \emptyset]$$

and

$$\langle g(\epsilon), \epsilon\rangle \in B \leftrightarrow \exists\eta\,[\langle g(\epsilon), \epsilon, \eta\rangle \in R_\epsilon \,\&\, K_\eta \cap A = \emptyset]$$

Proof. First suppose $\langle f(\epsilon), \epsilon\rangle \in A$. Let σ be such that $\langle f(\epsilon), \epsilon\rangle \in A^\sigma - A^{<\sigma}$. It follows that $f(<\tau, \epsilon) = f(\tau, \epsilon) = f(\epsilon)$ for all $\tau \geq \sigma$. Put $\nu = L(\sigma)$. Then $H(\sigma, \nu) \leq \epsilon < H(\sigma, \nu+1)$ and so $H(\tau, \mu) = H(\mu)$ for all $\mu \leq \nu$, $\tau \geq \sigma$. Since $A^\sigma \neq A^{<\sigma}$ we must have $g(\sigma, \epsilon') = \sigma + 1$ for all $\epsilon' \geq H(\sigma, \nu) = H(\nu)$. Furthermore there must be an η such that $\langle f(\epsilon), \epsilon, \eta\rangle \in R^\sigma_\epsilon$ and $K_\eta \cap B^{<\sigma} = \emptyset$. Note that $K_\eta \subseteq \eta < \sigma$. We claim that $K_\eta \cap B = \emptyset$. Suppose not, say $\langle g(\sigma', \epsilon'), \epsilon'\rangle \in K_\eta \cap (B^{\sigma'} - B^{<\sigma'})$, $\sigma' \geq \sigma$. Then $g(\sigma', \epsilon') < \sigma$ hence $g(\sigma', \epsilon') = g(\sigma, \epsilon')$ and $\epsilon' < H(\sigma, \nu) = H(\sigma', \nu)$. Put $\nu' = L(\sigma')$, then $H(\sigma', \nu') \leq \epsilon' < H(\sigma', \nu'+1)$ so $\nu' + 1 \leq \nu$. Hence $H(\sigma', \nu'+1) \leq H(\sigma', \nu) = H(\sigma, \nu) \leq \epsilon$. Therefore $f(\sigma', \epsilon) = \sigma' + 1 > \sigma$ a contradiction.

Now for the converse suppose $\langle f(\epsilon), \epsilon, \eta\rangle \in R^\sigma_\epsilon$ and $K_\eta \cap B = \emptyset$. Let σ be a stage such that $f(<\sigma, \epsilon) = f(\epsilon)$ and $\langle f(\epsilon), \epsilon, \eta\rangle \in R^\sigma_\epsilon$ and $H(\sigma, L(\sigma)) \leq \epsilon < H(\sigma, L(\sigma)+1)$. Then $\langle f(\epsilon), \epsilon\rangle \in A^\sigma \subseteq A$.

We have proved the $f(\epsilon)$ part of the lemma. The $g(\epsilon)$ part is similar. ■

4.8 Lemma. $A \not\leq_\alpha B$ *and* $B \not\leq_\alpha A$.

Proof. In the first place $\{\xi \mid K_\xi \subseteq B\}$ is α-r.e. since B is. Hence $A \leq_\alpha B$ would imply the existence of an $\epsilon < \alpha^*$ such that for all γ

$$\gamma \in A \leftrightarrow \exists\eta\,[\langle\gamma, \eta\rangle \in R_\epsilon \,\&\, K_\eta \cap B = \emptyset]\,.$$

But $\gamma = \langle f(\epsilon), \epsilon\rangle$ witnesses the contrary. ■

The proof of Theorem 4.1 is complete.

5. Further results

During 1970–1972 the theory of α-degrees has developed rapidly. Two of the prettiest theorems are to be found in the Ph.D. thesis of Richard Shore.

5.1 Theorem (Shore). *Let A be an α-regular, α-r.e. set and let B be a non-α-recursive, α-regular, α-r.e. set. Then there are hyperregular α-r.e. sets A_0, A_1 such that $A = A_0 \cup A_1$, $A_0 \cap A_1 = \emptyset$, $B \not\leq_\alpha A_0$, $B \not\leq_\alpha A_1$.*

Corollary. *Any non-zero α-r.e. degree is a nontrivial join of two hyperregular α-r.e. degrees.*

5.2 Theorem (Shore). *Let $\boldsymbol{a}$ and $\boldsymbol{c}$ be α-r.e. degrees such that $\boldsymbol{a} < \boldsymbol{c}$. Then there is an α-r.e. degree $\boldsymbol{b}$ such that $\boldsymbol{a} < \boldsymbol{b} < \boldsymbol{c}$. If $\boldsymbol{a}$ is hyperregular then $\boldsymbol{b}$ can be made hyperregular.*

Theorems 5.1 and 5.2 generalize two famous theorems of ordinary recursion theory due to Sacks. They are known respectively as the Splitting Theorem and the Density Theorem; for an exposition see Shoenfield's degree monograph [14]. The proofs of 5.1 and 5.2 are remarkable applications of Shore's "blocking" device duscussed in the Remark on page 182 above.

We offer the following analysis of the proof of 5.1. A typical goal or requirement of the construction is $c_B \neq [\epsilon]^{A_i}$ where $i < 2$, $\epsilon < \alpha^*$. In the special case $\alpha = \omega$, Sacks satisfies this requirement by preserving (with priority ϵ) those computations which make c_B and $[\epsilon]^{A_i}$ look equal on long initial segments of α. (Because of these preservations, Sacks is able to argue at the end that $c_B = [\epsilon]^{A_i}$ would imply that B is α-recursive.) In the case of an arbitrary admissible α, this will not work, because the preservations associated with an infinite, α-finite set of requirements can get out of hand if $\Delta_2 - \mathrm{cf}(\alpha) < \alpha^*$. Instead Shore breaks up the set of α^* requirements into $\Delta_2 - \mathrm{cf}(\alpha)$ blocks each of size less than α^*. A typical block $B_{\nu,i}$ is $\{c_B \neq [\epsilon]^{A_i} \mid H(\nu) \leq \epsilon < H(\nu+1)\}$ where $i < 2$, $\nu < \Delta_2 - \mathrm{cf}(\alpha)$. (See page 182.) Shore satisfies the requirements in block $B_{\nu,i}$ by preserving with priority ν those computations which make $\{\gamma < \alpha \mid c_B(\gamma)$ and $[\epsilon]^{A_i}(\gamma)$ look equal for *some* ϵ, $H(\nu) \leq \epsilon < H(\nu+1)\}$ contain long initial segments of α. Thus the block of requirements $B_{\nu,i}$ is handled as a single requirement, so the preservations associated with $B_{\nu,i}$ do not get out of hand.

In another line of investigation, M. Lerman and C.T. Chong have attempted to generalize the theorems of Lachlan's Lower Bounds paper [3] to α-recursion theory. The results of this attempt are as follows:

5.3 Theorem (Lerman). *If $\boldsymbol{a}$ and $\boldsymbol{b}$ are α-r.e. degrees with $\boldsymbol{a} \cap \boldsymbol{b} = \boldsymbol{0}$ then $\boldsymbol{a} \cup \boldsymbol{b} < \boldsymbol{0}'$.*

5.4 Theorem (Chong). *If $\boldsymbol{d}$ is any non-zero α-r.e. degree then there is a hyperregular α-r.e. degree $\boldsymbol{c}$ such that $\boldsymbol{0} < \boldsymbol{c} < \boldsymbol{d}$ and $\boldsymbol{c}$ is not a nontrivial meet of two α-r.e. degrees.*

Lerman and Sacks, with good reason, call an admissible ordinal α *refractory* if (i) there is a largest α-cardinal, call it $\aleph$; (ii) $\Delta_2 - \mathrm{cf}(\alpha) < \aleph$; and (iii) there is no Σ_2 function $f : \alpha \overset{1-1}{\to} \beta$ where $\beta < \aleph$. For example let α be the first constructibly uncountable admissible ordinal having constructible cofinality ω; then α is obviously refractory.

5.5 Theorem (Lerman–Sacks). *Suppose α is not refractory. Then there are nonzero hyperregular α-r.e. degrees $\boldsymbol{a}, \boldsymbol{b}$ such that $\boldsymbol{a} \cap \boldsymbol{b} = \boldsymbol{0}$.*

Of course one conjectures that the hypothesis of non-refractoriness can be eliminated, but this does not seem easy to do [11].

Another area where the results to date are fragmentary is the question of minimal α-degrees. An α-degree $\boldsymbol{m}$ is said to be *minimal* if $\boldsymbol{m} > \boldsymbol{0}$ and there is no α-degree $\boldsymbol{a}$ with $\boldsymbol{m} > \boldsymbol{a} > \boldsymbol{0}$.

5.6 Theorem (MacIntyre). *Let α be an admissible ordinal such that $|\alpha|$ is regular and every $X \subseteq \alpha$ of cardinality less than $|\alpha|$ is α-finite. Then there exists a regular, hyperregular, minimal α-degree.*

In particular, minimal α-degrees exist if α is countable. (For α = ω this is an old theorem of Spector.) MacIntyre has conjectured that regular, hyperregular, minimal α-degrees exist for all admissible ordinals α.

The proof of 5.6 is not a priority argument. By exploiting priorities à la Shoenfield [14 : pp. 54–56] and the "α-finite injury method" of Sacks–Simpson [13], one enlarges the class of admissible ordinals α for which minimal α-degrees are known to exist. Namely,

[11] Recently D. Posner has shown that for every admissible ordinal α there exist regular, hyperregular α-degrees $\boldsymbol{a}, \boldsymbol{b}$ such that $\boldsymbol{a} \cap \boldsymbol{b} = \boldsymbol{0}$.

5.7 Theorem (Shore). *Suppose* $\Delta_2 - \mathrm{cf}(\alpha) = \alpha$. *Then there is a regular, hyperregular, minimal α-degree* $\boldsymbol{m} < \boldsymbol{0}'$.

One can probably adapt the proofs of 5.6 and 5.7 to get finite distributive lattices as initial segments of α-degrees. However, to eliminate the special hypotheses on α seems a difficult problem indeed.

This completes our survey of the current theory of α-degrees. Apart from what we have reported here, most questions concerning α-degrees are virgin territory. It is not always easy to appropriately generalize the *statement* of a theorem of ordinary degree theory to α-degree theory, much less the proof. One obstacle is that the admissibility of α (i.e. of the structure $\langle L_\alpha, \in \rangle$) does not imply admissibility of the expanded structure $\langle L_\alpha, \in, C \rangle$ where $C \subseteq \alpha$, even if C is α-r.e. and α-regular. Therefore "relativization" to C (cf. Rogers [10: p. 257]) may be difficult or impossible.

This suggests a typical question: for an arbitrary admissible ordinal α, are there Σ_2 sets $A, B \subseteq \alpha$ such that $A \not\leq_\alpha B \oplus 0'$ and $B \not\leq_\alpha A \oplus 0'$? For $\alpha = \omega$ (indeed whenever $\Delta_2 - \mathrm{cf}(\alpha) = \alpha$) the answer is obviously yes by relativizing the Friedberg–Muchnik theorem to $0'$. For arbitrary admissible ordinals α, the answer is probably still yes but the proof may require an "infinite injury" argument.

Carl Jockusch has proved the following theorem of ordinary degree theory: there exists a degree $\boldsymbol{d}$ such that for all $\boldsymbol{b} \geq \boldsymbol{d}$ there is $\boldsymbol{a} < \boldsymbol{b}$ such that there is no $\boldsymbol{c}$ with $\boldsymbol{a} < \boldsymbol{c} < \boldsymbol{b}$. Jockusch's proof is unique in degree theory in that it employs the powerset axiom of ZFC (via a result of Paris concerning Gale–Stewart games). Therefore, the following question is of exceptional interest: for which admissible ordinals α does Jockusch's theorem generalize to α-degrees?

6. Appendix: the fine structure of L

The *constructible hierarchy* is defined by recursion on the ordinals as follows: $L_0 = \{\emptyset\}$; $L_{\sigma+1} = \{X \subseteq L_\sigma \mid X$ is first order definable over $\langle L_\sigma, \in \rangle$ allowing parameters from $L_\sigma\}$; $L_\lambda = \bigcup\{L_\sigma \mid \sigma < \lambda\}$ for limit ordinals λ; $L = \bigcup\{L_\sigma \mid \sigma$ an ordinal$\}$. Ronald Jensen [1] has made a detailed study of the fine structure of L. (Jensen then uses his results to settle a number of open questions in set theory under the assumption V = L.) There is a close connection between some of Jensen's ideas and some of the ideas in α-recursion theory. We now discuss this connection briefly.

First, a glossary comparing some of Jensen's terminology to some of ours. (The equivalence proofs would be tedious but straightforward.)

6.1. Let α be a limit ordinal. Then L_α is an *admissible set* iff α is an admissible ordinal.

6.2. Let α be an admissible ordinal and $A \subseteq \alpha$. Then
(i) A is $\Sigma_n(L_\alpha)$ iff A is Σ_n in our terminology;
(ii) $A \in L_\alpha$ iff A is α-finite;
(iii) The structure $\langle L_\alpha, \in, A \rangle$ is *amenable* iff A is α-regular.

6.3. Suppose α is admissible and $A \subseteq \alpha$ is α-regular and let f be a partial function from α into α. Then
(i) f is $\Sigma_1(L_\alpha, A)$ iff $f \leq_{w\alpha} A$;
(ii) the structure $\langle L_\alpha, \in, A \rangle$ is *admissible* iff A is α-hyperregular;
(iii) $B \subseteq \alpha$ is $\Delta_1(L_\alpha, A)$ iff $c_B \leq_{w\alpha} A$.

6.4. Suppose α is admissible and $A \subseteq \alpha$ is α-regular and α-hyperregular. Then $B \subseteq \alpha$ is $\Delta_1(L_\alpha, A)$ iff $B \leq_\alpha A$.

A decisive role in Jensen's theory is played by the notion of Σ_n projection. For α an admissible ordinal and $n \geq 1$, the Σ_n *projectum* of α is defined to be the least β such that there is a $\Sigma_n(L_\alpha)$ function from a subset of β onto α. Thus the Σ_1 projectum of α is just what we have been calling α^*. Jensen's main theorem reads as follows.

6.5. Theorem. *For $n \geq 1$, the Σ_n projectum of α is equal to the least β such that not every $\Sigma_n(L_\alpha)$ subset of β is α-finite.*

The proof of 6.5 is essentially model theoretic rather than recursion theoretic in nature. One needs a delicate refinement of the Skolem hull-condensation method first used by Gödel to prove the GCH assuming V = L. The reader is invited to study Jensen's proof in [1].

For α admissible and $n = 1$, Theorem 6.5 is trivial and we have used it repeatedly in the present paper. For $n = 2$ Theorem 6.5 is non-trivial but has been used elsewhere in α-recursion theory, e.g., in [7] and in Chapter 3 of

[17]. The interesting point is this: Theorem 6.5 for $n = 2$ can fail if one looks at admissible structures of the form $\langle L_\alpha, \in, A\rangle$. Namely, one can construct an admissible ordinal α and an α-regular, α-hyperregular set $A \subseteq \alpha$ such that (i) not every $\mathbf{\Delta}_2(L_\alpha, A)$ subset of ω is α-finite; (ii) there is no $\mathbf{\Sigma}_2(L_\alpha, A)$ function from a subset of ω onto α. This means that some arguments of α-recursion theory, e.g., the proof of Theorem 3.2 of [7], do not generalize straightforwardly to recursion theory on admissible structures $\langle L_\alpha, \in, A\rangle$. This is a sad but amusing state of affairs which deserves to be investigated further.

Bibliography

[1] R.B. Jensen, The fine structure of the constructible hierarchy, Annals of Math. Logic 4 (1972) 229–308.

[2] G. Kreisel, Some reasons for generalizing recursion theory, in: R.O. Gandy and C.E.M. Yates, eds., Logic Colloquium '69 (North-Holland, Amsterdam, 1971) 139–198.

[3] A.H. Lachlan, Lower bounds for pairs of recursively enumerable degrees, Proc. London Math. Soc. 16 (1966) 537–569.

[4] M. Lerman, Maximal α-r.e. sets, Trans. Amer. Math. Soc. (to appear).

[5] M. Lerman, Least upper bounds for minimal pairs of α-r.e. α-degrees (to appear).

[6] M. Lerman and G.E. Sacks, Some minimal pairs of α-recursively enumerable degrees, Annals of Math. Logic 4 (1972) 415–442.

[7] M. Lerman and S.G. Simpson, Maximal sets in α-recursion theory, Israel J. Math. (to appear).

[8] M. Machtey, Admissible ordinals and lattices of α-r.e. sets, Annals of Math. Logic 2 (1971) 379–417.

[9] J.M. MacIntyre, Minimal α-recursion theoretic degrees, J. Symb. Logic 38 (1973) 18–28.

[10] H. Rogers, Jr., Theory of Recursive Functions and Effective Computability (McGraw-Hill, 1967) 482 pp.

[11] G.E. Sacks, Post's problem, admissible ordinals, and regularity, Trans. Amer. Math. Soc. 124 (1966) 1–23.

[12] G.E. Sacks, Metarecursion theory, in: J.N. Crossley, ed., Sets, Models, and Recursion Theory (North-Holland, Amsterdam, 1967) 243–263.

[13] G.E. Sacks and S.G. Simpson, The α-finite injury method, Annals of Math. Logic 4 (1972) 343–367.

[14] J.R. Shoenfield, Degrees of Unsolvability (North-Holland, Amsterdam, 1971) 111 pp.

[15] R.A. Shore, Minimal α-degrees, Annals of Math. Logic 4 (1972) 393–414.

[16] R.A. Shore, Priority Arguments in α-recursion Theory, Ph.D. Thesis, M.I.T. 1972.

[17] S.G. Simpson, Admissible Ordinals and Recursion Theory, Ph.D. Thesis, M.I.T. 1971.

J.E.Fenstad, P.G.Hinman (eds.), Generalized Recursion Theory

MORE ON SET EXISTENCE

Françoise VILLE
University of Orleans

§ 1. Standard sets of a theory: definition and general results

The structure of the standard sets is, by definition, the structure $\langle V, \in_V \rangle$ of the cumulative hierarchy; it is extensional and well founded. The transitive closure of an element a of V is denoted by $[a]$.

Let us suppose that (T) is a theory formulated in a language $\mathcal{L}$, containing at least the binary relation symbol $\in$. Realisations of $\mathcal{L}$ have thus the form $\langle M, \in_M, ... \rangle$; if m is an element of M, $[m]_{\in_M}$ denotes the transitive closure of m with respect to $\in_M$.

To avoid confusion, membership in the metalanguage will be denoted ε.

Definition 1. Let $\mathcal{M}$ be a realisation of $\mathcal{L}$. An element a of V is called a standard set of $\mathcal{M}$ iff there is an m_a in M such that:

$$\langle [\{a\}], \in_V \upharpoonright [\{a\}] \rangle \text{ is isomorphic to } \langle [\{m_a\}]_{\in_M}, \in_M \upharpoonright [\{m_a\}]_{\in_M} \rangle$$

m_a is said to represent a in $\mathcal{M}$.

If $\langle M, \in_M \rangle$ is extensional, a standard set is represented by at most one element of M.

Definition 2. The element a of V is called a standard set of the theory (T) iff a is represented in every model of (T). $S(\mathrm{T})$ denotes the collection of all standard sets of (T).

Note that $\langle S(\mathrm{T}), \in_V \upharpoonright S(\mathrm{T}) \rangle$ is a transitive substructure of $\langle V, \in_V \rangle$.

When $(\mathrm{T}) \subset \mathcal{L}_{\mathrm{L}_\omega}$ (where $\mathrm{L}_\omega = \mathrm{V}_\omega = \mathrm{H}_\omega$ = the collection of well founded hereditarily finite sets) and so, each formula of (T) is of finite length, it can be shown by a compactness argument that every standard set of (T) is hereditarily finite: thus $S(\mathrm{T}) \subset \mathrm{L}_\omega$. If (T) is a denumerable set of formulas (of denumerable length) $\subset \mathcal{L}_{\omega_1,\omega}$ then one easily deduces from the Löwenheim–Skolem theorem that every element of $S(\mathrm{T})$ is hereditarily countable.

A brutal generalisation of these facts would be: "Let A be standard and admissible, and let (T) be a theory in the language $\mathcal{L}_A$ (associated with A, as in Kunen, Barwise etc.), then $S(\mathrm{T}) \subset A$".

But this cannot hold, as shown by the following example:

Take $A = \mathrm{L}_{\omega_1^{ck}}$, and the following theory (T):

Language of (T):

- the binary relation symbols $\in$ and $=$,
- for each integer n, a constant symbol c_n,
- two distinguished constant symbols c_W and c_ω.

Axioms of (T):

- extensionality
- $\forall x\, x \notin c_0$ and for each integer n:
  $\forall x\, (x \in c_{n+1} \leftrightarrow (x = c_n \vee x \in c_n))$
- $\forall x\, (x \in c_\omega \leftrightarrow \bigvee_{n \in \omega} x = c_n)$
- $\forall x\, (x \in c_W \rightarrow x \in c_\omega)$
- for each integer n:

$$c_n \in c_W \quad \text{iff} \quad n \in W$$

$$c_n \notin c_W \quad \text{iff} \quad n \notin W,$$

where $W \in \mathbf{\Pi}_1^1 - \mathbf{\Sigma}_1^1$, for example the set of Gödel numbers of recursive well orderings.

Clearly $(\mathrm{T}) \subset \mathcal{L}_{\mathrm{L}_{\omega_1^{ck}}}$ and $S(\mathrm{T}) = \omega \cup \{W\} \cup \{\omega\}$; but as is well known $W \notin \mathrm{L}_{\omega_1^{ck}}$, so $S(\mathrm{T}) \not\subset \mathrm{L}_{\omega_1^{ck}}$. Note that the theory (T) above is $\Pi_1^1 \cup \Sigma_1^1$ but not Π_1^1.

The principal aim of this paper is to show that $S(\mathrm{T}) \subset \mathrm{L}_{\omega_1^{ck}}$ is obtained by bounding the complexity of (T):

(∗) Provided that (T) has an extensional model, if (T) is an $\mathrm{L}_{\omega_1^{ck}}$-r.e. theory of $\mathcal{L}_{\mathrm{L}_{\omega_1^{ck}}}$, then $S(\mathrm{T}) \subset \mathrm{L}_{\omega_1^{ck}}$.

Below we shall use ω_1 to denote the least non recursive ordinal which we called ω_1^{ck} above. So L_{ω_1} will be the set of all sets constructible with recursive order. We shall also make the following assumptions on (T):

– there is at least one extensional model of (T)

– (T) is a theory with equality

In the following (T) *will be an* L_{ω_1}-r.e. theory of $\mathcal{L}L_{\omega_1}$.

§2. Properties definable on S(T)

In this paragraph we assume that D is a subset of S(T) which satisfies the following condition:

(D) Every element of D is defined in every model of (T) by a formula of $\mathcal{L}L_{\omega_1}$ that we shall denote by $\Phi_a(x)$.

So that for every a in D, for every model $\mathcal{M}$ of (T) there is an m_a in $\mathcal{M}$ which represents a in $\mathcal{M}$ such that $\forall x\,(x \in m_a \leftrightarrow \Phi_a(x))$ is true in $\mathcal{M}$.

Definition 3. a) A subset X of D is L_{ω_1}-definable on D in a given model $\mathcal{M}$ of (T) iff there is a formula $\Phi_{\mathcal{M}}(x, x_1, ..., x_n)$ of $\mathcal{L}L_{\omega_1}$ with $n+1$ free variables, and parameters $a_1, ..., a_n$ in $\mathcal{M}$ such that $X = \{a \in V : \langle m_a, a_1, ..., a_n\rangle$ satisfies $\phi_{\mathcal{M}}$ in $\mathcal{M}\} \cap$ D.

b) A subset X of D is L_{ω_1}-definable on D in (T) iff it is L_{ω_1}-definable on D in every model of (T).

Examples: a) for each element a of S(T), $a \cap$ D is defined on D in each model $\mathcal{M}$ of (T) by the formula $x \in x_1$ with parameter m_a.

b) If $L_\omega \in S$(T) and $L_\omega \subset$ D, then ω is definable on D in (T) as well as the graphs of + and $\times$, and every arithmetic subset of ω; here L_ω, or more precisely, its representative m_{L_ω}, can be used as a parameter.

Below we shall need the following analysis of L_{ω_1}-definability on D in terms of the consequence relation in (T):

Theorem 1. *Under the condition* (D), *if* $X \subset$ D *is* L_{ω_1}*-definable on* D *in* (T) *then there are formulas* $\psi_1(x)$ *and* $\psi_2(x)$ *in* $\mathcal{L}L_{\omega_1}$, *with only one free variable such that for every* a *in* D:

$$a \in X \leftrightarrow (\mathrm{T}) \vdash \exists x\,[\phi_a(x) \wedge \psi_1(x)]$$

$$a \in X \leftrightarrow (\mathrm{T}) \vdash \exists x\,[\phi_a(x) \wedge \psi_2(x)]\ .$$

Theorem 1, which can be proved via the omitting types theorem for denumerable admissible languages, is fundamental for the proof of (∗). The case for L_ω is treated in chap. 6 of [Kr.Kr].

We shall distinguish two cases according to whether (T) is an extension of KP [1] (§3) or not (§4).

§3. (T) is an extension of KP

In this case $S(T)$ which obviously contains every element of L_{ω_1} will provide, in every model of (T) all the parameters we need.

Every model of (T) is extensional; then by the remark following Definition 1, every element of $S(T)$ has at most one representative in every model of (T).

Every model of (T) is admissible; then if $S(T)$ has no infinite element, $S(T) = L_\omega$. If, on the contrary, there is an infinite element in $S(T)$ then $L_\omega \in S(T)$ and we have $L_{\omega_1} \subset S(T)$ (cf. [8] p. 243). Note that in each case:

Lemma 1. *There is an L_{ω_1}-recursive mapping $a \to \overline{\Phi}_a(x)$ from L_{ω_1} into $\mathcal{L}L_{\omega_1}$ such that for every a in $L_{\omega_1} \cap S(T)$, $\Phi_a(x)$ defines a in every model of* (T).

This is an easy consequence of the extensionality of every model of (T). Therefore L_{ω_1} satisfies condition (D) of §2.

Lemma 2. *Let a subset X of L_{ω_1} be L_{ω_1}-definable on L_{ω_1} in* (T). *Then X is L_{ω_1} recursive.*

As in Theorem 1 we prove that there are ψ_1, ψ_2 in $\mathcal{L}L_{\omega_1}$ such that for every a in L_{ω_1}:

$$a \in X \quad \text{iff} \quad (T) \vdash \exists x\, (\overline{\phi}_a(x) \wedge \psi_1(x))$$

$$a \notin X \quad \text{iff} \quad (T) \vdash \exists x (\overline{\phi}_a(x) \wedge \psi_2(x))\,.$$

[1] For the definition of KP see [8] p. 232.

The applications: $a \to \exists x(\overline{\phi}_a(x) \wedge \psi_1(x))$ and: $a \to \exists x(\overline{\phi}_a(x) \wedge \psi_2(x))$ are L_{ω_1}-recursive by Lemma 1; therefore, consequence in (T) being L_{ω_1}-r.e., X is an L_{ω_1}-recursive subset of L_{ω_1}:

Theorem 2. $\omega_1 \notin S(\mathrm{T})$.

Let us first prove one lemma about model of KP.

Lemma 3. *Let $\mathcal{M}$ be a model of* KP *and R a linear ordering of ω which is Δ_1-definable in $\mathcal{M}$. For every ordinal α of $\mathcal{M}$ if there is an isomorphism f of $\langle \alpha, \in_M \restriction \alpha \rangle$ onto an initial segment of the ordering R then f belongs to $\mathcal{M}$.*

f is an isomorphism from $\langle \alpha, \in_M \restriction \alpha \rangle$ into an initial segment of R (which we shall denote $<_R$) iff f satisfies the following equations, $I(f, \alpha)$:

$f(\emptyset) = a$ iff a is the least element in the ordering $<_R$,
and for all $\lambda < \alpha$:

$f(\lambda) = b$ iff b is the least element of R greater than any $f(\xi)$ for $\xi < \lambda$ in the ordering $<_R$.

Now we want to prove that we have in $\mathcal{M}$:

for every ordinal α, there is one and only one f such that $I(f, \alpha)$.

Note that there is a Σ_1 formula $Is(f, \alpha)$ which describes the relation f *is a function of domain* $\alpha \wedge I(f, \alpha)$ in $\mathcal{M}$ (we allow parameters) since R is Δ_1 in $\mathcal{M}$.

By induction on ordinals we establish that $\exists! f\, Is(f, \alpha)$ holds in $\mathcal{M}$ for every α. The uniqueness is easily established, so we have to establish the existence; we proceed by induction:

– if $\alpha = \emptyset$ then $f = \emptyset$ and is the only f satisfying $Is(f, \alpha)$ in $\mathcal{M}$.

– if $\alpha \neq \emptyset$ then by induction hypothesis there is a unique f, say f_ξ, which satisfies $Is(f_\xi, \xi)$ in $\mathcal{M}$ (for $\xi < \alpha$). Thus,

– if α is the successor of β, we put $f_\alpha = f_\beta \cup \langle \beta, b \rangle$ where $b \in \omega$ is chosen as described in I. As f_β belongs to $\mathcal{M}$, f_α obviously belongs to $\mathcal{M}$.

– if α is a limit ordinal, one can apply Σ_1-replacement, since $Is \in \Sigma_1$ and, by induction hypothesis

$$\mathcal{M} \models \forall \xi \in \alpha \, \exists ! f_\xi Is(f_\xi, \xi) .$$

Thus there is an f in $\mathcal{M}$, such that: $\mathcal{M} \models (f = f_\xi : \xi < \alpha)$. By $\mathbf{\Delta}_1$-comprehension $\mathbf{U}f$ belongs to $\mathcal{M}$. Putting $f_\alpha = \mathbf{U}f$ we have:

$$Is(f_\alpha, \alpha) .$$

Proof of Theorem 2. Let Q be a recursive linear ordering of ω with an initial segment I_Q of length ω_1 (for example the ordering in Gandy's [G]). Since Q is a recursive subset of ω^2, Q can be defined in every model of (T) by a $\mathbf{\Delta}_0$-formula, using ω as a parameter.

Suppose $\omega_1 \in S(\mathrm{T})$; and let $\mathcal{M}$ be any model of (T): ω_1 belongs to (is represented in) $\mathcal{M}$. By the choice of Q, there is an isomorphism f from $\langle \omega_1, \in_M \upharpoonright \omega_1 \rangle$ into $\langle I_Q, Q \rangle$. Since the hypothesis of Lemma 3 are satisfied, f belongs to $\mathcal{M}$. I_Q can then be defined by $\exists \xi \in \omega_1 \langle \xi, x \rangle \in f$ with x as the only free variable and the two parameters ω_1 and f.

So I_Q is defined on ω in (T) and, by Lemma 2, I_Q is L_{ω_1}-rec; and hence $\mathbf{\Delta}_1^1$, since $\mathbf{\Pi}_1^1$ over ω or over L_ω is equivalent to L_{ω_1}-r.e. (cf. [BGM]). But this contradicts the fact that I_Q is a $\mathbf{\Pi}_1^1$ subset of ω which is not $\mathbf{\Sigma}_1^1$.

Thus we cannot have $\omega_1 \in S(\mathrm{T})$.

Theorem 3. $S(\mathrm{T}) \subset \mathrm{L}_{\omega_1}$.

Lemma 4. *Let $\mathcal{M}$ be a model of* KP *and a be a subset of* L *which belongs to $\mathcal{M}$. Then there is an ordinal α of $\mathcal{M}$ such that $a \subset \mathrm{L}_\alpha$.*

Since $\mathcal{M}$ is admissible, the function od which associates to each constructible x its order ξ is expressible in $\mathcal{M}$ by a formula Od which is $\mathbf{\Sigma}_1$ over $\mathcal{M}$ and verifies $\mathcal{M} \models \forall x \in a \, \exists ! y \; Od(x,y)$.

By $\mathbf{\Sigma}_1$ replacement there is a b in $\mathcal{M}$ such that

$$\mathcal{M} \models b = \{Od(x) : x \in_M a\} .$$

Take α to be the least upper bound of b; it is an ordinal of $\mathcal{M}$ and $a \subset \mathrm{L}_\alpha$.

Proof of Theorem 3. Assume a is an element of $S(\mathrm{T})$ included in L_{ω_1}; the example a) following definition 3 shows that a is L_{ω_1}-definable on L_{ω_1} in (T) and Lemma 2 shows that a is L_{ω_1}-recursive.

In every model $\mathcal{M}$ of (T) there is, by Lemma 4, an α such that $a \subset L_\alpha$, and this α does obviously not depend on $\mathcal{M}$; therefore $\alpha \in S(T)$ and by Theorem 2, $\alpha < \omega_1$. So a is an L_{ω_1}-rec. subset of L_{ω_1} which is L_{ω_1} bounded, i.e. $a \in L_{\omega_1}$.

An induction shows that every element a of S(T) belongs to L_{ω_1}.

Theorem 4. $S(T) = L_\omega$ *or* $S(T) = L_{\omega_1}$.

Either S(T) has only finite elements and $S(T) = L_\omega$; or there is an infinite element in S(T). In this last case we saw that $L_{\omega_1} \subset S(T)$; by Theorem 3 $S(T) \subset L_{\omega_1}$, therefore $S(T) = L_{\omega_1}$.

Theorem 4 of Chapter 8 in [Kr.Kr] can now be generalised to $\mathcal{L}_{L_{\omega_1}}$.

Definition 4. A subset X of S(T) is uniformly L_{ω_1}-defined in (T) iff there is a formula Φ of $\mathcal{L}_{L_{\omega_1}}$ with a single free variable, such that for every model $\mathcal{M}$ of (T) $X = \{a \in M : \mathcal{M} \models \phi(a)\}$.

Theorem 5. *Every subset X of* S(T) *which is uniformly* L_{ω_1}*-definable in* (T) *is* L_{ω_1}*-finite.*

By Theorem 4 and Lemma 1 every element a of S(T) is defined in (T) by formula $\overline{\phi}_a(x)$ such that the mapping $a \nrightarrow \overline{\phi}_a(x)$ is L_{ω_1}-recursive. If $X \subset S(T)$ is uniformly L_{ω_1}-definable in (T), it is a fortiori L_{ω_1}-definable on L_{ω_1} in (T), and by Lemma 2 it is L_{ω_1}-recursive. Suppose ϕ is a formula of $\mathcal{L}_{L_{\omega_1}}$ which defines X uniformly in (T). Let c be a constant symbol not occurring in (T). The set of axioms

$$(T) \cup \{\phi(c)\} \cup \{\neg\overline{\phi}_a(c) : a \in X\}$$

is L_{ω_1}-r.e. and is obviously inconsistent.

By compactness there is a $b \in L_{\omega_1}$ such that

$$(T) \cup \{\phi(c)\} \cup \{\overline{\phi}_a(c) : a \in_V b\}$$

is consistent. Therefore

$$(\mathrm{T}) \vdash \phi(c) \rightarrow \bigvee_{a \in_{\mathrm{V}} b} \overline{\phi}_a(c)$$

and, as c does not occur in (T)

$$(\mathrm{T}) \vdash \forall x \quad \phi(x) \rightarrow \bigvee_{a \in_{\mathrm{V}} b} \overline{\phi}_a(x) \quad .$$

This means that X is bounded by b, which implies $X \varepsilon \mathrm{L}_{\omega_1}$.

§4. (T) is not an extension of KP

It is convenient to introduce auxiliary theories (T_1^+) and (E), formulated with an additional type of variables. (T_1^+) is needed to formalize the metamathematical notion of admissibility; let us call "metauniverse" the universe of models of (T_1^+). (E) relates standard sets of (T) to "real standard sets" i.e. standard sets of metauniverse.

We assume that (T) is consistent with extensionality.

Language of (T_1^+):

– its variables, of a type different of the types in (T) are denoted $X, Y, Z, \ldots$

– two binary relation symbols $\in'$ and $='$.

So the atomic formulas of this languages are of the form $X \in' Y, X =' Y \ldots$.

The axioms of (T_1^+) are those of KP formulated in this new language.

Language of (E):

– variables of the type of the arguments of $\in$ (the distinguished relation symbol of (T)) and of the type used in (T_1^+).

– a binary relation symbol E, as well as the symbols $\in$ and $\in'$.

The new atomic formulas have the form $x \mathrm{E} X$.

The axioms of (E) are:

(1) $\exists! X \forall x (\forall y\, y \notin x \rightarrow x \mathrm{E} X)$

(2) $\forall x (\forall y \in x \exists! Y\, y \mathrm{E} Y \rightarrow \exists! X\, x \mathrm{E} X)$

(3) $\forall x \forall y \forall X \forall Y (x \mathrm{E} X \wedge y \mathrm{E} Y \rightarrow (x \in y \leftrightarrow X \in' Y))$

(4) $\forall X (\forall Y \in' X \exists y\, y \mathrm{E} Y \rightarrow \exists x\, x \mathrm{E} X)$.

Let (T') be $(\mathrm{T}) \cup (\mathrm{T}_1^+) \cup (\mathrm{E})$.

Every realisation of the language of (T') has the form:

$$\langle M \cup M', \in_M, =_M, \in'_{M'}, ='_{M'}, \mathrm{E}_M, \ldots \rangle \, .$$

Definition 4. A standard set of $\langle M \cup M', ...\rangle$ is a standard set of the meta-universe, i.e. of $\langle M', \in'_{M'}, ='_{M'}\rangle$.

Lemma 5. *Let* $\mathfrak{M} \models$ (E) *and* $\langle M, \in_M\rangle$ *be extensional. Then the restriction of* E_M *to the well founded part* M_0 *of* M *is the graph of a function which maps representatives of standard sets in* $\langle M, \in_M\rangle$ *onto representatives of standard sets in* $\langle M', \in'_{M'}\rangle$.

We prove that for all x in M_0 $\mathfrak{M} \models \exists! X(x \mathrm{~E~} X)$ by induction along $\in_M \restriction M_0$ with the help of axioms (1) and (2). Thus there is a function, say φ, defined every where on M_0 whose graph is $\mathrm{E}_M \restriction M_0 \times M$. By Axiom (3) this function embeds $\langle M_0, \in_M \restriction M_0\rangle$ isomorphically in $\langle M', \in'_{M'}\rangle$; by Axiom (4), writing $\in_M$ for $\in_M \restriction M_0$

$$z \in'_{M'} \varphi(x) \leftrightarrow \exists y\,(y \in_{M_0} x \wedge z = \varphi y)\,.$$

Therefore if m_a represents the set a in $\langle M, \in_M, =_M\rangle$, $\phi(m_a)$ represents a in $\langle M', \in'_{M'}, ='_{M'}\rangle$.

Lemma 6. (T$'$) *is consistent.*

Let $\mathfrak{M} = \langle M, \in_M, ...\rangle$ be an extensional model of (T) and M_0 be its well founded part w.r.t. $\in_M$. Define by induction along $\in_{M_0}$ (i.e. $\in_M \mid M_0$) a function ψ from M_0 into V by:

$$\psi(x) = \{\psi(y) : y \in_{M_0} x\}$$

ψ is every where defined on M_0.

Put: $M_1 = \psi[M_0]$

M_1^+ = the least admissible containing M_1

$\in_1 = \in_V \restriction M_1^+$

$=_1$ = identity in M_1^+

E_1 = the graph of ψ

$\mathfrak{M}' = \langle M \cup M_1^+, \in_M, =_M, \in_1, =_1, \mathrm{E}_1, ...\rangle$.

$\mathfrak{M}'$ is a model of (T$'$) since:

– $\langle M, \in_M, =_M\rangle$ is by hypothesis a model of (T)

– $\langle M_1^+, \in_1, =_1\rangle$ is admissible and therefore a model of (T_1^+)

– $\mathfrak{M}'$ satisfies (E) by definition of ψ (cf the proof of Lemma 5).

Lemma 7. *Every standard set of* (T) *is a standard set of* (T′).

As (T) has an extensional model, (T) ∪ {Extensionality} is consistent; it is consistent with $(T_1^+) \cup (E)$. Let (T_1) be (T) ∪ {Extensionality}.

By Lemma 5 every standard set of (T_1) is a standard set of (T′): therefore as $(T) \subset (T_1)$, every standard set of (T) is a standard set of (T′).

Theorem 6. $\mathcal{S}(T) \subset L_{\omega_1}$.

(T′) is an L_{ω_1}-r.e. theory of $\mathcal{L}_{L_{\omega_1}}$ such that $(T') \supset KP$; by Theorem 4 $\mathcal{S}(T')$ is equal to either L_ω or L_{ω_1}. By Lemma 7, $\mathcal{S}(T) \subset \mathcal{S}(T')$. Therefore $\mathcal{S}(T) \subset L_{\omega_1}$.

This result evidently generalizes the results of [GKT] which establishes if a theory (T), formulated in the language of finite types, in $\mathbf{\Pi}_1^1$ then subsets of ω which appear in every ω-model of (T) are hyperarithmetic and every collection of subsets which appears in every ω model of (T) is L_{ω_1}-finite.

The only restriction we make on (T) is its consistency with extensionality. How to avoid this condition will be shown in a forthcoming paper.

References

[B] J. Barwise, Infinitary logic and admissible sets, JSL 34 (1969) 226–251.

[BGM] J. Barwise, R.O. Gandy and Y. Moschovakis, The next admissible set, JSL 36 (1971) 108–120.

[G] R.O. Gandy, Proof of Mostowski's conjecture, Bull. de l'Ac. Pol. des Sc. VIII 9 (1960).

[GKT] R.O. Gandy, G. Kreisel and W.W. Tait, Set existence, Bull. de l'Ac. Pol. des Sc. VIII, 9 (1960).

[Kr. Kr] G. Kreisel and J.L. Krivine, Elements of mathematical logic (North-Holland, Amsterdam, 1967).

PART III

INDUCTIVE DEFINABILITY

J.E.Fenstad, P.G.Hinman (eds.), Generalized Recursion Theory
© North-Holland Publ. Comp., 1974

INDUCTIVE DEFINITIONS AND THEIR CLOSURE ORDINALS

Stål AANDERAA *

IBM Thomas J. Watson Research Center, Yorktown Heights, New York

Abstract. We shall prove some theorems about inductive definitions and their closure ordinals. As corollaries we obtain the following results. $|\mathbf{\Delta}_1^1| < |\mathbf{\Pi}_1^1| < |\mathbf{\Sigma}_1^1| < |\mathbf{\Delta}_2^1\text{-mon}| = |\mathbf{\Delta}_2^1| = |\mathbf{\Sigma}_2^1\text{-mon}| < |\mathbf{\Pi}_2^1| < |\mathbf{\Delta}_3^1\text{-mon}|$. We also have that $|\mathbf{\Pi}_n^1| \neq |\mathbf{\Sigma}_n^1|$, $|\mathbf{\Pi}_n^1| \neq |\mathbf{\Sigma}_n^1\text{-mon}|$, $|\mathbf{\Sigma}_n^1| \neq |\mathbf{\Pi}_n^1\text{-mon}|$ for all n. Moreover if $n \geq 2$, then $|\mathbf{\Delta}_n^1\text{-mon}| = |\mathbf{\Delta}_n^1| \leq |\mathbf{\Sigma}_n^1\text{-mon}| < |\mathbf{\Sigma}_n^1| < |\mathbf{\Delta}_{n+1}^1\text{-mon}|$ and $|\mathbf{\Delta}_n^1\text{-mon}| = |\mathbf{\Delta}_n^1| \leq |\mathbf{\Pi}_n^1\text{-mon}| < |\mathbf{\Pi}_n^1| < |\mathbf{\Delta}_{n+1}^1\text{-mon}|$. Moreover, assuming the axiom of constructibility we have $|\mathbf{\Delta}_n^1| = |\mathbf{\Sigma}_n^1\text{-mon}| < |\mathbf{\Sigma}_n^1| < |\mathbf{\Pi}_n^1| < |\mathbf{\Delta}_{n+1}^1|$, for all $n \geq 2$. However, assuming that both Projective Determinacy (PD) and the axiom of Dependent Choices (DC) hold, then $|\mathbf{\Delta}_{2i}^1| = |\mathbf{\Sigma}_{2i}^1\text{-mon}| < |\mathbf{\Sigma}_{2i}^1| < |\mathbf{\Pi}_{2i}^1| < |\mathbf{\Delta}_{2i+1}| = |\mathbf{\Pi}_{2i+1}^1\text{-mon}| < |\mathbf{\Pi}_{2i+1}^1| < |\mathbf{\Sigma}_{2i+1}^1|$. Complete proofs are given for the most important results. Many less interesting results are stated without proofs, or with a sketch of a proof.

1. Introduction

Let ω denote the set of nonnegative integers which coincide with the ordinals less than ω.

Definition 1. By an *inductive definition* (abbreviated i.d.) we shall mean a mapping $\Gamma : P(\omega) \to P(\omega)$, where $P(\omega) = \{X \mid X \subseteq \omega\}$ = the set of all subsets of ω. ("Inductive definitions" is abbreviated i.ds.)

Definition 2. An i.d. Γ is *monotone* iff $X \subseteq Y \subseteq \omega$ implies $\Gamma(X) \subseteq \Gamma(Y)$.

Definition 3. Let Γ be an i.d. and let λ be an ordinal number. Then we define Γ^λ to be a set of integers, defined by transfinite recursion as follows:

$$\Gamma^0 = \emptyset .$$

* On leave from Institute of Mathematics, University of Oslo, Blindern, Oslo 3, Norway.

For $\lambda > 0$ we have

$$\Gamma^{\lambda} = \mathbf{U}\{\Gamma(\Gamma^{\xi}) \mid \xi < \lambda\} .$$

Definition 4. Let Γ be an i.d. The *closure ordinal* of Γ is the least ordinal λ such that $\Gamma^{\lambda} = \Gamma^{\lambda+1}$. We shall let $|\Gamma|$ denote the closure ordinal of Γ, and $\Gamma^{\infty} = \Gamma^{|\Gamma|}$ is the set *defined by* Γ.

Remark 1. $|\Gamma|$ exists for every i.e. Γ, and $|\Gamma|$ is a countable ordinal.

Definition 5. Let ${}^{\omega}\omega$ be the set of all functions of one number variable, i.e., the set of all total functions mapping ω into ω. Subsets of the product space

$$\chi = X_1 \times \ldots \times X_k$$
$$(X_i = \omega \text{ or } X_i = P(\omega) \text{ or } X_i = {}^{\omega}\omega \text{ for all } i = 1, 2, \ldots, k) ,$$

will be called *pointsets*. Sometimes we think of these as *relations* and we write interchangeably $x \in A \Leftrightarrow A(x)$.

Remark 2. It turns out that it is convenient for us to permit $X_i = P(\omega)$. In this way we deviate from the definition of pointsets in Moschovakis 1970 and 1972.

For $n, m \in \omega$, let $\langle n, m \rangle$ be a coding of pairs, say $\langle n, m \rangle = 1/2(n^2 + 2nm + m^2 + 3m + m)$ as in Rogers 1967, p. 64.

We shall use $f_0, f_1, f_2, \ldots$ to denote some functions throughout the paper according to the following definition.

Definition 6. For each $i \in \omega$, f_i is a mapping of ω into ω defined as follows: $f_i(x) = \langle i, x \rangle$, $x \in \omega$, and f_i is also a mapping of $P(\omega)$ into $P(\omega)$ defined as follows: $f_i(S) = \{f_i(x) \mid x \in S\}$. Moreover f_i^{-1} is a mapping of $P(\omega)$ into $P(\omega)$ defined as follows:

$$f_i^{-1}(S) = \{y \mid (\exists x)(x \in S \;\&\; x = f_i(y))\} .$$

Let $\alpha \in {}^{\omega}\omega$. Write $(\alpha)_i$ for the function $\lambda x \alpha(\langle i, x \rangle)$, i.e. $(\alpha)_i \in {}^{\omega}\omega$ and $(\alpha)_i(x) = \alpha(\langle i, x \rangle)$.

We do not need the full axiom of choice in this paper. At some places we need a weak axiom of choice:

Dependent Choices (DC). For each $A \subseteq {}^{\omega}\omega \times {}^{\omega}\omega$ we have

$$\text{(DC)} \qquad (\forall\alpha)(\exists\beta)A(\alpha,\beta) \Rightarrow \exists\alpha\forall n A((\alpha)_n, (\alpha)_{n+1}) .$$

This follows from the axiom of choice. In some cases a still weaker axiom is sufficient:

Countable Choices (CC). For each $A \subseteq \omega \times {}^{\omega}\omega$ we have:

$$\text{(CC)} \qquad (\forall x)(\exists\alpha)A(x,\alpha) \Rightarrow (\exists\alpha)(\forall x)A(x,(\alpha)_x) .$$

The axiom of Dependent Choices implies the axiom of Countable Choices. The axiom of Determinacy (AD) also implies CC.

Note that the axiom of CC has to be used in order to prove that the following prefix transformations are permissible:

$$\ldots \forall^0 \exists^1 \ldots \rightarrow \ldots \exists^1 \forall^0 \ldots$$

$$\ldots \exists^0 \forall^1 \ldots \rightarrow \ldots \forall^1 \exists^0 \ldots$$

(See Rogers 1967, p. 375 and Exercise 16–2, p. 446.)

Definition 7. A *point class* $\mathcal{C}$ is a class of pointsets, not necessarily all in the same product space.

It turns out to be convenient for us to represent each i.d. Γ by the pointset

$$\{(A,n) \mid A \subseteq \omega \text{ and } n \in \Gamma(A)\} \subseteq {}^{\omega}\omega \times \omega ,$$

instead of the pointset

$$\{(A,\Gamma(A)) \mid A\} \subseteq {}^{\omega}\omega \times {}^{\omega}\omega .$$

We shall identify an i.d. with its representation. Hence $\Gamma \in \mathcal{C}$ means

$$\{(A,n) \mid A \subseteq \omega \text{ and } n \in \Gamma(A)\} \in \mathcal{C} \ .$$

Given a pointset $\mathcal{C}$ we usually in this paper pay attention only to the subclass $({}^{\omega}\omega \times \omega) \cap \mathcal{C}$ of $\mathcal{C}$.

Definition 8. $\mathbf{\Sigma}^1_n$ is the class of all pointsets definable from recursive relations by $\mathbf{\Sigma}^1_n$ prefix. Similarly for $\mathbf{\Pi}^1_n$.

Definition 9. Let Γ be an i.d., then $\check{\Gamma} = ({}^{\omega}\omega \times \omega) - \Gamma$ (i.e. $\check{\Gamma}(S) = \omega - \Gamma(S)$ for all $S \subseteq \omega$).

Definition 10. Let $\check{\mathcal{C}}$ be a class of pointsets. Then $\check{\mathcal{C}} = \{\chi - A \mid A \subseteq \chi$ and $A \in \mathcal{C}\}$. (Here χ is a cartesian product as in Definition 5.)

Remark 3. Let Γ be an i.d. Then $\check{\Gamma} \in \mathbf{\Pi}^1_n$ iff $\Gamma \in \mathbf{\Sigma}^1_n$ and $\check{\mathbf{\Pi}}^1_n = \mathbf{\Sigma}^1_n$, $\check{\mathbf{\Sigma}}^1_n = \mathbf{\Pi}^1_n$.

Definition 11. Let $\mathcal{C}$ be a class of pointsets. Then the *supremum ordinal* for $\mathcal{C}$ is $|\mathcal{C}| = \sup\{|\Gamma| \mid \Gamma$ is an i.d. and Γ is an i.d. and $\Gamma \in \mathcal{C}\}$.

Definition 12. Let $A \subseteq \omega$ and let H be a pointset such that $H \subseteq X_1 \times X_2 \times \ldots \times X_k$ where the X_i's are as in Definition 5. Then $H^A = \{(x_1, x_2, \ldots, x_{k-1}) \mid (x_1, x_2, \ldots, x_{k-1}, A) \in H\}$ if $X_k = P(\omega)$ and $H^A = H$ otherwise. Let $\mathcal{C}$ be a class of pointsets. Then $H^A = \{H^A \mid H \in \mathcal{C}\}$.

Remark 4. Note that if $\mathcal{C} = \mathbf{\Sigma}^1_n$ then $\mathcal{C}^A = \mathbf{\Sigma}^{1,A}_n$, where $\mathbf{\Sigma}^{1,A}_n$ is defined as in Rogers 1967. (See pp. 409, 374 and 304.) In the same way, if $\mathcal{C} = \mathbf{\Pi}^1_n$ then $\mathcal{C}^A = \mathbf{\Pi}^{1,A}_n$.

Definition 13. Let $\mathcal{C}$ be a class of pointsets. The ordinal λ is called a *terminal ordinal* for $\mathcal{C}$ iff there exists an $A \subseteq \omega$ such that $\lambda = |\mathcal{C}^A|$. The set of terminal ordinals for $\mathcal{C}$ is denoted by $\mathrm{ter}(\mathcal{C})$.

Definition 14. Let $\mathcal{C}$ be a class of pointsets. The ordinal λ is a transit ordinal for $\mathcal{C}$ iff there exists an $A \subseteq \omega$ such that $\lambda < |\mathcal{C}^A|$ but $|\Gamma| \neq \lambda$ for all i.ds. $\Gamma \in \mathcal{C}$.

Definition 15. Let Λ and Γ be i.ds. Then Λ is *one-one reducible* to Γ *up to* λ

(notation: $\Lambda \leq_1{}^\lambda \Gamma$) iff there exists a one-one recursive function f such that $(\forall\xi)(\forall x)(\xi<\lambda \Rightarrow (x \in \Lambda^\xi \Leftrightarrow f(x) \in \Gamma^\xi))$. Λ is *one-one reducible to* Γ *at every ordinal* (notation: $\Lambda \leq_1^\infty \Gamma$) iff $\Lambda \leq_1^\lambda \Gamma$ where $\lambda = (\sup\{|\Lambda|, |\Gamma|\}) + 1$.

Definition 16. Let Λ and Γ be i.ds., and let λ be an ordinal; and let g be a mapping of ordinals into the ordinals such that $g(\xi)$ is defined if $\xi < \lambda$. Then Λ is *one-one reducible to the* g^{-1} *contraction* Γ up to λ (notation $\Lambda \leq_1^\lambda \leq_g \Gamma$) iff there exists a one-one recursive function f such that $(\forall\xi)(\forall x)(\xi<\lambda \Rightarrow (x \in \Lambda^\xi \Leftrightarrow f(x) \in \Gamma^{g(\xi)}))$. Λ is *one-one reducible to the* g^{-1}*-contraction of* Γ at every ordinal (notation: $\Lambda \leq_1^\infty \leq_g \Gamma$) iff $\Lambda \leq_1^\lambda \leq \Gamma$ where $\lambda = (\sup\{|\Lambda|, |\Gamma|\}) + 1$. (We shall write $\Lambda \leq_1^\lambda \leq \Gamma$ to mean that ($\Lambda \leq_1^\lambda \leq_g \Gamma$) holds for some g. $\Lambda \leq_1^\infty \leq \Gamma$ is interpreted similarly.

Definition 17. Let Λ and Γ be i.ds., then Λ is *one-one reducible to a reorganization of* Γ up to λ (notation $\Lambda \leq_1^\lambda \simeq \Gamma$) iff there exists a one-one recursive function f such that the following two conditions are satisfied, for all ordinals θ, ξ, where $\xi < \lambda$.

1. $\quad f^{-1}(\Gamma^\theta) \subseteq \Lambda^\xi \Rightarrow f^{-1}(\Gamma^{\theta+1}) \subseteq \Lambda^{\xi+1}$
2. $\quad f^{-1}(\Gamma^\theta) - \Lambda^\lambda \neq \emptyset \Rightarrow \Gamma^{\theta+1} - \Gamma^\theta \neq \emptyset$.

Λ is one-one reducible to a reorganization of Γ (notation: $\Lambda \leq_1^\infty \simeq \Gamma$), iff $\Lambda \leq_1^\lambda \simeq \Gamma$ where $\lambda = |\Lambda|$.

Remark 5. Let Λ and Γ be i.ds. and let h be a one-one recursive function. Then "$\Lambda \leq_1^\infty \Gamma$ via h" small mean that $\Lambda \leq_1^\infty \Gamma$ where the recursive one-one function f mentioned in Definition 15 can be chosen to be equal to h, i.e., $h^{-1}(\Gamma^\xi) = \Lambda^\xi$ for all $\xi \leq |\Lambda|$. Similarly for the following expressions: "$\Lambda \leq_1{}^\lambda \Gamma$ via h" "$\Lambda \leq_1{}^\lambda \leq_g \Gamma$ via h", "$\Lambda \leq_1^\infty \leq_g \Gamma$ via h", "$\Lambda \leq_1^\lambda \simeq \Gamma$ via h" and "$\Lambda \leq_1^\infty \simeq \Gamma$ via h".

Definition 18. Let $\mathcal{C}$ be a class of point sets, and let Γ be an i.d. Then Γ is *inductively* $\mathcal{C}$*-complete* iff $\Gamma \in \mathcal{C}$ and $(\forall\Lambda)(\Lambda \subseteq {}^\omega\omega \times \omega \;\&\; \Lambda \in \mathcal{C} \Rightarrow \Lambda \leq_1^\infty \Gamma)$.

Definition 19. Let $\mathcal{C}$ be a class of pointsets. Then the spectrum of $\mathcal{C}$ (denoted: $\mathrm{sp}(\mathcal{C})$) is the following set of ordinals: $\{|\Gamma| \mid \Gamma \in \mathcal{C}$ and Γ is an i.d.$\}$.

Remark 6. We shall from mow on assume that the axiom CC holds.

2.

We shall prove the following results.

Theorem 1. *Let $\mathcal{C} = \mathbf{\Sigma}_n^1$ or $\mathcal{C} = \mathbf{\Pi}_n^1$. Then there exist an i.d. Γ which is inductively $\mathcal{C}$-complete and such that $\Lambda^\lambda \leq_1 \Gamma^\infty$ for each i.d. Λ in $\mathcal{C}$, and for each ordinal λ. Hence $|\Gamma| = |\mathcal{C}| \in \mathrm{Sp}(\mathcal{C})$.*

Theorem 2. $|\mathbf{\Sigma}_n^1| \neq |\mathbf{\Pi}_n^1|$, *for every $n \geq 1$.*

Remark 7. It is proved in Aczel and Richter (1972) that $|\mathbf{\Sigma}_1^1| \neq |\mathbf{\Pi}_1^1|$; and they have later obtained a lot of results including proofs of theorems 1–4 in this paper.

Theorem 3. *Let $\mathcal{C} = \mathbf{\Pi}_n^1$ or let $\mathcal{C} = \mathbf{\Sigma}_n^1$ for some n. Suppose $|\breve{\mathcal{C}}| \leq |\mathcal{C}|$. Let Γ and Λ be inductively $\mathcal{C}$-complete and inductively $\breve{\mathcal{C}}$-complete, respectively. Let $g(\lambda) = \lambda + 1$ if λ is a successor ordinal, and let $g(\lambda) = \lambda$ otherwise. Then $\Lambda \leq_1^\infty \leq_g \Gamma$ and $\Gamma \leq_1^{|\Lambda|+1} \leq_g \Lambda$.*

Definition 20. Let K be a set of ordinals. An ordinal λ is a *point of accumulation of K* iff $(\forall\xi)(\xi < \lambda \Rightarrow ((\exists\eta)(\xi < \eta < \lambda \;\&\; \eta \in K))$, and λ is an *isolated point of K* iff $\lambda \in K$ and λ is not a point of accumulation of K. The set of points of accumulation of K is denoted by $\mathrm{Acc}(K)$ and the set of isolated points of K is denoted by $I_s(K)$.

Theorem 4. *Let $\mathcal{C}$ be a class of pointsets equal to $\mathbf{\Pi}_n^1$ or $\mathbf{\Sigma}_n^1$ for some $n \geq 1$. Then*

$$\mathrm{Acc}(\mathrm{tra}(\mathcal{C}) \cup \mathrm{ter}(\mathcal{C})) = \mathrm{Acc}(\mathrm{tra}(\breve{\mathcal{C}}) \cup \mathrm{ter}(\breve{\mathcal{C}}))$$

and

$$\mathrm{tra}(\mathcal{C}) - \mathrm{Acc}(\mathrm{tra}(\mathcal{C}) \cup \mathrm{ter}(\mathcal{C})) = \mathrm{ter}(\breve{\mathcal{C}}) - \mathrm{Acc}(\mathrm{tra}(\breve{\mathcal{C}}) \cup \mathrm{ter}(\breve{\mathcal{C}})).$$

Remark 8. $\mathrm{Tra}(\mathbf{\Pi}_1^1) = \emptyset$. Hence $\mathrm{Tra}(\mathbf{\Pi}_1^1) \neq \mathrm{Ter}(\mathrm{T}_1^1) = \mathrm{Ter}(\mathbf{\Sigma}_1^1)$. We do not know whether $\mathrm{Tra}(\mathbf{\Sigma}_1^1) = \mathrm{Ter}(\mathbf{\Pi}_1^1)$, but we would like to state that as a conjuncture. We do not know if it is consistent with set theory to assume $\mathrm{tra}(\mathbf{\Pi}_n^1) - \mathrm{ter}(\mathbf{\Sigma}_n^1) \neq \emptyset$ or $\mathrm{tra}(\mathbf{\Sigma}_n^1) - \mathrm{ter}(\mathbf{\Pi}_n^1) \neq \emptyset$ for some n. Theorem 4 will not be proved in this paper.

To state the next theorem we need the notions $\mathcal{C}$-norm and Prewell-ordering ($\mathcal{C}$).

Definition 21. A *norm* on a set p is a function $\phi : p \to$ ordinals; we call ϕ a $\mathcal{C}$-*norm* if there are relations $\underset{\bullet}{\leq}$ and $\overset{\bullet}{\leq}$ in $\mathcal{C}$ and $\breve{\mathcal{C}}$, respectively, such that $(\forall y)(y \in p \Rightarrow (\forall x)(x \underset{\bullet}{\leq} y \Leftrightarrow x \overset{\bullet}{\leq} y \Leftrightarrow [x \in p \;\&\; \phi(x) \leq \phi(y)]))$.

Definition 22. Prewellordering ($\mathcal{C}$) $\Leftrightarrow$ every pointset p in $\mathcal{C}$ admits a $\mathcal{C}$-norm.

Theorem 5. *Let Γ and $\hat{\Gamma}$ be i.ds., and let $\Gamma \in \mathcal{C}$ and $\hat{\Gamma} \in \breve{\mathcal{C}}$, where $\mathcal{C} = \mathbf{\Sigma}_n^1$ or $\mathcal{C} = \mathbf{\Pi}_n^1$ for some $n \geq 1$. Suppose $\Gamma(S) - S \neq \emptyset \Rightarrow \emptyset \neq \hat{\Gamma}(S) - S \subseteq \omega$. Then there exists an i.d. $\Lambda \in \mathcal{C}$ such that $\Gamma \leq_1^\infty \leq_g \Lambda$ and $|\Lambda| = |\Gamma| + 1$, where g is as in Theorem 3.*

Theorem 6. *Let $\mathcal{C} = \mathbf{\Sigma}_n^1$ or $\mathcal{C} = \mathbf{\Pi}_n^1$ for some $n > 1$, and suppose Prewell-ordering ($\mathcal{C}$). Then $|\mathcal{C}| < |\breve{\mathcal{C}}|$.*

Corollary 1. $|\mathbf{\Pi}_1^1| < |\mathbf{\Sigma}_1^1|$.

Corollary 2. $|\mathbf{\Sigma}_2^1| < |\mathbf{\Pi}_2^1|$.

Corollary 3. *Suppose that every element in ${}^\omega\omega$ is constructible in the sense of Gödel, then $|\mathbf{\Sigma}_n^1| < |\mathbf{\Pi}_n^1|$ for all $n \geq 2$.*

Corollary 4. *Assume Projective Determinacy (PD) and the axiom of Dependent Choices (DC) hold. Then $|\mathbf{\Pi}_{2i-1}^1| < |\mathbf{\Sigma}_{2i-1}^1|$ and $|\mathbf{\Sigma}_{2i}^1| < |\mathbf{\Pi}_{2i}^1|$ for all $i \geq 1$.*

These are the results which will be proved in this paper. At the end of the paper some recent results are stated which will be proved in a later paper.

2. Proof of Theorem 1.

Let $\mathcal{C} = \mathbf{\Sigma}_n^1$ or $\mathcal{C} = \mathbf{\Pi}_n^1$ for some $n \geq 1$. Then there exists a pointset $E \in \mathcal{C}$ such that $E \subseteq \omega \times \omega \times P(\omega)$, which enumerates all pointsets in $P(\omega \times P(\omega)) \cap \mathcal{C}$ in the sense that if $A \in \mathcal{C}$ and $A \subset \omega \times P(\omega)$, then there exists number i such that

$$A(x,S) \Leftrightarrow E(i,x,S) \text{ for all } x \in \omega \text{ and } S \subseteq \omega .$$

We shall write $x \in \Lambda_i(S)$ for $E(i,x,S)$. Then we have that for each i.d., $\Lambda \in \mathcal{C}$ there exists an $i \in \omega$ such that $\Lambda_i = \Lambda$. Let Γ be an i.d. defined as follows:

$$(1) \quad \Gamma(S) = \{x \mid (\exists y)(\exists z)(x = f_z(y) \;\&\; y \in \Lambda_z(f_z^{-1}(S)))\}$$
$$= \bigcup_{z=0}^{\infty} f_z(\Lambda_z(f_z^{-1}(S))).$$

(Recall that $f_z(x) = \langle z,x \rangle$, see Definition 6.)
Then $\Gamma \in \mathcal{C}$, and by induction on λ we can easily prove that $(\forall x)(x \in \Lambda_z^\lambda \Leftrightarrow f_z(x) \in \Gamma_z^\lambda)$. This proves that Γ is $\mathcal{C}$-complete since if Λ is an i.d., and $\Lambda \in \mathcal{C}$, then $\Lambda = \Lambda_i$ for some i and $\Lambda_i \leq_1^\infty \Gamma$ via f_i.

To prove the last part of the theorem we shall prove a lemma.

Lemma 1. *Let $\mathcal{C} = \mathbf{\Sigma}_n^1$ or $\mathcal{C} = \mathbf{\Pi}_n^1$ for some n. Let Γ be an i.d. in $\mathcal{C}$. Then there exists an i.d. Γ_0 in $\mathcal{C}$ such that $\Gamma \leq_1^\infty \Gamma_0$ and $\Gamma^\lambda \leq_1 \Gamma_0^\infty$ for each ordinal λ.*

Proof. Let Γ_0 be defined as follows:

$$\Gamma_0(S) = \{\langle 0,x \rangle \mid x \in \Gamma(f_0^{-1}(S)\}$$
$$\bigcup \{\langle z+1,x \rangle \mid (z \notin f_0^{-1}(S) \;\&\; x \in f_0^{-1}(S) \;\&\; z \in \Gamma(f_0^{-1}(S)))\}.$$

Then $\Gamma_0 \in \mathcal{C}$ and $(\forall x)(x \in \Gamma^\lambda \Leftrightarrow f_0(x) \in \Gamma_0^\lambda)$ for all ordinals λ. Moreover, if $\lambda \geq |\Gamma|$ then $(\forall x)(x \in \Gamma^\lambda \Leftrightarrow f_0(x) \in \Gamma_0^\infty)$, and if $\lambda < |\Gamma|$, choose $z \in \Gamma^{\lambda+1} - \Gamma^\lambda$. Then $(\forall x)(x \in \Gamma^\lambda \Leftrightarrow f_{z+1}(x) \in \Gamma_0^\infty)$. This proves Lemma 1.

We shall now complete the proof of Theorem 1.

Let Γ be defined by (1) as before and choose $\Gamma_0 \in \mathcal{C}$ such that $\Gamma \leq_1^\infty \Gamma_0$ and $\Gamma^\lambda \leq_1 \Gamma_0^\infty$ for all ordinals λ which is possible according to Lemma 1. Then $\Gamma_0 = \Lambda_j$ for some $j \in \omega$. Given $i \in \omega$, we have $\Lambda_i^\lambda \leq_1 \Gamma^\lambda <_1 \Lambda_j^\infty \leq_1 \Gamma^\infty$. Hence $\Lambda_i^\lambda \leq_1 \Gamma^\infty$ for all ordinals λ. This completes the proof of Theorem 1.

3. Proof of the Theorems 2 and 3.

To prove these theorems, we shall prove a lemma.

Lemma 2. *Let $\mathcal{C} = \mathbf{\Sigma}_n^1$ or $\mathcal{C} = \mathbf{\Pi}_n^1$, and let Λ and Γ be i.d.'s such that $\Lambda \in \check{\mathcal{C}}$ and $\Gamma \in \mathcal{C}$. Then there exists an i.d. $\Xi = \Xi^{\Lambda,\Gamma}$, such that $\Xi \in \check{\mathcal{C}}$ and*

$|\Xi| = |\Lambda|$ if $|\Lambda| \neq |\Gamma|$ and $|\Xi| = |\Lambda| + 1$ if $|\Lambda| = |\Gamma|$. Moreover, $\Lambda \leq_1^\infty \Xi$, and in addition we have $\Gamma \leq_1^\infty \leq_g \Xi$ if $|\Gamma| \leq |\Lambda|$ and $\Gamma \leq_1^{|\Lambda|+1} \leq_g \Xi$ if $|\Lambda| < |\Gamma|$, where g is the same as in Theorem 3.

Proof. We shall first outline the general idea of the proof, before giving the details of the construction. The main problem we have to solve is to construct $\Xi \in \breve{\mathcal{C}}$ from $\Lambda \in \breve{\mathcal{C}}$ and $\Gamma \in \mathcal{C}$ such that Γ^λ is coded in Ξ^λ in some way. First we note that $\breve{\Gamma} \in \breve{\mathcal{C}}$. Suppose that we can recover Γ^λ from Ξ^λ. We cannot get $\Gamma(\Gamma^\lambda)$ directly, but we can get $\breve{\Gamma}(\Gamma^\lambda) = \omega - \Gamma^{\lambda+1}$. If $\xi < \lambda$ then $\omega - \Gamma^\lambda \subseteq \omega - \Gamma^\xi$. Hence we cannot save $\omega - \Gamma^\lambda$ directly. But we can use the elements Λ^λ as indices. Hence we can try to choose Ξ such that $\Xi^\lambda = \bigcup_{\xi \leq \lambda} \Lambda^\xi \times (\omega - \Gamma^\xi)$. We shall use a somewhat more complicated construction. Let $I_\lambda = \{x+3 \mid x \in \Lambda^\lambda\}$. In order to generate the indices I_λ, code Λ^λ as $f_0(\Lambda^\lambda)$, and Γ^λ is coded as $\bigcup_{\xi \leq \lambda} (I_\xi \times (\omega - \Gamma^\xi))$. In this way, we can go on simulating both Λ and Γ as long as Λ generates new indices. (In the proof of Theorem 5 we shall use another construction to obtain new indices.)

In order to obtain $|\Xi| = |\Lambda| + 1$ if $|\Lambda| = |\Gamma|$ we add $f_1(0) = \langle 1, 0 \rangle$ when $\Gamma^{\lambda+1} = \Gamma^\lambda$. Moreover, in order to obtain $\Gamma \leq_1^{|\Gamma|+1} \leq_g \Xi$, we add $\langle 2, x \rangle$ to $\Xi^{\lambda+1}$ when $x \in \Gamma^\lambda$.

We shall now give the details of the construction.
Let $H(S) = \{u \mid (\exists v)(\langle v+3, u \rangle \notin S \;\&\; \langle 0, v \rangle \in S)\}$. We construct Ξ from Λ and Γ as follows.

$$\begin{aligned}\Xi(S) = {} & \{\langle 0, y \rangle \mid y \in \Lambda(f_0^{-1}(S))\} \\ & \cup \{\langle z+3, x \rangle \mid z \in \Lambda(f_0^{-1}(S)) \;\&\; x \notin \Gamma(H(S))\} \\ & \cup \{\langle 2, x \rangle \mid x \in H(S)\} \\ & \cup \{\langle 1, 0 \rangle \mid (\forall x)(x \in \Gamma(H(S)) \Rightarrow x \in H(S))\}.\end{aligned}$$

Then Ξ is an i.d. and $\Xi \in \breve{\mathcal{C}}$. We can easily prove by induction on λ.

i) $(\forall\lambda)(\forall x)(x \in \Lambda^\lambda \Leftrightarrow f_0(x) \in \Xi^\lambda)$
ii) $(\forall\lambda)(\lambda \leq |\Lambda| \Rightarrow \Gamma^\lambda = H(\Xi^\lambda))$
iii) $(\forall\lambda)(\lambda < |\Lambda| \Rightarrow \Gamma^\lambda = f_2^{-1}(\Xi^{\lambda+1}))$
iv) $(\forall\lambda)((\lambda \leq |\Lambda|$ and λ an limit ordinal$) \Rightarrow \Gamma^\lambda = f_2^{-1}(\Xi^\lambda))$.

Moreover, $|\Xi| = |\Lambda|$ except when $|\Lambda| = |\Gamma|$. If $|\Gamma| \leq |\Lambda|$ then $\langle 1, 0\rangle \in \Xi^{|\Gamma|+1} - \Xi^{|\Gamma|}$. Hence $|\Xi| = |\Gamma|+1 = |\Lambda|+1$ if $|\Lambda| = |\Gamma|$. Lemma 3 now follows easily and the proof of Lemma 3 is complete.

To prove Theorem 2, suppose $|\mathbf{\Sigma}_n^1| = |\mathbf{\Pi}_n^1|$ for some n. Choose $\Lambda \in \mathbf{\Sigma}_n^1$ and $\Gamma \in \mathbf{\Pi}_n^1$ such that $|\Lambda| = |\mathbf{\Sigma}_n^1|$ and $|\Gamma| = |\mathbf{\Pi}_n^1|$. Apply Lemma 2, and we get $\Xi \in \mathbf{\Sigma}_n^1$ such that $|\Xi| = |\Lambda|+1 = |\mathbf{\Sigma}_n^1|+1$, which is a contradiction. This proves Theorem 2.

To prove Theorem 3, let $\mathbf{\Sigma} = \mathbf{\Pi}_n^1$ or $\mathcal{C} = \mathbf{\Sigma}_n^1$ and suppose $|\breve{\mathcal{C}}| \leq |\mathcal{C}|$. Then $|\breve{\mathcal{C}}| < |\mathcal{C}|$ by Theorem 2. Let Γ and Λ be inductively $\breve{\mathcal{C}}$-complete and inductively $\mathcal{C}$-complete, respectively. Construct $\Xi_1 = \Xi^{\Lambda,\Gamma}$ and $\Xi_2 = \Xi^{\Gamma,\Lambda}$ as in Lemma 2. Then $\Xi_1 \in \breve{\mathcal{C}}$ and $\Xi_2 \in \mathcal{C}$ and $\Lambda \leq_1^\infty \Xi_1 \leq_1^\infty \Lambda$ and $\Gamma \leq_1^{|\Lambda|+1} \leq_g \Xi_1$. Hence $\Gamma \leq_1^{|\Lambda|+1} \leq_g \Lambda$. Moreover, $\Gamma <_1^\infty \Xi_2 \leq_1^\infty \Gamma$ and $\Lambda \leq_1^\infty \leq_g \Xi_2$. Hence $\Lambda <_1^\infty \leq_g \Gamma$. This proves Theorem 3.

We shall not prove Theorem 4 in this paper.

4. Proof of Theorems 5 and 6 and their corollaries

Let $\mathcal{C} = \mathbf{\Sigma}_n^1$ or $\mathcal{C} = \mathbf{\Pi}_n^1$ and let Γ and $\hat{\Gamma}$ satisfy the assumptions in Theorem 5. The proof is very similar to the proof of Lemma 2. The difference between the constructions in the proofs is the generation of indices I_λ. We shall now let $I_\lambda = \{x+3 \mid (\exists \xi)(\xi < \lambda \;\&\; x \in \hat{\Gamma}(\Gamma^\xi))\}$. We shall as before now code I_λ as $f_0(I_\lambda)$. Then $I_{\lambda+1} - I_\lambda \neq \emptyset$ if $\lambda < |\Gamma|$. As before, we let $\langle 1,0\rangle \in \Lambda^{|\Gamma|+1} - \Lambda^{|\Gamma|}$. The details of the construction are as follows. Let as before $f_i(y) = \langle i, y\rangle$ (see Definition 6) and let $H(S) = \{u \mid (\exists v)(\langle v+3, u\rangle \notin S \;\&\; \langle 0, v\rangle \in S)\}$, as in the proof of Lemma 2. Let $\Lambda(S)$ be defined as follows:

$$
\begin{aligned}
\Lambda(S) = \{\langle 0,x\rangle \mid x \in \hat{\Gamma}(H(S))\}&\\
\cup\ \{\langle z+3,x\rangle \mid z \in \hat{\Gamma}(H(S)) \;\&\; x \notin \Gamma(H(S))\}&\\
\cup\ \{\langle 1,0\rangle \mid (\forall x)(x \in \Gamma(H(S)) \Rightarrow x \in H(S))\}&\\
\cup\ \{\langle 2,x\rangle \mid x \in H(S)\}\,.&
\end{aligned}
$$

Then Λ is an i.d. and $\Lambda \in \check{\mathcal{C}}$. Moreover, we can easily prove by induction λ.

i. $\Gamma^\lambda = H(\Lambda^\lambda)$

ii. $\hat{\Gamma}(H(\Lambda^\lambda)) \subseteq \Gamma^{\lambda+1}$

iii. $\Gamma^{\lambda+1} - \Gamma^\lambda \neq \emptyset \Rightarrow \hat{\Gamma}(H(\Lambda^\lambda)) - H(\Lambda^\lambda) \neq \emptyset$

iv. $\Gamma^\lambda = f_2^{-1}(\Lambda^{\lambda+1})$

v. $\Gamma^\lambda = f_2^{-1}(\Lambda^\lambda)$ if λ is a limit ordinal.

Hence $|\Lambda| = |\Gamma| + 1$, since $|\Gamma| \leq |\Lambda|$ and $\langle 1,0\rangle \in \Lambda^{|\Gamma|+1} - \Lambda^{|\Gamma|}$. Moreover, by iv. and v. above we have $\Gamma \leq_1^\infty \leq_g \Lambda$. This completes the proof of Theorem 5.

We shall now prove Theorem 6. We need one more definition.

Definition 23. An i.d. Γ is *single-valued* iff $x \in \Gamma(S)$ and $y \in \Gamma(S)$ implies $x = y$. The set of single-valued $\mathbf{\Sigma}_n^1$ i.d.'s is denoted by ${}^{sv}\mathbf{\Sigma}_n^1$ and ${}^{sv}\mathbf{\Pi}_n^1$ denote the set of single-valued $\mathbf{\Pi}_n^1$ i.d.'s.

Theorem 6 is going to be an immediate consequence of Theorem 5 and the following lemma.

Lemma 3. *Let* $\mathcal{C} = \mathbf{\Sigma}_n^1$ *or* $\mathcal{C} = \mathbf{\Pi}_n^1$ *for some* $n \geq 1$, *and suppose Prewellordering* $(\mathcal{C})$. *Let* Γ *be an i.d. in* $\mathcal{C}$. *Then there exist i.d.'s* $\check{\Gamma}$ *in* $\mathcal{C}$ *and* $\hat{\Gamma}$ *in* $\check{\mathcal{C}}$, *such that* $\check{\Gamma}$ *is single-valued and if* $\Gamma(S) - S \neq \emptyset$ *then* $\emptyset \neq \hat{\Gamma}(S) = \check{\Gamma}(S) \subseteq \Gamma(S) - S$. *Moreover, if* $\Gamma(S) - S = \emptyset$, *then* $\hat{\Gamma}(S) = \omega$ *and* $\check{\Gamma}(S) = \emptyset$.

Proof. Let $\Gamma'(S) = \Gamma(S) - S$. Then $\Gamma' \in \mathcal{C}$. Consider the set $A = \{(x,S) \mid x \in \Gamma'(S)\}$. By Prewellordering $(\mathcal{C})$ we have that there exists $\mathcal{C}$-norm ϕ of A. Then there are relations $\underset{\bullet}{\leq}$ in $\mathcal{C}$ and $\dot{\leq}$ in $\check{\mathcal{C}}$ such that if $T \subseteq \omega$ and $y \in \omega$, and if $y \in \Gamma'(T)$, then for all $x \in \omega$ and all $S \subseteq \omega$ we have

$$(x,S) \dot{\leq} (y,T) \Leftrightarrow (x,S) \underset{\bullet}{\leq} (y,T) \Leftrightarrow (x \in \Gamma'(S) \;\&\; \phi((x,S)) \leq \phi((y,T)))$$

Then we define $\check{\Gamma}$ as follows.

$$\check{\Gamma}(S) = \{x \mid x \in \Gamma'(S) \;\&\; (\forall y)((y,S) \dot{\leq} (x,S) \Rightarrow ((x,S) \underset{\bullet}{\leq} (y,S) \;\&\; x \leq y))\}\,.$$

Then $\check{\Gamma} \in \mathcal{C}$. Given S, and suppose $\Gamma'(S) \neq \emptyset$. Let λ_S be the least ordinal λ such that $\lambda = \phi((x,S))$ and $x \in \Gamma'(S)$ for some x. Let x_S be the least $x \in \omega$

such that $\lambda_S = \phi((x,S))$. Then $\Gamma(S) = \{x_S\}$. If $\Gamma'(S) = \emptyset$ then $\Gamma(S) = \emptyset$. Hence Γ is single-valued. We now define Γ as follows

$$x \in \hat{\Gamma}(S) \Leftrightarrow (\forall y)(y \in \check{\Gamma}(S) \Rightarrow y = x) .$$

Then $\hat{\Gamma}(S) = \check{\Gamma}(S)$ if $\check{\Gamma}(S) \neq \emptyset$ and $\hat{\Gamma}(S) = \omega$ if $\check{\Gamma}(S) = \emptyset$. Lemma 3 now follows immediately.

Theorem 6 follows from Theorem 5, Theorem 1 and Lemma 3, by choosing $\Gamma \in \mathcal{C}$ such that $|\Gamma| = |\mathcal{C}|$. Then we obtain a $\Lambda \in \mathcal{C}$ such that $|\Lambda| = |\Gamma| + 1 = |\mathcal{C}| + 1$ by Lemma 3 and Theorem 5. This proves $|\mathcal{C}| < |\mathcal{C}|$, and the proof of Theorem 6 is complete.

Remark 9. We may obtain a slightly stronger result than Theorem 6 by using a weaker assumption than Prewellordering ($\mathcal{C}$). It is enough to assume that there exists a norm on A and a relation $R \subseteq \omega \times \omega \times P(\omega)$ such that if $(y,S) \in A$ (i.e., $y \in \Gamma'(A)$) then we have for all $x \in \omega$

$$R(x,y,S) \Leftrightarrow (x \in \Gamma'(S) \;\&\; \phi((x,S)) \leq \phi(y,S)) .$$

In this case we can define $\hat{\Gamma}$ as follows

$$\hat{\Gamma}(S) = \{x \mid (\forall y)(y \in \Gamma'(S) \Rightarrow R(x,y,S))\} .$$

Then $\Gamma(S) - S \neq \emptyset \Rightarrow \emptyset \neq \hat{\Gamma}(S) - S \subseteq \Gamma(S)$.

We shall now show that Corollaries 1–4 follow from Theorem 6.

We have that Prewellordering ($\mathbf{\Pi}_1^1$) and Prewellordering ($\mathbf{\Sigma}_2^1$) (see Moschovakis 1970, p. 33). Hence $|\mathbf{\Pi}_1^1| < |\mathbf{\Sigma}_1^1|$ and $|\mathbf{\Sigma}_2^1| < |\mathbf{\Pi}_2^1|$ by Theorem 6. This proves Corollaries 1 and 2.

According to Moschovakis 1970, p. 33, the arguments of Addison 1959a, suffice to show that if every real number is constructible in the sense of Gödel, then for each $k > 2$, Prewellordering ($\mathbf{\Sigma}_k^1$). By using this fact we obtain Corollary 3 from Theorem 5 immediately.

Finally, we have that PD and DC imply Prewellordering ($\mathbf{\Pi}_{2i-1}^1$) and Prewellordering ($\mathbf{\Sigma}_{2i}^1$). (See Moschovakis 1970, p. 33, or see Martin 1968.) Hence Corollary 4 follows.

5. Further results

We shall now state some results without proofs. A full treatment will be published elsewhere.

Theorem 7. $\omega_1 = |\boldsymbol{\Pi}_1^1\text{-mon}| < |\boldsymbol{\Delta}_1^1| = |{}^{\mathrm{sv}}\boldsymbol{\Sigma}_1^1| < |{}^{\mathrm{sv}}\boldsymbol{\Pi}_1^1| = |\boldsymbol{\Pi}_1^1| < |\boldsymbol{\Sigma}_1^1| < |\boldsymbol{\Delta}_2^1\text{-mon}| = |\boldsymbol{\Delta}_2^1| = |\boldsymbol{\Sigma}_2^1\text{-mon}| = |{}^{\mathrm{sv}}\boldsymbol{\Pi}_2^1| = |{}^{\mathrm{sv}}\boldsymbol{\Sigma}_2^1| < |\boldsymbol{\Sigma}_2^1| < |\boldsymbol{\Pi}_2^1|$.

Theorem 8. $|\boldsymbol{\Delta}_2^1| < |\boldsymbol{\Pi}_2^1\text{-mon}| < |\boldsymbol{\Pi}_2^1|$.

Remark 10. Let ω_1 be the first non-recursive ordinal. Spector showed that $|\boldsymbol{\Pi}_1^1\text{-mon}| = |\boldsymbol{\Pi}_1^0\text{-mon}| = \omega_1 < |\boldsymbol{\Delta}_1^1|$. According to T. Grilliot, $|\boldsymbol{\Sigma}_1^1\text{-mon}| = |\boldsymbol{\Sigma}_1^1|$.

Theorem 9. $|\boldsymbol{\Sigma}_n^1| \in \mathrm{Sp}(\boldsymbol{\Sigma}_n^1)$, $|\boldsymbol{\Pi}_n^1| \in \mathrm{Sp}(\boldsymbol{\Pi}_n^1)$, $|\boldsymbol{\Delta}_n^1| \notin \mathrm{Sp}(\boldsymbol{\Delta}_n^1)$, $|\boldsymbol{\Sigma}_n^1\text{-mon}| \in \mathrm{Sp}(\boldsymbol{\Sigma}_n^1\text{-mon})$, $|\boldsymbol{\Pi}_n^1\text{-mon}| \in \mathrm{Sp}(\boldsymbol{\Pi}_n^1\text{-mon})$. $|\boldsymbol{\Pi}_1^1| \in \mathrm{Sp}({}^{\mathrm{sv}}\boldsymbol{\Pi}_1^1)$, $|{}^{\mathrm{sv}}\boldsymbol{\Sigma}_2^1| \in \mathrm{Sp}({}^{\mathrm{sv}}\boldsymbol{\Sigma}_2^1)$. $|{}^{\mathrm{sv}}\boldsymbol{\Pi}_2^1| \notin \mathrm{Sp}({}^{\mathrm{sv}}\boldsymbol{\Sigma}_2^1)$.

Theorem 10. *Let $n \geq 2$. Then* $|{}^{\mathrm{sv}}\boldsymbol{\Sigma}_n^1| = |\boldsymbol{\Delta}_n^1\text{-mon}| = |\boldsymbol{\Delta}_n^1| \leq |\boldsymbol{\Sigma}_n^1\text{-mon}| < |\boldsymbol{\Sigma}_n^1| < |\boldsymbol{\Delta}_{n+1}^1\text{-mon}|$ *and* $|{}^{\mathrm{sv}}\boldsymbol{\Pi}_n^1| = |\boldsymbol{\Delta}_n^1\text{-mon}| = |\boldsymbol{\Delta}_n^1| \leq |\boldsymbol{\Pi}_n^1\text{-mon}| < |\boldsymbol{\Pi}_n^1| < |\boldsymbol{\Delta}_{n+1}^1\text{-mon}|$ *and* $|\boldsymbol{\Pi}_n^1\text{-mon}| \neq |\boldsymbol{\Sigma}_n^1|$ *and* $|\boldsymbol{\Pi}_n^1\text{-mon}| \neq |\boldsymbol{\Sigma}_n^1|$.

Theorem 11. $|{}^{\mathrm{sv}}\boldsymbol{\Pi}_n^1| \in \mathrm{Sp}({}^{\mathrm{sv}}\boldsymbol{\Pi}_n^1) \Rightarrow (|{}^{\mathrm{sv}}\boldsymbol{\Sigma}_n^1| \notin \mathrm{Sp}({}^{\mathrm{sv}}\boldsymbol{\Sigma}_n^1)$ *and* $|\boldsymbol{\Delta}_n^1| < |\boldsymbol{\Sigma}_n^1\text{-mon}|)$ *for all $n \geq 2$.*

Theorem 12. $|{}^{\mathrm{sv}}\boldsymbol{\Sigma}_n^1| \in \mathrm{Sp}({}^{\mathrm{sv}}\boldsymbol{\Sigma}_n^1) \Rightarrow (|{}^{\mathrm{sv}}\boldsymbol{\Pi}_n^1 \notin \mathrm{Sp}({}^{\mathrm{sv}}\boldsymbol{\Pi}_n^1)$ *and* $|\boldsymbol{\Delta}_n^1| < |\boldsymbol{\Pi}_n^1\text{-mon}|)$ *for all $n \geq 2$.*

Theorem 13. *Let $n \geq 2$. Suppose Prewellordering* $(\boldsymbol{\Pi}_n^1)$. *Then* $|\boldsymbol{\Delta}_n^1| = |\boldsymbol{\Pi}_n^1\text{-mon}| < |\boldsymbol{\Pi}_n^1| < |\boldsymbol{\Sigma}_n^1|$ *and* $|\boldsymbol{\Delta}_n^1| < |\boldsymbol{\Sigma}_n^1\text{-mon}| < |\boldsymbol{\Sigma}_n^1|$ *and* $|{}^{\mathrm{sv}}\boldsymbol{\Pi}_n^1| \in \mathrm{Sp}({}^{\mathrm{sv}}\boldsymbol{\Pi}_n^1)$ *and* $|{}^{\mathrm{sv}}\boldsymbol{\Sigma}_n^1| \notin \mathrm{Sp}({}^{\mathrm{sv}}\boldsymbol{\Sigma}_n^1)$.

Theorem 14. *Let $n \geq 2$. Suppose Prewellordering* $(\boldsymbol{\Sigma}_n^1)$. *Then* $|\boldsymbol{\Delta}_n^1| = |\boldsymbol{\Sigma}_n^1\text{-mon}| < |\boldsymbol{\Sigma}_n^1| < |\boldsymbol{\Pi}_n^1|$ *and* $|\boldsymbol{\Delta}_n^1| < |\boldsymbol{\Pi}_n^1\text{-mon}| < |\boldsymbol{\Pi}_n^1|$ *and* $|{}^{\mathrm{sv}}\boldsymbol{\Sigma}_n^1| \in \mathrm{Sp}({}^{\mathrm{sv}}\boldsymbol{\Sigma}_n^1)$ *and* $|{}^{\mathrm{sv}}\boldsymbol{\Pi}_n^1| \notin \mathrm{Sp}({}^{\mathrm{sv}}\boldsymbol{\Pi}_n^1)$.

Acknowledgement

I wish to thank Jens Erik Fenstad for having encouraged me to work on the problem of deciding the order relation between $|\mathbf{\Pi}_1^1|$ and $|\mathbf{\Sigma}_1^1|$.

I am also indebted to Peter Aczel and Wayne Richter for helpful comments on the preliminary draft of this paper.

I am also grateful to Leo Harrington for helpful discussions.

References

[1] Aczel, P. and W. Richter, Inductive definitions and analogues of large cardinals, in: Conference in Mathematical Logic, London, 1970. Lecture Notes in Mathematics Nr. 255 (Springer, Berlin, 1971) 1–9.

[2] Addison, T.W., Some consequences of the axiom of constructibility, Fund. Math. 46 (1959) 123–135.

[3] Addison, T.W. and Y.N. Moschovakis, Some consequences of the axiom of definable determinateness, Proc. Nat. Acad. Sci., U.S.A., 59 (1968) 708–712.

[4] Martin, D.A., The axiom of determinateness and reduction principles in the analytical hierarchy, Bull. Amer. Soc. 74 (1968) 687–689.

[5] Moschovakis, Y.N., Determinacy and Prewellordering of the continuum, in: Math. Logic and Foundations of Set Theory, Y. Bar Hillel (ed.) (North-Holland, Amsterdam, 1970) 24–62.

[6] Moschovakis, Y.N., Uniformization in a playful universe, Bull. Amer. Math. Soc. 77 (1971) 731–736.

[7] Rogers, Hartley, Jr., Theory of Recursive Functions and Effective Calculability (McGraw-Hill, New York, 1967).

[8] Spector, C., Inductively defined sets of natural numbers, in: Infinistic Methods (Pergamon Press, Oxford and PWN, Warsaw, 1961) 97–102.

J.E.Fenstad, P.G. Hinman (eds.), Generalized Recursion Theory
© North-Holland Publ. Comp., 1974

ORDINAL RECURSION AND INDUCTIVE DEFINITIONS

Douglas CENZER
University of Michigan and
University of Florida

1. Introduction

An inductive operator Γ over a set X is a map Γ from $P(X)$ to $P(X)$ such that for all $A \subseteq X, A \subseteq \Gamma(A)$. Γ determines a transfinite sequence $\{\Gamma^\alpha : \alpha \in \text{ORD (ordinals)}\}$, where $\Gamma^\alpha = \mathbf{U}\{\Gamma^\beta : \beta < \alpha\}$ for $\alpha = 0$ or α a limit and $\Gamma^{\alpha+1} = \Gamma(\Gamma^\alpha)$. Γ is monotone if, for all A, B in $P(X), A \subsetneq B$ implies $\Gamma(A) \subseteq \Gamma(B)$.

The closure ordinal $|\Gamma|$ of Γ is the least ordinal α such that $\Gamma^{\alpha+1} = \Gamma^\alpha$; clearly $|\Gamma|$ always has cardinality less than or equal to $\overline{\overline{X}}$. The closure $\overline{\Gamma}$ of Γ is $\Gamma^{|\Gamma|}$, the set inductively defined by Γ.

Inductive definitions are basic to the development of recursion theory. Following the methods of Kleene [11, 12], we will define ordinal recursion and recursion in a partial functional by means of inductive definitions.

Given a class C of inductive operators, one would like to characterize the closure ordinal $|C| = \sup\{|\Gamma| : \Gamma \in C\}$ and the closure algebra $\overline{C} = \{A : A$ is 1–1 reducible to $\overline{\Gamma}$ for some $\Gamma \in C\}$. We write C-mon for the class of monotone operators in C.

The following is a brief summary of results on inductive definitions over the natural numbers.

The first significant results on inductive definitions were obtained by Spector [24], who showed that $|\mathbf{\Pi}^0_1\text{-mon}| = \omega_1$, the first non-recursive ordinal, and that $\overline{\mathbf{\Pi}^0_1\text{-mon}} = \overline{\mathbf{\Pi}^1_1\text{-mon}} = \mathbf{\Pi}^1_1$. Gandy (unpublished) later showed that $|\mathbf{\Pi}^0_1| = \omega_1$ and $\overline{\mathbf{\Pi}^0_1} = \mathbf{\Pi}^1_1$. We make use of slightly generalized versions of Spector's results in section 3.

It is easily seen that $|\mathbf{\Pi}^1_1| > \omega_1$ and $\overline{\mathbf{\Pi}^1_1} \neq \mathbf{\Pi}^1_1$. Richter [19] demonstrated that even $|\mathbf{\Pi}^0_2|$ is a rather large admissible ordinal. Anderaa [1] recently proved that $|\mathbf{\Pi}^1_1| < |\mathbf{\Sigma}^1_1|$. On the other hand it follows from the work of Aczel [2] on

$\mathbf{\Sigma}_1^1$ operators that $|\mathbf{\Sigma}_1^1| = |\mathbf{\Sigma}_1^1\text{-mon}|$. Aczel and Richter [3, 4] have characterized $|\mathbf{\Pi}_1^1|$, $|\mathbf{\Sigma}_1^1|$, and, for all n, $|\mathbf{\Pi}_n^0|$ in terms of reflection principles in the constructible hierarchy.

Putnam [18] showed that $|\mathbf{\Delta}_2^1| = \delta_2^1$, the first non-$\mathbf{\Delta}_2^1$ ordinal; it is also known that $\overline{\mathbf{\Delta}_2^1} = \mathbf{\Delta}_2^1$. More generally, for all $n > 1$, $\overline{\mathbf{\Delta}_n^1} = \mathbf{\Delta}_n^1$, $\overline{\mathbf{\Pi}_n^1\text{-mon}} = \mathbf{\Pi}_n^1$, and $\overline{\mathbf{\Sigma}_n^1\text{-mon}} = \mathbf{\Sigma}_n^1$. (However, $\overline{\mathbf{\Sigma}_1^1\text{-mon}} \neq \mathbf{\Sigma}_1^1$.) Using the techniques of Lemma 9.12, we can now show that $|\mathbf{\Delta}_n^1| = \delta_n^1$ for all $n > 2$. (See Cenzer [8].)

2. Summary of results

In this paper we explore further the relation between non-monotone inductive definitions and ordinal recursion. As pointed out above, ordinal recursion can be defined by an inductive operator, and in return the theory of inductive definitions can be developed within the framework of ordinal recursion.

For any ordinal α, let α^+ be the least recursively regular (admissible) ordinal greater than α and let α^* be the least stable ordinal greater than α. Let S be the class of stable ordinals. (See section 3 for a brief development of ordinal recursion theory.)

Theorem A. (a). $|\mathbf{\Pi}_1^1|$ *is the least ordinal* α *which is not* α^+*-recursive*;
(b). $|\mathbf{\Pi}_1^1|$ *is the least ordinal* α *which is* α^+*-stable*;
(c). $|\mathbf{\Pi}_1^1|$ *is the least ordinal* α *such that* L_α *is a* $\mathbf{\Sigma}_1$*-elementary submodel of* L_{α^+}.

Part (b) was proven independently by Aczel and Richter [3, 4].

Theorem B. (a). $|\mathbf{\Sigma}_2^1|$ *is the least ordinal* α *which is not* α^**-recursive in* S;
(b). $|\mathbf{\Sigma}_2^1|$ *is the least ordinal* α *which is* α^**-stable in* S;
(c). $|\mathbf{\Sigma}_2^1|$ *is the least ordinal* α *such that* L_α *is a* $\mathbf{\Sigma}_2$*-elementary submodel of* L_{α^*}.

Theorem C. (a). $\overline{\mathbf{\Pi}_1^1}$ *is the class of sets of numbers which are* $|\mathbf{\Pi}_1^1|$*-semirecursive (the domain of a* $|\mathbf{\Pi}_1^1|$*-partial recursive function)*;
(b). $\overline{\mathbf{\Sigma}_2^1}$ *is the class of sets of numbers which are* $|\mathbf{\Sigma}_2^1|$*-semirecursive in S.*

We derive results similar to Theorem C regarding $\mathbf{\Sigma}_1^1$ and $\mathbf{\Pi}_2^1$ inductive operators.

We extend the above results to combinations of operators in which the components need be inductive. For example, if $(\mathbf{\Pi}_1^1)^2$ denotes the class of inductive operators which are the composition of two $\mathbf{\Pi}_1^1$ operators, then $|(\mathbf{\Pi}_1^1)^2|$ is the least ordinal α which is not α^{++}-recursive. Similar extensions of Theorems A, B, and C obtain for the classes $(\mathbf{\Pi}_1^1)^n$ and $(\mathbf{\Sigma}_1^1)^n$ for all n.

It is well known that for $A \subseteq \omega$ and α admissible, $A \in L_\alpha$ iff A is α-recursive. We generalize this result to the ordinal recursive arithmetic hierarchy and thus obtain results concerning constructibly analytic inductive operators. (A relation is constructibly analytic if definable by means of quantifiers restricted to the constructible reals.)

Analogously to the results of Aczel [2] on $E_1^\#$ and $\mathbf{\Sigma}_1^1$ inductive operators, we define a functional $G_1^\#$ and prove the following theorem.

Theorem D. (a). $|\mathbf{\Pi}_1^1| = \omega_1^{G^\#}$, *the least ordinal not recursive in* $G_1^\#$;
(b). $\mathbf{\Pi}_1^1$ *is the class of sets of numbers which are semirecursive in* $G_1^\#$.

3. Ordinal recursion

This section is intended to provide the necessary background of ordinal recursion theory. Proofs omitted here can be found in Cenzer [7].

Let $\langle\ \rangle$ be a natural sequencing function from $\bigcup_{n<\omega} \mathrm{ORD}^n$ to ORD. $(\langle\alpha_0, ..., \alpha_n\rangle)_i = \alpha_i$; $\ln(\langle\alpha_0, ..., \alpha_n\rangle) = n+1$; $\langle\alpha_0, ..., \alpha_n\rangle * \langle\beta_0, ..., \beta_n\rangle = \langle\alpha_0, ..., \alpha_n, \beta_0, ..., \beta_n\rangle$.

We give the inductive operator (actually a class of operators) which defines ordinal recursion in full detail here since we will later want to discuss two related operators with reference to this definition.

Definition 3.1. For any ordinal γ, any $l < \omega$, and any $f = (f_0, ..., f_{l-1})$, $\Omega_\gamma[f]$ is the monotone operator such that for all k and $n < \omega$, all $i < k$ and $j < l$, all $\boldsymbol{\alpha} = (\alpha_0, ..., \alpha_{k-1})$ with each $\alpha_i < \gamma$, all β, σ, τ, ξ, and $\zeta < \gamma$, and all $A \subseteq \mathrm{ORD}$:

(0) $\langle\langle 0, k, l, n\rangle, \boldsymbol{\alpha}, n\rangle \in \Omega_\gamma[f](A)$;

(1) $\langle\langle 1, k, l, i\rangle, \boldsymbol{\alpha}, \alpha_i\rangle \in \Omega_\gamma[f](A)$;

(2) $\langle\langle 2, k, l, i\rangle, \boldsymbol{\alpha}, \alpha_i + 1\rangle \in \Omega_\gamma[f](A)$;

(3) $\xi < \zeta$ implies $\langle\langle 3, k+4, l\rangle, \xi, \zeta, \sigma, \tau, \boldsymbol{\alpha}, \sigma\rangle \in \Omega_\gamma[f](A)$;
$\xi \geq \zeta$ implies $\langle\langle 3, k+4, l\rangle, \xi, \zeta, \sigma, \tau, \boldsymbol{\alpha}, \tau\rangle \in \Omega_\gamma[f](A)$;

(4) for all $m, b, c_0, ...,$ and $c_{m-1} < \omega$, and all $\tau_0, ..., \tau_{m-1} < \gamma$, if for all $i < m$,

$\langle c_i, \boldsymbol{\alpha}, \tau\rangle \in A$, and $\langle b, \boldsymbol{\tau}, \beta\rangle \in A$, then $\langle\langle 4, k, l, b, \boldsymbol{c}\rangle, \boldsymbol{\alpha}, \beta\rangle \in \Omega_\gamma[f](A)$;
(5) if $\forall \xi < \sigma\, \exists \zeta < \tau[\langle\langle 5, k+1, l, b\rangle, \xi, \boldsymbol{\alpha}, \zeta\rangle \in A]$, and
$\forall \zeta < \tau\, \exists \xi < \sigma\, \exists \rho < \tau[\rho > \zeta \wedge \langle\langle 5, k+1, l, b\rangle, \xi, \boldsymbol{\alpha}, \rho\rangle \in A]$, and
$\langle b, \tau, \sigma, \boldsymbol{\alpha}, \beta\rangle \in A$, then $\langle\langle 5, k+1, l, b\rangle, \sigma, \boldsymbol{\alpha}, \beta\rangle \in \Omega_\gamma[f](A)$;
(6) if $\forall \sigma < \beta\, \exists \tau > 0[\langle b, \sigma, \boldsymbol{\alpha}, \tau\rangle \in A]$, and $\langle b, \beta, \boldsymbol{\alpha}, 0\rangle \in A$, then
$\langle\langle 6, k, l, b\rangle, \boldsymbol{\alpha}, \beta\rangle \in \Omega_\gamma[f](A)$;
(7) if $\langle b, \boldsymbol{\alpha}, \beta\rangle \in A$, then $\langle\langle 7, k+1, l\rangle, b, \boldsymbol{\alpha}, \beta\rangle \in \Omega_\gamma[f](A)$;
(8) if $f_j(\alpha_i) \simeq \beta$, then $\langle\langle 8, k, l, i, j\rangle, \boldsymbol{\alpha}, \beta\rangle \in \Omega_\gamma[f](A)$;
(9) $\rho \in \Omega_\gamma[f](A)$ iff ρ is put in by one of clauses (0) to (8).

We write $\Omega_\gamma^\xi[f]$ for $(\Omega_\gamma[f])^\xi$ and $\overline{\Omega}_\gamma[f]$ for $\overline{\Omega_\gamma[f]}$; $\Omega_\infty^\xi[f]$ is $\mathbf{U}\{\Omega_\gamma^\xi[f] : \gamma \in \mathrm{ORD}\}$.

Definition 3.2. (a). $\{a\}_\gamma(\boldsymbol{\alpha}, f) \simeq \beta$ iff $\langle a, \boldsymbol{\alpha}, \beta\rangle \in \overline{\Omega}_\gamma[f]$;
(b). F is γ-recursive iff $\exists a < \omega . F = \{a\}_\gamma$;
(c). F is weakly γ-recursive iff $\exists a < \omega\; \exists \boldsymbol{\alpha} < \gamma \,.\, F = \lambda \boldsymbol{\sigma}, f \,.\, \{a\}_\gamma(\boldsymbol{\sigma}, \boldsymbol{\alpha}, f)$.
(F may be partial in the above definition.)

We point out that our definition of ω-recursion yields the usual ω-recursive functionals and that every ω-recursive functional is γ-recursive for all $\gamma > \omega$. Notice that for any a and any $\alpha < \beta$, $\{a\}_\alpha \subseteq \{a\}_\beta$. The following lemma is easily verified.

Lemma 3.3. *For all* $\gamma, f, \sigma, \boldsymbol{\alpha}; b, c_0, \ldots,$ *and* c_{m-1}:
(a) (*Composition*)
$\{\langle 4, k, l, b, \boldsymbol{c}\rangle\}_\gamma(\boldsymbol{\alpha}, f) \simeq \{b\}_\gamma(\{c_0\}_\gamma(\boldsymbol{\alpha}, f), \ldots, \{c_{m-1}\}_\gamma(\boldsymbol{\alpha}, f), f)$;
(b) (*Primitive Recursion*)
$\{\langle 5, k+1, l, b\rangle\}_\gamma(\sigma, \boldsymbol{\alpha}, f) \simeq \{b\}_\gamma(\sup_{\xi<\sigma}\{\langle 5, k+1, l, b\rangle\}_\gamma(\xi, \boldsymbol{\alpha}, f), \sigma, \boldsymbol{\alpha}, f)$;
(c) (*Search Operator*)
$\{\langle 6, k, l, b\rangle\}_\gamma(\boldsymbol{\alpha}, f) \simeq$ least $\beta_{\beta<\gamma} . \{b\}_\gamma(\beta, \boldsymbol{\alpha}, f) \simeq 0$;
(d) (*Enumeration*)
$\{\langle 7, k+1, l\rangle\}_\gamma(b, \boldsymbol{\alpha}, f) \simeq \{b\}_\gamma(\boldsymbol{\alpha}, f)$.

Definition 3.4. γ is recursively regular (RR(γ)) iff for all γ-recursive functions F, all $\boldsymbol{\alpha}$ and $\beta < \gamma$, if $\forall \sigma < \beta . F(\sigma, \boldsymbol{\alpha})\downarrow$, then $\sup_{\sigma<\beta} F(\sigma, \boldsymbol{\alpha}) < \gamma$.

As in recursion on the natural numbers we have the "recursion theorem".

Proposition 3.5. *For any $a<\omega$ there is an $\bar{e}$ such that for all $\gamma, \boldsymbol{\alpha}<\gamma$, and f: $\{e\}_\gamma(\boldsymbol{\alpha},f)\simeq\{a\}_\gamma(\bar{e},\boldsymbol{\alpha},f)$.*

In order to prove that various functionals are γ-recursive, we want to show that the set of γ-recursive functionals is closed under the following two schema:
(1) Strong Composition $(G, F_0, \ldots, F_{m-1}, G_0, \ldots, G_{n-1}) = H$ iff for all $\boldsymbol{\alpha}$ and all f:
$H(\boldsymbol{\alpha},f)\simeq F(F_0(\boldsymbol{\alpha},f), \ldots, F_{m-1}(\boldsymbol{\alpha},f), \lambda\beta\,.\,G_0(\beta,\boldsymbol{\alpha},f), \ldots, \lambda\beta\,.\,G_{n-1}(\beta,\boldsymbol{\alpha},f),f)$;
(2) Strong Primitive Recursion $(G) = F$ iff for all $\beta, \boldsymbol{\alpha}$, and f:
$F(\beta,\boldsymbol{\alpha},f)\simeq G(\beta,\boldsymbol{\alpha},(\lambda\sigma\,.\,F(\sigma,\boldsymbol{\alpha},f))\restriction\overset{\circ}{\beta},f)$,

where in general $g\restriction\overset{\circ}{\beta}(\sigma)\simeq\begin{cases} g(\sigma), & \text{if } \sigma<\beta,\\ 0\ , & \text{if } \sigma\geq\beta.\end{cases}$

The following two propositions, which demonstrate the desired closure, can be proven by means of the recursion theorem. See Cenzer [7], pp. 14–18, for details.

Proposition 3.6. *For all $m, n<\omega$, there is a primitive recursive function* $\mathrm{Cmp}^{m,n}$ *such that for all $a, b_0, \ldots, b_{m-1}, c_0, \ldots, c_{n-1}$, all γ, all $\boldsymbol{\alpha}<\gamma$, and all f: $\{\mathrm{Cmp}^{m,n}(a,\langle b\rangle,\langle c\rangle)\}_\gamma(\boldsymbol{\alpha},f)\simeq\{a\}_\gamma(\{b_0\}_\gamma(\boldsymbol{\alpha},f), \ldots$
$\ldots, \{b_{m-1}\}_\gamma(\boldsymbol{\alpha},f), \lambda\beta\,.\{c_0\}_\gamma(\beta,\boldsymbol{\alpha},f), \ldots, \lambda\beta\,.\,\{c_{n-1}\}_\gamma(\beta,\boldsymbol{\alpha},f),f)$.*

Proposition 3.7. *There is a primitive recursive function* Spr *such that for all $a<\omega$, all γ, all $\boldsymbol{\alpha},\beta<\gamma$, and all f:
$\{\mathrm{Spr}(a)\}_\gamma(\beta,\boldsymbol{\alpha},f)\simeq\{a\}_\gamma(\beta,\boldsymbol{\alpha},(\lambda\sigma\{\mathrm{Spr}(a)\}_\gamma(\sigma,\boldsymbol{\alpha},f))\restriction\overset{\circ}{\beta},f)$.*

Definition 3.8. $\mathrm{Sup}(\alpha,f)\simeq\beta$ iff f is total on α and $\beta=\sup\{f(\sigma):\ \sigma<\alpha\}$.

It is easily seen that the functional Sup is γ-recursive for any γ. For $\gamma>\omega$, $\omega = \text{least } \alpha<\gamma[\alpha\neq 0\wedge \mathrm{Sup}(\alpha,\lambda\beta\,.\,\beta+1)=\alpha]$, so ω is γ-recursive. Let $\{c_s\}=\mathrm{Sup}$ and $\{c_\omega\}=\omega$. We now distinguish an important class of functionals over ORD.

Definition 3.9. (a). POR is the smallest set of numbers containing c_s, c_ω, $\langle 0,k,l,n\rangle$, $\langle 1,k,l,i\rangle$, $\langle 2,k,l,i\rangle$, $\langle 3,k+4,l\rangle$, and $\langle 8,k,l,i,j\rangle$ for all $k,l,n,i<k$, and $j>l$, and closed under $\mathrm{Cmp}^{m,n}$ for all $m,n<\omega$ and Spr.

(b). F is primitive ordinal recursive (p.o.r.) iff $\exists a \in \mathrm{POR}.\ F = \{a\}_\infty$.

The usefulness of p.o.r. functionals lies in the following proposition, which can be proven by induction on POR.

Proposition 3.10. *For all* $a \in \mathrm{POR}$, *if* $F = \{a\}_\infty$, *then*
(a) *F is total on total functions*;
(b) *for any recursively regular* $\gamma > \omega$, *any* $\boldsymbol{\alpha} < \gamma$, *and any γ-recursive* f:
$F(\boldsymbol{\alpha}, f) \simeq \{a\}_\gamma(\boldsymbol{\alpha}, f)$.

We say that a relation or predicate is p.o.r. or α-recursive iff its characteristic function is. We list some properties of the p.o.r. relations and functions.

Proposition 3.11. (a). *The following functions and relations are* p.o.r.:
(1) *for all* i, $(\ )_i$, $\langle\ \rangle$, ln, *and* $*$;
(2) $<, \leq, \geq, >$, *and* $=$;
(3) $\lim(\alpha)$ *iff α is a limit ordinal*;
(4) *the operations* $+$, $\cdot$, *and* exp *of ordinal arithmetic*;
(5) $\dot{-}$, *defined by* $\alpha \dot{-} \beta = \sigma$ *iff* $\beta + \sigma = \alpha$ *or* $(\alpha \leq \beta \wedge \sigma = 0)$;
(6) *all arithmetic relations over* $\omega^k \times ({}^\omega\omega)^l$, *for all* $k, l < \omega$;
(b). *the p.o.r. functions and/or relations are closed under the following*:
(1) *union, intersection, and complementation*;
(2) *bounded quantification*;
(3) *definition by cases*;
(4) *bounded search operator* (least $\alpha_{\alpha \leq \gamma}.\ f(\alpha) = 0 \vee \alpha = \gamma$).

Our next goal is to define a p.o.r. T-predicate for ordinal recursion. First, we need to study p.o.r. inductive definitions (such as $\Omega_\gamma[f]$). If Γ is p.o.r., we want the function F, defined by $F(\xi, \alpha) \simeq \begin{cases} 1, & \text{if } \alpha \in \Gamma^\xi, \\ 0, & \text{if } \alpha \notin \Gamma^\xi, \end{cases}$ to be p.o.r.

Lemma 3.12. *If G is p.o.r. and F is defined by* $F(\xi, \alpha, \rho, \boldsymbol{\gamma}, f) \simeq \mathrm{Sup}(\xi, \lambda\sigma.\ F(\sigma, \alpha, \rho, \boldsymbol{\gamma}, f))$ *for* $\lim(\xi)$ *or* $\xi = 0$, *and* $F(\xi + 1, \alpha, \rho, \boldsymbol{\gamma}, f) \simeq G(\alpha, \rho, \boldsymbol{\gamma}, (\lambda\beta.\ F(\xi, \beta, \rho, \boldsymbol{\gamma}, f)) \upharpoonright \overset{0}{\rho}, f)$, *then F is also p.o.r.*

Proof. F can be defined by recursion on the lexicographic ordering of $\mathrm{ORD} \times \rho$.

Definition 3.13. For $f = (f_0, \ldots, f_{l-1})$, $T^l(\xi, \gamma, \tau, f)$ iff $\tau \in \Omega_\gamma^\xi[f]$.

Proposition 3.14. *T^l is p.o.r. for all l.*

Proof. $\Omega_\gamma[f]$ is an inductive definition over $\rho = \sup_{n<\omega} \overbrace{\langle \gamma, \ldots, \gamma \rangle}^{n}$, $\Omega_\gamma[f]$ is seen to be p.o.r. by inspection of Definition 3.1. It follows from Lemma 3.12 that T is p.o.r.

Proposition 3.15. (a) *For all recursively regular γ, all $\boldsymbol{\alpha}, \beta < \gamma$, all $a < \omega$, and all γ-recursive f, $\{a\}_\gamma(\boldsymbol{\alpha}, f) \simeq \beta$ iff $\exists \xi, \sigma < \gamma \cdot T^l(\xi, \sigma, \langle a, \boldsymbol{\alpha}, \beta \rangle, f)$ iff $T^l(\gamma, \gamma, \langle a, \boldsymbol{\alpha}, \beta \rangle, f)$.*
(b) *for any ordinal γ and any f, γ is recursively regular in f iff $\Omega_\gamma^{\gamma+1}[f] = \Omega_\gamma^\gamma[f]$.*

Proof. We can show by inspection of Definition 3.1 that $\bigcup_{\sigma<\gamma} \bigcup_{\xi<\gamma} \Omega_\sigma^\xi[f]$ is closed under $\Omega_\gamma[f]$, and therefore equals $\overline{\Omega}[f]$; (a) and half of (b) follows from this. If γ is not regular in f then we have a functional $\{a\}_\gamma$ and $\boldsymbol{\alpha}, \beta$ with $\sup_{\sigma<\beta}\{a\}_\gamma(\sigma, \boldsymbol{\alpha}, f) \simeq \gamma$; it is then not difficult to construct a functional $\{c\}_\gamma$ with $\langle c, \boldsymbol{\alpha}, \beta, 0 \rangle \in \overline{\Omega}_\gamma[f] - \Omega_\gamma^\gamma$.

Corollary 3.16. RR *is p.o.r.*

Proof. Let $h(\gamma) = \sup_n \overbrace{\langle \gamma, \ldots, \gamma \rangle}^{n}$. h is p.o.r. and RR(γ) iff $\forall \tau < h(\gamma)[T^l(\gamma+1, \gamma, \tau) \to T^l(\gamma, \gamma, \tau)]$.

4. Recursive analogues of large cardinals

The class of recursively regular ordinals is intended as a recursive analogue of the class of regular cardinals; ω_1 obviously corresponds to $\aleph_1$.

Definition 4.1. (a) ω_α is the α'th ordinal which is recursively regular or a limit of recursively regular ordinals;
(b) $\alpha^+ = \text{least}\, \beta\, [\beta > \alpha \wedge \text{RR}(\beta)]$, the "next" recursively regular.

Notice that $\omega_0 = \omega$, $(\omega_\alpha)^+ = \omega_{\alpha+1}$, and for limit ordinals α $\sup_{\beta<\alpha} \omega_\beta = \omega_\alpha$.

Definition 4.2. (a) α is recursively inaccessible (RI (α)) iff RR (α) and α is a limit of recursively regulars;

(b) α is recursively Mahlo (RM (α)) iff RR (α) and every normal weakly α-recursive function f from α to α has a recursively regular fixed point (f is normal iff strictly increasing and continuous at limit ordinals).

The following is easily verified (part (c) using Propositions 3.15 and 3.16.)

Proposition 4.3. (a) RI (α) *iff* RR $(\alpha) \wedge \omega_\alpha = \alpha$;
(b) RM $(\alpha) \to$ RI (α):
(c) RI *and* RM *are p.o.r.*

We could defined properties like recursively hyper-Mahlo and hyper inaccessible, but our interest here is in stronger notions of recursive largeness.

Definition 4.4. (a) α is absolutely projectible to β iff there is a function f α-recursive in parameters $\xi < \beta$ mapping α 1-1 into β;
(b) α is projectible to β iff there is a weakly α-recursive function f mapping α 1-1 into β;
(c) α is non-absolutely projectible (NAP (α)) iff RR (α) and there is no $\beta < \alpha$ such that α is absolutely projectible to ω_β;
(d) α is non-projectible (NP (α)) iff RR (α) and there is no $\beta < \alpha$ such that α is projectible to ω_β.

The following concept turns out to be very useful in the study of recursively large ordinals and particularly relevant to the two notions of projectibility.

Definition 4.5. For all ordinals α and all $\boldsymbol{\sigma}$ and $\beta < \alpha$:
(a) β is α-recursive in $\boldsymbol{\sigma}$ iff there is a α-recursive function F such that $\beta = F(\boldsymbol{\sigma})$;
(b) β is α-recursive iff there is an index a such that $\{a\}_\alpha \simeq \beta$.

We derive the following equivalences for projectibility.

Proposition 4.6. *For all recursively regular ordinals* α *and all* $\beta < \alpha, \beta$ *closed under* $\langle\ \rangle$;
(a) α *is absolutely projectible to* β *iff*
$\forall \tau < \alpha\ \exists \sigma < \beta [\tau$ *is* α*-recursive in* $\sigma]$;
(b) α *is projectible to* β *iff*
$\exists \gamma < \alpha\ \forall \tau < \alpha\ \exists \sigma < \beta [\tau$ *is* α*-recursive in* $\gamma, \sigma]$.

Proof. We give the proof for (a); (b) is similar. (→) Suppose we have $\xi<\beta$ and F such that $\lambda\tau\,.\,F(\tau,\xi)$ is a projection of α to β. Given $\tau<\alpha$, let $\sigma = F(\tau,\xi)$; then τ = least $\zeta<\alpha\,.\,F(\zeta,\xi)\simeq\sigma$, and τ is α-recursive in $\langle\sigma,\xi\rangle$ (←) Suppose $\forall\tau<\alpha\,\exists\sigma<\beta[\tau$ is α-recursive in $\sigma]$. Then β in particular is α-recursive in parameters less than β. Let $f_0(\tau)$ = least $\langle a,\sigma,\xi\rangle<\alpha\,[\langle a,\sigma\rangle<\beta\wedge T^0(\xi,\xi,\langle a,\sigma,\tau\rangle)]$, and let $f(\tau)$ = $\langle(f_0(\tau))_0,(f_0(\tau))_1\rangle$. This f is the canonical projection of α to β.

Corollary 4.7. (a) NP(α) *iff* NAP$(\alpha)\wedge$ RI(α);
(b) NP *and* NAP *are p.o.r.*

In order to prove that ordinals exist which are recursively Mahlo or non-projectible, we first discuss the concept of stability.

Definition 4.8. (a) α is γ-stable in f iff α is closed under all functions γ-recursive in f.
(b) α is stable in f iff α is ∞-stable in f;
(c) α is stable $(S(\alpha))$ iff α is stable in $\emptyset$.

We need the following lemma and proposition to prove that stable ordinals exist. See Cenzer [7], pp. 47–49, for a proof of Lemma 4.9.

Lemma 4.9. *For any ordinal γ, any $D\subseteq\gamma$, and any f from D to D such that D is closed under all functions γ-recursive in f, if π is the collapsing function mapping D 1-1 onto an ordinal in an order-preserving fashion, then for any $\alpha,\beta\in D$ and any γ-recursive functional F,*

$$F(\alpha,f)\simeq\beta \text{ implies } F(\pi(\alpha),f)\simeq\pi(\beta).$$

Proposition 4.10. *For any ordinal γ, any $\alpha<\gamma$, and any f from α to α, $\{F(\xi,f):\ \xi<\alpha, F\ \gamma\text{-recursive}\} = D$ is an initial segment of γ.*

Proof. Notice that $\alpha\subseteq D$; it follows from Lemma 4.9 that for $\xi<\alpha$, $F(\xi,f)\simeq\beta$ implies $F(\xi,f)\simeq\pi(\beta)=\beta$. Now given any $\beta\in D$, $\beta=F(\xi,f)$ for some $\xi<\alpha$, and therefore $\pi(\beta)=\beta$.

Corollary 4.11. (a) *For any ordinal α and any function f from α to α, there is an ordinal δ stable in f with $\overline{\overline{\delta}}=\overline{\overline{\alpha}}$.*

(b) *For any f from ω to ω, there is a countable δ stable in f.*

We can use a different technique to find ordinals stable in functions from ORD to ORD.

Proposition 4.12. *For any* f: ORD → ORD *and any* α, *there is an ordinal* $\beta > \alpha$ *such that* β *is stable in* f.

Proof. Let $D_0 = \{F(\xi, f) : \xi < \alpha$ and F is ∞-recursive$\}$ and $\sigma_0 = \operatorname{Sup} D_0$. For all $n < \omega$, let $D_{n+1} = \{F(\xi, f) : \xi < \sigma_n$ and F is ∞-recursive and $\sigma_{n+1} = \operatorname{Sup} D_{n+1}$. Then $\beta = \operatorname{Sup}_{n<\omega} \sigma_n$ is stable in f.

Definition 4.13. (a) δ^A_α = the α'th ordinal stable in χ_A;
(b) $\delta^* = \text{least } \beta\, [\beta > \delta \wedge S(\beta)]$;
(c) $\delta_\alpha = \delta^\emptyset_\alpha$.

We point out that δ_1 is the least non-∞-recursive ordinal and for all n, δ_{n+1} is the least ordinal which is not ∞-recursive in δ_n.

Proposition 4.14. (a) *For any* β *and any* $\gamma > \beta$, *if* β *is* γ-*stable, then* β *is recursively Mahlo*;
(b) β *is non-projectible iff* β *is a limit of ordinals which are* β-*stable*;
(c) *the relation* RS, *defined by* RS(α,β) *iff* α *is* β-*stable, is* p.o.r.;
(d) S *is not* ∞-*recursive.*

Proof. (a) Given a normal weakly β-recursive function f, let

$$g(\alpha) = \begin{cases} \operatorname{Sup}(\alpha, \lambda\sigma\,.f(\sigma)), & \text{for } \lim(\alpha); \\ f(\alpha), & \text{otherwise.} \end{cases}$$

Then g is γ-recursive, $g(\alpha) \simeq f(\alpha)$ for all $\alpha < \beta$, and $g(\beta) \simeq \beta$. But then $\alpha_0 = \text{least } \alpha < \gamma\, [\mathrm{RR}(\alpha) \wedge g(\alpha) \simeq \alpha] < \beta$ by stability, so that f has a recursively regular fixed point.
(b) If β is non-projectible, then for any $\alpha < \beta$, $\sigma_\alpha = \operatorname{Sup}\{F(\xi), \xi \leq \alpha \wedge F\ \beta\text{-recursive}\}$ is β-stable and $\alpha < \sigma_\alpha$; if β is projectible to $\alpha < \beta$, then by Proposition 4.6, $\exists \gamma < \beta\, \{F(\sigma, \gamma) : \sigma < \alpha\} = \beta$. It is clear that there can be no β-stable ordinal greater than $\max(\alpha, \gamma)$.
(c) This follows directly from Proposition 3.15.
(d) If S were ∞-recursive, then $\delta_1 = \text{least } \delta\,.S(\delta)$ would be ∞-recursive, contradicting its stability.

Corollary 4.15. (a) NP(α) *implies* RM(α);
(b) *there are non-projectibles less than* δ_1.

Given any p.o.r. relation R on ORD, ρ = least α . $R(\alpha)$ is ∞-recursive, but further, ρ = least $\alpha < \rho^+$. $R(\alpha)$, and so ρ^+-recursive. For example, ω_1 = least $\alpha < \omega_2$ [RR$(\alpha) \wedge \alpha \neq \omega$]. We have the following.

Proposition 4.16. *If* α *is any of the following*: ω_1, *the least recursively inaccessible, the least recursively Mahlo, the least non-absolutely projectible, the least non-projectible, then* α *is* α^+*-recursive.*

Definition 4.17. (a) $\alpha^{(0)} = \alpha; \alpha^{(n+1)} = (\alpha^{(n)})^+$;
(b) γ_n = least α . α not $\alpha^{(n)}$-recursive.

Since for any stable ordinal δ, δ is not $\delta^{(n)}$-recursive, it is clear that the γ_n exist and are all less than δ_1.

Proposition 4.18. *For all* $n > 0$, (a) RR(γ_n) *and* RI(γ_n);
(b) $\forall \tau < \gamma_n$. τ *is* γ_n*-recursive*;
(c) $\forall \sigma, \gamma_n \leq \sigma < \gamma_n^{(n)}$, σ *is not* $\gamma_n^{(n)}$*-recursive.*

Proof. We give the proof for $n = 1$.
(a) If $\neg$RR(γ_1), then $\gamma_1 = \sup_{\sigma<\beta} F(\sigma, \xi)$, for some β, $\xi < \gamma_1$ and γ_1-recursive F; but then β, ξ are γ_1^+-recursive (β is β^+-recursive and $\beta^+ < \gamma_1^+$) and so is F. This implies that γ_1 is γ_1^+-recursive, a contradiction. If $\neg$RI(γ_1), then $\gamma_1 = \beta^+$, for some $\beta < \gamma_1$; but then γ_1 = least $\gamma < \gamma_1^+$ [RR$(\gamma) \wedge \beta < \gamma$].
(b) $\tau < \gamma_1$ implies τ τ^+-recursive, but since RI(γ_1), $\tau^+ < \gamma_1$.
(c) If σ is γ_1^+-recursive and $\gamma_1 \leq \sigma < \gamma_1^+$, then
γ_1 = least $\gamma < \gamma_1^+$. $\forall \tau \leq \sigma$[RR$(\tau) \rightarrow \tau \leq \gamma$].

The following is easily verified.

Proposition 4.19. (a) *For all* $n > 0$, γ_n = *least* γ . γ is $\gamma^{(n)}$*-stable*;
(b) γ_1^+ *is the least non-absolutely projectible ordinal.*

Corollary 4.20. (a) *For all* $n > 0$, RM(γ_n);
(b) *the least recursively Mahlo ordinal is less than* γ_1.

The following characterization of γ_n will be useful.

Proposition 4.21. *For all $n > 0$, γ_n is the least ordinal γ such that*
$\forall a < \omega [\{a\}_{\gamma^{(n)}}(\gamma)\downarrow \rightarrow \exists \alpha < \gamma \,.\, \{a\}_{\alpha^{(n)}}(\alpha)\downarrow]$.

Proof. ($\geq$) Suppose $\{a\}_{\gamma_1^+}(\gamma_1)\downarrow$. By 4.18c,
$\rho = \text{least } \beta_{\beta < \gamma_1^+} [T^0((\beta)_2, (\beta)_2, \langle a, (\beta)_1, (\beta)_0\rangle) \wedge (\beta)_2 < (\beta)_1^+]$ is less than γ_1.
(Notice that $\sigma < \tau^+$ iff $\forall \xi \leq \sigma [\mathrm{RR}(\xi) \rightarrow \xi \leq \tau]$ is p.o.r.) Let $\alpha = (\rho)_1$;
$\{a\}_{\alpha^+}(\alpha)\downarrow$ and $\alpha < \gamma_1$.
($\leq$) Given $\gamma < \gamma_1$, we know that $\gamma \simeq \{a\}_{\gamma^+}$ for some $a < \omega$. Let
$\{b\}_{\sigma^+}(\sigma) \simeq \text{least } \xi_{\xi < \sigma^+} \,.\, \{a\}_{\sigma^+} \simeq \xi \wedge \xi = \sigma]$. Clearly, $\{b\}_{\gamma^+}(\gamma)\downarrow$ and for any
$\alpha < \gamma$, $\{b\}_{\alpha^+}(\alpha)\uparrow$.
The proof for $n > 1$ is similar.

Corollary 4.22. *For any $n > 0$ and any $i < \omega$:*
(a) *if $\{a\}_{\gamma_n^{(n)}}(\gamma_n) \simeq i$, then $\exists \alpha < \gamma_n \,.\, \{a\}_{\alpha^{(n)}}(\alpha) \simeq i$;*
(b) *if $\forall \alpha < \gamma_n \,.\, \{a\}_{\alpha^{(n)}}(\alpha) \simeq i$ and if $\{a\}_{\gamma_n^{(n)}}(\gamma_n)\downarrow$, then $\{a\}_{\gamma_n^{(n)}}(\gamma_n) \simeq i$.*

In section 6 we make use of the foregoing material to prove that $|\mathbf{\Pi}_1^1| = \gamma_1$. First it is necessary to study the class of $\mathbf{\Pi}_1^1$ relations with reference to ordinal recursion.

5. $\mathbf{\Pi}_1^1$ relations

In this section we prove the following two generalizations of the Kreisel–Sacks [14] result that for $A \subseteq \omega$, $A \in \mathbf{\Pi}_1^1$ iff A is ω_1-semirecursive.

Proposition 5.1. *If $Q \subseteq \omega \times P(\omega)$ is $\mathbf{\Pi}_1^1$, then there is a p.o.r. functional F with* $\mathrm{rg}(F) \subseteq \{0, 1\}$ *such that for all m and A: $Q(m, A)$ iff*
$\exists \sigma \,.\, F(\sigma, m, \chi_A) \simeq 1$ *iff* $\exists \sigma < \omega_1^A \,.\, F(\sigma, m, \chi_A) \simeq 1$.

Proposition 5.2. *The relation K, defined by $K(\langle a, \boldsymbol{m}, n\rangle, A)$ iff $\{a\}_{\omega_1^A}(\boldsymbol{m}) \simeq n$, is $\mathbf{\Pi}_1^1$.*

We want to code ordinals into the natural numbers. Let $\{a\}^A$ be the a'th function partial recursive in A.

Definition 5.3. (a) $W(\phi)$ iff ϕ is the characteristic function of a non-strict well-ordering of a subset of ω;
(b) Field $(\phi) = \{s : \phi(s,s) = 1\}$; if $W(\phi)$, $|\phi|$ is the ordinal isomorphic to the well-ordering defined by ϕ and, for $s \in \text{Field}(\phi)$, $|s|_\phi$ is the image of s under the map from Field(ϕ) to $|\phi|$; $s \leq_\phi t$ iff $|s|_\phi \leq |t|_\phi$ iff $\phi(s,t) = 1$.

Lemma 5.4. (a) *W is $\mathbf{\Pi}_1^1$;*
(b) *there are $\mathbf{\Sigma}_1^1$ relations M and M' such that for all ϕ and all ψ such that $W(\psi)$, $M(\phi,\psi)$ iff $W(\phi) \wedge |\phi| \leq |\psi|$, and $M'(\phi,\psi)$ iff $W(\phi) \wedge |\phi| < |\psi|$;*
(c) *there are $\mathbf{\Pi}_1^1$ relations L and L' such that for all ϕ and ψ: $L(\phi,\psi)$ iff $W(\phi) \wedge W(\psi) \wedge |\phi| \leq |\psi|$, and $L'(\phi,\psi)$ iff $W(\phi) \wedge W(\psi) \wedge |\phi| < |\psi|$.*

Proof. (a) $W(\phi)$ iff ϕ is a linear ordering $\wedge \forall\theta\, \exists p \rightarrow (\theta(p+1) <_\phi \theta(p))$.
(b) $M(\phi,\psi)$ iff $\exists\theta\, \forall m\, \forall n.\ \phi(\langle m,n\rangle) = \psi(\langle\theta(m),\theta(n)\rangle)$; M' similar.
(c) $L(\phi,\psi)$ iff $W(\phi) \wedge W(\psi) \wedge \neg M'(\psi,\phi)$; L' similar.

The following lemma is an easy relativization of the standard result; see Shoenfield [22], p. 184, Cenzer [7], pp. 29–31.

Lemma 5.5. (a) *For all $A \subseteq \omega$, $\{a\colon W(\{a\}^A)\}$ is $\mathbf{\Pi}_1^1$–A complete;*
(b) *(Boundedness Principle) For any $A \subseteq \omega$ and any $\mathbf{\Sigma}_1^1$ $V \subseteq \{a\colon W(\{a\}^A)\}$, there is a u such that for all v in V, $L(\{v\}^A, \{u\}^A)$.*

We generalize Spector's results [24] on $\mathbf{\Pi}_1^1$ monotone inductive definitions; see Cenzer [7], pp. 32–34 for proofs.

Proposition 5.6. *If K is a $\mathbf{\Pi}_1^1$ relation such that for all A, Γ_A, defined by $\Gamma_A(B) = \{m\colon K(m,B,A)\}$, is a monotone inductive operator, then:*
(a) *the relation P, defined by $P(m,u,A)$ iff $m \in \Gamma_A^{|\{u\}^A|}$, is $\mathbf{\Pi}_1^1$;*
(b) *the relation S, defined by $S(m,A)$ iff $m \in \Gamma^{\omega_1^A}$, is $\mathbf{\Pi}_1^1$.*
(c) *for every $A \subseteq \omega$, $|\Gamma_A| \leq \omega_1^A$;*
(d) *the relation Q, defined by $Q(m,A)$ iff $m \in \overline{\Gamma_A}$, is $\mathbf{\Pi}_1^1$.*

The following lemma completes the preparation for the proof of Proposition 5.1.

Lemma 5.7. *If Q is a $\mathbf{\Pi}_1^1$ relation on $\omega \times P(\omega)$, then there is a $\mathbf{\Pi}_1^0$ relation K*

such that for all A, Γ_A, *defined by* $\Gamma_A(B) = \{m\colon K(m,B,A)\}$, *is a monotone inductive operator such that for all* m, $Q(m,A)$ *iff* $\langle m,1\rangle \in \overline{\Gamma}_A$.

Proof. Suppose $Q(m,A)$ iff $\forall\phi\,\exists p\cdot R(m,\bar{\phi}(p),A)$, where for all $s,t,R(m,s,A) \wedge s \subseteq t \to R(m,t,A)$. Let $K(\langle m,s\rangle,B,A)$ iff $\langle m,s\rangle \in B \vee R(m,s,A) \vee \forall p\,.\,\langle m,s*\langle p\rangle\rangle \in B$, and let Γ_A be defined from K as above. It is easily seen that for all m, s, and A, $\langle m,s\rangle \in \overline{\Gamma}_A$ iff $\forall\phi(s \subseteq \phi \to \exists p\,.\,R(m,\bar{\phi}(p),A)$, so that $Q(m,A)$ iff $\langle m,1\rangle \in \overline{\Gamma}_A$ (1 being the empty sequence).

Proposition 5.1. *If* $Q \subseteq \omega \times P(\omega)$ *is* $\mathbf{\Pi}^1_1$, *then there is a p.o.r. functional* F *with* $\mathrm{rg}(F) \subseteq \{0,1\}$ *such that for all* m *and* A: $Q(m,A)$ *iff* $\exists\sigma\,.\,F(\sigma,m,\chi_A) \simeq 1$ *iff* $\exists\sigma < \omega_1^A\,.\,F(\sigma,m,\chi_A) \simeq 1$.

Proof. Let K and Γ_A be defined from Q as in Lemma 5.7. Then $Q(m,A)$ iff $\exists\sigma\,.\,m \in \Gamma_A^\sigma$ iff $\exists\sigma < \omega_1^A\,.\,m \in \Gamma_A^\sigma$, where the latter equivalence follows from Proposition 4.6c. Since K is $\mathbf{\Pi}^0_1$, it is p.o.r. by Proposition 3.11. If we set $F(\sigma,m,\chi_A) \simeq 1$ iff $\langle m,1\rangle \in \Gamma_A^\sigma$, then F is p.o.r. by Lemma 3.12.

To prove Proposition 5.2, we code up the ordinal recursive functions on ω_1^A in a $\mathbf{\Pi}^1_1 - A$ fashion.

Definition 5.8. $N(u,m,A)$ iff $W(\{u\}^A) \wedge |\{u\}^A| = m$.

The relation N is easily seen to be arithmetic.

Definition 5.9. $K_A^l[\mathbf{\Phi}] = \{\langle a,u,v\rangle\colon T^l(\omega_1^A,\omega_1^A,\langle a,|\{u\}^A|,|\{v\}^A|\rangle,\mathbf{\Phi})$.

Proposition 5.10. *For all* $l > \omega$, *the relation* R^l, *defined by* $R^l(m,\mathbf{\Phi},A)$ *iff* $m \in K_A^l[\mathbf{\Phi}]$, *is* $\mathbf{\Pi}^1_1$.

Proof. $K_A^l[\mathbf{\Phi}]$ is defined by a $\mathbf{\Pi}^1_1 - \mathbf{\Phi},A$ monotone inductive operator $\Omega_A[\mathbf{\Phi}]$ which parallels $\Omega_{\omega_1^A}[\mathbf{\Phi}]$. From a slight generalization of Proposition 5.6b it follows that $K_A^l[\mathbf{\Phi}]$, which equals $\Omega_A[\mathbf{\Phi}]^{\omega_1^A}$ is $\mathbf{\Pi}^1_1 - \mathbf{\Phi},A$ uniformly in $\mathbf{\Phi},A$. We give some cases of the definition of $\Omega_A[\mathbf{\Phi}]$:
(0) $\langle\langle 0,k,l,n\rangle, u_0, \ldots, u_{k-1}, v\rangle \in \Omega_A[\mathbf{\Phi}](B)$ iff $W(\{u_0\}^A) \wedge \ldots \wedge W(\{u_{k-1}\}^A) \wedge N(v,n,A)$;

(5) $\langle\langle 5,k+1,l,b\rangle,s,\boldsymbol{u},v\rangle \in \Omega_A[\boldsymbol{\phi}](B)$ iff $\exists t[W(\{t\}^A) \wedge$
$\forall x(M'(\{x\}^A,\{s\}^A) \to \exists z(L'(\{z\}^A,\{t\}^A) \wedge \langle\langle 5,k+1,l,b\rangle,x,\boldsymbol{u},z\rangle \in B)) \wedge$
$\forall z(M'(\{z\}^A,\{t\}^A) \to \exists y\,\exists x(L'(\{x\}^A,\{s\}^A) \wedge L'(\{y\}^A,\{t\}^A) \wedge$
$L'(\{z\}^A,\{y\}^A) \wedge \langle\langle 5,k+1,l,b\rangle,x,\boldsymbol{u},y\rangle \in B)) \wedge \langle b,t,s,\boldsymbol{u},v\rangle \in B]$;
(7) $\langle\langle 7,k+1,l\rangle,y,\boldsymbol{u},v\rangle \in \Omega_A[\boldsymbol{\phi}](B)$ iff $\exists b(N(y,b,A) \wedge \langle b,\boldsymbol{u},v\rangle \in B)$;
(8) $\langle\langle 8,k,l,i,j\rangle,\boldsymbol{u},v\rangle \in \Omega_A[\boldsymbol{\phi}](B)$ iff $\exists m\,\exists n(\phi_j(m) = n \wedge N(u_i,m,A) \wedge$
$N(v,n,A))$.

Now let $l = 1$ and $\boldsymbol{\phi} = \chi_A$.

Proposition 5.11. (a) $K_A^1[\chi_A]$ *is* $\boldsymbol{\Pi}_1^1$ *uniformly in* A;
(b) $K_A^1[\chi_A] = \{\langle a,\boldsymbol{u},v\rangle : \{a\}_{\omega_1^A}(|\{\boldsymbol{u}\}^A|,\chi_A) \simeq |\{v\}^A|$;
(c) ω_1^A *is recursively regular in* A.

Proof. (a) This follows from Proposition 5.10.
(b) Let $\omega_1^A = \alpha$; by Proposition 5.6c, $|\Omega_A[\chi_A]| \leq \alpha$. Since $\Omega_\alpha[\chi_A]$ parallels $\Omega_A[\chi_A]$, we have $\Omega_\alpha^\alpha[\chi_A] = \overline{\Omega_\alpha[\chi_A]}$, so $\{a\}_\alpha(|\{\boldsymbol{u}\}^A|,\chi_A) \simeq |\{v\}^A|$ iff $T^1(\alpha,\alpha,\langle a,|\{\boldsymbol{u}\}^A|,|\{v\}^A|\rangle,\chi_A)$.
(c) This follows from (b) above and Proposition 3.14b.

Proposition 5.2 is a corollary to 5.11.

Proposition 5.2. *The relation* K_1, *defined by* $K_1(\langle a,\boldsymbol{m},n\rangle,A)$ *iff* $\{a\}_{\omega_1^A}(\boldsymbol{m}) \simeq n$, *is* $\boldsymbol{\Pi}_1^1$.

Proof. $K_1(\langle a,\boldsymbol{m},n\rangle,A)$ iff $\exists u,v[\langle a,\boldsymbol{u},v\rangle \in K_A^1[\chi_A] \wedge N(u_0,m_0,A) \wedge \ldots$
$\ldots \wedge N(u_{k-1},m_{k-1},A) \wedge N(v,n,A)]$.

Proposition 5.12. (a) *For any functions* f, *any* α *recursively regular in* f, *and any function* ϕ *weakly* α*-recursive in* f, *if* $W(\phi)$, *then* $|\phi| < \alpha$;
(b) *the relation* W_0, *defined by* $W_0(\alpha,\phi)$ *iff* $W(\phi) \wedge |\phi| = \alpha$, *is p.o.r.*

Proof. (a) Define H by recursion on σ so that $H(\sigma,\phi) \simeq$ the unique $m\,.\,|m|_\phi = \sigma$.

$$\text{Let } G(m,\phi) \simeq \begin{cases} \text{least } \sigma_{\sigma<\alpha}\,.\,H(\sigma,\phi) \simeq m, & \text{if } \phi(m,m) = 1 \\ 0, & \text{if } \phi(m,m) = 0. \end{cases}$$

Then H and G are α-recursive and $|\phi| = \sup_{m<\omega} G(m,\phi) < \alpha$ by regularity.
(b) The graphs of G and H above are p.o.r. in α,ϕ.

Corollary 5.13. (a) *For all $A \subseteq \omega$, ω_1^A is the least ordinal $\neq \omega$ recursively regular in A;*
(b) *for any recursively regular $\alpha > \omega$ and any $A \subseteq \omega$, if χ_A is weakly α-recursive, then $\omega_1^A \leq \alpha$.*

Proof. Suppose α is recursively regular in A and $\omega < \alpha < \omega_1^A$. Then there is a well-ordering ϕ ω-recursive in A of type α. But ϕ must be weakly α-recursive in A, since $\alpha > \omega$, and then by 5.12a $|\phi| < \alpha$.
(b) Any well-ordering ϕ recursive in A must be weakly α-recursive, but then by 5.12a, $|\phi| < \alpha$.

6. Π_1^1 inductive definitions

In this section we prove

Theorem 6.1. (a) $|\Pi_1^1| = \gamma_1$;
(b) *for all $A \subseteq \omega$, $A \in \Pi_1^1$ iff A is γ_1-semirecursive.*

Definition 6.2. Given a Π_1^1 operator Γ, we say that F and I, rg $\subseteq \{0, 1\}$, are p.o.r. functionals associated with Γ if for all m, A, β, τ,
(a) $m \in \Gamma(A)$ iff $\exists \sigma \,.\, F(\sigma, m, \chi_A) \simeq 1$ iff $\exists \sigma < \omega_1^A \,.\, F(\sigma, m, \chi_A) \simeq 1$ (as in 5.2)
(b) $I(\tau, m, \beta) \simeq \mathrm{Sup}(\tau, \lambda\xi \,.\, I(\xi, m, \beta))$, if $\lim(\tau)$ or $\tau = 0$;
$I(\tau{+}1, m, \beta) \simeq \mathrm{Sup}(\beta, \lambda\sigma \,.\, F(\sigma, m, \lambda n \,.\, I(\tau, n, \beta)))$.

We see by inspection that if Γ_β is defined by $m \in \Gamma_\beta(A)$ iff $\exists \sigma < \beta \,.\, F(\sigma, m, \chi_A) \simeq 1$, then $m \in \Gamma_\beta^\tau$ iff $I(\tau, m, \beta) \simeq 1$. For sufficiently large β (e.g., $\beta = \aleph_1$), $\Gamma_\beta = \Gamma$. We get a better bound on β by looking at the Γ^τ individually.

Lemma 6.3. *For any Π_1^1 inductive operator Γ with associated I, and for all ordinals τ:*
(a)$_\tau$ *For all m and all $\beta \geq \omega_\tau$, $\chi_{\Gamma^\tau}(m) \simeq I(m, \tau, \beta)$;*
(b)$_\tau$ $\omega_1^{\Gamma^\tau} \leq \omega_{\tau+1}$;

Proof. We break the proof into four parts:
(1) $(a)_0$ and $(b)_0$ are trivial.

(2) $(a)_\tau \wedge (b)_\tau \rightarrow (a)_{\tau+1}$ by Definition 6.2.
(3) $(a)_\tau \rightarrow (b)_\tau$ by Corollary 5.13b.
(4) $[\lim(\tau) \wedge \forall\sigma<\tau\,.\,(a)_\sigma] \rightarrow (a)_\tau$ by Definition 6.2.

Proposition 6.4. *For any* $\mathbf{\Pi}_1^1$ *inductive operator* Γ, $|\Gamma| \leq \gamma_1$.

Proof. For any γ, $\mathrm{RI}(\gamma)$ implies by 6.3 that for all m, $\chi_{\Gamma^\gamma}(m) \simeq I(\gamma, m, \gamma)$ and $m \in \Gamma^{\gamma+1}$ iff $\exists\sigma<\gamma^+\,.\,F(\sigma, m, \lambda p\,.\,I(\gamma,p,\gamma) \simeq 1$. Given $m \in \Gamma^{\gamma_1+1}$, choose a so that $\{a\}_\tau(\alpha) \simeq$ least $\sigma<\alpha[F(\sigma,m,\lambda p\,.\,I(\alpha,p,\alpha)) \simeq 1 \wedge \mathrm{RI}(\alpha)]$. Then $\{a\}_{\gamma_1^+}(\gamma_1)\downarrow$, so by Proposition 4.21, there is an $\alpha<\gamma_1$ such that $\{a\}_{\alpha^+}(\alpha)\downarrow$, so that $m \in \Gamma^{\alpha+1} \subseteq \Gamma^{\gamma_1}$.

We next define a $\mathbf{\Pi}_1^1$ inductive operator Λ_1 such that $|\Lambda_1| = \gamma_1$. Recall from Proposition 5.2 that the relation $K(\langle a,m,n\rangle, A)$ iff $\{a\}_{\omega_1^A}(m) \simeq n$ is $\mathbf{\Pi}_1^1$.

Definition 6.5. $\Lambda_1(A) = A \cup \{\langle a,m,n\rangle : \{a\}_{\omega_1^A}(m) \simeq n\}$.

Lemma 6.6. *For all* $\tau \leq \gamma_1 + 1$,
$(a)_\tau$ $\Lambda_1^0 = \emptyset$; $\tau > 0 \rightarrow \Lambda_1^\tau = \mathbf{U}_{\sigma<\omega_\tau} \{\langle a,m,n\rangle : \{a\}_\sigma(m) \simeq n\} = A_\tau$;
$(b)_\tau$ for $\tau \leq \gamma_1, \omega_1^{\Gamma^\tau} = \omega_{\tau+1}$.

Proof. The proof is broken up into four parts:
(1) $(a)_0$ and $(b)_0$ are trivial.
(2) $(a)_\tau \wedge (b)_\tau \rightarrow (a)_{\tau+1}$ by the definition of Λ_1 and the regularity of $\omega_{\tau+1}$.
(3) $(a)_\tau \rightarrow (b)_\tau$ for all $\tau \leq \gamma_1$, as follows: for any $\tau \leq \gamma_1$ and any $\xi < \omega_\tau$, there exists some $a<\omega$ and $\sigma<\omega_\tau$ such that $\xi \simeq \{a\}_\sigma$. Choose b so that $\{b\}_\alpha(m,n) \simeq 0$ iff $\{m\}_\alpha \leq \{n\}_\alpha$. Then $\{\langle m,n\rangle : \langle b,m,n,0\rangle \in A_\tau\}$ is a pre-well-ordering of length ω_τ. Refine this to a well-ordering by choosing a least member of each equivalence class to obtain a well-ordering recursive in A_τ of type ω_τ; it follows that $\omega_1^{A_\tau} \geq \omega_{\tau+1}$. On the other hand, it follows from Proposition 3.15 that A_τ is p.o.r. in ω_τ and therefore weakly $\omega_{\tau+1}$-recursive; then by Corollary 5.13(b), $\omega_1^A \leq \omega_{\tau+1}$.
(4) $(\lim(\tau) \wedge \forall\sigma<\tau\,.\,(a)_\sigma) \rightarrow (a)_\tau$ is trivial.

Corollary 6.7. $|\Lambda_1| = \gamma_1$.

Proof. $|\Lambda_1| \leq \gamma_1$ by Proposition 6.4. By Proposition 4.18(b), for any $\tau<\gamma_1$, there is an m such that $\{m\}_{\gamma_1} \simeq \omega_{\tau+1}$, so that $\{m\}_\tau\uparrow$. Let b be as in (3) of the proof of Lemma 6.6 above: then $\langle b,m,m,0\rangle \in \Lambda_1^{\gamma_1} - \Lambda_1^\tau$, so that $|\Lambda_1| \geq \gamma_1$.

This completes the proof of Theorem 6.1(a). We prove 6.1(b) (that $\mathbf{\Pi}_1^1 =$ γ_1-semirecursive over $P(\omega)$) as follows:
($\subseteq$) For any $\mathbf{\Pi}_1^1$ inductive operator Γ, by Lemma 6.3(a), $\overline{\Gamma} = \{m\colon \exists\sigma < \gamma_1 [I(\sigma,m,\sigma) \simeq 1 \wedge \mathrm{RI}(\sigma)]\}$ and is therefore γ_1-semirecursive. Any $A \subseteq \omega$ such that $A \leq_{1-1} \overline{\Gamma}$ is also γ_1-semirecursive.
($\supseteq$) Suppose that $m \in A$ iff $\{a\}_{\gamma_1}(m)\downarrow$. Choose b so that $\{b\}_{\gamma_1}(m) \simeq 0 \cdot \{a\}_{\gamma_1}(m)$. Then by Lemma 6.6(a), $m \in A$ iff $\langle b,m,0\rangle \in \overline{\Lambda_1}$.

7. A functional $G_1^\#$ with $\omega_1^{G_1^\#} = \gamma_1$

Recall that for any partial functional F, ω_1^F is the least ordinal not equal to $|\phi|$ for any well-ordering ϕ recursive in F. It is easily seen that for any F, there is an α such that $\omega_1^F = \omega_\alpha$, although ω_α need not be recursively regular, as indicated by the remark following Lemma 7.8.

Hinman has defined the functional $E_1^\#$ from ω^ω to $\{0,1\}$.

Definition 7.1. $E_1^\#(f) \simeq \begin{cases} 0, & \text{if } \forall\Phi\exists p\,.\, f(\overline{\phi}(p)) \simeq 0;\\ 1, & \text{if } \exists\phi\forall p\,.\, f(\overline{\phi}(p)) > 0.\end{cases}$

The following results are proved by Aczel [2]:

Theorem 7.2. (a) $\omega_1^{E_1^\#} = |\mathbf{\Sigma}_1^1\text{-mon}|$;
(b) *for all* $A \subseteq \omega$, $A \in \mathbf{\Sigma}_1^1$-mon *iff* A *is* $E_1^\#$-*semirecursive.*

Grilliot first pointed out that, using $E_1^\#$, one can prove that $|\mathbf{\Sigma}_1^1\text{-mon}| = |\mathbf{\Sigma}_1^1|$.

Definition 7.3. $G_1^\#(f) \simeq n$ iff $\{f(0)\}_{(\omega_1^{f})^+}(f(1)) \simeq n$.

In this section we prove the following theorem.

Theorem 7.4. (a) $\omega_1^{G_1^\#} = |\mathbf{\Pi}_1^1| = \gamma_1$;
(b) *for all* $A \subseteq \omega$, $A \in \mathbf{\Pi}_1^1$ *iff* A *is* $G_1^\#$-*semirecursive*
iff A *is* γ_1-*semirecursive.*

We define recursion in $G_1^\#$ as in the definition of ordinal recursion (3.1); it is possible to consolidate the first seven clauses into five, since we are doing ω-recursion. Clause (8) is replaced by the following to define $\Omega[G_1^\#]$:
$(8)^\#$ for any $f \in \omega^\omega$, if for all s and t, $f(s) \simeq t$ implies $\langle b, s, \boldsymbol{m}, t\rangle \in A$ and $G_1^\#(f) \simeq n$, then $\langle 6, k, b\rangle, \boldsymbol{m}, n\rangle \in \Omega[G_1^\#](A)$.

Lemma 7.5. (a) $G_1^\#$ *is consistent, that is, for any f and g in ω^ω and any $n < \omega$,* $(f \subseteq g \wedge G_1^\#(f) \simeq n) \to G_1^\#(g) \simeq n$.
(b) *for any a and $\boldsymbol{m}$, there is at most one n such that* $\langle a, \boldsymbol{m}, n\rangle \in \overline{\Omega[G_1^\#]}$.

Proof. (a) Given $f \subseteq g$ and $\{f(0)\}_{(\omega_1^f)^+}(f(1)) \simeq n$.we have $g(0) \simeq f(0)$, $g(1) \simeq f(1)$, and $\omega_1^g \geq \omega_1^f$.
(b) This follows directly from (a).

Definition 7.6. (a) $\{a\}^{G_1^\#}(\boldsymbol{m}) \simeq n$ iff $\langle a, \boldsymbol{m}, n\rangle \in \overline{\Omega[G_1^\#]}$;
(b) $f \in \omega^\omega$ is $G_1^\#$-recursive iff there is an $a < \omega$ such that $f = \{a\}^{G_1^\#}$.

We need the following definition and lemma to prove that $\omega_1^{G_1^\#} \geq \gamma_1$.

Definition 7.7. $F_\alpha(a, m) \simeq \beta$ iff $\{a\}_\alpha(m) \simeq \beta \wedge \beta < \omega$.

Lemma 7.8. *For any recursively regular α, with $\omega < \alpha < \gamma_1$, $\omega_1^{F_\alpha} = \alpha$.*

Proof. ($\leq$) Since F_α is α-recursive, it follows as in Corollary 5.13(b) that $\omega_1^{F_\alpha} \leq \alpha$.
($\geq$) As in (3) of the proof of Lemma 6.6, we have for any $\omega_\tau < \alpha$ a well-ordering ϕ of type ω_τ which is $\omega_{\tau+1}$-recursive and therefore α-recursive. But then for some a, $\phi = \lambda t\,.\,F_\alpha(a, t)$ and is therefore recursive in F_α.

We remark that for non-regular limits α of recursively regulars, if one sets $F_\alpha(m) \simeq n$ iff $\exists \sigma < \alpha.\ \{a\}_\sigma(m) \simeq n$, then $\omega_1^{F_\alpha} = \alpha$ as in Lemma 7.8.

Proposition 7.9. $\omega_1^{G_1^\#} \geq \gamma_1$.

Proof. Suppose $\omega_1^{G_1^\#} = \tau < \gamma_1$. Let $\theta(a, m, t) \simeq n$ iff $(t = 0 \wedge n = a) \vee (t = 1 \wedge n = m) \vee \exists c, p(t = \langle c, p\rangle \wedge \{c\}^{G_1^\#}(p) \simeq n)$. Then θ is $G_1^\#$-recursive and for all a, m, $\omega_1^{\lambda t\,.\,\theta(a,m,t)} = \tau$. Hence for all a, m,

$G_1^{\#}(\lambda t\,.\,\theta(a,m,t)) \simeq F_{\tau^+}(a,m)$. But then F_{τ^+} is $G_1^{\#}$-recursive and by Lemma 7.8, $\omega_1^{G_1^{\#}} \geq \tau^+$, a contradiction.

Our next goal is to show that $|\Omega[G_1^{\#}]| \leq \gamma_1$. We need to be able to compute ω_1^f from the graph of f in order to construct associated functionals F and I analoguous to those of §6. The following lemma is a rather technical application of Proposition 5.12(b); we refer the reader to Cenzer [7] for the proof.

Lemma 7.10. *The relation R_0, defined by $R_0(\alpha,\phi)$ iff there is an $f \in \omega^\omega$ such that $\phi = \chi_{\mathrm{graph}(f)}$ and $\omega_1^f = \alpha$, is p.o.r.*

We now define a p.o.r. functional F associated with $\Omega[G_1^{\#}]$.

Lemma 7.11. *There is a p.o.r. functional F with $\mathrm{rg}(F) \subseteq \{0,1\}$ such that for all ordinals β and all $A \subseteq \omega$, if $(\omega_1^A)^+ \leq \beta$, then for all s, $s \in \Omega[G_1^{\#}](A)$ iff $F(\beta,s,\chi_A) \simeq 1$.*

Proof. F is defined by cases and the only interesting one is, of course, case $(8)^{\#}$, the application of $G_1^{\#}$. Let

$$J(b,p,\langle m\rangle,\theta) \simeq \begin{cases} 1, & \text{if } \exists s,t(p=\langle s,t\rangle \wedge \theta(\langle b,s,m,t\rangle) = 1; \\ 0, & \text{otherwise.} \end{cases}$$

Then for any b,m,f, and A, if $f(s) = t$ iff $\langle b,s,m,t\rangle \in A$, then $\lambda p \cdot J(b,p,\langle m\rangle,\chi_A) = \chi_{\mathrm{graph}(f)}$. We now define $F(\beta,\langle\langle 8,k,b\rangle,m,n\rangle,\theta) \simeq 1$ iff $\theta(\langle\langle 8,k,b\rangle,m,n\rangle,\theta) \simeq 1 \vee$ $\exists\xi,\sigma < \beta\, \exists a,t[R_0(\sigma,\lambda p\,.\,J(b,p,\langle m\rangle,\theta)) \wedge \xi < \sigma^+ \wedge \theta(\langle b,0,m,a\rangle) = \theta(\langle b,1,m,t\rangle) = 1 \wedge T^0(\xi,\xi,\langle a,t,n\rangle)]$. If $R_0(\sigma,\lambda p\,.\,J(b,p,\langle m\rangle,\chi_A))$, then $\sigma \leq \omega_1^A$, so taking $\beta \geq (\omega_1^A)^+$ is sufficient.

We next define a p.o.r. functional I associated with $\Omega[G_1^{\#}]$.

Definition 7.12. $I(\tau+1,m,b) \simeq F(\beta,m,\lambda n\,.\,I(\tau,n,\beta))$;
$I(\tau,m,\beta) \simeq \mathrm{Sup}\,(\tau,\lambda\sigma.\,I(\sigma,m,\beta))$, if $\lim(\tau)$ or $\tau = 0$.

Proposition 7.13. *For all $m,n < \omega$, all limit ordinals α, all $\tau = \alpha+n$, and all $\beta \geq \omega_{\alpha+2n}$:*
$(a)_\tau$ $m \in \Omega^\tau[G_1^{\#}]$ iff $I(\tau,m,\beta) \simeq 1$;

$(b)_\tau\ \omega_1^{\Omega^\tau[G_1^\#]} \leq \omega_{\alpha+2n+1}$.

Proof. As usual, we break the proof into four parts:
(1) $(a)_0$ and $(b)_0$ are trivial.
(2) $(a)_\tau \wedge (b)_\tau \rightarrow (a)_{\tau+1}$ by Lemma 7.11.
(3) $(a)_\tau \rightarrow (b)_\tau$ follows from Corollary 5.13(b).
(4) $(\lim(\tau) \wedge \forall \sigma < \tau\,.(a)_\sigma) \rightarrow (a)_\tau$ is trivial.

Corollary 7.14. (a) *For all recursively inaccessible ordinals* α, *all* $b, m < \omega$, *and all* $f \in \omega^\omega$ *such that* $f(s) \simeq u$ *iff* $\langle b, s, m, u\rangle \in \Omega^\alpha[G_1^\#]$, *f is* α*-recursive*;
(b) *for all recursively inaccessible* α *and all* $m < \omega$, $m \in \Omega^{\alpha+1}[G_1^\#]$ *iff* $I(\alpha+1, m, \alpha^+) \simeq 1$.

Proof. (a) $f(s) \simeq (\text{least } \rho < \alpha\,.\,I((\rho)_1, \langle b, s, m, (\rho)_0\rangle, (\rho)_2) \simeq 1 \wedge (\rho)_2 \geq \omega_{(\rho)_1+\omega})]_0$.
(b) Now whenever f comes from $\Omega^\alpha[G_1^\#]$ as above, $\omega_1^f \leq \alpha$, so that as in Lemma 7.11, taking $\beta \geq \alpha^+$ suffices.

Proposition 7.15. (a) $|\Omega[G_1^\#]| \leq \gamma_1$;
(b) *for all* $f \in \omega^\omega$, *if f is* $G_1^\#$*-recursive, then f is* γ_1*-recursive*;
(c) $\omega_1^{G_1^\#} \leq \gamma_1$.

Proof. (a) Suppose $q \in \Omega^{\gamma_1+1}[G_1^\#] - \Omega^{\gamma_1}[G_1^\#]$; the difficult case is $q = \langle\langle 8, k, b\rangle, m, n\rangle$. By Corollary 7.14(b), for all recursively inaccessible α, $q \in \Omega^{\alpha+1}[G_1^\#]$ iff $\exists \xi, \sigma, a, t < \alpha^+ [R_0(\sigma, \lambda p\,.\,J(b, p, \langle m\rangle, \lambda s\,.\,I(\alpha, s, \alpha))) \wedge \xi < \sigma^+ \wedge I(\alpha, \langle b, 0, m, a\rangle, \alpha) \simeq I(\alpha, \langle b, 1, m, t\rangle, \alpha) \simeq 1 \wedge T^0(\xi, \xi, \langle a, t, n\rangle)]$. We can write $\xi < \sigma^+$ as $\forall \tau \leq \xi(\mathrm{RR}(\tau) \rightarrow \tau \leq \sigma)$; thus the inside of the brackets is a p.o.r. relation, say, $P(\xi, \sigma, a, t, \alpha)$. Choose c so that for all α, $\{c\}_{\alpha^+}(\alpha) \simeq \text{least } \langle \xi, \sigma, a, t\rangle < \alpha^+ [\mathrm{RI}(\alpha) \wedge P(\xi, \sigma, a, t, \alpha)]$. Since $q \in \Omega^{\gamma_1+1}[G_1^\#]$, $\{c\}_{\gamma_1^+}(\gamma_1)\downarrow$, and therefore by Proposition 4.21 there is an $\alpha < \gamma_1$ such that $\{c\}_{\alpha^+}(\alpha)\downarrow$. Then $q \in \Omega^{\alpha+1}[G^\#]\ \Omega^{\gamma_1}[G^\#]$, contradicting the hypothesis and completing the proof of (a).
(b) By (a), $\overline{\Omega[G^\#]} = \Omega^{\gamma_1}[G^\#]$, so by Corollary 7.14(a), every $G_1^\#$-recursive function is γ_1-recursive.
(c) Any $G_1^\#$-recursive well-ordering ϕ is now γ_1-recursive so that $|\phi| < \gamma_1$ by Proposition 5.12(a).

Combining (c) with Proposition 7.9, we have

Corollary 7.16. $\omega_1^{G_1^\#} = \gamma_1$.

Finally, we obtain the other directions of (a) and (b) of Proposition 7.15.

Proposition 7.17. (a) $|\Omega[G_1^\#]| = \gamma_1$;
(b) *for all* $f \in \omega^\omega$, *f is* $G_1^\#$*-recursive iff f is* γ_1*-recursive*;
(c) *for all* $A \subseteq \omega$, *A is* $G_1^\#$*-semirecursive iff A is* γ_1*-semirecursive.*

Proof. (a) Suppose $|\Omega[G_1^\#]| \leq \alpha < \gamma_1$; without loss of generality we assume that α is recursively inaccessible. Then it follows as in Proposition 7.15(c) that $\omega_1^{G_1^\#} \leq \alpha < \gamma_1$, contradicting Proposition 7.9.
(b) By Corollary 7.16, we have $\omega_1^{G_1^\#} = \gamma_1$; as in the proof of Proposition 7.9, it follows that the function $F_{\gamma_1^+}$ is $G_1^\#$-recursive. Now for any partial γ_1-recursive function $f \in \omega^\omega$, for some a, $f(m) \simeq \{a\}_{\gamma_1}(m) \simeq \{a\}_{\gamma_1^+}(m) \simeq F_{\gamma_1^+}(a, m)$ for all m.
(c) This follows directly from (b).

This completes the proof of Theorem 7.4.

8. Extensions of the class of $\mathbf{\Pi}_1^1$ operators

Richter [19] defines a method of combining operators as follows.

Definition 8.1. (a) $[\Gamma_0, \Gamma_1](A) = \begin{cases} \Gamma_0(A), & \text{if } \Gamma_0(A) \neq A; \\ \Gamma_1(A), & \text{if } \Gamma_0(A) = A; \end{cases}$
(b) $[\Gamma_0, \ldots, \Gamma_{n+1}] = [[\Gamma_0, \ldots, \Gamma_n], \Gamma_{n+1}]$;
(c) $[C_0, \ldots, C_n] = \{[\Gamma_0, \ldots, \Gamma_n] : \Gamma_i \in C_i\}$.

The following results, due to Richter, are of interest.

Theorem 8.2. (a) $|[\mathbf{\Pi}_1^1, \mathbf{\Pi}_0^0]|$ *is the least recursively inaccessible ordinal,* $|[\mathbf{\Pi}_1^0, \mathbf{\Pi}_0^0, \mathbf{\Pi}_0^0]|$ *is the least recursively hyper-inaccessible ordinal, and similarly for* $|[\mathbf{\Pi}_1^0, \mathbf{\Pi}_0^0, \ldots, \mathbf{\Pi}_0^0]|$;
(b) $|[\mathbf{\Pi}_1^0, \mathbf{\Pi}_1^0]|$ *is the least recursively Mahlo ordinal,* $|[\mathbf{\Pi}_1^0, \mathbf{\Pi}_1^0, \mathbf{\Pi}_1^0]|$ *is the least recursively hyper-Mahlo ordinal, and similarly for* $|[\mathbf{\Pi}_1^0, \ldots, \mathbf{\Pi}_1^0]|$.

Using the techniques of 6, it is not difficult to prove

Proposition 8.3. *For all n,* $|[\overbrace{\Pi^1_1, \ldots, \Pi^1_1}^{n}]| < \gamma_2$.

We obtain a result analoguous to Theorem 8.2; the proof is omitted since it is similar to that in Richter [19].

Proposition 8.4. (a) $|[\Pi^1_1, \Pi^0_0]|$ *is the least recursively regular ordinal* γ *such that* γ *is not* γ^+*-recursive and* γ *is a limit of ordinals* α *which are recursively regular and not* α^+*-recursive*;
(b) $|[\Pi^1_1, \Pi^0_1]|$ *is the least recursively regular ordinal* γ *such that* γ *is not* γ^+*-recursive and such that every normal weakly* γ*-recursive function has a fixed point* α *which is recursively regular and not* α^+*-recursice.*

While the ordinals of Theorem 8.2 present a nice hierarchy of recursively large ordinals, those of Proposition 8.4 and its obvious extensions seem of less interest. The natural ordinals to consider above γ_1 would appear to be γ_2, γ_3, and so forth. It turns out that there are natural classes of inductive operators with these as closure ordinals.

Definition 8.5. (a) $(\Gamma_0 \cdot \Gamma_1)(A) = \Gamma_1(\Gamma_0(A))$;
(b) $\Gamma_0 \cdot \ldots \cdot \Gamma_{n+1} = (\Gamma_0 \cdot \ldots \cdot \Gamma_n) \cdot \Gamma_{n+1}$;
(c) $C_0 \cdot \ldots \cdot C_n = \{\Gamma : \exists \Gamma_i \in C_i\, (\Gamma = \Gamma_0 \cdot \ldots \cdot \Gamma_n)\}$;
(d) $(C)^n = \overbrace{C \cdot \ldots \cdot C}^{n}$.

For $\Gamma = \Gamma_0 \cdot \ldots \cdot \Gamma_n$, we do not require each operator Γ_i to be inclusive, but only the composition Γ. For example, $\Sigma^1_1 \subseteq (\Pi^1_1)^2$, since for any Σ^1_1 operator Γ, if $\Gamma_0(A) = \omega - \Gamma(A)$ and $\Gamma_1(A) = \omega - A$, then $\Gamma = \Gamma_0 \cdot \Gamma_1$.

We now state the primary result of this section.

Theorem 8.6. *For any* $n > 0$,
(a) $|(\Pi^1_1)^n| = \gamma_n$;
(b) *for all* $A \subseteq \omega$, $A \in \overline{(\Pi^1_1)^n}$ *iff* A *is* γ_n*-semirecursive.*

Theorem 8.6 is proven in a manner similar to the proof of Theorem 6.1.

Proposition 8.7. *For all* $n > 0$, $|(\Pi^1_1)^n| \leq \gamma_n$.

Proof. We sketch the proof for $(\boldsymbol{\Pi}_1^1)^2$. Given $\Gamma = \Gamma_0 \cdot \Gamma_1 \in (\boldsymbol{\Pi}_1^1)^2$, we have by Proposition 5.1(a), p.o.r. functionals F_0 and F_1 with $\mathrm{rg}(F_i) \subseteq \{0,1\}$ such that for all m and A: $m \in \Gamma_i(A)$ iff $\exists\sigma\,.F_i(\sigma,m,\chi_A) \simeq 1$ iff $\exists\sigma < \omega_1^A\,.F_i(\sigma,m,\chi_A) \simeq 1$. We define p.o.r. functions H and I by:
$I(\tau,m,\beta) \simeq \mathrm{Sup}\,(\tau,\lambda\xi.I(\xi,m,\beta))$, if $\lim(\tau)$ or $\tau = 0$;
$H(\tau+1,m,\beta) \simeq \mathrm{Sup}\,(\beta,\lambda\sigma.F_0(\sigma,m,\lambda p.I(\tau,p,\beta)))$;
$I(\tau+1,m,\beta) \simeq \mathrm{Sup}\,(\beta,\lambda\sigma.F_1(\sigma,m,\lambda p.I(\tau,p,\beta)))$.

The following lemma is proven in the same manner as Lemma 6.3.

Lemma 8.8. *For all ordinals τ and β, all $\alpha < \tau$ and $n < \omega$, if* $\lim(\alpha)$, $\tau = \alpha+n$, *and* $\beta \geq \omega_{\alpha+2n}$, *then*
(a)$_\tau$ for all m, $\chi_{\Gamma^\tau}(m) \simeq I(\tau,m,\beta))$, *and*
(b)$_\tau$ $\omega_1^{\Gamma^\tau} \leq \omega_{\alpha+2n+1}$.

Now for recursively inaccessible ordinals α, $m \in \Gamma^\alpha$ iff $I(\alpha,m,\alpha) \simeq 1$, and $m \in \Gamma^{\alpha+1}$ iff $\exists\sigma < \alpha^{++}F_1(\sigma,m,\lambda p\,.H(\alpha+1,p,\alpha^+)) \simeq 1$. As in Proposition 6.4, it follows that if $m \in \Gamma^{\gamma_2+1}$, then for some $\alpha < \gamma_2$, $m \in \Gamma^{\alpha+1}$, so $m \in \Gamma^{\gamma_2}$; thus $|\Gamma| \leq \gamma_2$.

Next we construct operators $\Lambda_n \in (\boldsymbol{\Pi}_1^1)^n$ such that $|\Lambda_n| = \gamma_n$.

Definition 8.9. $\Lambda(A) = \{\langle x,y\rangle : L(\{x\}^A, \{y\}^A)\}$; $\Lambda_n = (\Lambda)^{n-1}.\,\Lambda_1$.

The following lemma is easily verified.

Lemma 8.10. *For any* $A \subseteq \omega$, *let* $\omega_1^A = \alpha$; *then*
(a) *$\Lambda(A)$ is a pre-well-ordering of length* α;
(b) *if A is p.o.r. in* α, *then* $\omega_1^{\Lambda^n(A)} = \alpha^{(n)}$;
(c) *if A is p.o.r. in* α, *then* $\Lambda_{n+1}(A) = \{\langle a,m,t\rangle : \{a\}_{\alpha^{(n)}}(m) \simeq t\}$.

Applying this lemma and proceeding as in the proof of Lemma 6.6, we have

Lemma 8.11. *For all k, all* $n > 0$, $\tau < \gamma_n$, *and* $\alpha < \tau$, *if* $\lim(\alpha)$ *and* $\tau = \alpha + k$, *then*
(a)$_\tau$ $\Lambda_n^0 = \emptyset$; $\Lambda_n^\tau = \mathbf{U}_{\sigma<\omega_{\alpha+kn}} \{\langle a,m,t\rangle : \{a\}_\sigma(m) \simeq t\}$ *for* $\tau > 0$;
(b)$_\tau$ $\omega_1^{\Lambda_n^\tau} = \omega_{\alpha+kn+1}$;
(c) $\omega_1^{(\Lambda)^{n-1}(\Lambda_n^\tau)} = \omega_{\alpha+n\,(k+1)}$.

Proposition 8.12. *For all $n > 0$,*
(a) $\overline{\Lambda_n} = \{\langle a, m, t\rangle : \{a\}_{\gamma_n}(m) \simeq t\}$;
(b) $|\Lambda_n| = \gamma_n$.

It follows from Proposition 8.7 that $\overline{\Lambda_n} = \Lambda_n^{\gamma_n}$, which equals $\{\langle a, m, t\rangle : \{a\}_{\gamma_n}(m) \simeq t$ by Lemma 8.11.
(b) This follows from (a).

This completes the proof of Theorem 8.6(a); 8.6(b) $((\mathbf{\Pi}_1^1)^n = \gamma_n$-semirecursive) is easily obtained.
($\subseteq$) This follows from Lemma 8.8(a) and Proposition 8.7.
($\supseteq$) This follows directly from Proposition 8.12(a).

Finally, we can extend the results of §7 to $(\mathbf{\Pi}_1^1)^n$ and γ_n.

Definition 8.13. $G_n^{\#}(f) \simeq t$ iff $\{f(0)\}_\alpha(f(1)) \simeq t < \omega$ for $\alpha = (\omega_1^f)^{(n)}$.

Following the development of §7, we can prove

Proposition 8.14. *For all $n > 0$ and all $k < \omega$,*
(a) $\omega_1^{G_n^{\#}} = \gamma_n$;
(b) *for all $f: \omega^k \to \omega$, f is $G_n^{\#}$-recursive iff f is γ_n-recursive*;
(c) *for all $A \subseteq \omega$, A is $G_n^{\#}$-semirecursive iff A is γ_n-semirecursive.*

9. Σ_1^1 inductive definitions

The main result of this section is a characterization of $|\mathbf{\Sigma}_1^1|$ which is the dual of the characterization of $|\mathbf{\Pi}_1^1| = \gamma_1$ given by Proposition 4.21 (in combination with Theorem 6.1).

Theorem 9.1. $|\mathbf{\Sigma}_1^1|$ *is the least ordinal σ such that for all $a < \omega$,* $(\forall \tau < \sigma \,.\, \{a\}_{\tau^+}(\tau)\downarrow) \to \{a\}_{\sigma^+}(\sigma)\downarrow$.

We also prove an analogue to Theoren 6.1(b).

Theorem 9.2. *For all $A \subseteq \omega$, $A \in \overline{\mathbf{\Sigma}_1^1}$ iff A is $|\mathbf{\Sigma}_1^1|$-semirecursive.*

Anderaa [1] has recently obtained a significant result regarding dual classes of inductive operators which implies that $|\mathbf{\Pi}_1^1| < |\mathbf{\Sigma}_1^1|$ and $|\mathbf{\Sigma}_2^1| < |\mathbf{\Pi}_2^1|$. Combining the former inequality with the techniques involved in Theorems 9.1 and 9.2, we are able to characterize the spectra of the two classes.

Definition 9.3. For any class C of inductive operators, Spectrum $(C) = \{|\Gamma| : \Gamma \in C\}$.

Theorem 9.4. (a) Spectrum $(\mathbf{\Sigma}_1^1) = \{\alpha : \alpha \leq \gamma_1\}$;
(b) Spectrum $(\Sigma_1^1) = \{\alpha \leq |\Sigma_1^1| : \alpha$ is α^+-recursive$\}$.

We begin by applying Proposition 5.1 to Σ_1^1 inductive operators.

Lemma 9.5. *For any* Σ_1^1 *operator* Γ, *there is a p.o.r. functional* F *with* $\mathrm{rg}(F) \subseteq \{0,1\}$ *such that for all* m *and* A,
$m \in \Gamma(A)$ *iff* $\forall \sigma \,.\, F(\sigma, m, \chi_A) \simeq 0$ *iff* $\forall \sigma < \omega_1^A \,.\, F(\sigma, m, \chi_A) \simeq 0$.

Lemma 9.6. *For any* Γ *and* F *as in Lemma 9.5, if* I *is defined by*
$I(\tau, m, \beta) \simeq \mathrm{Sup}\,(\tau, \lambda\xi \,.\, I(\xi, m, \beta))$, *if* $\lim(\tau)$ *or* $\tau = 0$, *and*
$I(\tau+1, m, \beta) \simeq 1 - \mathrm{Sup}\,(\beta, \lambda\sigma \,.\, F(\sigma, m, \lambda n \,.\, I(\tau, n, \beta)))$, *then*
(a) *for all* m, *all* τ *and all* $\beta \geq \omega_\tau$, $\chi_{\Gamma^\tau}(m) \simeq I(\tau, m, \beta)$;
(b) *for all* τ, $\omega_1^{\Gamma^\tau} \leq \omega_{\tau+1}$.

Lemma 9.6 is proven after the pattern of Lemma 6.3.

Proposition 9.7. *For any* Σ_1^1 *inductive operator* Γ, *if* $\gamma = |\Gamma|$ *is recursively inaccessible, then* γ *is* γ^+*-recursive.*

Proof. For all inaccessible γ, $m \in \Gamma^\gamma$ iff $I(\gamma, m, \gamma) \simeq 1$ and $m \in \Gamma^{\gamma+1}$ iff $\forall \sigma < \gamma^+ \,.\, F(\sigma, m, \lambda n \,.\, I(\gamma, n, \gamma)) \simeq 0$. Hence $\Gamma^{\gamma+1} = \Gamma^\gamma$ iff $\forall m < \omega [I(\gamma, m, \gamma) \simeq 0 \rightarrow \exists \sigma < \gamma^+ \,.\, F(\sigma, m, \lambda n \,.\, I(\gamma, n, \gamma)) \simeq 1]$ iff $\exists \tau < \gamma^+ \forall m < \omega [I(\gamma, m, \gamma) \simeq 0 \rightarrow \exists \sigma < \tau \,.\, F(\sigma, m, \lambda n \,.\, I(\gamma, n, \gamma)) \simeq 1]$, the last equivalence by the regularity of γ^+. It is clear that if γ is the least such that $\Gamma^{\gamma+1} = \Gamma^\gamma$, then γ is γ^+-recursive.

This gives half of Theorem 9.4(b). Since, as we pointed out in §8, $\Sigma_1^1 \subseteq (\mathbf{\Pi}_1^1)^2$, we also have the following.

Three rather technical lemmas are needed for the proof of Theorem 9.1. We refer the reader to Cenzer [7] for proofs of these lemmas.

Lemma 9.9. *For any non-recursively-inaccessible ordinal* $\alpha < \gamma$,
(a) *if* α *is a limit of inaccessibles, then* α *is* α^+*-recursive*;
(b) *if there is a largest recursively inaccessible* $\beta < \alpha$, *then* α^+*-recursive in* β.

Lemma 9.10. *There is a* $\mathbf{\Pi}_1^1$ *relation* K^* *such that for all* $a < \omega$, *for all* α, *and all* $A \subseteq \omega$ *such that* A *is a prewellordering of length* α, $K^*(a, A)$ *iff* $\{a\}_{\alpha^+}(\alpha)\downarrow$.

Definition 9.11. For any operator Γ, let $\hat{\Gamma}$ be defined by
$\hat{\Gamma}(A) = \{\langle m,n\rangle : m,n \in \Gamma(\{p : \langle p,p\rangle \in A\}) \wedge \langle n,n\rangle \notin A\} \cup A$.

Lemma 9.12. *For any inductive operator* Γ, *any ordinal* τ,
(a) $\hat{\Gamma}^\tau = \{\langle m,n\rangle : m,n \in \Gamma^\tau \wedge (\textit{least } \sigma \,.\, m \in \Gamma^\sigma) \leq (\textit{least } \sigma \,.\, n \in \Gamma^\sigma)\}$;
(b) $|\hat{\Gamma}| = |\Gamma|$.
(c) *if* Γ *is* $\mathbf{\Sigma}_1^1(\mathbf{\Pi}_1^1)$, *then so is* $\hat{\Gamma}$.

We are now ready to prove Theorem 9.1, that
$|\mathbf{\Sigma}_1^1| = \text{least } \sigma \,.\, \forall a < \omega[(\forall \tau < \sigma \,.\, \{a\}_{\tau^+}(\tau)\downarrow) \rightarrow \{a\}_{\sigma^+}(\sigma)\downarrow]$.
($\leq$) For any $\sigma < |\mathbf{\Sigma}_1^1|$, we find an index $\bar{a}$ such that $\forall \tau < \sigma \,.\, \{\bar{a}\}_{\tau^+}(\tau)\downarrow$, but $\{\bar{a}\}_{\sigma^+}(\sigma)\uparrow$. The proof splits into two parts.
(a) Let α be recursively inaccessible. We have a $\mathbf{\Sigma}_1^1$ operator Γ with associated F and I and an $m \in \Gamma^{\alpha+1} - \Gamma^\alpha$. It follows from Lemma 9.6 that there is an index $\bar{a}$ such that for all τ, $\{\bar{a}\}_{\tau^+}(\tau)\downarrow$ iff $\neg\, \mathrm{RI}(\tau) \vee m \notin \Gamma^{\tau+1}$.
(b) It follows from Lemma 9.9 that there is an index a such that for any $\alpha < |\mathbf{\Sigma}_1^1|$, at least one of the following holds:
(i) $\{a\}_{\alpha^+}(\alpha) \simeq 0$ and $\mathrm{RI}(\alpha)$;
(ii) $\{a\}_{\alpha^+}(\alpha) \simeq c$ and $\{c\}_{\alpha^+} \simeq \alpha$;
(iii) $\{a\}_{\alpha^+}(\alpha) \simeq \langle \beta, b\rangle$, β is the largest recursively inaccessible ordinal less than α, and $\{b\}_{\alpha^+}(\beta) \simeq \alpha$.
It is important to note that the function $\lambda\alpha \,.\, \{a\}_{\alpha^+}(\alpha)$ is 1–1 on the set of non-recursively-inaccessibles less than $|\mathbf{\Sigma}_1^1|$. If (i) holds, we apply (a) above. If (ii) holds, choose $\bar{a}$ so $\{\bar{a}\}_{\tau^+}(\tau) \simeq \text{least } \xi < \tau^+[\{a\}_{\tau^+}(\tau) \neq c]$. If (iii) holds, apply (a) to β to find $\bar{b}$ such that $\forall \tau < \beta \,.\, \{\bar{b}\}_{\tau^+}(\tau)\downarrow$ but $\{\bar{b}\}_{\beta^+}(\beta)\uparrow$; then let

$$\{\bar{a}\}_{\tau^+}(\tau) \simeq \begin{cases} 0, & \text{if } \{a\}_{\tau^+}(\tau) < \omega \vee (\{a\}_{\tau^+}(\tau))_1 \neq b; \\ \{\bar{b}\}_{\tau^+}((\{a\}_{\tau^+}(\tau))_0), & \text{otherwise.} \end{cases}$$

($\geq$) Let $\sigma = |\Sigma_1^1|$ and suppose that for some a, $\{a\}_{\tau^+}(\tau)\downarrow$ for all $\tau < \sigma$ but $\{a\}_{\sigma^+}(\sigma)\uparrow$. Let Γ be a Σ_1^1 operator such that $|\Gamma| = \sigma$, and define Γ_1 as follows, using Lemmas 9.10 and 9.12: $m \in \Gamma_1(A)$ iff $m \in \hat{\Gamma}(A) \vee (m = 0 \wedge \neg K^*(a, A))$. Γ_1 is Σ_1^1 and $0 \in \Gamma_1^{\sigma+1} - \Gamma_1^{\sigma}$, so that $|\Gamma_1| > \sigma = |\Sigma_1^1|$, a contradiction. (Notice that since $\hat{\Gamma}^{\sigma}$ contains only pairs, $0 \notin \hat{\Gamma}^{\sigma}$.)

This completes the proof of Theorem 9.1; we have as a corollary to the proof the following.

Proposition 9.13. $|\Sigma_1^1|$ *is recursively inaccessible.*

Proof. If not, then part (b) (ii–iii) of ($\leq$) will provide a contradiction to the conclusion in ($\geq$).

We now complete the proof of Theorem 9.4.

Proposition 9.14. (a) *For any* $\sigma < |\Sigma_1^1|$, *if* σ *is* σ^+*-recursive, then* $\sigma \in \text{Spectrum}(\Sigma_1^1)$;
(b) *for any* $\sigma < |\Pi_1^1|$, $\sigma \in \text{Spectrum}(\Pi_1^1)$.

Proof. (a) Given $\sigma \simeq \{c\}_{\sigma^+}$, choose b so that $\{b\}_{\tau^+}(\tau) \simeq$ least $\xi < \tau^+ . \{c\}_{\tau^+} \simeq \tau$. Then $\forall \tau < \sigma . \{b\}_{\tau^+}(\tau)\uparrow$ but $\{b\}_{\sigma^+}(\sigma)\downarrow$. Let Γ be a universal Σ_1^1 operator; then by Lemmas 9.10 and 9.12, $\forall \tau < \sigma . \neg K^*(b, \hat{\Gamma}^{\tau})$ but $K^*(b, \hat{\Gamma}^{\sigma})$. We define Γ_0 with $|\Gamma_0| = \sigma$ by $\Gamma_0(A) = \{m : m \in \hat{\Gamma}(A) \wedge \neg K^*(b, A)\} \cup A$.
(b) Since $|\Pi_1^1| < |\Sigma_1^1|$, there is an index a such that $\forall \tau < \sigma . \{a\}_{\tau^+}(\tau)\downarrow$, but $\{a\}_{\sigma^+}(\sigma)\uparrow$. Let Γ be a universal Π_1^1 operator and let $\Gamma_1(A) = \{m : m \in \hat{\Gamma}(A) \wedge K^*(a, A)\}$; then $|\Gamma_1| = \sigma$.

For the proof of Theorem 9.2 we need the following lemma.

Lemma 9.15. *There is a* Σ_1^1 *relation* $\bar{K}$ *such that for all* $a < \omega$, *all* α, *and all* $A \subseteq \omega$ *such that* A *is a prewellordering of length* α, $\bar{K}(t, A)$ *iff* $T^0(\alpha, \alpha, t)$.

Proof. Since T is p.o.r., this is an easy consequence of Lemma 9.10.

We now prove Theorem 9.2, that $\overline{\Sigma_1^1} = |\Sigma_1^1|$-semirecursive. Let $\sigma = |\Sigma_1^1|$.
($\subseteq$) Given a Σ_1^1 operator Γ, we have by Lemma 9.6 and Theorem 9.1,

$m \in \overline{\Gamma}$ iff $\exists \alpha < \sigma\,(\mathrm{RI}(\alpha) \wedge I(\alpha, m, \alpha) \simeq 1)$.
($\supseteq$) Let Γ be a universal $\mathbf{\Sigma}_1^1$ operator and let Γ_1 be defined by
$\Gamma_1(A) = A \cup \{\langle 0, s\rangle : s \in \hat{\Gamma}(\{p : \langle 0, p\rangle \in A\})\}$
$\qquad \cup \{\langle 1, t\rangle : \bar{K}(t, \{p : \langle 0, p\rangle \in A\})\}$.
We have $|\Gamma_1| = \sigma$ and for all a, m, and $n < \omega$, $\{a\}_\sigma(m) \simeq n$ iff $\langle 1, \langle a, m, n\rangle\rangle \in \overline{\Gamma_1}$.

10. Σ_2^1 relations, stable ordinals, and $\aleph_1$-recursion

In §4 we defined, for $A \subseteq \mathrm{ORD}$, the ordinal δ_1^A and showed in Corollary 4.11 that for $A \subseteq \omega$, $\delta_1^A < \aleph_1$ and $\aleph_1$ is stable in A. It follows that, for any A and B in $P(\omega)$, B is ∞-semirecursive in A iff B is $\aleph_1$-semirecursive in A iff B is δ_1^A-semirecursive in A. Recall that $\delta_2^1\!-\!A$ is the least ordinal not isomorphic to a well-ordering $\mathbf{\Delta}_2^1$ in A. In this section we present the following results, parallel to those in §5 regarding $\mathbf{\Pi}_1^1$ and ω_1^A.

Proposition 10.1. *If $Q \subseteq \omega \times P(\omega)$ is $\mathbf{\Sigma}_2^1$, then there is a p.o.r. functional F with $\mathrm{rg}(F) \subseteq \{0, 1\}$ such that for all m and A: $Q(m, A)$ iff $\exists \alpha\,.\,F(\alpha, m, \chi_A) \simeq 1$ iff $\exists \alpha < \aleph_1\,.\,F(\alpha, m, \chi_A) \simeq 1$ iff $\exists \alpha < \delta_1^A\,.\,F(\alpha, m, \chi_A) \simeq 1$.*

Proposition 10.2. *The relation K_2, defined by $K_2(\langle a, m, n\rangle, A)$ iff $\{a\}_{\delta_1^A}(m, \chi_A) \simeq n$, is $\mathbf{\Sigma}_2^1$.*

Proposition 10.3. *For all $A \subseteq \omega$, $\delta_1^A = \delta_2^1\!-\!A$.*

The proof of Proposition 10.1 is basically an adaptation of Shoenfield's [21] Absoluteness Theorem to ordinal recursion theory with a set parameter; the proof follows:

Proof of Proposition 10.1: The second and third equivalences follow by stability; we prove the first. Suppose we have $Q(m, A)$ iff $\exists \phi\, \forall \psi\, \exists p\,.\,R(m, \bar{\phi}(p), \bar{\psi}(p), A)$, where R is recursive and $p < t \wedge R(..p..) \to R(..t..)$. Let $S_{m,\phi,A} = \{s : \mathrm{seq}(s) \wedge \neg R(m, \bar{\phi}(\mathrm{ln}(s)), s, A)\}$; let $\circledless$ be the Kleene–Brouwer sequence ordering. Then, as in the proofs cited for Lemma 5.5, $Q(m, A)$ iff $\exists \phi\,[S_{m,\phi,A}$ is well-ordered by $\circledless]$ iff $\exists \phi\, \exists \tau\, \exists f (f : \omega \to \tau)\, \forall s\, \forall t\, [s, t \in S_{m,\phi,A} \wedge s \circledless t \to f(s) < f(t)]$ iff

$\exists\sigma\exists g(g\colon\omega\to\sigma)\forall p\,.\,G(m,\bar g(p),\chi_A)\simeq 1$, where G is p.o.r. with $\mathrm{rg}(G)\subseteq\{0,1\}$. The direction $(\to)$ of the last equivalence involves letting g code up a given ϕ and f and taking $\sigma=\langle\omega,\tau\rangle$ for a given τ. Let $h(\sigma)=\mathrm{Sup}(\omega,\lambda n\,.\langle\overbrace{\sigma,\ldots,\sigma}^{n}\rangle)$ so that for any $g\colon\omega\to\sigma$ and any p, $\bar g(p)<h(\sigma)$. As in Lemma 5.7 there is a p.o.r. relation K such that for all ordinals σ and all $A\subseteq\omega$, $\Gamma_{\sigma,A}$, defined by $\tau\in\Gamma_{\sigma,A}(C)$ iff $K(\tau,\sigma,A,C)$, satisfies $\langle m,1\rangle\in\overline{\Gamma_{\sigma,A}}$ iff $\forall g(g\colon\omega\to\sigma)\exists p\,.\,G(m,\bar g(p),\chi_A)\simeq 0$. Then $Q(m,A)$ iff $\exists\sigma\,.\,\langle m,1\rangle\notin\overline{\Gamma_{\sigma,A}}$ iff $\exists\sigma\exists\tau[\Gamma^{\tau+1}_{\sigma,A}=\Gamma^{\tau}_{\sigma,A}\wedge\langle m,1\rangle\notin\Gamma^{\tau}_{\sigma,A}]$.

$$\text{Let } F(\alpha,m,\chi_A)\simeq\begin{cases}1, & \text{if } \Gamma^{(\alpha)_1+1}_{(\alpha)_0,A}=\Gamma^{(\alpha)_1}_{(\alpha)_0,A}\wedge\langle m,1\rangle\notin\Gamma^{(\alpha)_1}_{(\alpha)_0,A};\\ 0, & \text{otherwise.}\end{cases}$$

F is easily seen to be p.o.r. and $Q(m,A)$ iff $\exists\alpha\,.\,F(\alpha,m,\chi_A)\simeq 1$.

This completes the proof of Proposition 10.1; next we prove Proposition 10.2.

Proof of Proposition 10.2: Since δ^A_1 is stable in A, $K_2(\langle a,m,n\rangle,A)$ iff $\{a\}_{\aleph_1}(m,\chi_A)\simeq n$ iff $\exists B\,.\,\{a\}_{\omega^B_1}(m,\chi_A)\simeq n$ iff $\exists B\exists\boldsymbol{u},v(N(\boldsymbol{u},\boldsymbol{m},B)\wedge N(v,n,B)\wedge\langle a,\boldsymbol{u},v\rangle\in K^1_B[\chi_A])$, where N and K_B are taken from Definitions 5.8 and 5.9. It is clear from Proposition 5.10 that this is a $\boldsymbol{\Sigma}^1_2$ relation.

The proof of Proposition 10.3 depends heavily on the Novikoff, Kondo [13], Addison Uniformization Theorem: (See Shoenfield [21], p. 188.)

Theorem 10.4. *For any* $\boldsymbol{\Pi}^1_1$ *relation* P *there is a* $\boldsymbol{\Pi}^1_1$ *relation* Q *such that for all* $\boldsymbol{m},\boldsymbol{\phi},\boldsymbol{\psi}$, *and* A:
(a) $Q(\boldsymbol{m},\boldsymbol{\phi},\boldsymbol{\psi},A)\to P(\boldsymbol{m},\boldsymbol{\phi},\boldsymbol{\psi},A)$;
(b) $\exists\boldsymbol{\phi}.P(\boldsymbol{m},\boldsymbol{\phi},\boldsymbol{\psi},A)\leftrightarrow\exists!\,\boldsymbol{\phi}.Q(\boldsymbol{m},\boldsymbol{\phi},\boldsymbol{\psi},A)$.

Corollary 10.5. (a) *If* Q *is* $\boldsymbol{\Pi}^1_1$, *then for all* $\boldsymbol{m}$, $\boldsymbol{\psi}$, *and* A:
$\exists\boldsymbol{\phi}.Q(\boldsymbol{m},\boldsymbol{\phi},\boldsymbol{\psi},A)$ *iff* $\exists\boldsymbol{\phi}\in\boldsymbol{\Delta}^1_2-A\,.\,Q(\boldsymbol{m},\boldsymbol{\phi},\boldsymbol{\psi},A)$;
(b) *If* Q *is* $\boldsymbol{\Sigma}^1_2$, *then for all* $\boldsymbol{m}$, $\boldsymbol{\psi}$, *and* A:
$\exists\boldsymbol{\phi}.Q(\boldsymbol{m},\boldsymbol{\phi},\boldsymbol{\psi},A)$ *iff* $\exists\boldsymbol{\phi}\in\boldsymbol{\Delta}^1_2-A\,.Q(\boldsymbol{m},\boldsymbol{\phi},\boldsymbol{\psi},A)$.

Proof of Proposition 10.3, that $\delta^A_1=\delta^1_2-A$ *for all* $A\subseteq\omega$:
$(\leq)$ We show that δ^1_2-A is stable in A. Suppose $\{a\}_\infty(\alpha,\chi_A)\downarrow$ for some a and some $\alpha<\delta^1_2-A$; choose $\phi\in\boldsymbol{\Delta}^1_2-A$ such that $|\phi|=\alpha$. Then $\exists B\exists u,v(|\phi|=|\{u\}^B|\wedge\langle a,u,v\rangle\in K^1_B[\chi_A])$; applying Corollary 10.5 there is a B which is

there is a B which is Δ^1_2 in A. It follows that $\{a\}_\infty(\alpha, \chi_A) < \delta^1_2 - A$.
($\geq$) For any $\sigma < \delta^1_2 - A$, we show that σ is not stable in A. Given $\sigma < \delta^1_2 - A$, we have a $\Delta^1_2 - A$ well-ordering ϕ of type σ. It follows from Proposition 10.1 that ϕ is ∞-recursive in A, so if σ is stable in A, then ϕ is σ-recursive in A. But then by Proposition 5.12, $|\phi| < \sigma$, a contradiction.

We remark that Propositions 10.1 and 10.2 could be combined and restated as follows:

Proposition 10.6. *A relation over natural numbers and sets of natural numbers is Σ^1_2 iff it is ∞-semirecursive iff it is $\aleph_1$-semirecursive.*

11. Stability and the ordinal arithmetic hierarchy

A crucial point in the study of Π^1_1 inductive definitions in §6 is the fact that the relations RR and RI are p.o.r. A comparison of §5 with §10 makes it clear that in the study of Σ^1_2 inductive definitions the relation "stable" will have a large part.

Definition 11.1. (a) $S^1(\alpha)$ iff α is stable;
(b) for all $n > 0$, $S^{n+1}(\alpha)$ iff α is stable in S^n.

We say that α is n-stable if $S^n(\alpha)$. By Proposition 4.12, n-stables exist for all $n > 0$. It is interesting to note, however, that for $n > 1$ there need not be any countable n-stables and uncountable cardinals need not always be n-stable. Let S^0 be all of the ordinals for the sake of simplicity. Recursion in the S^n is closely related to the ordinal arithmetic hierarchy, defined similarly to the usual arithmetic hierarchy.

Definition 11.2. $R \subseteq \mathrm{ORD}^k \times (\mathrm{ORD}^{\mathrm{ORD}})$ is $\gamma - \Sigma_n$ iff there is a p.o.r. relation P and an alternating sequence $\exists\beta_1 < \gamma \ldots Q_n\beta_n < \gamma$ of ordinal quantifiers such that for all $\boldsymbol{\alpha}$ and f: $R(\boldsymbol{\alpha}, f)$ iff $\exists\beta_1 < \gamma \ldots Q_n\beta_n < \gamma \,.P(\boldsymbol{\beta}, \boldsymbol{\alpha}, f)$; $\gamma - \boldsymbol{\Pi}_n$ and $\boldsymbol{\gamma} - \boldsymbol{\Sigma}_n$ are defined analogously.

These γ-arithmetic classes are comparable to the usual arithmetic hierarchy for sufficiently regular γ.

Proposition 11.3. (a) *For any f, any γ recursively regular in f, and any partial function F, F is γ-recursive in f iff graph(F) is $\gamma - \Sigma_1$ in f;*

(b) *For any regular cardinal* κ (*or* $\kappa = \infty$) *and any partial functional F, F is* κ-*recursive iff graph* (F) *is* $\kappa - \mathbf{\Sigma}_1$.

We need a notion of relative n-stability.

Definition 11.4. α is $n-\beta$-stable ($\mathrm{RS}^n(\alpha,\beta)$) iff $\mathrm{RR}(\beta)$ and for all $\beta - \mathbf{\Sigma}_n$ relations R and all $\boldsymbol{\tau} < \alpha$, $\exists\gamma < \beta \,.\, R(\boldsymbol{\tau},\gamma) \rightarrow \exists\gamma < \alpha \,.\, R(\boldsymbol{\tau},\gamma)$.

Lemma 11.5. *For all* $n > 0$, RS^n *is p.o.r.*

Proof. This is an easy application of Propositions 3.14 and 3.15.

In contrast to Lemma 11.5 is the following result.

Lemma 11.6. *For all n,* S^{n+1} *is not* ∞-*recursive in* S^n.

Proof. If S^{n+1} were ∞-recursive in S^n, then least $\alpha \,.\, S^{n+1}(\alpha)$ would also be ∞-recursive in S^n, contradicting its $n+1$-stability.

We can now prove an ordinal arithmetic "Hierarchy Theorem".

Theorem 11.7. *For all* $n > 0$,
(a) *for all* α, $S^n(\alpha)$ *iff for any* $\infty - \mathbf{\Sigma}_n$ *relation R and any* $\boldsymbol{\tau} < \alpha$,
$\exists\beta \,.\, R(\boldsymbol{\tau},\beta) \leftrightarrow \exists\beta < \alpha \,.\, R(\boldsymbol{\tau},\beta)$;
(b) *for any n-stable ordinal* β *and any* $\alpha < \beta$, $S^n(\alpha)$ *iff* $\mathrm{RS}^n(\alpha,\beta)$;
(c) S^n *is* $\infty - \mathbf{\Pi}_n$ *but not* $\infty - \mathbf{\Sigma}_n$;
(d) *for any* $R \subseteq \mathrm{ORD}^k$, *R is* $\infty - \mathbf{\Sigma}_{n+1}$ *iff R is* $\infty - \mathbf{\Sigma}_1$ *in* S^n.

Proof. Let $n = 1$; for $n > 1$ the proof is similar but more involved.
(a) This follows from Proposition 11.3(a).
(b) This is immediate from (a).
(c) $S^1(\alpha)$ iff $\forall\gamma\forall\sigma\forall\,\boldsymbol{\tau} < \alpha\forall a < \omega[T^0(\sigma,\sigma,\langle a,\boldsymbol{\tau},\gamma\rangle) \rightarrow \gamma < \alpha]$; if S^1 were also $\infty - \mathbf{\Sigma}_1$, it would be ∞-recursive, contradicting Lemma 11.6.
(d) ($\rightarrow$) $\exists\beta\forall\gamma \,.\, P(\beta,\gamma,\boldsymbol{\alpha})$ iff $\exists\beta\exists\sigma[S^1(\sigma) \wedge \langle\boldsymbol{\alpha},\beta\rangle < \sigma \wedge \forall\gamma < \sigma \,.\, P(\beta,\gamma,\boldsymbol{\alpha})]$.
($\leftarrow$) $\exists\beta \,.\, \{a\}_\infty(\beta,\boldsymbol{\alpha},S^1) \simeq 1$ iff $\exists\beta\exists\sigma[S^1(\sigma) \wedge \langle\boldsymbol{\alpha},\beta\rangle < \sigma \wedge$ $T^1(\sigma,\sigma,\langle a,\beta,\boldsymbol{\alpha},1\rangle,\lambda\tau \,.\, \mathrm{RS}^1(\tau,\sigma))]$, which is $\infty - \mathbf{\Sigma}_2$ since by (c) S^1 is $\infty - \mathbf{\Pi}_1$. (We identify S^n and RS^n with their characteristic functions for the sake of simplicity.)

We can relativize Theorem 11.7 (d) to large ordinals.

Proposition 11.8. *For all $n > 0$, all ordinals α such that α is a limit of n-stables, and all $R \subseteq \mathrm{ORD}^k$, R is $\alpha - \boldsymbol{\Sigma}_{n+1}$ iff R is $\alpha - \boldsymbol{\Sigma}_1$ in S^n.*
In particular, R is $\aleph_1 - \boldsymbol{\Sigma}_2$ iff R is $\aleph_1 - \boldsymbol{\Sigma}_1$ in S.

Although the S^n are not $\infty - \boldsymbol{\Pi}_n$ complete (for example, since for any ∞-recursive f and any m, $f(m) < \delta_1$ and is therefore not stable, the only $A \subseteq \omega$ reducible to S by an ∞-recursive function is the empty set), they play the role in Theorem 11.7 of the n'th "jump" of $\emptyset$. We can think of stability as a jump operator in the following sense (proven as 11.7 (c)).

Proposition 11.9. *For any $A \subseteq \mathrm{ORD}$, $\{\alpha: \alpha$ is stable in $A\}$ is $\infty - \boldsymbol{\Pi}_1$ in A but not $\infty - \boldsymbol{\Sigma}_1$ in A.*

For the remainder of the section we discuss S^1 or S for short.

Definition 11.10. (a) $\overline{\mathrm{RR}}(\alpha, f)$ iff α is recursively regular in f;
(b) α is inaccessibly stable (IS (α)) iff $\overline{\mathrm{RR}}(\alpha, S)$ and α is a limit of stable ordinals.

Proposition 11.11. (a) IS (α) *iff* $\overline{\mathrm{RR}}(\alpha, S) \wedge \alpha = \delta_\alpha$;
(b) $\overline{\mathrm{RR}}$ *is p.o.r. and* IS *is p.o.r. in* S;
(c) *for all α, $\delta_{\alpha+1}$ is recursively regular in S;*
(d) *α stable in S implies α is recursively regular in S.*

Proof. (a), (b), and (d) are similar to results on RR, RI, and stability. To prove (c), notice that by Theorem 11.7 (b), $S(\beta) \leftrightarrow \beta = \delta_\alpha \vee \mathrm{RS}^1(\beta, \delta_\alpha)$ for $\beta < \delta_{\alpha+1}$, so that $S \upharpoonright \delta_{\alpha+1}$ is weakly $\delta_{\alpha+1}$-recursive; (c) now follows by the regularity of $\delta_{\alpha+1}$.

It is clear that δ_1^S must be inaccessible stable, but as δ_1^S need not be countable (see §13), we need something else.to construct a countable inaccessible stable ordinal.

Lemma 11.12. *For all ordinals β, $S(\beta)$ iff for all $\boldsymbol{\alpha} < \beta$ and all ∞-recursive F, $F(\boldsymbol{\alpha}) \not\simeq \beta$.*

Proof. By Proposition 4.10, $D_\beta = \{F(\boldsymbol{\alpha}) : \boldsymbol{\alpha} < \beta \wedge F \text{ is } \infty\text{-recursive}\}$ is a countable initial segment of ORD such that Sup (D_β) is the least stable ordinal greater than or equal to β. The lemma follows directly from this fact.

Proposition 11.13. *There are countable inaccessibly stables.*

Proof. The least ordinal which is not ∞-recursive in S is clearly regular in S and will be stable by Lemma 11.12.

We are interested in ordinals much larger than the first inaccessibly stable because of the following result, which is parallel to Proposition 4.16.

Proposition 11.14. *If α is any of the following: δ_1, the least inaccessibly stable ordinal, the least hyper-inaccessibly stable ordinal, then α is α^*-recursive in S.*

Proof. For example, δ_1 = least $\alpha < \delta_2 \,.\, S(\alpha)$.

For any α, let $\alpha^{*n} = \overbrace{\alpha^{*\cdots *}}^{n}$.

Definition 11.15. β_n = least β. β is not β^{*n}-recursive in S.

It is clear that the β_n exist and are less than the least ordinal not ∞-recursive in S, and therefore countable. The β_n are large with respect to stability as the γ_n of §4 are with respect to regularity.

Proposition 11.16. *For all $n > 0$, β_n is inaccessibly stable.*

Proof. For any $\boldsymbol{\alpha} < \beta_n$, each α_i is α_i^{*n}-recursive in S and therefore β_n^{*n}-recursive in S; any ∞-recursive function F is equivalent on β_n to a β_n^{*n}-recursive function F_0 by the stability of β_n^{*n}. By the definition of β_n, $F(\boldsymbol{\alpha}) \simeq F_0(\boldsymbol{\alpha}) \not\simeq \beta_n$. Hence by Lemma 11.12, β_n is stable. The proof that β_n is regular in S and is a limit of stables is parallel to the proof of Proposition 4.18(a).

We can characterize the β_n with a proposition similar to Proposition 4.21. We state our result for $n = 1$.

Proposition 11.17. *β_1 is the least ordinal β such that for all $a < \omega$, $\{a\}_{\beta*}(\beta, S)\downarrow \to \exists\alpha < \beta \,.\, \{a\}_{\alpha*}(\alpha, S)\downarrow$.*

Corollary 11.18. *For any a and $i < \omega$:*
(a) *if $\{a\}_{\beta_1^*}(\beta_1, S) \simeq i$, then $\exists\alpha < \beta_1 \,.\, \{a\}_{\alpha*}(\alpha, S) \simeq i$;*
(b) *if for all $\alpha < \beta_1$, $\{a\}_{\alpha*}(\alpha, S) \simeq i$ and if $\{a\}_{\beta_1^*}(\beta_1, S)\downarrow$, then $\{a\}_{\beta_1^*}(\beta_1, S) \simeq i$.*

12. Σ_2^1 inductive definitions

In this section we present the following theorems.

Theorem 12.1. *For all $n > 0$,* (a) $|(\Sigma_2^1)^n| = \beta_n$;
(b) *for all $A \subseteq \omega$, $A \in \overline{(\Sigma_2^1)^n}$ iff A is β_n-semirecursive.*

Theorem 12.2. *$|\Pi_2^1|$ is the least ordinal β such that for all $a < \omega$, if $\forall\alpha < \beta \,.\, \{a\}_{\alpha*}(\alpha, S)\downarrow$, then $\{a\}_{\beta*}(\beta, S)\downarrow$.*

Theorem 12.3. *For all $A \subseteq \omega$, $A \in \overline{\Pi_2^1}$ iff A is $|\Pi_2^1|$-semirecursive in S.*

Theorem 12.4. (a) Spectrum$(\Sigma_2^1) = \{\alpha : \alpha < \beta_1\}$;
(b) Spectrum$(\Pi_2^1) = \{\alpha < |\Pi_2^1| : \alpha \text{ is } \alpha^*\text{-recursive in } S\}$.

We begin by showing that $\Gamma \in \Sigma_2^1$ implies $|\Gamma| \leq \beta_1$. For any Σ_2^1 operator Γ, we have by Proposition 10.1 a p.o.r. F so that $m \in \Gamma(A)$ iff $\exists\alpha \,.\, F(\alpha, m, \chi_A) \simeq 1$ iff $\exists\alpha < \delta_1^A \,.\, F(\alpha, m, \chi_A) \simeq 1$. Let I be defined from F as in Definition 6.2. Parallel to Lemma 6.3, we have

Lemma 12.5. (a) *For all m and α, $\chi_{\Gamma^\alpha}(m) \simeq I(\alpha, m, \delta_\alpha)$;*
(b) *for all α, $\delta_1^{\Gamma^\alpha} \leq \delta_{\alpha+1}$.*

Now for any inaccessibly stable ordinal β and any m: $m \in \Gamma^{\beta+1}$ iff $\exists\alpha < \beta^* \,.\, F(\alpha, m, \lambda n \,.\, I(\beta, n, \beta)) \simeq 1$. It follows from Proposition 11.17 that for any m, β_1 can not be the least β with $m \in \Gamma^{\beta+1}$, so that $|\Gamma| \leq \beta_1$.

To show that $|\Sigma_2^1| = \beta_1$, we define a Σ_2^1 inductive operator Υ such that for all $\alpha \leq \beta_1$, $\Upsilon^\alpha = \bigcup_{\sigma < \delta_\alpha} \{\langle a, m, n\rangle : \{a\}_\sigma(\boldsymbol{m}, S) \simeq n\} = B_\alpha$.

We need the following rather technical lemma.

Lemma 12.6. *There is an index s such that for all ordinals β, if all $\alpha < \delta_\beta$ are δ_β-recursive in S (for example, any $\beta \leq \beta_1$) or $\beta = 0$, then for all $\alpha < \delta_{\beta+1}$, $\{s\}_{\delta_{\beta+1}}(\alpha, \chi_{B_\beta}) \simeq S(\alpha)$.*

Proof. As in the proof of Lemma 6.6, there is a well-ordering $\{d\}^{B_\beta}$ (d for short) of length δ_β such that for b in the field of d, $\{b\}_{\delta_\beta}(S) \simeq |b|_d$. Choose c so that for all a and τ, $\{c\}_\tau(a,S) \simeq 0$ iff $\{a\}_\tau(S)$ is stable. Recall from Proposition 5.12 the relation W_0 such that $W_0(\sigma,\phi)$ iff $W(\phi) \wedge |\phi| = \sigma$. Choose s so that

$$\{s\}_{\delta_{\beta+1}}(\alpha, \chi_{B_\beta}) \simeq \begin{cases} 1, \text{ if } W_0(\alpha, d) \text{ or } \exists a[\langle c, a, 0\rangle \in B_\beta \wedge \\ \qquad W_0(\alpha, d \restriction_0 \{b : d(b,a) \simeq 1 \wedge a \neq b\}]; \\ 0, \text{ otherwise.} \end{cases}$$

Proposition 12.7. *There is a Σ^1_2 relation K^* such that $K^*(\langle a,m,n\rangle, B_\beta)$ iff $\{a\}_{\delta_{\beta+1}}(m,S) \simeq n$,* for all *$a,m,n$, and β.*

Proof. Begin with K_2 from Proposition 10.2 and apply Lemma 12.6 and the recursion theorem (Proposition 3.5).

Definition 12.8. $\Upsilon(A) = A \cup \{s : K^*(s,A)\}$.

The following proposition is easily verified and completes the proof of Theorem 12.1(a) for $n = 1$.

Proposition 12.9. (a) *For all $\alpha \leq \beta_1$, $\delta_1^{\Upsilon^\alpha} = \delta_{\alpha+1}$;*
(b) *for all $\alpha \leq \beta_1$, $\Upsilon^\alpha = B_\alpha$;*
(c) $|\Upsilon| = \beta_1$.

It is now easy to see that $\overline{\Sigma^1_2} = \beta_1$-semirecursive:
($\subseteq$) We have $\overline{\Upsilon} = \{\langle a,m,n\rangle : \{a\}_{\beta_1}(m,S) \simeq n\}$ by Proposition 12.9.
($\supseteq$) For any Σ^1_2 operator Γ with associated I, by Lemma 12.5, $m \in \overline{\Gamma}$ iff $\exists \alpha < \beta_1 [\mathrm{IS}(\alpha) \wedge I(\alpha,m,\alpha) \simeq 1]$.

The proof of Theorem 12.1 for $n > 1$ is straightforward except for the construction of an operator Υ_n with $|\Upsilon_n| = \beta_n$. We need the following lemma, a fairly difficult corollary to the Uniformization Theorem (10.4). (See Cenzer [7] for a proof.)

Lemma 12.10. *There is a Σ^1_2 relation L_2 such that for all $A \subseteq \omega$,*

$\{\langle u,v\rangle : L_2(u,v,A)\}$ *is a well-ordering of type* δ_1^A.
Let $\Gamma_{i+1}(A) = \{\langle\langle i,\langle u,v\rangle\rangle\rangle : L_2(u,v,A)\} \cup A$.

We can generalize Lemma 12.6.

Lemma 12.11. *For all* $n > 0$, *there is an index* s_n *such that for all p and all limit ordinals* $\tau < \beta_n$, *if* $A = (\Gamma_1 \cdot ... \cdot \Gamma_{n-1})(B_{\tau+np})$, *then for all* $\alpha < \delta_{\tau+n(p+1)}$, $\{s_n\}_{\delta_{\tau+n(p+1)}}(\alpha, \chi_B) \simeq S(\alpha)$.

Definition 12.12. For all $n > 1$, all $A \subseteq \omega$,
(a) $\Lambda_n^*(A) = \{\langle a,m,t\rangle : \{a\}_{\delta_1^A}(m, \lambda\alpha\,.\,\{s_n\}_{\delta_1^A})) \simeq t\}$;
(b) $\Upsilon_n = \Gamma_1 \cdot ... \cdot \Gamma_{n-1} \cdot \Lambda_n^*$.

As in Proposition 12.7 each Λ_n^* is $\mathbf{\Sigma}_2^1$, so $\Upsilon_n \in (\mathbf{\Sigma}_2^1)^n$. It is not difficult to check that for all $n > 1$, $|\Upsilon_n| = \beta_n$.

Parallel to Propositions 8.3 and 8.4, we have:

Proposition 12.13. (a) $|[\mathbf{\Sigma}_2^1, ..., \mathbf{\Sigma}_2^1]| < \beta_2$;
(b) $|[\mathbf{\Sigma}_2^1, \mathbf{\Pi}_0^0]|$ *is the least ordinal* β *which is not* β^**-recursive in S and which is a limit of ordinals with the same property.*

The results regarding $\mathbf{\Pi}_2^1$ inductive definitions are parallel to those in §9 on $\mathbf{\Sigma}_1^1$ inductive definitions. Since all proofs are basically adaptations of the techniques of §9, we omit them.

An interesting open problem in this area is whether or not $|\mathbf{\Pi}_2^1\text{-mon}| = |\mathbf{\Pi}_2^1|$. The fact that $\mathbf{\Pi}_2^1\text{-mon} = \mathbf{\Pi}_2^1$, whereas $\mathbf{\Pi}_2^1$ is much larger suggests that $|\mathbf{\Pi}_2^1\text{-mon}| < |\mathbf{\Pi}_2^1|$. On the other hand, $|\mathbf{\Sigma}_1^1\text{-mon}| = |\mathbf{\Sigma}_1^1|$ and, although we have no functional for $\mathbf{\Pi}_2^1$ corresponding to $E_1^\#$, many other results on $\mathbf{\Pi}_1^1$ and $\mathbf{\Sigma}_1^1$ operators carry over to $\mathbf{\Sigma}_2^1$ and $\mathbf{\Pi}_2^1$ operators, so perhaps the two ordinals are equal.

13. Ordinal recursion and the constructible hierarchy

Our definition of α-recursive is equivalent to the Kripke [15] – Platek [17] definition of $\mathbf{\Sigma}_1$-definable (without parameters) over L_α. In this section we explore the relationship between ordinal recursion and constructibility.

Let Φ and Ψ denote formulas of the language of ZF (Zermelo-Fraenkel set theory – see Shoenfield [21] for details). We follow Levy in classifying formulas as $\mathbf{\Delta}_0$, $\mathbf{\Sigma}_1$, and so forth.

Proposition 13.1. *For all recursively regular ordinals $\alpha > \omega$ (or $\alpha = \infty$), all k and $n > 0$, and all $R \subseteq \alpha^k$, R is $\alpha - \mathbf{\Sigma}_n$ iff R is $\mathbf{\Sigma}_n$ definable over L_α (without parameters).*

Proof. ($\rightarrow$) By induction on the class of p.o.r. functionals we see that all p.o.r. relations are $\mathbf{\Delta}_1$-definable over L_α; the result follows by adjoining quantifiers on each side.
($\leftarrow$) We first prove a lemma; let $\{F(\sigma) : \sigma \in \mathrm{ORD}\}$ be Gödel's [9] enumeration of the constructible sets.

Lemma 13.2. *The relations E, defined by $E(\sigma,\tau)$ iff $F(\sigma) \in F(\tau)$, and C, defined by $C(\sigma,\tau)$ iff $\tau = F(\sigma)$, are p.o.r.*

Proof. E and C are defined by an induction similar to that which defines F. The result follows from this lemma by induction over formulas.

For structures $\mathcal{A}$ and $\mathcal{B}$ for the language of **ZF**, we write $\mathcal{A} \prec \mathcal{B}$ iff $\mathcal{A}$ is an elementary submodel of $\mathcal{B}$ and $\mathcal{A} \prec_{\Sigma_n} \mathcal{B}$ iff $\mathcal{A} \subseteq \mathcal{B}$ and for all $\mathbf{\Sigma}_n$ formulas Φ and all $\boldsymbol{a} \in |\mathcal{A}|$, $\mathcal{A} \models \Phi[\boldsymbol{a}]$ implies $\mathcal{L} \models \Phi[\boldsymbol{a}]$.

Proposition 13.3. *For all α, β, and n,*
(a) *α is n–β-stable iff $L_\alpha \prec_{\Sigma_n} L_\beta$;*
(b) *$S^n(\alpha)$ iff $L_\alpha \prec_{\Sigma_n} L$.*

(See Levy [16] for a definition of satisfaction for $\mathbf{\Sigma}_n$ formulas in proper classes like L.)

(a) This follows from Proposition 13.1.
(b) This follows from Theorem 11.7(a) and Proposition 13.1.

Combining Proposition 13.3 with Theorems 6.1 and 12.1, we obtain new characterizations for $|\mathbf{\Pi}_1^1|$ and $|\mathbf{\Sigma}_2^1|$.

Proposition 13.4. (a) $|\mathbf{\Pi}_1^1|$ *is the least* α *such that* $L_\alpha \prec_{\Sigma_1} L_{\alpha^+}$;
(b) $|\mathbf{\Sigma}_2^1|$ *is the least* α *such that* $L_\alpha \prec_{\Sigma_2} L_{\alpha^*}$.

There is no comparable characterization of $|\mathbf{\Sigma}_1^1|$ or $|\mathbf{\Pi}_2^1|$, since for any α, $L_\alpha \prec_{\Pi_1} L_{\alpha^+}$ iff $L_\alpha \prec_{\Delta_0} L_{\alpha^+}$, which is true for any regular ordinal α and $L_\alpha \prec_{\Pi_2} L_{\alpha^*}$ iff $L_\alpha \prec_{\Sigma_1} L_{\alpha^*}$, which is true for any stable ordinal α.

There is a characterization for each of the four closure ordinals in terms of certain reflection principles.

Definition 13.5. (a) (α, α^+) is $\mathbf{\Sigma}_1(\mathbf{\Pi}_1)$ reflecting iff for any $\mathbf{\Sigma}_1(\mathbf{\Pi}_1)$ formula Φ, $L_{\alpha^+} \models \Phi[\alpha] \to \exists\beta < \alpha \,.\, L_{\beta^+} \models \Phi[\beta]$;
(b) (α, α^*) is $\mathbf{\Sigma}_2(\mathbf{\Pi}_2)$ reflecting iff for any $\mathbf{\Sigma}_2(\mathbf{\Pi}_2)$ formula Φ, $L_{\alpha^*} \models \Phi[\alpha] \to \exists\beta < \alpha \,.\, L_{\beta^*} \models \Phi[\beta]$.

Combining Proposition 13.1 with Propositions 4.21 and 11.17 and Theorems 9.1 and 12.2 we obtain

Proposition 13.6. (a) $|\mathbf{\Pi}_1^1|$ *is the least ordinal* α *such that* (α, α^+) *is* $\mathbf{\Sigma}_1$ *reflecting*;
(b) $|\mathbf{\Sigma}_1^1|$ *is the least ordinal* α *such that* (α, α^+) *is* $\mathbf{\Pi}_1$ *reflecting*;
(c) $|\mathbf{\Sigma}_2^1|$ *is the least ordinal* α *such that* (α, α^*) *is* $\mathbf{\Sigma}_2$ *reflecting*;
(d) $|\mathbf{\Pi}_2^1|$ *is the least ordinal* α *such that* (α, α^*) *is* $\mathbf{\Pi}_2$ *reflecting*;

There is an obvious extension of this result to classes like $(\mathbf{\Pi}_1^1)^n$ which we leave to the reader. In §14 we will obtain similar characterizations of $|\mathbf{\Sigma}_n^1 - L|$ and $|\mathbf{\Pi}_n^1 - L|$ for $n > 2$.

We indicated in §11 that δ_1^S need not be countable; our next goal is to show that in fact it is a rather large constructible cardinal. The following result is very helpful.

Proposition 13.7. *There is a p.o.r. function* F_L *such that for all ordinals* α, L
$$^\alpha\alpha = \{\lambda\beta \,.\, F_L(\beta, \sigma, \alpha) : \sigma \in \mathrm{ORD}\}$$
$$= \{\lambda\beta \,.\, F_L(\beta, \sigma, \alpha) : \bar{\bar{\sigma}} \leq \bar{\bar{\alpha}}\}.$$

Proof. Let $\mathrm{FN}(\sigma, \alpha)$ iff $F(\sigma) \in {}^\alpha\alpha$ iff $\forall\beta < \alpha \,.\, \exists\gamma < \alpha \,.\, \exists\tau < \sigma[C(\tau, \langle\beta, \gamma\rangle) \wedge E(\tau, \sigma)]$. Define F_L so that $F_L(\beta, \sigma, \alpha) = (F(\sigma))(\beta)$ if $F(\sigma) \in {}^\alpha\alpha$:
$$F_L(\beta, \sigma, \alpha) \simeq \begin{cases} \text{least } \gamma < \alpha \,.\, \exists\tau < \sigma[C(\tau, \langle\beta, \gamma\rangle) \wedge E(\tau, \sigma)], \text{ if } \mathrm{FN}(\sigma, \alpha)]; \\ 0, \text{ otherwise.} \end{cases}$$

The second equality follows from a collapsing argument using Lemma 4.9.

Definition 13.8. (a) LC(α) iff $L \models \alpha$ is a cardinal;
(b) $\aleph^L_\alpha$ is the α'th infinite constructible cardinal.

Proposition 13.9. (a) LC *is* $\infty - \mathbf{\Pi}_1$;
(b) LC *is* ∞*-recursive in* S:
(c) $\lambda\alpha \, . \, \aleph^L_\alpha$ *is* ∞*-recursive in* S.

Proof. (a) LC (α) iff $\forall\sigma \cdot \rightarrow \lambda\beta \cdot F_L(\beta, \sigma, \alpha)$ maps α 1–1 into some $\tau < \alpha$.
(b) and (c) follow from Theorem 11.7 (d).

Corollary 13.10. *For any ordinal* α,
(a) $S^2(\alpha) \rightarrow \mathrm{LC}(\alpha)$; (b) $S^2(\alpha) \rightarrow \alpha = \aleph^L_\alpha$.

Now if $V = L$, then the least 2-stable is a rather large uncountable cardinal. Andreas Blass has pointed out in conversation that it can be proven by a simple forcing argument that CON(**ZFC**) $\rightarrow$ CON(**ZFC**+ $\exists$ countable 2-stable).

Since $S^n \cap \aleph^L_1 = \emptyset$ for $n > 1$, $\aleph^L_1$-recursion in S^n is the same as $\aleph^L_1$-recursion; in § 14 we obtain results connecting the $\mathbf{\Sigma}_{n+1} - L$ relations over ω with $n - \aleph^L_1$-stability.

It is interesting to note that although by Proposition 4.12 $\delta^{S^n}_1$ exists for each n, the two-place relation $S^n(\alpha)$ is not definable in **ZF** (being in fact "equivalent" to a satisfaction relation for L), so that the existence of an ordinal which is n-stable for all n is independent of **ZF**.

14. The constructible analytical hierarchy

It follows from Proposition 13.1 that for any α and any n, all $\alpha - \mathbf{\Sigma}_n$ relations are constructible. On the other hand, it is consistent with **ZFC** that there be a $\mathbf{\Delta}^1_3$ non-constructible set of natural numbers (see Jensen–Solovay [10]). Therefore it need not be the case that every $\mathbf{\Sigma}^1_3$ relation over ω be $\infty - \mathbf{\Sigma}_n$ for any n. However, we are able to generalize Proposition 10.6 if we restrict the discussion to constructibly $\mathbf{\Sigma}^1_n$ relations.

Definition 14.1. A relation R is constructibly $\mathbf{\Sigma}^1_n$ ($R \in \mathbf{\Sigma}^1_n - L$) iff R can be defined by a $\mathbf{\Sigma}_n$ formula with function quantifiers restricted to $L \cap {}^\omega\omega$. $\mathbf{\Pi}^1_n - L$ and other constructible definability classes are similarly defined.

Theorem 14.2. *For all n, k and all $R \subseteq \omega^k$, R is $\boldsymbol{\Sigma}^1_{n+1} - L$ iff R is $\aleph_1^L - \boldsymbol{\Sigma}_n$.*

Proof. (→) Begin with Proposition 10.1 and replace additional function quantifiers by ordinal quantifiers using Proposition 13.7.
(←) This is proven from Proposition 5.10 as was Proposition 10.2.

We need some terminology for $n - \aleph_1^L$-stability.

Definition 14.3. (a) $S_1^n(\alpha)$ iff $\mathrm{RS}^n(\alpha, \aleph_1^L)$ iff α is $n - \aleph_1^L$ stable;
(b) $\delta_n - A$ is the least ordinal $n - \aleph_1^L$ stable in A;
(c) $\delta_{n,\alpha}$ is the α'th $n - \aleph_1^L$ stable ordinal;
(d) $sc^n(\alpha)$ is the least $n - \aleph_1^L$-stable greater than α.

We generalize Theorem 14.2(→) by adding a set parameter.

Proposition 14.4. *For any $\boldsymbol{\Sigma}^1_{n+2} - L$ relation Q on $\omega \times P(\omega)$, there is an $\aleph_1^L - \boldsymbol{\Sigma}_n$ relation P such that for all m and A, Q(m, A) iff $\exists \alpha < \aleph_1^L \,.\, P(\alpha, m, A)$ iff $\exists \alpha < \delta_{n+1} - A \,.\, P(\alpha, m, A)$.*

We obtain further results parallel to those in §10 from the following corollary to Addison's [5] uniformization theorem for $\boldsymbol{\Pi}^1_n - L$.

Proposition 14.5. *For any $n > 2$, any $\Sigma^1_n - L$ relation Q, any $\boldsymbol{m}$ and any constructible $A \subseteq \omega$, $\exists \phi \in L \,.\, Q(\boldsymbol{m}, \phi, A)$ iff $\exists \phi (\phi$ is $\boldsymbol{\Delta}^1_n - L$ in $A) \,.\, Q(\boldsymbol{m}, \phi, A)$.*

Parallel to Proposition 10.3, we have

Proposition 14.6. *For all $n \geq 1$ and all constructible $A \subseteq \omega$, $\delta_n - A = (\delta_{n+1} - A)^L$.*

Theorem 11.7 can be relativized directly to $n - \aleph_1^L$-stability and $\aleph_1^L - \boldsymbol{\Sigma}_n$ relations. We leave the details to the reader.

Definition 14.7. For any $n \geq 1$, σ_n is the least ordinal σ which is not $\mathrm{sc}^n(\sigma)$-recursive in S_1^n.

Notice that σ_1 is the ordinal β_1 defined in §11. The following results are proven as in §11.

Proposition 14.8. *For all* $n \geq 1$,
(a) $S_1^n(\sigma_n)$ *and* σ_n *is a fixed point of the* $n - \aleph_1^L$ *stables*;
(b) σ_n *is the least ordinal* σ *such that* σ *is* $n+1-\mathrm{sc}^n(\sigma)$*-stable*;
(c) σ_n *is the least ordinal* σ *such that for all* $a < \omega$,
$\{a\}_{\mathrm{sc}^n(\sigma)}(\sigma, S_1^n)\downarrow \to \exists \tau < \sigma\,.\,\{a\}_{\mathrm{sc}^n(\tau)}(\tau, S_1^n)\downarrow$.

We can now prove the main result of the section.

Theorem 14.9. *For all* $n \geq 1$, $|\Sigma^1_{n+1}| = \sigma_n$.

Sketch of proof: Let $n = 2$ for simplicity.
($\leq$) For any $\Sigma^1_3 - L$ operator Γ, we have by Proposition 14.4 a p.o.r. functional F with $\mathrm{rg}(F) \subseteq \{0,1\}$ such that for all m, A: $m \in \Gamma(A)$ iff $\exists \alpha < \delta_n - A\,.\,\forall \beta < \delta_n - A\,.\,F(m, \alpha, \beta, \chi_A) \simeq 1$. Defining I by bounding the quantifiers we have as in Lemma 12.5:

Lemma 14.10. (a) *For all* m *and* α, $\chi_{\Gamma^\alpha}(m) \simeq I(\alpha, m, \delta_{2,\alpha}, \delta_{2,\alpha})$;
(b) *for all* α, $\delta_2 - \Gamma^\alpha \leq \delta_{2,\alpha+1}$;
(c) *for all* $\alpha = \delta_{2,\alpha}$,
$m \in \Gamma^{\alpha+1}$ *iff* $\exists \sigma < \mathrm{sc}^2(\alpha)\,.\,\forall \tau < (\langle \sigma, \alpha \rangle)^*\,.\,F(m, \sigma, \tau, \lambda t\,.\,I(\alpha, m, \alpha, \alpha)) \simeq 1$.

It is now easy to check that for any m, $m \in \Gamma^{\sigma_2+1} \to m \in \Gamma^{\sigma_2}$.
($\geq$) As in the proof of Theorem 12.1, we can define a $\Sigma^1_3 - L$ operator Υ such that for all $\alpha \leq \sigma_2$, $\delta_2 - \Upsilon^\alpha = \delta_{2,\alpha+1}$ and
$\Upsilon^\alpha = \bigcup_{\sigma < \delta_{2,\alpha}} \{\langle a, m, n \rangle : \{a\}_\sigma(m, S_1^2) \simeq n\}$.
Combining Propositions 13.3(a) and 14.8(b), we have

Proposition 14.11. *For all* $n \geq 1$, $|\Sigma^1_{n+1} - L|$ *is the least ordinal* α *such that* $L_\alpha \prec_{\Sigma_{n+1}} L_{\mathrm{sc}^n(\alpha)}$.

We can prove directly an extension of Theorem 12.2.

Theorem 14.12. *For all* $n \geq 1$, $|\Pi^1_{n+1} - L|$ *is the least ordinal* α *such that for all* α, *if* $\forall \tau < \alpha\,.\,\{a\}_{\mathrm{sc}^n(\tau)}(\tau, S_1^n)\downarrow$, *then* $\{a\}_{\mathrm{sc}^n(\alpha)}(\alpha, S_1^n)\downarrow$.

Definition 14.13. For any $n \geq 1$, $(\alpha, \mathrm{sc}^n(\alpha))$ is $\Sigma_n(\Pi_n)$ reflecting iff for any $\Sigma_n(\Pi_n)$ formula Φ, $L_{\mathrm{sc}^n(\alpha)} \models \Phi[\alpha] \to \exists \beta < \alpha\,.\,L_{\mathrm{sc}^n(\beta)} \models \Phi[\beta]$.

Parallel to Proposition 13.6 we have

Proposition 14.14. *For all* $n \geq 1$,
(a) $|\mathbf{\Sigma}^1_{n+1} - L|$ *is the least ordinal* α *such that* $(\alpha, \mathrm{sc}^n(\alpha))$ *is* $\mathbf{\Sigma}_{n+1}$*-reflecting*;
(b) $|\mathbf{\Pi}^1_{n+1} - L|$ *is the least ordinal* α *such that* $(\alpha, \mathrm{sc}^n(\alpha))$ *is* $\mathbf{\Pi}_{n+1}$*-reflecting.*

The results of this section can be extended in an obvious fashion to obtain characterizations for $|(\mathbf{\Sigma}^1_n - L)^k|$ and $|(\mathbf{\Pi}^1_n - L)^k|$.

References

[1] S. Aanderaa, this volume.
[2] P. Aczel, Representability in some systems of second order arithmetic, Israel J. Math. 8 (1970) 309–328.
[3] P. Aczel and W. Richter, Inductive definitions and analogues of large cardinals, Proc. Conf. Math. Logic London 70, Springer Lecture Notes #255.
[4] P. Aczel and W. Richter, this volume.
[5] J.W. Addison, Some consequences of the axiom of constructibility, Fund. Math. 46 (1959) 337–357.
[6] J. Barwise, R.O. Gandy and Y.N. Moschovakis, The next admissible set, J. Symbolic Logic 36 (1971) 108–120.
[7] D. Cenzer, Ordinal recursion and inductive definitions, Ph.D. Thesis, University of Michigan, 1972.
[8] D. Cenzer, Analytic inductive definitions, to appear.
[9] K. Gödel, The Consistency of the Axiom of Choice and of the Generalized Continuum Hypothesis with the Axioms of Set Theory (Princeton Univ. Press, Princeton, 1958).
[10] R.B. Jensen and R.M. Solovay, Some applications of almost disjoint sets, in: Y. Bar-Hillel (ed.) Mathematical Logic and Foundations of Set Theory (North-Holland, Amsterdam, 1970) pp. 88–104.
[11] S.C. Kleene, Recursive functionals and quantifiers of finite types, I, Trans. Amer. Math. Soc. 91 (1959) 1–52.
[12] S.C. Kleene, Recursive functionals and quantifiers of finite types, II, Trans. Amer. Math. Soc. 108 (1963) 106–142.
[13] M. Kondo, Sur l'uniformization des complementaires analytiques et les ensembles projectifs de la seconde classe, Japanese J. Math. 15 (1938) 197–230.
[14] G. Kreisel and G. Sacks, Metarecursive sets, J. Symbolic Logic 30 (1965) 318–338.
[15] S. Kripke, Transfinite recursion, constructible sets, and analogues of large cardinals, in: Lecture notes prepared in connection with the Summer Institute on Axiomatic Set Theory held at UCLA, July–August, 1967.
[16] A. Levy, A hierarchy of formulas in set theory, Mem. Amer. Math. Soc. No. 57, 1965.
[17] R.A. Platek, Foundations of recursion theory, Ph.D. Thesis, Stanford University, 1966.

[18] H. Putnam, On hierarchies and systems of notations, Proc. Amer. Math. Soc. 15 (1964) 44–50.

[19] W. Richter, Recursively Mahlo ordinals and inductive definitions, in: R.O. Gandy and C.E.M. Yates (eds.) Logic Colloquium '69 (North-Holland, Amsterdam, 1971) pp. 273–288.

[20] H. Rogers, Theory of Recursive Functions and Effective Computability (McGraw-Hill, New York, 1967).

[21] J.R. Shoenfield, The problem of predicativity, in: Y. Bar-Hillel (ed.) Essays on the Foundations of Mathematics (The Magnes Press, Jerusalem, 1961 and North-Holland, Amsterdam, 1962).

[22] J.R. Shoenfield, Mathematical Logic (Addison–Wesley, Reading, Mass., 1967).

[23] C. Spector, Recursive well-orderings, J. Symbolic Logic 20 (1955) 151–163.

[24] C. Spector, Inductively defined sets of natural numbers, in: Infinitistic Methods (Pergamon, Oxford, 1961) pp. 97–102.

J.E.Fenstad, P.G. Hinman (eds.), Generalized Recursion Theory

INDUCTIVE DEFINITIONS

Robin O. GANDY
Mathematical Institute, Oxford University

§0. Introduction

Mathematical Logic is certainly permeated with inductive definitions. Here are some examples of concepts which are usually or readily defined in this way. In syntax; the notions of *expression, well-formed formula, proof, theorem*. In semantics and model theory; *the satisfaction relation, validity*, Morley's notion of *rank*. In set theory; *well-founded set, ordinal, constructible set*, the *forcing relation, Borel set*. In recursion theory the question is rather: are there any fundamental notions which are not inductively defined? All this suggests that a study of inductive definitions in general should produce interesting and applicable results. Of course it could be that it is always *particular* features of the definitions which are significant, so that a general study will only yield trivial results. But in fact this is not the case. One example is Barwise's completeness and compactness theorems: the theorems are consequences of the *form* of the inductive definition of *derivation*, not of its particular details. (This is discussed in §2 below.) Another example is Moschovakis' notion of *hyper-projective*; the original definition was by rather elaborate schemata. Under the new title *hyper-elementary* Moschovakis (in [4]) has reworked the material as a study of first-order positive inductive definitions; the proofs are more general, shorter, and more transparent. A final example is the study of extended systems of notations for ordinals which flourished in the 1950's. The authors (no names, no pack-drill!) were often at pains to verify, case by weary case, that their systems of notation had certain simple properties (e.g. of belonging to $\mathbf{\Delta}^1_2$). But *this* verification was quite unnecessary; all that was needed was to observe that the definitions were built up using arithmetical (not necessarily monotonic) clauses, and then to apply a trivial theorem about such definitions.

General studies, then, are worth pursuing. And once this has been accepted, it would be unreasonably Draconian to deny them autonomy. The view taken here is that inductive definitions are interesting in their own right. Of course we are also interested in applications; but we do not have to back up each line of enquiry with a promise of applicability.

My original intention was to give in this paper a fairly systematic account of first-order inductive definitions on an admissible set. But the recent work of Moschovakis [4], Aczel [1] and Barwise (this volume) made my account obsolete. So what is presented here consists, in effect, of remarks and reflections.

In §1, I discuss the various methods which have been used to investigate certain particular classes of inductive definitions. Section §2 represents the residue of the first draft; it describes the method of semantic tableau which may have further uses. The results of 2.3. are certainly, the result of 2.4. possibly, more expeditiously proved by other methods. The results of 2.5. about $\mathbf{\Pi}_1$ positive inductive definitions are new; but it is not clear if that class is significant. In §3 an effort is made to present Kleene's theory of recursion in a type 2 object as a branch of inductive theory. In so far as the effort is successful (when '$\mathbf{E}$ is recursive in $\mathbf{F}$') it gives a clear indication (already discussed by Aczel in [1]) of how to set up the theory for structures other than the natural numbers. There is some inconclusive discussion of the contrary case. In §4 I draw attention to some of the problems which were ignored in §1–3, and make propaganda for the investigation of the *forms* of inductive definition which occur in proof theory. (This propaganda is directed as much at recursion-theorists as it is at proof-theorists.)

§1. Preliminaries and a discussion of methods

1.1. Let $\mathfrak{A}$ be an arbitrary first-order structure with domain A. An n-place inductive operator Φ is a map $\mathbf{P}({}^nA) \to \mathbf{P}({}^nA)$ ($\mathbf{P}$ for power-set, ${}^nA$ for $A \times \ldots \times A$). For simplicity we shall always suppose that Φ is *progressive*, i.e., $R \subseteq \Phi R$; if not, replace Φ by Φ', where $\Phi' R = R \cup \Phi R$. The *α-th iterate*, $\Phi^\alpha(R_0)$, of Φ applied to R_0 is defined by

$$\Phi^\alpha(R_0) = R_0 \cup \mathbf{U}_{\beta<\alpha} \Phi(\Phi^\beta(R_0)) .$$

If $R_0 = \emptyset$ we write simply Φ^α. The *closure ordinal* $|\Phi(R_0)|$ of Φ from R_0 is defined by

$$|\Phi(R_0)| = (\mu\beta)(\Phi^{\beta+1}(R_0) = \Phi^\beta(R_0)) .$$

The *closure*, $\Phi^\infty(R_0)$ of Φ from R_0 is simply $\Phi^\alpha(R_0)$ where $\alpha = |\Phi(R_0)|$. If $a \in \Phi^\alpha(R_0)$, the *stage* $|a|_{\Phi(R_0)}$ of a is defined by

$$|a|_{\Phi(R_0)} = (\mu\beta)(a \in \Phi^{\beta+1} - \Phi^\beta) .$$

It is often convenient to set $|a|_{\Phi(R_0)} = |\Phi(R_0)|$ for $a \notin \Phi^\infty(R_0)$. From now on we suppose $R_0 = \emptyset$.

1.2. We shall be interested in these things as Φ ranges over some given collection $\mathcal{C}$. So we define

$$|\mathcal{C}| = \mathrm{Sup}\{|\Phi| : \Phi \in \mathcal{C}\} .$$

The class $\mathcal{C}^\infty$ of *$\mathcal{C}$-fixed points* is defined by $\mathcal{C}^\infty = \{\Phi^\infty : \Phi \in \mathcal{C}\}$. Of greater interst is the class $\mathcal{C}^{(\infty)}$ of *$\mathcal{C}$-inductive* relations:

$$\mathcal{C}^{(\infty)} = \{R : (\exists \bar{a} \in {}^nA)(\exists \Phi \in \mathcal{C})(\forall \bar{x} \in {}^mA)(R\bar{x} \leftrightarrow \Phi^\infty \overline{a}\,\overline{x})\}$$

(where R is m-place and Φ is $n+m$-place). A relation is *$\mathcal{C}$-co-inductive* ('$\in \mathcal{C}^{(-\infty)}$') iff its complement is $\mathcal{C}$-inductive. It is *$\mathcal{C}$-bi-inductive* ('$\in \mathcal{C}^{(\pm\infty)}$') iff it is both $\mathcal{C}$-inductive and $\mathcal{C}$-co-inductive.

1.3. One is normally interested in inductive operators which can be defined in some language $\mathbf{L}$. Naturally $\mathbf{L}$ must contain variables ranging over A, and a relation symbol $\dot{R}$. *We do not automatically assume that* $\mathbf{L}$ *contains* '=', *nor that identity is a relation of* $\mathfrak{A}$. Then Φ is defined by a formula $\varphi(\mathrm{x}, \dot{R})$ of $\mathbf{L}$ (which shall not contain any free variables other than $\mathrm{x}\,(= \mathrm{x}_1, ..., \mathrm{x}_n)$) iff

(1) $$\Phi R = \{\bar{a} \in {}^nA : (\mathfrak{A}, ..., R) \models \varphi(\bar{a}, \dot{R})\} ;$$

here ... indicates whatever enlargement of $\mathfrak{A}$ is necessary to give a realisation of $\mathbf{L}$. We adopt the convention that $\varphi, \Phi; \psi, \Psi; ...,$ are always related as in (1).

Note that this useful convention may sometimes conceal the true nature of Φ. For example, if

$$\Phi R = \{\tau(\bar{a}) \colon \bar{a} \in {}^m A \ , (\mathfrak{A}, R) \models \psi(\bar{a}, \dot{R})\}$$

where R is one-place and ψ is quantifier-free and τ is a term then the corresponding φ is *not* quantifier-free. In what follows, however, we shall mostly be concerned with rather broad classes $\mathcal{C}$ for which this difficulty does not arise, and we shall use syntactic classes $\mathcal{C}$ of formulae to characterise the corresponding classes of inductive operators. If $\mathcal{F}$ is a class of formulae we shall be particularly concerned with the class $\mathcal{F}+$ of inductive operators defined by formulae of $\mathcal{F}$ in which $\dot{R}$ occurs only positively, and the class $\mathcal{F}m$ of inductive operators which are defined by formulae of $\mathcal{F}$ and which are also *monotonic*; i.e. which satisfy $R \subseteq S \to \Phi R \subseteq \Phi S$.

There are two classic cases; much of the recent work on inductive definitions stems from trying to understand them and to generalise them. In both, the underlying structure is $\mathfrak{N} = \langle N, 0, S, = \rangle$.

(R) (Post–Smullyan). If $\mathcal{C}$ lies in the range from Rud + to $\Sigma_1^0 m$ then $\mathcal{C}^{(\infty)} = \Sigma_1^0$ and $\mathcal{C}^{(\pm\infty)}$ = Recursive. (Here Rud is Smullyan's 'rudimentary' (= constructive arithmetic)).

(H) (Kleene–Spector). If $\mathcal{C}$ lies in the range from Π_1^0+ to $\Pi_1^1 m$ then $\mathcal{C}^{(\infty)} = \Pi_1^1$ and $\mathcal{C}^{(\pm\infty)}$ = Hyperarithmetic.

1.4. The standard problems of inductive theory for a given $\mathfrak{A}$, $\mathcal{C}$ are to determine $|\mathcal{C}|$, to characterise $\mathcal{C}^{(\infty)}$, to determine the closure properties of $\mathcal{C}^{(\infty)}$ and $\mathcal{C}^{(\pm\infty)}$, and to uncover any additional structure which these sets may have. Further problems arise from relativisation – that is by considering Φ which depend on parameters. The methods which have been most used may be summarised as follows.

1.4.1. *Direct methods.* One proceeds by constructing particular inductive definitions. A good example is the definition and use of O in hyperarithmetic theory. This way of proceeding appeals naturally to the purist. For many recursion theorists it also has a psychological attraction: there is a pleasure in working out the details of an intricate recursion which is akin to the pleasure of constructing a tangible object. And for investigating the fine structure of

$\mathcal{C}^{(\infty)}$ it seems to be the only method available. In classifying the r.e. sets, for example, one has actually to construct simple sets, maximal sets, and so on, in order to prove their existence and discover their properties.

1.4.2. *Use of higher type recursion*. In case (H) it is possible to consider $\mathcal{C}^{(\pm\infty)}$ as the class of relations which are recursive in the jump operator. This was shown by Kleene in [1]. The first study of the generalisation of (H) (by Moschovakis in [1–3]) was based on this method. But it has turned out that the results are more easily obtained by other methods (in particular 1.4.1. and 1.4.4.).

1.4.3. *The use of normal forms*. In case (H), for example, many of the closure properties of $\mathcal{C}^{(\infty)}$ and $\mathcal{C}^{(\pm\infty)}$ and a certain amount of the structure of these classes can be most easily derived from the fact that $\mathcal{C}^{(\infty)} = \mathbf{\Pi}^1_1$. Recently (in [4]) Moschovakis has obtained a normal form for the generalisation of case (H) to arbitrary structures by using a 'game quantifier' (the idea underlying this is discussed in §2.3 below). But I think it would be a mistake to place too much reliance on this method. For I believe that the future development of the subject will be concerned with finer classes $\mathcal{C}$. These will not have the sort of broad and simple syntactic characterisation of the classes so far considered; it is not to be expected then that $\mathcal{C}^{(\infty)}$ will have a simple syntactic form.

1.4.4. *The method of embedding*. We can enlarge a first order structure $\mathfrak{A}$ to a structure $(\mathfrak{A}, S, \in)$, where S is a subset of the cumulative hierarchy of types V_A formed with the elements of A as individuals (*urelements*). I.e., $V_A = \mathbf{U}(V^\alpha_A : \alpha \in \mathrm{On}\}$ where $V^\alpha_A = \mathbf{U}\{\mathbf{P}(V^\beta \cup A) : \beta < \alpha\}$. In the language of $(\mathfrak{A}, S, \in)$ we can use Levy's classification of set-theoretic formulae. If $\mathcal{F}$ is a class of formulae in this language, then we denote the class of relations on A which can be defined in $(\mathfrak{A}, S, \in)$ by formulae of $\mathcal{F}$, by $\mathcal{F}/(\mathfrak{A}, S, \in)$.

We give three examples in which, given $\mathcal{C}$, S can be chosen so that

(1) $$\mathcal{C}^{(\infty)} = \mathbf{\Sigma}_1/(\mathfrak{A}, S, \in)\,.$$

Example (A):

$\mathcal{C} = \mathbf{\Sigma}^0_1+$ (i.e., first-order existential positive operators),
S = the set of hereditarily finite members of V_A.

Example (B):

$\mathcal{C} = \mathbf{\Delta}_0^1+$ (i.e., first-order positive operators),

S = the smallest subset of V_A such that $(\mathfrak{A}, S, \in)$ is a model for the axioms KPU^+ of Kripke–Platek set theory over a *set* of urelements.

(For the description of KPU^+ see Barwise's paper in this volume; also 2.4.1. below.) In both these examples some conditions must be placed on $\mathfrak{A}$ in order to ensure that RHS (1) $\subseteq$ LHS (1). A sufficient condition is that $\mathfrak{A}$ should have only finitely many functions and relations, and be such that $\mathcal{C}^{(\pm\infty)}$ contains a pairing function for $\mathfrak{A}$. Barwise gives beautiful, short proofs of both these results, and example (B) is his theorem. (These proofs will appear in some lecture notes which Barwise is preparing, see the introduction to Barwise [3].) Barwise–Gandy–Moschovakis had previously proved it when $\mathfrak{A}$ itself is the structure of an admissible set. I think example (A) should be credited to Grilliot [1]; but the class had been investigated from various points of view in Moschovakis [1], Montague [1] and Gordon [1]. What is so fine about Barwise's proof is that one does not need to prove first any intricate results (e.g. the stage comparison and pre-well-ordering theorems) about $\mathcal{C}^{(\infty)}$; but these theorems are easily proved once (1) has been established. In Moschovakis [4] the reader will find statements and proofs of these theorems, valid even when (1) fails, and some discussion of this case. So far as I know the following problems are open for both examples.

Problem 1. Find minimal conditions on $\mathfrak{A}$ for (1) to hold.

Problem 2. Are there any examples where (1) fails, but $\mathcal{C}^{(\infty)}$ is an interesting class in its own right. (In the simplest failure of (1) for example (B), $\mathcal{C}^{(\infty)}$ is merely the class of $\mathfrak{A}$-definable relations.)

Problem 3. If the answer to problem 2 is 'yes', is it possible to find a restriction of the RHS which makes (1) true?

Example (C). Here $\mathfrak{A}$ is $\mathfrak{N}$ and we identify N with $\omega \in V_A$. For $\mathcal{C}$ we take one of the classes $\mathbf{\Pi}_{n+1}^0$, $\mathbf{\Sigma}_{n+1}^0$, $\mathbf{\Pi}_1^1$, $\mathbf{\Sigma}_1^1$ of non-monotonic operators considered by Aczel and Richter in [1]. Then $S = \mathrm{L}_{|\mathcal{C}|}$. Of course this does not solve the problem (as they do) of characterising $|\mathcal{C}|$. It would be good to have a Barwise type proof of their results. As an interim measure we mention that the structure of $L_{|\mathcal{C}|}$ can be coded in $\mathcal{C}^{(\pm\infty)}$; this does simplify the Aczel–Richter proofs.

To sum up: the wearisome feature of so much earlier work on generalised (and even ordinary) recursion theory has been the excessive use of notations and codings. The details of these were frequently irrelevant to the results; but without keeping track of the details one could not prove the results. This criticism also holds, I think, for those more elegant versions (e.g. Richter [1]) where, rather than some particular system of coding, one deals with a type of system. The method described in this section avoids these *longeurs*. One passes as rapidly as possible to an equation like (1); thereafter one can use the actual objects (ordinals, cumulative sets) instead of their codes.

1.4.5. *Invariant definability*. For the most studied cases, $\mathcal{C}^{(\infty)}$ has received elegant characterisations in terms of invariant definability. The classic references are Grzegorczyk, Mostowski and Ryll-Nardzewski [1], Mostowski [1] and Kreisel [1]. More recent work is in Kunen [1], Moschovakis [3], Grilliot [2], Barwise, Gandy and Moschovakis [1], and the papers by Barwise and Ville in this volume. Just as the direct methods appeal to those who like to think in terms of constructions (Pascal's '*espirit geometrique*') so do these to those people who prefer to think in terms of structures for a language or theory ('*espirit analytique*'). But, so far as I know, there is no general approach to the problem of characterising a given $\mathcal{C}^{(\infty)}$ in terms of invariant definability. Two particular problems might serve as first steps.

Problem 4. Give an invariant definability characterisation of $(\mathbf{\Pi}_1^0+)^{(\infty)}$ for any arbitrary structure, or for an admissible set. (Grilliot in [1] has shown that for acceptable $\mathfrak{A}$, this is the class of relations which are semi-prime-computable in $\mathbf{E}$.)

Problem 5. Characterise by means of inductive definitions the classes defined by Grilliot in his paper in this volume.

It is also relevant to seek invariant definability characterisations of the classes $\mathcal{C}$ of inductive operators. The only work known to me along this line is Feferman [1]. He showed that $\mathbf{\Sigma}_1^0+$ is the class of operators which are monotonic both with respect to the relation argument and with respect to end-extensions of the structure $\mathfrak{A}$. This suggests:

Problem 6. Develop a theory of the connection between invariant definability characterisations of $\mathcal{C}$ and those of $\mathcal{C}^{(\infty)}$.

§2. The method of semantic tableaux

The results we prove here are not essentially new. The method is an obvious one, but I have not seen it used elsewhere; it is described in the hope that it may prove useful in other contexts. And we take this opportunity of showing how easy it is to work in the system of set theory with a structural collection of urelements (as introduced by Barwise in his paper in this volume). Given a structure $(\mathfrak{A}, S, \in)$ of the kind described in 1.4.4., we shall use a, b, c to range over A, r, s, t to range over S, and x, y, z, u, v, to range over $A \cup S$.

2.1. Definition. The class $\in$-Prim $(\mathfrak{A})$ (or $\mathcal{P}$ for short) of $\in$-primitive recursive functions over $\mathfrak{A}$ is defined to be the last class $\mathcal{P}$ of functions over $A \cup V_A$ satisfying the following conditions.

(i) If F is a function of $\mathfrak{A}$, and

$$F'\bar{x} = F\bar{x} \quad \text{if } \bar{x} \in {}^nA\,,$$
$$= \emptyset \qquad \text{otherwise,}$$

then $F' \in \mathcal{P}$.

(ii) If R is a relation of $\mathfrak{A}$ and

$$R^*\bar{x} = \emptyset \qquad \text{if } \bar{x} \in {}^nA \text{ and } R\bar{x}$$
$$= \{\emptyset\} \text{ otherwise,}$$

then $R^* \in \mathcal{P}$.

(iii) The following functions $\in \mathcal{P}$.

$$(x;y) =_{\mathrm{Df}} \{z : z \in x \vee z = y\}.$$

$$\mathbf{U}x \;\; =_{\mathrm{Df}} \{y : (\exists u \in x)(y \in u)\}.$$

$$\mathbf{C}xyuv =_{\mathrm{Df}} x \text{ if } u \in v,$$
$$=_{\mathrm{Df}} y \text{ otherwise.}$$

(iv) $\mathcal{P}$ is closed under explicit definition.
(v) $\mathcal{P}$ is closed under $\in$-recursion; i.e. if $G, H \in \mathcal{P}$ and

$$\begin{aligned} Fx\bar{y} &= Gx\bar{y} && \text{if } x \in A, \\ &= H(F \upharpoonright x, \bar{y})x\bar{y} && \text{otherwise} \end{aligned}$$

then $F \in \mathcal{P}$, (where $F \upharpoonright x, \bar{y} = \{\langle u, \bar{y}, z\rangle : u \in x \wedge z = Fu\bar{y}\}$).

Mutatis mutandis all the results proved in Jensen & Karp [1] hold also for $\mathcal{P}$. In particular the characteristic function of any $\mathbf{\Delta}_0$ relation (without parameters) lies in $\mathcal{P}$, and the graph of any function in $\mathcal{P}$ is $\mathbf{\Delta}_1$. Also, $\mathbf{U} \in \mathcal{P}$ and (v) can be dropped in favour of (v′) which is obtained from (v) by substituting the equations:

$$\begin{aligned} Fx\bar{y} &= Gx\bar{y} && \text{if } x \in A, \\ &= H(\mathbf{U}\{Fu\bar{y} : u \in x\})x\bar{y} && \text{otherwise.} \end{aligned}$$

Note. With the definition of $\mathbf{C}$ given above, identity on A becomes a primitive recursive relation. I do not know if a satisfactory theory of $\mathcal{P}$ can be developed which avoids this.

2.2. *Notations and hypotheses.* Let $\mathfrak{D} = (\mathfrak{A}, S, \in)$ be such that D $(= A \cup S)$ is closed under $\mathcal{P}$. Since D is closed under pairing we need only consider inductively defined subclasses of D (we reserve 'set' for members of S); X ranges over $\mathbf{P}(D)$, and we introduce a 1-place predicate symbol '$\in \dot{X}$'. Let $\mathbf{L} = \mathbf{L}_{S,\omega}(\dot{X})$ be the infinitary language for $(\mathfrak{D}, X)$. We use a, b, ..., x, y, ..., r, s, t, to denote variables of $\mathbf{L}$ of the appropriate sorts. We suppose that $\exists, \forall, \bigwedge, \bigvee$, are all primitives of $\mathbf{L}$, so that we may suppose without loss of generality that *negation is applied only to atomic formulae*. A *basic* formula is an atomic formula or the negation of one. We suppose that $\mathbf{L}$ contains constants for all elements of D. We suppose that a 1:1 coding function $g : \mathbf{L} \to D$ has been defined, and we identify formulae with their codes. We further suppose that g has been chosen so that all the functions of elementary syntax, and the truth definition for atomic sentences not containing $\dot{X}$, belong to $\mathcal{P}$. (This imposes some limitation on $\mathfrak{A}$; it is certainly possible if $\mathfrak{A}$ has only finitely many functions and relations.) The only realisations of $\mathbf{L}$

we consider are $(\mathfrak{D}, X)$, so the truth of a sentence varies only with X.

We denote the class of predicates definable by formulae in which $\forall$ (resp. $\exists$) does not occur by $\boldsymbol{\Sigma}_D$ (resp. $\boldsymbol{\Pi}_D$). We use $\boldsymbol{\Delta}_0$, $\boldsymbol{\Sigma}_1$, ..., to refer to predicates definable using conjunctions and disjunctions over finite sets. We introduce $\boldsymbol{\nabla}_1$ for the closure of $\boldsymbol{\Sigma}_1$ under (finitary) boolean operations. Finally, as in § 1, a '+' indicates that '$\in \dot{X}$' only occurs unnegated.

2.3.1. Definition. We define *the semantic tableau* for a sentence φ by induction on the level n.

(1) At level 0 there is a single point P, the *vertex* and the sentence at P is φ.

(2) Let P be a point of level n, and let the sentence at P be ψ.

  (i) If ψ is basic, P is a *tip* and there are no points below it.

  (ii) If ψ is $\mathbf{W}s$, or $(\exists \mathrm{x})\theta(\mathrm{x})$ then P is disjunctive.

  (iii) If ψ is $\mathbf{M}s$, or $(\forall \mathrm{x})\theta(\mathrm{x})$ then P is conjunctive.

  (iv) If ψ is $\mathbf{W}s$ or $\mathbf{M}s$, then for each $\theta \in s$ there is a point Q_θ of level $n + 1$ immediately below P at which the sentence is θ.

  (v) If ψ is $(\exists \mathrm{x})\theta(\mathrm{x})$ or $(\forall \mathrm{x})\theta(\mathrm{x})$, then for each $y \in D$ there is a point $Q_{\theta(y)}$ of level $n + 1$ immediately below P at which the sentence is $\theta(y)$.

  (vi) 'below' is the transitive closure of 'immediately below'.

Evidently a semantic tableau is well-founded and every 'branch' through it comes to a tip.

2.3.2. Definition. Let an assignment X to $\dot{X}$ be given; we define the *grounded* points of a tableau T for φ and their ordinals as follows.

  (i) If P is a tip, it is grounded iff the sentence at P is true, and in this case $|P| = 0$.

  (ii) If P is a disjunctive point it is grounded iff some point immediately below it is grounded, and in this case

$$|P| = \mathrm{Min}\,\{|Q| : Q \text{ immediately below } P\} + 1.$$

  (iii) If P is a conjunctive point it is grounded iff all points immediately below it are grounded and in this case

$$|P| = \mathrm{Sup}^+\,\{|Q| : Q \text{ immediately below } P\}$$

(where $\mathrm{Sup}^+ Y = \mathrm{Sup}\,\{\alpha + 1 : \alpha \in Y\}$).

(iv) The tableau T is grounded iff its vertex V is, and then $|T| = |V|$.

From this there follows straightforwardly:

2.3.3. Lemma. *Under the assignment X for $\dot{X}$, φ is true iff the tableau T for φ is grounded.*

2.3.4. Now let $\Phi \in$ **L**+ be a positive inductive operator. The *complete* tableau $T(x, \Phi)$ for $x \in \Phi^{(\infty)}$ is defined by modifying the definition of the tableau for $\Phi(x, \dot{X})$. Clause (2) (i) of 2.3.1. is altered to:–

(i) (a) If ψ is basic and does not contain $\dot{X}$, then P is a tip

(i) (b) If ψ is $y \in \dot{X}$, then P is (conventionally) disjunctive; there is just one point immediately below P, and the sentence there is $\varphi(y, \dot{X})$.

The definition of grounded for points of the complete tableau for $x \in \Phi^\infty$ is again 2.3.2. Notice that $\dot{X}$ does not occur in the formula at a tip, so that this definition does not depend on an assignment for $\dot{X}$.

2.3.5. Lemma. $x \in \Phi^\infty$ *iff its complete tableau is grounded.*

This is readily proved, by lemma 2.3.3., using transfinite induction on $|y|_\Phi$ and $|P|$.

2.3.6. *Remarks.* (1). I believe the notion of trees with both conjunctive and disjunctive points was first introduced by Beth in [1].

(2) As Moschovakis first pointed out, and has greatly exploited, the notion of 'grounded' can be given a very intuitive explication in terms of game theory. Players **Λ** and **V** choose in succession a sequence of points on a branch through the complete tableau, starting at the vertex. Suppose $P_1, \ldots, P_n$ have been played and P_n is not a tip; if P_n is conjunctive, **Λ** must play next, otherwise **V**. In either case the appropriate player must choose for P_{n+1} a point immediately below P_n. If P_n is a tip then the game is finished; it is a win for **V** if the sentence there is true, for **Λ** if it is false. Then *player* **V** *has a winning strategy just in case the tableau is grounded*. From this it is plain that $x \in \Phi^\infty$ can be expressed using an ω-sequence of quantifiers. If we code consecutive turns by the same player into a single turn, then the ω-sequence becomes an alternating sequence of $\forall$'s and $\exists$'s.

(3) It is almost obvious that 'grounded' has been given a $\mathbf{V}_1+$ inductive definition. This is verified in the proof of the following:

2.3.7. Theorem. *There is a* $J \in (\mathbf{V}_1+)^{(\infty)}$ *such that for any* $\Phi \in \mathbf{L}+$, *any* $x \in D$,

$$x \in \Phi^\infty \leftrightarrow \langle \varphi, x \rangle \in J .$$

Thus J is universal for $(\mathbf{L}+)^{(\infty)}$. The theorem generalises to the infinitary language theorem 6 of Moschovakis [2], and its proof derives from his.

Proof. We code a point P on a complete tableau by a finite sequence $u(P) = \langle u_0, ..., u_{R-1} \rangle$; the u_i are just the sentences (θ, $\theta(y)$ or $\varphi(y)$) of 2.3.1. (iv), (v) or 2.3.4. (ib) chosen to lead one from the vertex (coded by $\langle\ \rangle$) to the point P.

Let $R\varphi vz$ be the relation which holds just in case either (i) v is a sentence of the form $\mathbf{W}s$ or $\mathbf{M}s$ and $z \in s$, or (ii) v is a sentence of the form $y \in \dot{X}$ and z is $\varphi(y)$, or (iii) v is a sentence of the form $(\exists y)\,\theta(y)$ and z is of the form $\theta(y)$. Then $R \in \mathcal{P}$ by the stipulation of 2.2. To see that this is true in case (iii), observe that either $\theta(y)$ does not contain the variable y free, or the constant y belongs to the transitive closure of $\theta(y)$. Using R it is easy to construct a function $G \in \mathcal{P}$ such that $G\varphi xu = 1$, 2 or 3 if u codes a tip, a conjunctive point or a disjunctive point on $T(x, \Phi)$, and $= 0$ otherwise (this shall include the case that φ is not a formula of L+ with at most one free variable). Let F be defined by

(1) $F\varphi x\langle\ \rangle = \varphi(x)$ if $G\varphi x\langle\ \rangle \neq 0$,
$F\varphi xu = (u)_{\mathrm{lh}u-1}$ if $G\varphi xu \neq 0$ and $\mathrm{lh}u > 0$,
$= \emptyset \in \emptyset$ otherwise.

Then if u codes a point on $T(x, \varphi)$, $F\varphi xu$ is the sentence at that point.

Now we give an inductive definition of a class K.

(2) $G\varphi xu = 1 \wedge F\varphi xu$ is a true basic sentence not containing $\dot{X}$
$\rightarrow \langle \varphi, x, u \rangle \in K$,
$G\varphi xu = 2 \wedge (\forall z)(G\varphi x(u * z) = 0 \vee \langle \varphi, x, u * z \rangle \in K)$
$\rightarrow \langle \varphi, x, u \rangle \in K$,
$G\varphi xu = 3 \wedge (\exists z)(G\varphi x(u * z) \neq 0 \wedge \langle \varphi, x, u * z \rangle \in K)$
$\rightarrow \langle \varphi, x, u \rangle \in K$,

where $u * z$ codes the sequence got by adjoining z at the end of the sequence coded by u.

Note that if $G\varphi xu = 0$, then $\langle\varphi xu\rangle \notin K$. It is a straightforward matter to verify that

(3) $\langle\varphi,x,u\rangle \in K \leftrightarrow u$ is a grounded point of $T(x,\varphi)$.

Evidently $w_0 = \langle\emptyset \notin \emptyset, \emptyset, \langle\ \rangle\rangle \in K$

and $w_1 = \langle\emptyset \in \emptyset, \emptyset, \langle\ \rangle\rangle \notin K$.

Then there are $H_1, H_2 \in \mathcal{P}$ which satisfy

(4) $H_1\langle\varphi,x,u\rangle z = \langle\varphi,x,u*z\rangle$ if $G\varphi xu = 2$ and $G\varphi x(u*z) \neq 0$,
$= w_0$ otherwise,
$H_2\langle\varphi,x,u\rangle z = \langle\varphi,x,u*z\rangle$ if $G\varphi xu = 3$ and $G\varphi x(u*z) \neq 0$,
$= w_1$ otherwise.

So we can rewrite (2) in the form:

(5) $Q(w) \vee (\forall z)(H_1wz \in K) \vee (\exists z)(H_2wz \in K) \rightarrow w \in K$

where, since the truth function for basic sentences $\in \mathcal{P}$, $Q \in \mathcal{P}$. But $\mathcal{P} \subseteq \mathbf{\Delta}_1$; thus we see that $K \in (\mathbf{\nabla}_1^+)^\infty$.

Finally we set

$$J = \{\langle\varphi,x\rangle : \langle\varphi,x\langle\ \rangle\rangle \in K\}.$$

By (3) and 2.5.3., J satisfies the conditions of the theorem.

QED

2.3.8. *Remarks* (a) Besides providing a universal set for $(\mathbf{L}+)^{(\infty)}$, K allows us to compare ordinals using $(\mathbf{\nabla}_1+)^{(\infty)}$ relations. More precisely the relations

$$w_1 <_K w_2 \leftrightarrow_{\mathrm{Df}} |w_1|_K < |w_2|_K$$

$$w_1 \leqslant_K w_2 \leftrightarrow_{\mathrm{Df}} w_1 \in K \;\&\; |w_1|_K \leqslant |w_2|_K$$

are both $(\mathbf{\nabla}_1^+)^{(\infty)}$. For, from 2.3.3. and (5) we have

$$Q(w) \rightarrow |w|_K = 0\,,$$

$$G(w)_0(w)_1(w)_2 = \left.\begin{matrix}2\\ \\ 3\end{matrix}\right\} \rightarrow |w|_K = \begin{cases}\operatorname{Sup}^+_z |H_1wz| \\ \\ \operatorname{Min}_z |H_2wz| + 1\,.\end{cases}$$

And from these it is a straightforward matter to write down a $\mathbf{\nabla}_1+$ simultaneous inductive definition for $<_K, \leqslant_K$. (For details see Moschovakis [2].)

Further, if one of $w_1, w_2 \in K$, then either $w_1 < w_2$ or $w_2 \leqslant w_1$. Hence there is a partial function (or selection operator) with graph in $(\mathbf{\nabla}_1)^{(\infty)}$ which is defined if at least one of its two arguments is in K, and which will then select one of its arguments which is in K. For a general discussion of the principles involved see Grilliot [3]. For more refined theorems of the same kind see Moschovakis [4].

(b). The arguments in remark (a) depended on the fact that $w \in K$ if it is an initial member, or if all or some of its 'predecessors' $H_i wz$ belong to K. Recently Aczel [1] has shown that the introduction of a special inductive class with this property is *not* essential. If $\mathcal{C}$ is sufficiently closed, then $\leqslant_\Phi$, $<_\Phi$ are $\mathcal{C}$-inductive for any $\Phi \in \mathcal{C}$. And, of course, for the case we are actually discussing, Barwise's method (cf. 1.4.4.) can be applied.

(c). The inductive set K has a similar role to O in hyperarithmetic theory. Of course $\mathrm{O} \in (\mathbf{\Pi}_1^0+)^{(\infty)}$. The realisation that in general one needs $(\mathbf{\nabla}_1^0+)^{(\infty)}$ to get a universal set for $(\mathbf{\Delta}_0^1+)^{(\infty)}$ is due to Moschovakis. We discuss the problem of when $\mathbf{\nabla}_1^0+$ can be replaced by $\mathbf{\Pi}_1^0+$ in §2.5 below.

2.4. For our next application of tableaux we need:

2.4.1. Definition. (Barwise). $\mathfrak{D} = (\mathfrak{A}, S, \in)$ is *admissible* iff it satisfies the axioms of (KPU). It is an admissible *beyond* $\mathfrak{A}$ iff it satisfies the axioms of (KPU$^+$). We give an alternative characterisation:

(i) $\mathfrak{D}$ is admissible iff (a) it is closed under $\mathcal{P}$, and
(b) it satisfies the axiom of $\mathbf{\Delta}_0$-collection:–

$$(\forall x \in s)(\exists y)\varphi \rightarrow (\exists t)(\forall x \in s)(\exists y \in t)\varphi$$

where φ is any $\mathbf{\Delta}_0$ formula.

(ii) $\mathfrak{D}$ is admissible beyond $\mathfrak{A}$ if it is admissible and $A \in S$.

It is easily shown that an admissible $\mathfrak{D}$ satisfies the axioms of $\mathbf{\Delta}_1$-separation and $\mathbf{\Sigma}_1$-collection. We write $o(\mathfrak{D})$ for $S \cap \mathrm{On}$.

2.4.2. Theorem. *If* $\mathfrak{D}$ *is admissible then*

$$|\mathbf{\Sigma}_D| = o(\mathfrak{D}) \text{ and } (\mathbf{\Sigma}_D+)^{(\infty)} = \mathbf{\Sigma}_D .$$

This theorem makes plain why admissible sets carry a recursion theory similar to ordinary recursion theory. It is inherent in the original development of admissibility theory by Kripke and Platek, but I believe it was first stated (for Σ_1) in lectures which I gave in Manchester and UCLA in 1968.

Proof. By a *subtableau* with vertex V of a complete tableau $T = T(x, \varphi)$ we mean a subclass Y of the class of points of T such that:

(i) $V \in Y$ and all other points of Y lie below V;

(ii) if $Q \in Y$ is conjunctive, then all the points of T immediately below Q belong to Y;

(iii) if $Q \in Y$ is disjunctive then at least one of the points of T immedialy below Q belongs to Y.

A subtableau Y is *well-founded* if:

(iv) the sentence at every tip of Y is true

(v) the relation 'below' on Y is well-founded.

2.4.3. Lemma. *A subtableau is well-founded iff all its points are grounded points of T.*

For 'if' use induction on $|P|$; for only if use induction on 'below'.

Corollary 1. *The union of a collection of well-founded subtableaux with vertex V is itself a well-founded tableau with vertex V.*

For the union will be a class of grounded points of T which satisfies (i)–(iii).

Corollary 2. *The sentence at the vertex of a well-founded subtableau is true under the assignment of* Φ^∞ *to* $\ddot{X}$.

2.4.4. Lemma. *For* $\Phi \in L+$ *the relation 'Y is a well-founded subtableau with vertex* v_0 *of the complete tableau for* $x \in \Phi^\infty$*' is* Σ_1 *in* Y, v_0, x, φ.

We refer, of course, to the coding of points introduced in the proof of theorem 2.3.7. Observe that the relation 'u is below v' is simply $u \subset v \leftrightarrow_{\mathrm{Df}} \mathrm{lh}(u) < \mathrm{lh}(v) \wedge (\forall i < \mathrm{lh}(u))((u)_i = (v)_i)$. Using the functions G and F it is easy to construct a primitive recursive predicate $L(Y, v_0, x, \varphi)$ which expresses conditions (i)–(iv), for the quantifiers in those conditions are all restricted to Y. And (v) can be expressed by:

$$M(Y) \leftrightarrow_{\mathrm{Df}} (\exists H)\,(H \text{ is a function with domain } Y \text{ and}$$
$$\text{range} \subseteq \mathrm{On} \wedge (\forall u,v \in Y)\,(u \subset v \rightarrow Hv < Hu)).$$

This is obviously $\mathbf{\Sigma}_1$, and so

$$N(\varphi,x,v_0,Y) \leftrightarrow_{\mathrm{Df}} L(Y,v_0,x,\varphi) \wedge M(Y)$$

gives the required relation.

Now let $\Phi \in \mathbf{\Sigma}_{D^+}$. The proof of the theorem rests on the crucial.

2.4.5. Lemma. *If v_0 is a grounded point of the complete tableau for $x \in \Phi^\infty$ then there exists a well-founded subtableau Y with vertex v_0 such that Y is a* set (*i.e., $Y \in D$ or Y is 'D-finite'*).

The proof is by induction on $|v_0|_{x,\Phi}$. If v_0 is a tip of T the required subtableau is $\{v_0\}$. If v_0 is a disjunctive point, then by IH we have a well-founded subtableau y for some point immediately below v_0 and $(y;v_0)$ is the required subtableau.

If v_0 is conjunctive, then the sentence at v_0 is $\bigwedge s$, say, and, by IH,

$$(\forall \theta \in s)(\exists z) N(\varphi,x,v_0 * \theta,z)\,.$$

Hence, since D satisfies $\mathbf{\Sigma}_1$-collection, there is a $t \in D$ for which

$$(\forall \theta \in s)(\exists z \in t) N(\varphi,x,v_0 * \theta,z)\,.$$

Now set

$$Y = \mathbf{U}_{\theta \in s}\, \mathbf{U}\{(z;v_0) : z \in t \wedge N(\varphi,x,v_0 * \theta,z)\}\,.$$

For any $\theta \in s$

$$Y_\theta = \mathbf{U}(z \in t \wedge N(\varphi,x,v_0 * \theta,z)\}$$

is a well-founded subtableau with vertex $v_0 * \theta$, by Corollary 1 to 2.4.3. But then Y is evidently a well-founded subtableau with vertex v_0. Finally since D satisfies the $\mathbf{\Delta}_1$-separation axiom and is closed under $\mathcal{P}$, $Y \in D$ as required.[1]

[1] See footnote on page 299.

This completes the proof of the lemma.

Finally, to prove the theorem, we claim:

$$x \in \Phi^\infty \leftrightarrow (\exists y) N(\varphi, x, \langle\ \rangle, y) .$$

For if $x \in \Phi^\infty$, $\langle\ \rangle$ is grounded on $T(x, \Phi)$ and the RHS follows by 2.4.4. Conversely if the RHS holds, then $\varphi(x, \Phi^\infty)$ by corollary 2 of 2.4.3.

QED

2.4.6. Corollary. *With D, Φ as in 2.4.2., if we put*

$$\psi(x, \dot{X}) \leftrightarrow (\exists z)(z \subseteq \dot{X} \wedge \varphi(x, z)) ,$$

then $\Psi^\infty = \Phi^\infty$.

For, since the subtableau Y of 2.4.4. belongs to D, so does

$$z = \{y : y \in \dot{X} \text{ occurs on } Y\} .$$

2.4.7. *Remarks.* (1) A corollary of this theorem is lemma 2.5. of Barwise [1], which may be stated thus: the class of derivable sequents of $L_{D,\omega}$, where D is admissible is $\boldsymbol{\Sigma}_1/D$. A comparison of Barwise's proof with ours shows, I think, the advantages both of working on *forms* of inductive definition, and of using the complete tableau. We have only 4 cases to consider (tips, $\exists, \mathbf{W}, \mathbf{M}$) against Barwise's 10. And because the complete tableau contains all the different possible justifications of $x \in \Phi^\infty$ we have avoided having to deal with derivations which contain, hereditarily, sets of derivations as subderivations.

(2) P.W. Grant has given (unpublished) a rather neat proof of the theorem for the case $(\boldsymbol{\Sigma}_1/D)^{(\infty)}$ based on the second recursion theorem for D-recursive functions.

(3) Suppose we consider a relativised inductive operator $\varphi(\overset{+}{B})$ where B is a given subset of D and atoms $z \in B$ occur only positively in $\varphi(B)$. In general lemma 2.4.4. fails: we can only assert that Y belongs to some extension D' of D in which the $\boldsymbol{\Delta}_0(\overset{+}{B})$ axiom of collection holds. [This is in contrast to the particular case when D is the hereditarily finite structure over A.] However, if $B \in \Sigma_D$, then 2.4.4. still holds, and so, if $x \in (\Phi(B))^\infty$ $b = \{y : y \in B$ occurs

on the subtableau for $x \in \Phi^\infty\}$ belongs to D. I.e. if $x \in (\Phi(\overset{+}{B}))^\infty$ then $x \in (\Phi(b))^\infty$ for some 'D-finite' $b \in D$. The Barwise compactness theorem is just a special case of this fact. Also, in this case, if $X \in (\Phi(\overset{+}{B}))^{(\pm\infty)}$, then X is weakly D-metarecursive in B. This suggests

Problem 7. Are there easily characterised subclasses $\mathcal{C}(B)$ of $\mathbf{\Sigma}_A+(B)$ such that

$$(\mathcal{C}(B))^{(\pm\infty)} = \{X : X \text{ is weakly (strongly) } D\text{-metarecursive in } B\}?$$

2.5. In this section we consider $(\mathbf{\Pi}_D+)^\infty$. First we note that by using the proof of 2.3.7. it is easy to show:

2.5.1. Theorem. *If D satisfies the stipulations of 2.2., then*

$$(\mathbf{\Pi}_D+)^{(\infty)} = (\mathbf{\Pi}_1+/D)^{(\infty)}, \quad |\mathbf{\Pi}_D+| = |\mathbf{\Pi}_1+|\,.$$

Now we prove our main result, which concerns those D all of whose members are countable.

2.5.1. Theorem. *Let v be a function such that for all $s \in D$, $\{vsn : 0<n<\omega\} = s$. Then there is a relation R', primitive recursive in v, such that, for any $\Phi \in \mathbf{\Pi}_D+$,*

$$x \in \Phi^\infty \leftrightarrow \{\langle y,z\rangle : R'\varphi xyz\} \text{ is well-founded.}$$

Proof. We use the coding for points on the complete tableau T for $x \in \Phi^\infty$ used in 2.3. Since x, Φ remain fixed throughout the argument, we omit all further mention of them. We use p, q to range over the set Seq of codes for finite sequences from ω. For definiteness we suppose 1 is the code for the empty sequence, and we suppose $p \subset q$ implies $p < q$. Let ρ : Seq $\to D$; we establish a mapping u_ρ from Seq into T as follows:

(1a) $\quad u_\rho(1) = \langle\ \rangle \quad$ (the vertex of T);

(1b) if $u_\rho(p)$ is a conjunctive point of T

$$u_\rho(p*0) = u_\rho(p) * \rho(p*0)\ ;$$

(1c) if $u_\rho(p)$ is a disjunctive point of T, at which the sentence has the form $\mathbf{W}s$, and if $n > 0$

$$u_\rho(p*n) = u_\rho(p) * \nu sn\ ;$$

(1d) if $u_\rho(p)$ is a point of T at which $y \in \dot{X}$ stands

$$u_\rho(p*1) = u_\rho(p) * \varphi(y)\ ;$$

(1e) in all other cases

$$u_\rho(p*n) = u_\rho(p)\ .$$

We consider a tree Γ whose infinite branches correspond to different choices of ρ. A point $\bar{\rho}(p)$ of this tree is determined by the values of ρ up to *and including* the value at p:

$$\bar{\rho}(p) = \langle \rho(1), ..., \rho(q), ..., \rho(p) \rangle \qquad (q \in \mathrm{Seq},\ 1 \leqslant q \leqslant p).$$

We say the branch ρ is *secured* at p if $u_\rho(p)$ is a tip of T. Let

$$\Gamma' = \{\bar{\rho}(p) : (\forall q < p)(\bar{\rho}(q) \text{ is not immediately secured})\}\ ,$$

be the tree of non-past secured points of Γ. Γ' is well-founded if every branch is secured at some point. And this is so iff the relation

$$(2) \qquad R(\bar{\rho}(p), \bar{\rho}(q)) \leftrightarrow_{\mathrm{Df}} \bar{\rho}(q) \subseteq \bar{\rho}(q)\ ,$$

restricted to Γ', is well-founded.

2.5.2. Lemma. *If T is grounded, Γ' is well-founded.*

Let T be grounded. For a given ρ consider the sequence

$$p_0 = \langle\ \rangle$$

$$p_{i+1} = p_i * 0 \quad \text{if } u_\rho(p_i) \text{ is conjunctive,}$$
$$= p_i * (\mu m > 0)(u_\rho(p_i * m) \text{ is grounded})$$
$$\text{if } u_\rho(p_i) \text{ is disjunctive,}$$
$$= p_i \quad \text{if } u_\rho(p_i) \text{ is a tip.}$$

Since T is grounded, p_i is always defined, and $u_\rho(p_0), u_\rho(p_1), \ldots$ is a sequence of grounded points of T, each one immediately below the one before. But this sequence must terminate at a tip, $u_\rho(p_k)$ say, and then ρ is secured at p_k.

2.5.3. Lemma. *If Γ' is well-founded, then T is grounded.*

Suppose Γ' is well-founded; we prove the following by induction up Γ':–
(3) If $\bar{\rho}(p) \in \Gamma'$, there is a $q \leqslant p$ such that $u_\rho(q)$ is a grounded point of T.

If $\bar{\rho}(p)$ is immediately secured, then $u_\rho(p)$ is a tip and so is grounded. If not, let the points of Γ' immediately below $\bar{\rho}(p)$ be $\bar{\rho}(p) * \rho(p' * m)$, where $p' \leqslant p$, and $\rho(p' * m)$ ranges over D. By IH, for each of these points there is a $q \leqslant p' * m$, with $u_\rho(q)$ grounded. If for any of them this q is $\leqslant p$, then (3) holds for $\bar{\rho}(p)$. If not, $u_\rho(p' * m)$ is grounded, where ρ takes the values assigned by $\bar{\rho}(p)$ at arguments $\leqslant p$, and can take any value at argument $p' * m$. By (1b), (1c) or (1d) this means that all the points of T immediately below $u_\rho(p)$ are grounded points of T. Hence $u_\rho(p)$ is a grounded point of T; so (3) is proved.

But then, since $\bar{\rho}(1) \in T, u_\rho(1)$ ($= \langle\ \rangle$) is a grounded point of T, and so T is grounded.

To prove the theorem we have to show that R (as given by (2)) restricted to Γ' is primitive recursive in ν. Since obviously, $R \in \mathcal{P}$, we have to show that '$\in \Gamma'$' is primitive recursive in ν.

But, given $\bar{\rho}(p)$, equations (1) determine $u_\rho(q)$ for $q \leqslant p$ by an ω-recursion which uses ν and the G and F of 2.3. This concludes the proof of 2.5.1.

Remark. An alternative proof could be constructed using semantic tableau rules similar to (but simpler than) those used by Lopez-Escobar in [1].

To see the interest of this theorem, let

$$\omega_1^D = \text{Sup}^+ \{R : R \text{ is a } \mathfrak{D}\text{-recursive well-ordering whose field} \subseteq D\}.$$

We consider only admissible D of the form L_α (so we are taking $A = \emptyset$). Let η be the least admissible ordinal such that for some $x \in L_\eta$, x is not ω-enumerated by any L_η-recursive function. Certainly η is quite a large ordinal. For example η is much greater than β_0 the closure ordinal for ramified analysis.

2.5.4. Theorem. *If α is an admissible ordinal less than η then*

$$|\mathbf{\Pi}_{L_\alpha}{}^+| = \omega_1^{L_\alpha}.$$

Proof. LHS $\geqslant$ RHS follows from considering the inductive definition for the initial well-ordered segment of an L_α-recursive ordering. For RHS $\geqslant$ LHS we apply 2.5.1. Since an L_α-recursive ω-enumeration of any $s \in L_\alpha$ is L_α bounded and so belongs to L_α, we can find it by search; thus there is an L_α-recursive ν as in 2.5.1. And since there is an L_α-recursive well-ordering of L_α, we can pass from the well-founded relation R' to a well-ordering relation Q with $|Q| \geqslant |R'|$.

Further, it is straightforward to verify the following by induction on the ordinals concerned.

(1) If $\bar{\rho}(p)$ and q are related as in (3), then

$$|\bar{\rho}(p)|_{\Gamma'} \geqslant |u_\rho(q)|_T\,;$$

Hence

$$|R'| \geqslant |\langle\ \rangle|_T\,.$$

(2) $$|\langle\ \rangle|_{T(x,\Phi)} \geqslant |x|_\Phi\,.$$

Thus $|Q| \geqslant |x|_\Phi$ (recall that R', Q depend on x, φ). This suffices to prove RHS $\geqslant$ LHS in 2.5.4.

It is known that there are admissible ordinals $< \beta_0$ for which

$$|\mathbf{L}_{L_\alpha}{}^+| > \omega_1^{L_\alpha}$$

(see Gostanian [1]; he calls them *bad* ordinals). Thus there are certainly ordinals α for which $|\mathbf{V}_1 + /L_\alpha| > |\mathbf{\Pi}_1 + /L_\alpha|$. I have found it hard to get much insight into $(\mathbf{\Pi}_D)^{(\infty)}$. Here are two problems which may well be easy, but which I could not solve.

Problem 8. Are there admissible D (preferably with $A = \emptyset$) for which

$$|\mathbf{\Pi}_D + | > \omega_1^D \ ?$$

Problem 9. Is it true that

$$\mathbf{\Sigma}_1/D \subseteq (\mathbf{\Pi}_D +)^{(\infty)} \ ?$$

§3. Inductive definitions with type 2 parameters

3.1. The natural way of relativising a class $\mathcal{C}$ of inductive operators over A to a given total or partial function $\zeta : A \to A$ is well-known and well understood: one enlarges $\mathcal{C}$ by allowing atoms $\hat{\zeta}xy$ to occur positively in the defining formulae for $\mathcal{C}$, where $\hat{\zeta}$ is the graph of ζ. If $\mathcal{C}$ satisfies some simple closure conditions all goes very smoothly, and the theorems about $\mathcal{C}^{(\infty)}$ are easily relativised to theorems about $\mathcal{C}(\zeta)$. We begin an investigation here of relativisation to a type 2 functional $\mathbf{F} : A^A \to A$; in particular we wish to connect this relativisation with the recursion theories of Kleene and Platek. We consider only the case where $\mathcal{C}$ is $\mathbf{\Delta}_0^1+$ (first-order, positive) or $\mathbf{\Delta}_0^0+$ (quantifier free, positive), or $\mathbf{\Sigma}_1^0+$. Further we assume that A contains a pairing function $\langle\ \rangle$ with projection functions $(\ )_0, (\ )_1$, and a copy of the natural numbers (so that it is an expansion of an 'acceptable' structure).

3.1.1. *Notations.* Capital italics stand for subsets of ${}^nA$. To avoid a plethora of qualifying marks we shall not always distinguish between a relation over A and its coding as a subset of A. Thus R, $\{(x,y) : xRy\}$, $\{\langle x,y\rangle : xRy\}$ may all be denoted by R. The variables ζ, η, range over $(A \to_p A)$, the set of *partial* functions; $\hat{\zeta}, \hat{\eta}$ denote their graphs in the above ambiguous sense. For $Y \subseteq A$, $\langle x, Y\rangle$ denotes $\{\langle x,y\rangle : y \in Y\}$, and conversely, Y_x, the x-th component of Y is $\{y : \langle x,y\rangle \in Y\}$. Letters $\mathbf{F}, \mathbf{G}$ stand for partial functionals from $(A \to_p A)$ to A. In order to code their graphs we define, for a binary relation R,

$[R : y] = \langle 0,R\rangle \cup \{\langle 1,y\rangle\}$, and then set

$$\hat{\mathbf{F}} = \{[\hat{\zeta} : \mathbf{F}\zeta] : \zeta \in \mathrm{Dom}\,\mathbf{F}\}\,.$$

F is *consistent* iff

$$\zeta \in \mathrm{Dom}\,\mathbf{F} \wedge \hat{\zeta} \subseteq \hat{\eta}. \rightarrow \mathbf{F}\eta = \mathbf{F}\zeta\,.$$

3.1.2. Let $\mathbf{Z} \in \mathbf{P}(\mathbf{P}(A))$, and let $\dot{\mathbf{Z}}$ be a formal variable for it. Let the class of formulas $\mathcal{C}$ be given by initial and closure conditions; we enlarge it to $\mathcal{C}(\dot{\mathbf{Z}})$ as follows:

(i) $\varphi \in \mathcal{C} \rightarrow \varphi \in \mathcal{C}(\dot{\mathbf{Z}})$;

(ii) if $\varphi \in \mathcal{C}(\dot{\mathbf{Z}})$, $\{x : \varphi\}$ is an *abstract* of $\mathcal{C}(\dot{\mathbf{Z}})$;

(iii) if τ is an abstract of $\mathcal{C}(\dot{\mathbf{Z}})$, $\tau \in \dot{\mathbf{Z}}$ is a formula of $\mathcal{C}(\dot{\mathbf{Z}})$;

(iv) other closure conditions as for $\mathcal{C}$.

Now let $\mathcal{C}$ be included in a language for positive inductive definitions, and consider a progressive operator Φ given by $\Phi X = \{u : \varphi(x, X, \mathbf{Z})\}$ where $\varphi(\mathrm{x}, \dot{X}, \dot{\mathbf{Z}}) \in \mathcal{C}(\dot{\mathbf{Z}})$, and has no free variables other than x. Aczel (in [1]) has pointed out that if $\mathbf{Z}$ is monotonic (i.e. $X \subseteq Y \wedge X \in \mathbf{Z}. \rightarrow Y \in \mathbf{Z}$) then Φ is a monotonic operator. Further, for $\mathcal{C} = \Delta_0^1 +$ much of the theory of $\mathcal{C}^{(\infty)}$ can be lifted up to $(\mathcal{C}(\mathbf{Z}))^{(\infty)}$; in particular Aczel's proof of the stage comparison theorems goes through. Aczel makes use of the fact that $\mathbf{Z}$ can be treated as a quantifier: $(\mathbf{Z}x)\varphi$ holds just in case $\{x : \varphi\} \in \mathbf{Z}$. We add the remark that provided $\mathbf{Z}$ is monotonic we may introduce $\mathbf{Z}$-rules into the calculus of sequents:

$|\mathbf{Z}$: from $\{\Gamma \vdash \varphi(x), \Delta \;: x \in Y\}$ for some $Y \in \mathbf{Z}$
infer $\Gamma \vdash (\mathbf{Z}x)\,\varphi(x), \Delta$

$\mathbf{Z}|$: from $\{\Gamma, \varphi(x) \vdash \Delta : x \in Y\}$ for some Y which meets every $X \in \mathbf{Z}$
infer $\Gamma, (\mathbf{Z}x)\,\varphi(x) \vdash \Delta$.

These rules permit cut-elimination. This should make it possible to extend Barwise's method (cf. 1.4.4) to this case.

3.1.3. Provided **F** is consistent, it makes sense to introduce the monotonic set of sets $\mathbf{F}^\circ$ defined by:

$$\mathbf{F}^\circ = \{Y : (\exists X \in \hat{\mathbf{F}})(X \subseteq Y)\}\,.$$

And then

(1) $$\mathbf{F}\zeta = y \leftrightarrow [\hat{\zeta} : y] \in \mathbf{F}^{\circ} .$$

From now on we always assume that **F** *is consistent.*

We note that a consequence of the definition of $\mathbf{F}^{\circ}$ is:

(2) $$y \in \operatorname{Rge} \mathbf{F} \leftrightarrow [A \times A : y] \in \mathbf{F}^{\circ} .$$

3.2. We use Kleene's schemata S1–S8 to define the class $\mathcal{R}(\mathbf{F})$ of A-theoretic functions partial-recursive in **F**. We take $\mathfrak{A} = \mathfrak{N}$, and for simplicity ignore relativisation to a function – that is we drop S.7. We do not insist that **F** be defined only for total functions, so that S.8. reads:

$$\{z\}(\bar{u}, \mathbf{F}) \simeq \mathbf{F}(\lambda t \cdot \{z_1\}(\bar{u}, t, \mathbf{F})) ,$$

where $\bar{u} = u_1, \ldots, u_m$, $z = \langle 8, \langle m, 0, 1 \rangle, 2, z_1 \rangle$; both sides are defined iff

$$\lambda t \cdot \{z_1\}(\bar{u}, t, \mathbf{F}) \in \operatorname{Dom}(\mathbf{F}) .$$

For the case considered this is, in effect, Platek's definition of 'recursive in **F**' and we shall refer to it as 'Kleene–Platek', or 'K–P' recursion. If the domain of **F** consists exactly of the total functions we say that **F** is *clean*, and write $\mathbf{F} \in \mathcal{K}$:

We write '$\{z\}(a)\uparrow$' for '$\{z\}(a)$ is defined'; (Moschovakis introduced '$\downarrow$' for this purpose, but a thumbs up sign seems more appropriate). We set

$$D = D(\mathbf{F}) = \{\langle z, \bar{u} \rangle : \{z\}(\bar{u}, \mathbf{F})\uparrow\} ,$$

$$V = V(\mathbf{F}) = \{\langle\langle z, \bar{u} \rangle, y \rangle : \{z\}(\bar{u}, \mathbf{F}) = y\} .$$

The class $\mathcal{R}^{+}(\mathbf{F})$ of sets *semi-recursive* in **F** is defined by

$$\mathcal{R}^{+}(\mathbf{F}) = \{Y : (\exists z)(Y = D_z)\} .$$

Similarly

$$\mathcal{R}^{\pm}(\mathbf{F}) = \{Y : Y, N - Y \in \mathcal{R}^{+}(\mathbf{F})\} .$$

(This would be pronounced as 'bi-semi-recursive in **F**'.) It is known that not every non-empty set of $\mathcal{R}^+(\mathbf{F})$ is the range of a total function recursive in **F**; that is why, following a warning from Kreisel in [2], we avoid using 'r.e. in **F**' for $\mathcal{R}^+(\mathbf{F})$. Also, in general, $\mathcal{R}^\pm(\mathbf{F}) \neq \mathcal{R}(\mathbf{F})$.

3.3. We now seek connections between the definitions in 3.1. and those in 3.2.

3.3.1. Theorem. *For any consistent* **F** *over* $\mathfrak{N}$

(i) $\qquad \mathcal{R}^+(\mathbf{F}) \subseteq (\Sigma^0_1{+}(\mathbf{F}^\circ))^{(\infty)}$,

(ii) $\qquad (\Sigma^0_1{+}(\mathbf{F}^\circ))^{(\infty)} = (\Delta^0_0{+}(\mathbf{F}^\circ))^{(\infty)}$.

Remark. We separate out the two parts because (i) can be generalised to other structures, but (ii) depends essentially on the fact that $\mathfrak{N}$ is finitely generated.

The proof of (i) consists of writing out formally an inductive definition Φ which follows Kleene's definition of $D(\mathbf{F})$ (for $\Phi_0^{(\ast)}$). An existential quantifier is essential in the clause corresponding to substitution (S.4.). Further details will be found in Grilliot [1], which also introduced the 'counting down' technique necessary for the proof of (ii). We illustrate this by a simple example. Suppose an inductive definition for X has just one existential clause:

$$(\exists y)(\langle x,y\rangle \in X) \to x \in X\,.$$

We replace X by the component X_0 in the other clauses, and replace the above clause by:

$$\langle x,y\rangle \in X_0 \vee \langle x,Sy\rangle \in X_1 \to \langle x,y\rangle \in X_1\,,$$

$$\langle x,0\rangle \in X_1 \to x \in X_0\,.$$

3.3.2. Theorem. *If* **F** *satisfies the condition* $(*)$ *below, in particular if* $\mathbf{F} \in \mathcal{K}$, *then*

$$(\Delta^1_0{+}(\mathbf{F}^\circ))^{(\infty)} = (\Delta^0_0{+}(\mathbf{F}^\circ))^{(\infty)}\,.$$

(∗) *There is a total function* $\eta \in (\Sigma_1^0(\mathbf{F}^\circ))^{(\infty)}$ *in the domain of* $\mathbf{F}$, *such that none of the partial functions* $\eta \restriction \{i : i < n\}$ $(n = 0, 1, 2, \ldots)$ *are in the domain of* $\mathbf{F}$.

This theorem is implicit in Grilliot [4]. We illustrate the method of proof by a particular example. Let $\Phi \in \mathbf{\Pi}_1^0 + (F^\circ)$ be associated with the formula $(\forall t)\varphi_1(\mathrm{t}, \mathrm{x}, \dot{X}, \dot{\mathbf{Z}})$, and let $\eta = \Theta^\infty$ with $\Theta \in \mathbf{\Sigma}_1^0 + (\mathbf{F}^\circ)$; (in general one should consider $\eta = \Theta_0^\infty$). Consider the inductive operator $\Psi \in \mathbf{\Sigma}_1^0 + (\mathbf{F}^\circ)$ given by:

$$\varphi_1(t, x, X_0, \mathbf{F}^\circ) \to \langle t, x\rangle \in X_1 ,$$

$$\theta(z, X_2, \mathbf{F}^\circ) \to z \in X_2 ,$$

$$\langle 0, x\rangle \in X_1 \wedge \langle 0, v\rangle \in X_2 \,.\!\!\to \langle x, \langle 0, v\rangle\rangle \in X_3 ,$$

$$(\exists v')(\langle x, \langle t, v'\rangle\rangle \in X_3) \wedge \langle St, x\rangle \in X_1 \wedge \langle St, v\rangle \in X_2$$
$$.\!\!\to \langle x, \langle St, v\rangle\rangle \in X_3 ,$$

$$(\exists y)([X_{3,x} : y] \in \mathbf{F}^\circ) \to x \in X_0 ,$$

(where $X_{3,x}$ denotes $\{z : \langle 3, \langle x, z\rangle\rangle \in X\}$). We claim that $\Psi_0^\infty = \Phi^\infty$. To see that this is so it is sufficient to observe the following facts.

(a) For some $\beta \leqslant |\Psi|, \Psi_2^\beta = \eta$.
(b) If $x \in \Phi^\infty$, then for some $\gamma \leqslant |\Psi|, \Psi_{3,x}^\gamma = \eta$.
(c) If $n = (\mu t) \to \varphi_1(t, x, \Phi^\infty, \mathbf{F}^\circ)$, then

$$\Psi_{3,x}^\infty = \eta \restriction \{i : i < n\} .$$

For the general case, one applies the above trick repeatedly, starting from the inside, so as to get rid of all the ∀'s from φ. The ∃'s can be got rid of by 3.3.2.(ii).

QED

3.4. Now we would like to be able to establish a converse to 3.3.1. Under any reasonable definition of 'function recursive in $\mathbf{F}$' however, we would not expect Rge $\mathbf{F}$ to be recursive in $\mathbf{F}$. But by 3.1.3.(2) Rge $\mathbf{F}$ is definable from $\mathbf{F}^\circ$! Because we required $\mathbf{F}^\circ$ to be monotonic, we have allowed $[R : y] \in \mathbf{F}^\circ$ even when the relation R is not functional. But in the proof of 3.3.1. and 3.3.2. we never needed this extension from $\hat{\mathbf{F}}$ to $\mathbf{F}^\circ$. In fact those theorems remain true if $\hat{\mathbf{F}}$ be substituted for $\mathbf{F}^{\circ}$ throughout. The disadvantage is, of course, that the operators in $\hat{\mathbf{F}}$ are not monotonic, and we cannot apply Aczel's results to them. So we make the following definition:

3.4.1. Definition. The operator $\Phi(\mathbf{F}^{\circ})$ determined by substituting $\mathbf{F}^{\circ}$ for $\dot{\mathbf{Z}}$ in the formula $\varphi(\mathrm{x}, \dot{X}, \dot{\mathbf{Z}})$ is said to be *functional* (at $\mathbf{F}$) iff

$$(\Phi(\mathbf{F}^{\circ}))^{\infty} = (\Phi(\hat{\mathbf{F}}))^{\infty} .$$

As the example of $V(\mathbf{F})$ and $D(\mathbf{F})$ shows, it may be obvious from the form of φ that $\Phi(\mathbf{F}^{\circ})$ is functional at all $\mathbf{F}$. We write $\Phi \in \mathrm{Fn}-\mathcal{C}(\mathbf{F}^{\circ})$ to mean that $\Phi \in \mathcal{C}(\mathbf{F}^{\circ})$ and is functional at $\mathbf{F}$.

We recall that $\mathbf{E} \in \mathcal{K}$ and satisfies $\mathbf{E}\zeta = 0$ if $(\exists x)(\zeta x = 0)$, $\mathbf{E}\zeta = 1$ otherwise.

3.4.2. Theorem. *If* $\mathbf{F} \in \mathcal{K}$, *then*

(i) $$(\mathrm{Fn} - \Delta_0^0 + (\mathbf{F}^{\circ}))^{(\infty)} = \mathcal{R}^{+}(\mathbf{F}, \mathbf{E}) = (\mathrm{Fn} - \Delta_0^1 + (\mathbf{F}^{\circ}))^{(\infty)}$$

(ii) $$(\mathrm{Fn} - \Delta_0^0 + (\mathbf{F}^{\circ}))^{(\pm\infty)} = \mathcal{R}(\mathbf{F}, \mathbf{E}) = (\mathrm{Fn} - \Delta_0^1 + (\mathbf{F}^{\circ}))^{(\pm\infty)} .$$

The equality of the extreme members is given by 3.3.2. For (i), $\mathcal{R}^{+}(\mathbf{F}, \mathbf{E}) \subseteq$ RHS is proved by a simple extension of the proof of 3.3.1. To prove LHS (i) $\subseteq \mathcal{R}^{+}(\mathbf{F}, \mathbf{E})$ we apply the first recursion theorem. We first show that for $\varphi \in \Delta_0^0 + (\dot{\mathbf{Z}})$ there is a functional J_φ, with index j_φ, defined by S1–S9 such that if $\mathrm{Rge}\,\zeta \subseteq \{0\}$ then

(1) $(\forall \bar{u})(\varphi(\bar{u}, \mathrm{Dom}\,\zeta, \hat{\mathbf{F}}) \rightarrow J_\varphi(\bar{u}, \zeta, \mathbf{F}, \mathbf{E}) = 0)$, and

(2) $(\forall \bar{u})(J_\varphi(\bar{u}, \zeta, \mathbf{F}, \mathbf{E}) = 0 \rightarrow \varphi(\bar{u}, \mathrm{Dom}\,\zeta, \mathbf{F}^{\circ}))$.

We proceed by induction on the construction of φ; we only exhibit relevant occurrences of variables and constants.

Case (a). $\varphi(\bar{u})$ is basic, not containing $\dot{X}$ nor $\dot{\mathbf{Z}}$.

Then $J_\varphi(\bar{u}) = 0$ if $\varphi(\bar{u})$ and is undefined otherwise.

Case (b). $\varphi(\bar{u})$ is $\tau(\bar{u}) \in X$.

Then $J_\varphi(\bar{u}) \simeq \zeta(\tau(\bar{u}))$.

Case (c). $\varphi(\bar{u})$ is $\{z : \psi(\bar{u}, z)\} \in \dot{\mathbf{Z}}$.

Then $J_\varphi(\bar{u}) \simeq J_\psi(\bar{u}, \langle 1, \mathbf{F}(\lambda t \cdot (\mu v)(J_\psi(\bar{u}, \langle 0, \langle t, v \rangle\rangle) = 0))\rangle)$.

Suppose $\{z : \psi(\bar{u}, z)\} \in \hat{\mathbf{F}}$. Then

$$(\forall t)(\exists ! v)\,\psi(\bar{u}, \langle 0, \langle t, v \rangle\rangle) .$$

So, assuming (1) holds with ψ for φ, the λ term in the definition of J_φ is a total function, η say, and $\psi(\bar{u},\langle 1,\mathbf{F}\eta\rangle)$ is true. But then $J_\varphi(\bar{u}) = 0$, so (1) holds.

Conversely, suppose $J_\varphi(\bar{u}) = 0$. Then the λ term must define a total function η, and so by (2) with ψ for φ, $\langle 0,\hat{\eta}\rangle \subseteq \{z : \psi(\bar{u},z)\}$. Similarly $\langle 1,\mathbf{F}\eta\rangle \in \{z : \psi(\bar{u},z)\}$. Hence $\{z : \psi(\bar{u},z)\} \in \mathbf{F}^\circ$; i.e. (2) holds.

Case (d). φ is $\psi_1 \wedge \psi_2$.

Then $\quad J_\varphi(\bar{u}) \simeq J_{\psi_1}(\bar{u}) \cdot J_{\psi_2}(\bar{u})$.

Case (e). φ is $\psi_1 \vee \psi_2$.

Then $\quad J_\varphi(\bar{u}) \simeq \{h\}(j_{\psi_1}, j_{\psi_2}, \bar{u}, \zeta, \mathbf{F}, \mathbf{E})$

where h is an index such that the RHS is defined with value 0 just in case

$$(\exists i<2)(\{j_{\psi_i}\}(\bar{u},\zeta,\mathbf{F},\mathbf{E}) = 0)\,.$$

Such an index h exists by Gandy's selection operator theorem. (For a proof see Moschovakis [5].)

Now we can apply the first recursion theorem (for a proof see Platek [1]) to obtain a partial function $\zeta \in \mathcal{R}(\mathbf{F},\mathbf{E})$ which is the minimal solution of

$$\zeta(x) \simeq J_\varphi(x,\zeta,\mathbf{F},\mathbf{E})\,.$$

But then, by (1), $(\Phi(\hat{\mathbf{F}}))^\infty \subseteq \mathrm{Dom}\,\zeta$, and by (2) $\mathrm{Dom}\,\zeta \subseteq (\Phi(\mathbf{F}^\circ))^\infty$. So, if Φ is functional at $\mathbf{F}$, $(\Phi(\mathbf{F}^\circ))^\infty = \mathrm{Dom}\,\zeta$, which suffices to prove (i).

To prove (ii) we note that we can strengthen the result about h (which shows that $\mathcal{R}(\mathbf{F},\mathbf{E})$ is closed under the 'strong or') to get a k (corresponding to 'strong definition by cases') satisfying

$$\begin{aligned}\{k\}(j_{\psi_1}, j_{\psi_2}, -) &= 0 \ \text{ if } \ \{j_{\psi_1}\}(-)\uparrow;\ \text{if not}\\ &= 1 \ \text{ provided } \ \{j_{\psi_2}\}(-)\uparrow\,.\end{aligned}$$

This completes the proof of the theorem.

3.5. *Discussion.* I believe I was the first to realize that $\mathcal{R}(\mathbf{F},\mathbf{E})$ (or $\mathcal{R}(\mathbf{F})$ if $\mathbf{E} \in \mathcal{R}(\mathbf{F})$) had a 'nice' recursion theory. Since then Sacks and his colleagues at MIT have carried it to stratospheric levels of sophistication (see, e.g. Sacks [1] and MacQueen [1]). Theorem 3.4.2. represents an attempt to explain why this was possible. It also suggests a different way of developing the theory. Namely one starts from $(\mathrm{Fn} - \mathbf{\Delta}_0^1 + (\mathbf{F}^\circ)^{(\infty)}$ and then follows the lines suggested in 3.1., checking functionality where necessary.

The theorem is not easily weakened or generalised. If we drop the requirement $\mathbf{F} \in \mathcal{K}$ then we cannot any longer use $\mathbf{E}$ to compare ordinals of computations. However, the use of Hinman's $\mathbf{E}^{\#}$ ($\mathbf{E}^{\#}\zeta = 0$ if $(\exists x)(\zeta x = 0)$, $= 1$ if $(\forall x)(\zeta x > 0)$, undefined otherwise) will yield an h as in (c) above, so that (i) holds with $\mathbf{E}^{\#}$ replacing $\mathbf{E}$. But, as Platek pointed out, a functional $\mathbf{K}$ which satisfies $\mathbf{K}\zeta\eta x = 0$ if $\zeta(x) = 0$, $= 1$ if $\eta(x) = 0$ and $\zeta(x)$ undefined, is not consistent. So it seems that to restore (ii) in this case one would need to introduce a multi-valued search operator. One would then be operating with relations rather than functions and it would seem more plausible to take the inductive definitions as primitive rather than to base them on an artificial notion of computability or recursiveness cooked up to make (ii) true.

Let us consider now extending the theorem to structures other than $\mathfrak{N}$. Theorem 3.3.1. (ii) will fail, so the inductive definitions to consider will be $\mathrm{Fn} - \mathbf{\Sigma}_1^0 + (\mathbf{F}^\circ)$. As Grilliot has shown in [1] the existential quantifier in the inductive definitions calls for a search operator in the recursion theory. One might expect to restore the theorem then, by replacing $\mathbf{\Delta}_0^0$ by $\mathbf{\Sigma}_1^0$, and interpreting $\mathcal{R}(\mathbf{F})$ as 'search-computable in $\mathbf{F}$'. But this means that $\mathbf{F}$ must act on many-valued functions – i.e. relations, and the last sentence of the previous paragraph may be re-applied.

Suppose now we take $\mathcal{R}$ as given, and seek for an appropriate class of inductive definitions. In particular we consider trying to find $\mathcal{C}$ so that for $\mathbf{F} \in \mathcal{K}$

$$(1) \qquad (\mathcal{C}(\mathbf{F}^\circ))^{(\infty)} = \mathcal{R}^+(\mathbf{F})\,.$$

Grilliot [4] has shown that $\mathcal{R}^+(\mathbf{F})$ is not closed under union, so $\mathcal{C}$ cannot be closed under $\vee$. At first sight this seems to make the case hopeless, as one needs $\vee$ to connect separate clauses. However, examination of the clauses in the inductive definitions of $D(\mathbf{F})$ and $V(\mathbf{F})$ show that they are deterministic in the sense that if e.g., $\langle z, \bar{x}\rangle \in D$ then the relevant clause is determined by z. Indeed Kleene has chosen the definition of 'index' so that this shall be so. But, alas, there is still a non-deterministic feature. The clause for S4 (substitution) has the form

$$(2) \qquad (\exists y)(\langle\langle z_1, \bar{u}\rangle, y\rangle \in V \wedge \langle z_2, \bar{u}, y\rangle \in D) \to \langle z, \bar{u}\rangle \in D$$

where z_1 and z_2 are determined by projection functions from z. We cannot eliminate the $\exists$ by counting down, since this uses a non-deterministic $\vee$.

Compare (1) with

(3) $(\exists i<2)(\langle z,i,\bar{u}\rangle \in D) \rightarrow \langle z,\bar{u}\rangle \in X$.

It seems almost impossible to imagine a criterion which would permit (2) but reject (3). On the other hand (3) appears to lead one outside $\mathcal{R}^+(\mathbf{F})$. To be more precise, what Grilliot shows is that there is no method *uniform in* **F** for presenting the union of two members of $\mathcal{R}^+(\mathbf{F})$ as a member of $\mathcal{R}^+(\mathbf{F})$. So one poses:

Problem 10. Is there an **F** such that $\mathcal{R}^+(\mathbf{F})$ is not closed under union?

To sum up. For an arbitrary structure the classes $(\mathrm{Fn} - \mathbf{\Delta}_0^1 + (\mathbf{F}^\circ))^{(\infty)}$ $(\mathrm{Fn} - \mathbf{\Sigma}_1^0 + (\mathbf{F}^\circ))^{(\infty)}$ promise to have nice closure properties, and be amenable to a variety of techniques. For $\mathfrak{N}$ when **E** is recursive in $\mathbf{F} \in \mathcal{X}$, the classes coincide with $\mathcal{R}^+(\mathbf{F})$. When **E** is not recursive in **F** the techniques cannot be applied and $\mathcal{R}^+(\mathbf{F})$ appear to behave very differently. (I think that all that is known is contained in Grilliot [6]). It is not yet clear whether in this case $\mathcal{R}^+(\mathbf{F})$ should be regarded as merely pathological.

§4. Discussion

A number of important or interesting questions received scant or zero attention in §1–3. We indicate some of these by listing further problems. Then we discuss the possibility of applying the theory of inductive definitions to proof theory.

Problem 11. Find conditions which ensure that $(\mathcal{C}m)^{(\infty)} = (\mathcal{C}+)^{(\infty)}$.

Spector showed that for case (H) of 1.3. the equation holds for all relevant $\mathcal{C}$. I had hoped that semantic tableaux would provide a proof for the case considered there; maybe they can be forced to do this. An entirely different approach would be along the lines suggested by Feferman (cf. 1.4.5.). One considers $\mathcal{C}$ and $\mathcal{C}^{(\infty)}$ not just for $\mathfrak{A}$, but for a class of models for the theory of $\mathfrak{A}$. For this class one can apply Lyndon's theorem to get $\mathcal{C}m = \mathcal{C}+$. One then has to work one's way back – via semi-invariant definability – to $(\mathcal{C}+)^{(\infty)}$ for $\mathfrak{A}$ itself.

The next problem is concerned with generalising the boundeness theorems of hyperarithmetic and hyperprojective theory. Let

$$\mathcal{C}^{(<\infty)} = \{R : (\exists \bar{a} \in {}^nA)(\exists \Phi \in \mathcal{C})(\exists \alpha < |\mathcal{C}|)$$
$$(\forall \bar{x} \in {}^mA)(R\bar{x} \leftrightarrow \Phi^\alpha \bar{a}\bar{x})\} .$$

Problem 12. Under what conditions is $\mathcal{C}^{(<\infty)} = \mathcal{C}^{(\pm\infty)}$?

Moschovakis [4] shows, for arbitrary $\mathfrak{A}$, that the equation holds for $\mathcal{C} = \mathbf{\Delta}_0^1+$. It is obviously not true (in general) if $\mathcal{C} = \mathbf{\Sigma}_1^0+$. The problem is important because when the equation holds one has a natural definition for 'finiteness' in $\mathcal{C}^{(\infty)}$.

Problem 13. Give a direct proof that, for countable admissible A, $(\mathbf{\Sigma}_1+/A)^{(\infty)} = (s - \mathbf{\Pi}_1^1+/A)^{(\infty)}$.

Here '$s - \mathbf{\Pi}_1^1$' is 'strict $\mathbf{\Pi}_1^1$' as introduced by Barwise in [2] (i.e., formulae which can be put in the form $(\forall X \subseteq A)\varphi$ with $\varphi \in \mathbf{\Sigma}_1$). Present proofs use the fact that the validity predicate is $s - \mathbf{\Pi}_1^1$ and then apply the completeness theorem for $\mathbf{L}_{\omega_1,\omega}$. It is even possible that the above equation holds for uncountable A; although then it is known that in general $|s - \mathbf{\Pi}_1^1+|$ is much greater than $|\mathbf{\Sigma}_1+|$. If the method of semantic tableaux could be extended to $s - \mathbf{\Pi}_1^1$ it should provide answers to this and other problems.

In §1–3 we used languages which contained constants for all members of A. There are many cases in which this is much too crude and will prevent interesting distinctions from being made. (For some examples, see Hinman & Moschovakis [1].) In [1] and [2] Moschovakis systematically allowed for the use of parameters *only* from a given $B \subseteq A$. If one uses the method of embedding (1.4.4.) then one would also restrict the use of parameters in the axioms of KPU. There is not exactly a problem here; just a line to pursue. Another very obvious line is to investigate what happens to the results and problems of §3 if one relativises to an '$\mathbf{F}$' of type higher than 2. One should then consider not only inductively defined subsets of A ('1-sections'), but inductively defined subsets of $\mathsf{P}^m(A)$. (See Grilliot [4] and [5].)

There is one part of mathematical logic which is wholly concerned with inductively defined sets; namely that part of proof theory which is concerned with the study of formal proofs. At a first encounter it may look as if no kind of general study of the *sorts* of inductive definition used will be of much use here: one studies a particular formal system, and evolves methods which may be specific to it. And it could be that a theory of inductive definitions has little or nothing to contribute. One does not, after all, expect such a theory to tell one anything interesting about a particular inductively defined subset of N such as the prime numbers.

But I think recent work does suggest that there are certain *patterns* or *forms* of inductive definition waiting to be discovered. The intuitive definitions I have in mind are: (a) of cut-free (infinite) proof trees (see, for examples, Schütte [1]); (b) of reducible or computable proofs (ranging from Gentzen's original consistency proof to e.g., Martin-Löf [1]); (c) of ordinal notations and their ordering relations. It should be observed that even when the end product of the definition is in fact a recursively enumerable or recursive set, the inductive definitions used are usually Π^0_1 (not necessarily monotonic – see, for example, Martin-Löf's definition of 'computable'). What one might hope from a theory of forms is that inductive definitions of the same form would have the same closure ordinal, and would endow it with the same structure (e.g., by exhibiting the same pattern of principal sequences or by defining the same functions on it). The evidence for optimism is as follows. First, it is often easy to spot the irrelevant elements in a given definition, and also to discover which are the clauses which really 'do the work'. Second, the same ordinals with, at least approximately, the same structure turn up in numerous different contexts (e.g. ϵ_0, Γ_0, $F(\epsilon^{\Omega_1+1})$). One has the feeling, often, that this is no accident, without quite being able to discern the cause. Third, the method originated by Bachmann [1] and since extensively elaborated (for example, in Pfeiffer [1] and Isles [1]) establishes natural-seeming connections between various inductively defined sets of notations and sets (in particular the finite number classes) for which there is an obvious classification. (A simplification of Bachmann's method, due to Feferman and Aczel, has been explored, described and extended in Bridge [1].) A further, and less arbitrary connection with a known classification is provided by Martin-Löf's conjecture (in [1]) that the provable well-orderings of his *formal* system for n-fold iterated Π^0_1+ inductive definitions have precisely those ordinals for which Pfeiffer *et al.* provided notations using the $n+1$-th number class. Now the closure ordinal of n-fold iterated Π^0_1+ inductive definitions is precisely the $n+1$-st admissible ordinal, which is the 'recursive' analogue of $\aleph_n$.

Martin-Löf's system and his normalisation procedure have a naturalness and transparency (as compared, say, with work on Π_1^1-analysis) which suggest that it should be possible to find the *reason* for the connection.

Lastly, I quote an example where a 'form' of inductive definition can actually be given, rather than merely hinted at. Richter [1] introduced the operation $[\Phi_1, \Phi_2]$; this acts like Φ_1 on sets which are not closed under Φ_1, and like Φ_2 on those which are. The idea can readily be extended to *finite* sequences: $[\Phi_1, \Phi_2, ..., \Phi_n]$. If the Φ_i are Σ_1^0+, then the closure ordinal of the resulting Π_1^0+ operation is $\leqslant \omega^n$. Schmidt, in her dissertation [1], has shown how ϵ_0 can be characterised using *transfinite* sequences of Σ_1^0+ operations, labelled by previously introduced notations. The resulting 'form' is not ideal, since it does not cover other Π_1^0+ inductive definitions which are known to close off at ϵ_0. But it suggests a promising direction for further investigation.

References

P. Aczel [1], Stage comparison theorems and game playing with inductive definitions (to appear).

P. Aczel and W. Richter [1], Inductive definitions and analogues of large cardinals, in: Conference in Mathematical Logic, London '70, Springer Lecture Notes in Maths. 255 (1972) 1–9.

H. Bachmann [1], Die Normalfunktionen und das Problem der ausgezeichneten Folgen von Ordnungzahlen, Vierteljschr. Naturforsch. Ges. Zürich 95 (1950) 115–147.

K.J. Barwise [1], Infinitary Logic and admissible sets, J. Symbolic Logic 34 (1969) 226–252.

K.J. Barwise [2], Implicit definability and compactness in infinitary languages, in: The Syntax and Semantics of Infinitary Languages, Springer Lecture Notes in Maths. 72 (1968) 1–34.

K.J. Barwise [3], this volume.

K.J. Barwise, R.O. Gandy and Y.N. Moschovakis [1], The next admissible set, J. Symbolic Logic 36 (1971) 108–120.

E.W. Beth [1], Semantic construction of intuitionistic logic, Mededelingen der Kon. Ned. Akad. v. Wet., new series 19 (1950) no. 11.

J.E. Bridge [1], Some problems in mathematical logic (systems of ordinal functions and ordinal notations) D. Phil thesis, Oxford, 1972.

S. Feferman [1], Uniform inductive definitions and generalized recursion theory, ASL meeting, Cleveland, Ohio, April 30, 1969.

R.O. Gandy [1], General recursive functions of finite type and hierarchies of functions, Annales de la Faculté des Sciences de l'Université de Clermont, Maths. 4 Fascicule (1967) 5–24.

C.E. Gordon [1], A comparison of abstract computability theories, Ph.D. dissertation, UCLA, 1968.

R. Gostanian [1], The next admissible ordinal, Ph.D. dissertation, N.Y. University, 1971.
T.J. Grilliot [1], Inductive definitions and computability, Trans. Amer. Math. Soc. 158 (1971) 309–317.
T.J. Grilliot [2], Implicit definability and hyperprojectivity, Scripta Mathematica, to appear.
T.J. Grilliot [3], Selection functions for recursive functionals, Notre Dame Journal of Formal Logic 10 (1969) 225–234.
T.J. Grilliot [4], Recursive Functions of Finite Higher Types, Ph.D. Dissertation, Duke University, 1967.
T.J. Grilliot [5], Hierarchies based on objects of finite type, J. Symbolic Logic 34 (1969) 177–182.
T.J. Grilliot [6], On effectively discontinuous type-2 objects, J. Symbolic Logic 36 (1971) 245–248.
T.J. Grilliot [7], Abstract recursion theory: a summary, this volume.
A. Grzegorczyk, A. Mostowski and C. Ryll-Nardzewski [1], The classical and the ω-complete arithmetic, J. Symbolic Logic 23 (1958) 188–206.
D. Isles [1], Regular ordinals and normal forms, in: Intuitionism and Proof Theory, A. Kino et al. (eds) (North-Holland, Amsterdam, 1970) 339–361.
R.B. Jensen and C. Karp [1], Primitive recursive set functions, Proc. Symposium Pure Math. Amer. Math. Soc. 13, Part I (1971) 143–176.
P.G. Hinman and Y.N. Moschovakis [1], Computability over the continuum, in: Logic Colloquium '69, R.O. Gandy, C.E.M. Yates (eds.) (North-Holland, Amsterdam, 1971) 77–105.
S.C. Kleene [1], Recursive functionals and quantifiers of finite types I, Trans. Amer. Math. Soc. 91 (1959) 1–52.
G. Kreisel [1], Model theoretic invariants, in: The Theory of Models, J. Addison et al. (eds.) (North-Holland, Amsterdam, 1965) 190–205.
G. Kreisel [2], Some reasons for generalising recursion theory, in: Logic Colloquium '69, R.O. Gandy, C.E.M. Yates (eds.) (North-Holland, Amsterdam, 1971) 139–198.
K. Kunen [1], Implicit definability and infinitary languages, J. Symbolic Logic 33 (1968) 446–451.
E.G.K. Lopez-Escobar [1], An interpolation theorem for denumerably long formulas, Fundamenta Math. 58 (1965) 254–272.
D.B. MacQueen [1], Post's problem for recursion in higher types, Ph.D. dissertation, M.I.T. 1972.
P. Martin-Löf [1], Hauptsatz for the intuitionistic theory of iterated inductive definitions, in: Proc. 2nd Scandinavian Logic Symposium, J.-E. Fenstad (ed.) (North-Holland, Amsterdam, 1971) 179–216.
R. Montague [1], Recursion theory as a branch of model theory, in: Logic, Methodology and Philosophy of Science III, B. van Rootselaar, J. Staal (eds.) (North-Holland, Amsterdam, 1968) 63–86.
Y.N. Moschovakis [1], Abstract first order computability I, Trans. Amer. Math. Soc. 138 (1969) 427–464.
Y.N. Moschovakis [2], Abstract first order computability II, Trans. Amer. Math. Soc. 138 (1969) 465–504.
Y.N. Moschovakis [3], Abstract computability and invariant definability, J. Symbolic Logic 34 (1969) 605–633.
Y.N. Moschovakis [4], Elementary induction on abstract structures (North-Holland, Amsterdam, 1973).

Y.N. Moschovakis [5], Hyperanalytic predicates, Trans. Amer. Math. Soc. 129 (1967) 249–282.
A. Mostowski [1], Representability of sets in formal systems, Proc. Symposium Pure Math. Amer. Math. Soc. 5 (1962) 29–48.
H. Pfeiffer [1], Ausgezeichnete Folgen für gewisse Abschnitte der zweiten und weiterer Zahlklassen, Dissertation, Hannover, 1964.
R. Platek [1], Foundations of Recursion Theory, Ph.D. Dissertation, Stanford University, 1966.
W. Richter [1], Recursively Mahlo ordinals and inductive definitions, in: Logic Colloquium '69, R.O. Gandy, C.E.M. Yates (eds.) (North-Holland, Amsterdam, 1971) 273–288.
G.E. Sacks [1], this volume.
D. Schmidt [1], Topics in mathematical logic (characterisations of small constructuve ordinals; constructive finite number classes) D. Phil. thesis, Oxford, 1972.
K. Schütte [1], Beweistheorie (Springer, Berlin, 1960).
R.M. Smullyan [1], Theory of formal systems (Princeton University Press, 1961).
C. Spector [1], Inductively defined sets of natural numbers, in: Infinitistic Methods (Pergamon, Oxford, 1961) 97–102.

Note added in proof. Moschovakis has recently made a very considerable extension of the method of embedding. He has shown that if $A \subseteq V$ is a 'nice' transitive set, and if $\mathcal{C}$ is any class of inductive operators over A satisfying certain rather weak closure conditions, then there is an admissible set A^* such that

$$\mathcal{C}^{(\infty)} = \Sigma_1/A^* .$$

[1] Note that the existential quantifer in N should also be exposed, and a bound (for $\theta \in s$) assigned to it.

J.E.Fenstad, P.G.Hinman (eds.), Generalized Recursion Theory
© North-Holland Publ. Comp., 1974

INDUCTIVE DEFINITIONS AND REFLECTING PROPERTIES OF ADMISSIBLE ORDINALS

Wayne RICHTER
The University of Minnesota

and

Peter ACZEL
University of Oslo and Manchester University

Contents

Introduction

An operator or inductive definition (i.d.) $\Gamma : P(\omega) \to P(\omega)$ determines a transfinite sequence $\langle \Gamma^{\xi} : \xi \in \mathrm{ON} \rangle$ of subsets of ω, where $\Gamma^{\lambda} = \mathbf{U}\{\Gamma(\Gamma^{\xi}) : \xi < \lambda\}$. The *closure ordinal* $|\Gamma|$ of Γ is the least ordinal λ such that $\Gamma^{\lambda+1} = \Gamma^{\lambda}$. The set *defined by* Γ is $\Gamma^{\infty} = \Gamma^{|\Gamma|}$. Γ is *monotone* if $\Gamma(X) \subseteq \Gamma(Y)$ whenever $X \subseteq Y \subseteq \omega$. For monotone Γ we have

$$\Gamma^{\infty} = \bigcap \{X : \Gamma(X) \subseteq X\} .$$

Monotone inductive definitions have long been used in logic and in particular in recursion theory. For example the definitions of the terms, formulas and theorems of predicate logic may be naturally formulated as monotone inductive definitions. More generally Post's production systems give a wide class of monotone inductive definitions, for defining sets of strings of symbols in a finite alphabet. These lead to a natural characterisation of the class of recursively enumerable sets of integers. All these inductive definitions have closure ordinal $\leq \omega$. But inductive definitions with larger closure ordinals may also be considered, and they determine notation systems for ordinals in the following way. For each $x \in \Gamma^{\infty}$ let $|x|_{\Gamma}$ be the least ordinal λ such that $x \in \Gamma^{\lambda+1}$. Then $\langle \Gamma^{\infty}, | \; |_{\Gamma} \rangle$ is a notation system for the ordinal $|\Gamma|$. Note that because $| \; |_{\Gamma}$ maps Γ^{∞} onto $|\Gamma|$, $|\Gamma|$ must be a countable ordinal. For example let Λ be the i.d.:

$$\Lambda(X) = \{1\} \cup \{2^x : x \in X\} \cup \{3.5^e : \forall n [e](n) \in X\} ,$$

where $[e]$ is the e'th primitive recursive function, in a standard recursive enumeration of them. Then $\langle \Lambda^{\infty}, | \; |_{\Lambda} \rangle$ is a slightly modified version of Kleene's system of notations for the recursive ordinals, i.e. Λ^{∞} is a complete $\mathbf{\Pi}_1^1$ set such that $|\Lambda| = \omega_1$, the first non-recursive ordinal. Note that Λ is monotone. Certain monotone i.d.'s are basic to Kleene's definition of recursion in higher type objects, [9]. Also monotone i.d.'s are extensively investigated in [13].

As ω_1 is a constructive analogue of the first uncountable ordinal, it was natural to try to formulate constructive analogues for larger initial ordinals by constructing systems of notations for them. This led to the use of non-monotone i.d.'s. (See [14] and [15].) An independent development led to the Kripke–Platek theory of recursion on admissible ordinals. These ordinals are a constructive analogue of the regular ordinals, the first two admissible ordinals being ω and ω_1.

The main aim of this paper is to formulate constructive analogues for large regular ordinals, and to obtain notation systems for them using non-monotone inductive definitions.

Our results will be concerned with classes $\mathcal{C}$ of those i.d.'s that are definable in a certain way. Thus we say that the i.d. Γ is $\mathbf{\Pi}^n_m$ if $\{(x,X) \in \omega \times P(\omega) : x \in \Gamma(X)\}$ is definable by a $\mathbf{\Pi}^n_m$ formula in the language of finite types over arithmetic. Similarly we define the classes of $\mathbf{\Sigma}^n_m$ and $\mathbf{\Delta}^n_m$ i.d.'s. For example the i.d.'s involved in Post's production systems are all $\mathbf{\Sigma}^0_1$ when coded on ω. The operator Λ, above, is an example of a $\mathbf{\Pi}^0_1$ monotone i.d. We shall write $\mathbf{\Pi}^n_m$-mon for the class of monotone $\mathbf{\Pi}^n_m$ i.d.'s. Similarly for $\mathbf{\Sigma}^n_m$-mon and $\mathbf{\Delta}^n_m$-mon. Given a class $\mathcal{C}$ of i.d.'s we will be interested in $|\mathcal{C}| = \mathrm{Sup}\{|\Gamma| : \Gamma \in \mathcal{C}\}$ and $\mathrm{Ind}(\mathcal{C}) = \{X \subseteq \omega : X \leq_m \Gamma^\infty$ for some $\Gamma \in \mathcal{C}\}$. Here $X \leq_m Y$ means that X is many-one reducible to Y.

In many cases $|\mathcal{C}|$ can be compared with $\omega(\mathcal{R})$ for a suitably chosen class $\mathcal{R}$ of relations on ω. $\omega(\mathcal{R})$ is defined to be the sup. of the order types of the well-ordering relations in $\mathcal{R}$. Thus it is well-known that $\omega_1 = \omega(\mathbf{\Delta}^0_1) = \omega(\mathbf{\Delta}^1_1)$.

We next list some of the earlier results on the ordinals of i.d.'s.

Proposition. (i) $|\mathbf{\Pi}^0_0| = |\mathbf{\Sigma}^0_1| = \omega$;

(ii) (Spector [20]) $|\mathbf{\Pi}^0_1\text{-mon}| = |\mathbf{\Pi}^1_1\text{-mon}| = \omega_1$;

(iii) (Gandy, unpublished) $|\mathbf{\Pi}^0_1| = \omega_1$;

(iv) (Richter, [16]) $|\mathbf{\Pi}^0_2|$ is a large admissible ordinal; e.g. much larger than the first recursively Mahlo ordinal;

(v) (Putnam, [14]) $|\mathbf{\Delta}^1_2| = \omega(\mathbf{\Delta}^1_2)$;

(vi) (Gandy, unpublished) $|\mathbf{\Sigma}^1_2\text{-mon}| = \omega(\mathbf{\Delta}^1_2)$.

We now summarise our results. π^n_m is the least ordinal λ such that $\langle L_\lambda, \in\rangle$ reflects every $\mathbf{\Pi}^n_m$ sentence. σ^n_m is defined using $\mathbf{\Sigma}^n_m$ sentences. For a precise definition see § 1.

Theorem A. $|\mathbf{\Pi}^0_m| = |\mathbf{\Sigma}^0_{m+1}| = \pi^0_{m+1} = \sigma^0_{m+2}$.

Theorem B. $|\mathbf{\Pi}^1_1| = \pi^1_1$ *and* $|\mathbf{\Sigma}^1_1| = \sigma^1_1$.

In general the characterisations of the closure ordinals of i.d.'s must be more complicated. If A is a relation on ordinals let $\pi^n_m(A)$ be the least ordinal λ such that $\langle L_\lambda[A], \in, A\rangle$ reflects every $\mathbf{\Pi}^n_m$ sentence. Similarly for $\sigma^n_m(A)$. Let $\pi^n_m(\Gamma) = \pi^n_m(A_\Gamma)$ where $A_\Gamma = \{(n,\alpha) : \alpha \in \mathrm{ON}\ \&\ n \in \Gamma^\alpha\}$. Also let $\sigma^n_m(\Gamma) = \sigma^n_m(A_\Gamma)$.

Theorem C. *For $m, n > 0$*

(i) $|\boldsymbol{\Pi}^n_m| = \pi^n_m(\Gamma)$ *where* Γ *is complete* $\boldsymbol{\Pi}^n_m$.

(ii) $|\boldsymbol{\Sigma}^n_m| = \sigma^n_m(\Gamma)$ *where* Γ *is complete* $\boldsymbol{\Sigma}^n_m$.

The proofs of the above characterisations actually give much more information. In each case, as well as characterising $|\mathcal{C}|$ we may also characterise $\text{Ind}(\mathcal{C})$. In the next result we use the notion of a "closed" class $\mathcal{C}$. Each $\boldsymbol{\Pi}^n_m$ and $\boldsymbol{\Sigma}^n_m$ is a closed class for $m > 0$ and every closed class $\mathcal{C}$ has a "-complete" element. See §8 for a definition of this notion.

Theorem D. *If* $\mathcal{C}$ *is a closed class* $\supseteq \boldsymbol{\Pi}^0_1$, Γ *is* $\mathcal{C}$*-complete and* $\lambda = |\mathcal{C}|$ *then*

(i) λ *is admissible relative to* A_Γ;

(ii) λ *is projectible to* ω *relative to* A_Γ;

i.e. there is a λ*-recursive in* A_Γ *injection* $f : \lambda \to \omega$;

(iii) $\text{Ind}(\mathcal{C}) = \{X \subseteq \omega : X \text{ is } \lambda\text{-r.e. relative to } A_\Gamma\}$.

When $\mathcal{C}$ is "sufficiently absolute" then $A_\Gamma \restriction \lambda$ is λ-recursive, so that the relativisation to A_Γ may be omitted in the statement of Theorem D. This is the case when $\mathcal{C}$ is $\boldsymbol{\Pi}^0_{m+1}, \boldsymbol{\Sigma}^0_{m+2}, \boldsymbol{\Pi}^1_1$ or $\boldsymbol{\Sigma}^1_1$.

Results along the lines of Theorems C and D have been recently obtained, independently, by Moschovakis. Moreover he has generalised them to classes of inductive definitions on arbitrary abstract structures.

In the next result we locate the ordinals of inductive definitions in relation to the ordinals of certain wellorderings. Part (i) has been independently obtained by Cenzer (see [5]).

Theorem E. *Let* $m, n > 0$.

(i) *If* $m+n > 2$ *then* $|\boldsymbol{\Delta}^n_m| = \omega(\boldsymbol{\Delta}^n_m)$.

$$\text{(ii)}\ |\boldsymbol{\Delta}^n_{m+1}| \left\{ \begin{array}{l} > |\boldsymbol{\Pi}^n_m| \geq \pi^n_m \geq \omega(\boldsymbol{\Sigma}^n_m) \geq \\ \\ > |\boldsymbol{\Sigma}^n_m| \geq \sigma^n_m \geq \omega(\boldsymbol{\Pi}^n_m) \geq \end{array} \right\} \omega(\boldsymbol{\Delta}^n_m).$$

Note that $m+n > 2$ is essential in (i) as $|\boldsymbol{\Delta}^1_1| \geq |\boldsymbol{\Pi}^0_2| > \omega_1 = \omega(\boldsymbol{\Delta}^1_1)$. When $m+n > 2$ Sacks has shown in [18] that $\omega(\boldsymbol{\Delta}^n_m)$ is a stable ordinal, so that $|\boldsymbol{\Delta}^n_m|$ is stable. It might be conjectured from this that $|\boldsymbol{\Delta}^1_1|$ is at least admissible. But we have:

Theorem F. $|\boldsymbol{\Delta}_1^1|$ *is not admissible.*

The diagram in (ii) of Theorem E leaves open the order relationship between several pairs of ordinals of the diagram.

The next result, obtained independently by Aanderaa in [1], gives us some more information.

Theorem G. *If* $m, n > 0$ *then* $|\boldsymbol{\Pi}_m^n| \neq |\boldsymbol{\Sigma}_m^n|$.

When $m = n = 1$ this result was first proved by showing directly that $\pi_1^1 \neq \sigma_1^1$. But we do not know if $\pi_m^n \neq \sigma_m^n$ for $n + m > 2$.

The proof of Theorem G is symmetric between $\boldsymbol{\Pi}_m^n$ and $\boldsymbol{\Sigma}_m^n$ and hence gives no information on the relative magnitudes of the two ordinals.

This is explained by the following result of Aanderaa, which we state here for completeness. (See [1].) PW($\mathcal{C}$) denotes that $\mathcal{C}$ has the pre-wellordering property. See [1] for a precise definition.

Theorem (Aanderaa). *If* $m, n > 0$ *then*

(i) $\mathrm{PW}(\boldsymbol{\Pi}_m^n) \Rightarrow |\boldsymbol{\Pi}_m^n| < |\boldsymbol{\Sigma}_m^n|$,

(ii) $\mathrm{PW}(\boldsymbol{\Sigma}_m^n) \Rightarrow |\boldsymbol{\Sigma}_m^n| < |\boldsymbol{\Pi}_m^n|$.

The following summarises what is known about when the pre-wellordering property holds.

Proposition. (i) $\mathrm{PW}(\boldsymbol{\Pi}_1^1)$ *and* $\mathrm{PW}(\boldsymbol{\Sigma}_2^1)$,

(ii) V = L *implies* $\mathrm{PW}(\boldsymbol{\Sigma}_m^1)$ *for* $m > 2$,

(iii) PD *implies* $\mathrm{PW}(\boldsymbol{\Pi}_{2m+1}^1)$ *and* $\mathrm{PW}(\boldsymbol{\Sigma}_{2m+2}^1)$ *for* $m > 0$.

Here PD denotes the axiom of projective determinacy. It follows that $|\boldsymbol{\Pi}_1^1| < |\boldsymbol{\Sigma}_1^1|$ and $|\boldsymbol{\Sigma}_2^1| < |\boldsymbol{\Pi}_2^1|$.

This should be compared with the following. (See [21] for more details.)

Proposition. (i) $\omega(\boldsymbol{\Sigma}_1^1) = \omega(\boldsymbol{\Delta}_1^1) = \omega_1$ *and* $\omega(\boldsymbol{\Pi}_1^1) = \omega_1^+$, *where* α^+ *is the first admissible ordinal* $> \alpha$;

(ii) $\omega(\boldsymbol{\Pi}_2^1) = \omega(\boldsymbol{\Delta}_2^1)$ *and* $\omega(\boldsymbol{\Delta}_2^1) < \omega(\boldsymbol{\Sigma}_2^1) < \omega(\boldsymbol{\Delta}_2^1)^+$;

(iii) V = L *implies* $\omega(\boldsymbol{\Delta}_m^1) = \omega(\boldsymbol{\Pi}_m^1) < \omega(\boldsymbol{\Sigma}_m^1)$ *for* $m > 2$.

These results lead to an improvement of Theorem E(ii) in certain cases. For example we have $\sigma_1^1 = |\mathbf{\Sigma}_1^1| > |\mathbf{\Pi}_1^1| = \pi_1^1 > |\mathbf{\Delta}_1^1| > \omega(\mathbf{\Pi}_1^1) > \omega(\mathbf{\Sigma}_1^1) = \omega(\mathbf{\Delta}_1^1)$, $|\mathbf{\Pi}_2^1| > |\mathbf{\Sigma}_2^1| \geq \sigma_2^1 > \omega(\mathbf{\Sigma}_2^1) > \omega(\mathbf{\Pi}_2^1) = \omega(\mathbf{\Delta}_2^1) = |\mathbf{\Delta}_2^1|$, and $|\mathbf{\Pi}_2^1| \geq \pi_2^1 > \omega(\mathbf{\Sigma}_2^1)$.

Whether $|\mathbf{\Sigma}_2^1| = \sigma_2^1$ or $|\mathbf{\Pi}_2^1| = \pi_2^1$ remains open. Also, the relationship between π_2^1 and $|\mathbf{\Sigma}_2^1|$ or σ_2^1 is not known.

The paper is divided into two parts. In Part I we give alternative characterisations of some of the reflecting properties, compare them with the reflecting properties for the indescribable cardinals, and investigate their relative magnitudes. See § 1 for a survey of the definitions and results of Part I. This part makes no use of inductive definitions and may be read without reference to Part II.

In Part II we prove the results stated in this introduction. Most of this part depends only on § 1 of Part I, so that the reader mainly interested in inductive definitions can probably omit the other sections of Part I on a first reading.

In § 7 we examine first order inductive definitions. In particular we give upper bounds to their ordinals, proving half of Theorem A. For the other half we need the construction introduced in § 8. In this section we formulate the notion of a closed class of operators. The construction of the notation systems $\mathcal{M}^{\Theta}$ and the associated coding lemma are the key to getting lower bounds for the ordinals of inductive definitions, and to proving Theorem D. The coding lemma is proved in the appendix. The proof of Theorem A is completed in § 9. Theorems B and C are proved in § 10, while § 11 has proofs of Theorems E, F and G. Many of the results in this paper were first announced in [2].

PART I. REFLECTING PROPERTIES

§ 1. Summary of definitions and results

In this part we shall study some of the classes of ordinals that will be used to characterise the ordinals associated with inductive definitions. These classes of ordinals will be defined in terms of certain "reflecting properties" closely analogous to those used in defining the indescribable cardinals of Hanf and Scott. (See [7] and also Lévy's [11] for a detailed discussion.) In order to bring out this analogy we shall start by considering the indescribable cardinals.

We shall use a perhaps excessively large language $\mathcal{L}$ within which we can conveniently formulate all our reflection properties. $\mathcal{L}$ has the usual propositional connectives, and has variables and quantifiers for all finite types (variables of type 0 range over individuals, those of type 1 range over sets of individuals, etc.). $\mathcal{L}$ also has a name (individual constant) for each set and a name (relation symbol) for each relation on sets. (We will use the same symbol for the object and its name.) In particular $\in$ will denote the membership relation between sets. The restricted quantifiers $(\forall x \in y), (\exists x \in y)$ are defined in the usual way. Formulae of $\mathcal{L}$ may be classified according to their prenex form. When doing this we shall follow Lévy in ignoring restricted quantifiers that do not bound unrestricted quantifiers. A formula is $\mathbf{\Pi}^n_m$ ($\mathbf{\Sigma}^n_m$) if it is logically equivalent to a formula in prenex form which first has m alternating blocks of type n universal and existential quantifiers starting with a block of universal (existential) quantifiers and then has quantifiers of types $<n$ and restricted quantifiers. The allowance for restricted quantifiers is of course only significant when $n = 0$. If a $\mathbf{\Pi}^n_m$ formula φ contains no constants then we call it a $\mathbf{\Pi}^{n-}_m$ formula. Similarly for $\mathbf{\Sigma}^{n-}_m$.

If $R_1, ..., R_n, a_1, ..., a_m$, are the relation symbols and individual constants occurring in a sentence φ of $\mathcal{L}$ let $A \models \varphi$ denote that A is a non-empty set such that $a_1, ..., a_m \in A$ and φ is true in the structure $\langle A, R_1 \restriction A, ..., R_n \restriction A, a_1, ...$ $..., a_m \rangle$.

We can now define the (weak) indescribables.

1.1. Definition. Let $X \subseteq \mathrm{On}$ and $\alpha \in \mathrm{On}$. If φ is a sentence of $\mathcal{L}$ then α *reflects* φ *on* X if

$$\alpha \models \varphi \Rightarrow (\exists \beta \in X \cap \alpha)\, \beta \models \varphi\, .$$

(Note that On is the class of all ordinals and that we indentify an ordinal with its set of predecessors.) α *reflects* φ if α reflects φ on On.

α *is* $\mathbf{\Pi}^n_m$ ($\mathbf{\Sigma}^n_m$)*-indescribable* [*on* X] if α reflects [*on* X] every $\mathbf{\Pi}^n_m$ ($\mathbf{\Sigma}^n_m$) sentence of $\mathcal{L}$.

Some properties of the indescribable cardinals are summarised in the following theorems. Proofs of most of these may be found in [11].

1.2. Theorem. α *is* $\mathbf{\Pi}^0_2$*-indescribable* $\Longleftrightarrow \alpha > \omega$ *is regular.*

Let $\mathrm{Rg} = \{\alpha > \omega : \alpha \text{ is regular}\}$. The ordinal α is *Mahlo on* X if for every $f : \alpha \to \alpha$ there is a $\beta > 0$ closed under f such that $\beta \in X \cap \alpha$.

1.3. Theorem. (i) *the following are equivalent*

a) α *is* $\mathbf{\Pi}^0_0$*-indescribable on* X,

b) α *is* $\mathbf{\Sigma}^0_2$*-indescribable on* X,

c) $\alpha = \mathrm{Sup}\,(X \cap \alpha)$.

(ii) *the following are equivalent*

a) α *is* $\mathbf{\Pi}^0_2$*-indescribable on* X

b) α *is* $\mathbf{\Pi}^1_0$*-indescribable on* X

c) α *is Mahlo on* X.

(iii) α *is* $\mathbf{\Pi}^1_n$*-indescribable on* $X \Leftrightarrow \alpha$ *is* $\mathbf{\Sigma}^1_{n+1}$*-indescribable on* X.

(iv) *If* $n > 0$ *or* $m > 2$ ($n > 0$ *or* $m > 3$) *then* α *is* $\mathbf{\Pi}^n_m$ ($\mathbf{\Sigma}^n_m$)*-indescribable on* $X \Leftrightarrow \alpha$ *is* $\mathbf{\Pi}^n_m$ ($\mathbf{\Sigma}^n_m$)*-indescribable on* $X \cap \mathrm{Rg}$.

Hierarchies of classes of large cardinals have been obtained by iterating such operators as L and M where for $X \subseteq \mathrm{On}$:

$$L(X) = \{\alpha \in X : \alpha = \mathrm{Sup}\,(X \cap \alpha)\}$$

$$M(X) = \{\alpha \in X : \alpha \text{ is Mahlo on } X\}\,.$$

Iterations of an operator F are defined by transfinite induction on λ:

$$F^{\lambda}(X) = X \cap \bigcap_{\mu < \lambda} F(F^{\mu}(X))\,.$$

The elements of $L^{\lambda}(\mathrm{Rg})$ are the (weak) λ*-hyperinaccessibles*, while the elements of $M^{\lambda}(\mathrm{Rg})$ are the (weak) λ*-hyperMahlo ordinals.* Let $H_1(X) = \{\alpha \in X : \alpha \text{ is } \mathbf{\Pi}^0_1\text{-indescribable on } X\}$ and let $H_{n+2}(X) = \{\alpha \in X : \alpha \text{ is } \mathbf{\Pi}^1_n\text{-indescribable on } X\}$. Then by Theorem 1.3 $H_1 = L$ and $H_2 = M$.

The relative magnitudes of the ordinals in $H_n(\mathrm{Rg})$ may be indicated by using the following diagonalisation of iterations:

$$F^{\Delta}(X) = \{\alpha > 0 : \alpha \in F^{\alpha}(X)\}\,.$$

1.4. Theorem (Lévy). *If* $n > 0$ *then* $H_{n+1}(\mathrm{Rg}) \subseteq H_n^{\Delta}(\mathrm{Rg}), (H_n^{\Delta})^{\Delta}(\mathrm{Rg})$, *etc.*

Let us now turn to the strongly indescribable cardinals. These are defined using reflecting properties of the cumulative hierarchy of sets. Let $R(\alpha) = \bigcup_{\beta<\alpha} P(R(\beta))$ for all $\alpha \in \mathrm{On}$. ($P(x)$ is the power set of x).

1.5. Definition. $R(\alpha)$ *reflects* φ *on* $X \subseteq \mathrm{On}$ if

$$R(\alpha) \models \varphi \Rightarrow (\exists \beta \in X \cap \alpha) R(\beta) \models \varphi .$$

$R(\alpha)$ *reflects* φ if $R(\alpha)$ reflects φ on On. α *is strongly* $\mathbf{\Pi}^n_m$ $(\mathbf{\Sigma}^n_m)$*-indescribable* [*on* X] if $R(\alpha)$ reflects [on X] every $\mathbf{\Pi}^n_m$ $(\mathbf{\Sigma}^n_m)$ sentence of $\mathcal{L}$.

The properties of the notions of Definition 1.5 closely resemble those of Definition 1.1. The strong $\mathbf{\Pi}^0_2$-indescribables coincide with the strongly inaccessible ordinals. For $n > 0$ an ordinal is strongly $\mathbf{\Pi}^n_m$ $(\mathbf{\Sigma}^n_m)$-indescribable if and only if it is strongly inaccessible and is $\mathbf{\Pi}^n_m$ $(\mathbf{\Sigma}^n_m)$-indescribable. So, assuming the GCH, the two notions coincide when $n > 0$.

Let L_α be the set of constructible sets of order $< \alpha$, (i.e. $\mathrm{L}_\alpha = \bigcup_{\beta<\alpha} \mathrm{Def}(\mathrm{L}_\beta)$ where $\mathrm{Def}(x)$ is the set of subsets of x definable in $\langle x, \in\restriction x, a\rangle_{a\in x}$).

1.6. Definition. L_α *reflects* φ *on* X if

$$\mathrm{L}_\alpha \models \varphi \Rightarrow (\exists \beta \in X \cap \alpha)\, \mathrm{L}_\beta \models \varphi .$$

L_α *reflects* φ if L_α reflects φ on On.

If this definition is used as in Definition 1.5 the resulting indescribability notions may easily be seen to coincide with those of Definition 1.1.

In order to obtain the classes of ordinals that we are interested in we restrict the language $\mathcal{L}$. Let $\mathcal{L}_\in$ be the sublanguage of $\mathcal{L}$ obtained by only allowing $\in$ as a relation symbol.

1.7. Definition. α *is* $\mathbf{\Pi}^n_m$ $(\mathbf{\Sigma}^n_m)$*-reflecting* [*on* X] if L_α reflects [on X] every $\mathbf{\Pi}^n_m$ $(\mathbf{\Sigma}^n_m)$ sentence of $\mathcal{L}_\in$.

Some properties of this definition are summarised in the following theorems, which should be compared with Theorems 1.2 and 1.3.

1.8. Theorem. *α is Π^0_2-reflecting iff α is an admissible ordinal $> \omega$.*

This result and Theorem 1.9 below will be proved in §2.

Let $\mathrm{Ad} = \{\alpha > \omega : \alpha \text{ is admissible}\}$. $\alpha \in \mathrm{Ad}$ is *recursively Mahlo* if for every α-recursive function $f : \alpha \to \alpha$ there is an ordinal $\beta > 0$ closed under f such that $\beta \in X \cap \alpha$.

1.9. Theorem. (i) *The following are equivalent*

a) *α is Π^0_0-reflecting on X*

b) *α is Σ^0_2-reflecting on X*

c) $\alpha = \mathrm{Sup}\,(X \cap \alpha)$.

(ii) *α is Π^0_2-reflecting on $X \Leftrightarrow \alpha$ is recursively Mahlo on X.*

(iii) *α is Π^0_n-reflecting on $X \Leftrightarrow \alpha$ is Σ^0_{n+1}-reflecting on X.*

(iv) *If $n > 0$ or $m > 2$ ($n > 0$ or $m > 3$) then α is Π^n_m (Σ^n_m)-reflecting on $X \Leftrightarrow \alpha$ is Π^n_m (Σ^n_m)-reflecting on $X \cap \mathrm{Ad}$.*

As it is often easier to work with ordinals rather than the constructible hierarchy the following characterisations will be useful. Let $\mathcal{L}_p$ be the sub-language of $\mathcal{L}$ that has relation symbols only for the primitive recursive relations on sets (see [8] for the properties of this notion).

1.10. Theorem. *α is Π^n_m (Σ^n_m)-reflecting* [*on X*] *if and only if α reflects* [*on X*] *every Π^n_m (Σ^n_m) sentence of $\mathcal{L}_p$.*

The primitive recursive relations in the language $\mathcal{L}_p$ are needed for reflecting properties on ordinals in order to compensate for the richness of the $\in$ relation for reflecting properties on the constructible hierarchy. Theorem 1.10 will be proved in §3.

$\mathrm{L}^\lambda(\mathrm{Ad})$ is the class of *λ-recursively inaccessible ordinals*, while if $\mathrm{RM}(X) = \{\alpha \in X : \alpha \text{ is recursively Mahlo on } X\}$ then $\mathrm{RM}^\lambda(\mathrm{Ad})$ is the class of *λ-recursively Mahlo* ordinals. Let $M_n(X) = \{\alpha \in X : \alpha \text{ is } \Pi^0_n\text{-reflecting on } X\}$. Then $M_0 = M_1 = \mathrm{L}$ and $M_2 = \mathrm{RM}$. The next result indicates the relative magnitudes of the ordinals in $M_n(\mathrm{Ad})$ and should be compared with Theorem 1.4.

1.11. Theorem. *If $n > 0$ then*

$$M_{n+1}(\mathrm{Ad}) \subseteq M_n^\Delta(\mathrm{Ad}), (M_n^\Delta)^\Delta(\mathrm{Ad}), \textit{etc. }.$$

This will be proved in §4.

1.12. Definition. Let π^n_m (σ^n_m) be the least $\mathbf{\Pi}^n_m$ ($\mathbf{\Sigma}^n_m$)-reflecting ordinal.

By 1.9 $\pi^0_0 = \pi^0_1 = \omega$ and $\pi^0_2 = \omega_1$ are the recursive analogues of the first two regular cardinals. What can we say about π^0_3? By 1.9 and 1.11 π^0_3 is greater than the least recursively Mahlo ordinal, the least recursively hyper-Mahlo ordinal etc. In fact π^0_3 appears to be greater than any "reasonable" iteration into the transfinite of this diagonalisation process. When one thinks of a corresponding cardinal in set theory (with "recursively Mahlo" now replaced by "Mahlo") the cardinal which comes to mind is the least $\mathbf{\Pi}^1_1$-indescribable cardinal. We shall now try and justify the view that $\mathbf{\Pi}^0_3$-reflection is the recursive analogue of $\mathbf{\Pi}^1_1$-indescribability. The same ideas with some additional notational complexity provide an analogy between $\mathbf{\Pi}^0_{n+2}$-reflection and $\mathbf{\Pi}^1_n$-indescribability for all $n > 0$, but we shall concentrate on the case $n = 1$.

The analogy is obtained as follows. A class of cardinals, called the 2-regular cardinals, is defined, as well as a recursive analogue of this class whose members are called 2-admissible. We then show that a cardinal is 2-regular if and only if it is strongly $\mathbf{\Pi}^1_1$-indescribable, and an ordinal is 2-admissible if and only if it is $\mathbf{\Pi}^0_3$-reflecting.

Certain properties of infinity can be stated in terms of fixed points of operations. For example $\kappa > \omega$ and κ is regular if and only if:

(1) for every $f: \kappa \to \kappa$ there is some $0 < \alpha < \kappa$ such that $f''\alpha \subseteq \alpha$. (We say α is a *witness* for f.)

If we modify (1) by requiring that the witness be regular, we obtain the Mahlo cardinals, etc.

An alternative way of modifying (1) is by using higher type operations on κ. Let $F: {}^\kappa\kappa \to {}^\kappa\kappa$. F is *κ-bounded* if for every $f: \kappa \to \kappa$ and $\xi < \kappa$, the value $F(f)(\xi)$ is determined by less than κ values of f. More precisely, F is κ-bounded if

$$\forall f\, \exists \gamma < \kappa\, \forall g [g \restriction \gamma = f \restriction \gamma \Rightarrow F(f)(\xi) = F(g)(\xi)] \,.$$

$0 < \alpha < \kappa$ is a *witness for* F if for every $f: \kappa \to \kappa$,

$$f''\alpha \subseteq \alpha \Rightarrow F(f)''\alpha \subseteq \alpha.$$

1.13. Definition. $\kappa > 0$ is *2-regular* if every κ-bounded $F : {}^\kappa\kappa \to {}^\kappa\kappa$ has a witness.

1.14. Theorem. *κ is 2-regular iff κ is strongly $\mathbf{\Pi}^1_1$-indescribable.*

We now look at a recursive analogue of 2-regularity. Roughly speaking the following definition of 2-admissible is obtained by replacing in the definition of 2-regular, "bounded" by "recursive" and the functions by their Gödel numbers. In the following definition we write $\{\xi\}_\kappa : \kappa \to \kappa$ to mean that $\{\xi\}_\kappa$ is total on κ.

1.15. Definition. (i) Let $\kappa \in \mathrm{Ad}$ and $\xi < \kappa$. $\{\xi\}_\kappa$ *maps κ-recursive functions to κ-recursive functions if*

$$\forall \beta < \kappa\, [\{\beta\}_\kappa : \kappa \to \kappa \Rightarrow \{\{\xi\}_\kappa(\beta)\}_\kappa : \kappa \to \kappa]\ ;$$

(ii) Suppose $\{\xi\}_\kappa$ maps κ-recursive functions to κ-recursive functions. $\alpha \in \kappa \cap \mathrm{Ad}$ is a *witness for* ξ if $\xi < \alpha$ and $\{\xi\}_\alpha$ maps α-recursive functions to α-recursive functions.

(iii) $\kappa \in \mathrm{Ad}$ is 2-admissible if every $\xi < \kappa$ such that $\{\xi\}_\kappa$ maps κ-recursive functions to κ-recursive functions has a witness.

1.16. Theorem. *κ is 2-admissible iff κ is $\mathbf{\Pi}^0_3$-reflecting.*

Theorems 1.14 and 1.16 will be proved in §5.

Certain classes of ordinals, defined in terms of reflecting properties, also have characterisations in terms of stability properties. Let $A \prec_{\mathbf{\Sigma}^0_1} B$ if A and B are transitive sets such that $A \subseteq B$ and $B \models \varphi \Rightarrow A \models \varphi$ for every $\mathbf{\Sigma}^0_1$ sentence φ of $\mathcal{L}_\in$ that only has constants for elements of A. Kripke has defined the notion of an ordinal α being β-stable (see [10]). His definition used his systems of equations for defining recursion on ordinals. For admissible β he gave the following characterisation, which we shall take as a definition:

1.17. Definition. α *is β-stable* if $\alpha < \beta$ and $\mathrm{L}_\alpha \prec_{\mathbf{\Sigma}^0_1} \mathrm{L}_\beta$.

When β is not admissible, this notion may well diverge from Kripke's original one.

1.18. Theorem. *α is Π_0^1-reflecting if and only if α is $\alpha+1$-stable.*

1.19. Theorem. *For countable α, α is Π_1^1-reflecting if and only if α is α^+-stable, where α^+ is the first admissible ordinal $> \alpha$.*

These results will be proved in §6.

Given $A \subseteq {}^n\mathrm{ON}$ all of our definitions and results will relativise to A. As we shall need the relativisations in Part II we spell out exactly what this means.

Definition 1.6 is relativised by using $\langle \mathrm{L}_\alpha[A] : \alpha \in \mathrm{ON}\rangle$ instead of $\langle \mathrm{L}_\alpha : \alpha \in \mathrm{ON}\rangle$. Here $\mathrm{L}_\alpha[A] = \bigcup_{\beta<\alpha} \mathrm{Def}_A(\mathrm{L}_\beta[A])$ where $\mathrm{Def}_A(x)$ is the set of subsets of x definable in $\langle x, \in \restriction x, A \restriction x, a\rangle_{a\in x}$. The language $\mathcal{L}_\in$ must be replaced by the language $\mathcal{L}_\in(A)$ which is $\mathcal{L}_\in$ with an added n-ary relation symbol to denote A. Definition 1.7 becomes: α is $\Pi_m^n(A)$*-reflecting* [*on* X] if $\mathrm{L}_\alpha[A]$ reflects [on X] every Π_m^n sentence of $\mathcal{L}_\in(A)$. Similarly for $\Sigma_m^n(A)$-reflecting.

Theorems 1.8 and 1.9 relativise in the obvious way. Ad must be replaced by $\mathrm{Ad}(A) = \{\alpha > \omega \mid \alpha \text{ is admissible relative to } A \restriction \alpha\}$.

The language $\mathcal{L}_p(A)$ is defined by allowing relation symbols for all relations primitive recursive in A. Most of the proofs relativise in a routine way.

§2. Elementary facts

In order to prove our theorems we shall need to assume some familiarity with the notions of primitive recursive set function, admissible class, admissible ordinal and ordinal recursion on an admissible ordinal. We shall use [8] as our basic reference and will usually follow the terminology they use. We shall also need to refer to [6] when we use Jensen's notion of a rudimentary set function.

The notion of a primitive recursive function with domain M has been formulated for various classes M e.g. ${}^m\omega$, ${}^m\mathrm{ON}$ and ${}^m\mathrm{V}$. As shown in [8] all these notions turn out to be special cases of the following: $F : M \to \mathrm{V}$ is *primitive recursive* if M is a primitive recursive function with domain M has been to M of a primitive recursive set function. In [8] a transitive prim closed class M is defined to be admissible if M satisfies the Σ_1^0*-collection principle* (there called Σ_1^0-reflection principle) which we shall formulate as follows:

For every prenex Σ_1^0 formula θ of $\mathcal{L}_\in$ if $M \models \forall x \in a\theta$ then

$M \models \forall x \in a \theta^b$ for some $b \in M$, where if θ is $\exists y_1 \ldots \exists y_k \Psi$, with Ψ Σ_0^0, then θ^b is $\exists y_1 \in b \ldots \exists y_k \in b \Psi$.

We shall find it more useful to use the characterisation in [6].

2.1. Definition. The transitive class M is *admissible* if M is rud closed and satisfies Σ_1^0-collection.

This definition is relativized by replacing Σ_1^0-collection by $\Sigma_1^0(A)$-collection, obtained by using $\mathcal{L}_\in(A)$ instead of $\mathcal{L}_\in$, and adding the condition that $a \in M \Rightarrow A \cap a \in M$.

A relation R on a transitive set M is Σ_1^0 *on* M if R is defined on M by a Σ_1^0 formula of $\mathcal{L}_\in$. A partial function with arguments and values in M is Σ_1^0 on M if its graph is. We shall assume some familiarity with the closure properties of these relations and functions on an admissible M, as presented for example in [8]. In particular we shall need the following:

2.2. Proposition (Definition by Σ_1^0-recursion). *Let M be an admissible set. Let G be a function such that $G \restriction M : M \times M \to M$ and $G \restriction M$ is Σ_1^0 on M. Let*

$$F(x) = G(x, F \restriction x) .$$

Then $F \restriction M : M \to M$ and $F \restriction M$ is Σ_1^0 on M. Moreover the Σ_1^0 definition of $F \restriction M$ depends only on the Σ_1^0 definition of $G \restriction M$ (and not on M).

Usually we will only be interested in $F \restriction M \cap \mathrm{ON}$.

For the notion of an admissible ordinal α and α-recursion we shall follow [8]. An ordinal α is *admissible* if L_α is admissible. $f : {}^n\alpha \to \alpha$ is *α-recursive* if it is Σ_1^0 on L_α.

The following lemma will be useful and the proof will illustrate some of the techniques of α-recursion.

2.3. Lemma. *If $\alpha > \omega$ is an admissible ordinal and $f : \alpha \to \alpha$ is α-recursive then there are arbitrarily large limit ordinals $< \alpha$ that are closed under f.*

Proof. Let $\alpha > \omega$ be admissible and let $f : \alpha \to \alpha$ be α recursive. Define $g : \alpha \to \alpha$ by $g(x) = \mathrm{Max}\,(x+1, \mathrm{Sup}_{y \leq x} f(y))$. Then g is α-recursive, $x < g(x)$ and $f(x) \leq g(y)$ for $x \leq y < \alpha$. Given $\gamma_0 < \alpha$ let $\gamma_n = g^n(\gamma_0)$. Then

$\gamma_0 < \gamma_1 < ... < \alpha$ and $x \leq \gamma_n \Rightarrow f(x) \leq \gamma_{n+1}$. Let $\gamma = \text{Sup}_{n<\omega}\gamma_n$. Then $\gamma \leq \alpha$ is a limit ordinal such that $\gamma_0 < \gamma$ and γ is closed under f as

$$x < \gamma \Rightarrow x \leq \gamma_n \quad \text{for some } n$$

$$\Rightarrow f(x) \leq \gamma_{n+1} < \gamma .$$

So it only remains to show that $\gamma < \alpha$. For this we need 2.2. Let $F(x) = G(x, F \upharpoonright x)$ where $G(x,y) = g(z)$ if x is a successor ordinal, y is a function such that $y(x-1)$ is defined with value $z < \alpha$, and $G(x,y) = \gamma_0$ otherwise.

Then it is not hard to see that $\gamma_n = F(n)$ for each $n \in \omega$, and that as $G \upharpoonright L_\alpha : L_\alpha \times L_\alpha \rightarrow L_\alpha$ and is Σ_1^0 on L_α it follows that $F \upharpoonright \alpha$ is α-recursive and hence $\gamma = \text{Sup}_{n<\omega}\gamma_n = \text{Sup}_{n<\omega}F(n) < \alpha$.

Proof of Theorem 1.8. Let α be Π_2^0-reflecting. If $a < \alpha$ then $L_\alpha \models \neg(a \in a)$. Hence there is a $\beta < \alpha$ such that $L_\beta \models \neg(a \in a)$; i.e. $a < \beta < \alpha$. Hence α is a limit number. So $L_\alpha \models \forall x \exists y (x \in y)$, which implies that there is a $\beta < \alpha$ such that $L_\beta \models \forall x \exists y (x \in y)$. Hence α is a limit number $> \omega$. Using Lemma 6 of [6] it is not hard to show that L_α is rud closed for any limit ordinal α. Hence it remains only to show that L_α satisfies Σ_1^0-collection. So let $L_\alpha \models \forall x \in a \theta$ where θ is a Σ_1^0 formula of $\mathcal{L}_\in$. Then by Π_2^0-reflection there is a $\beta < \alpha$ such that $L_\beta \models \forall x \in a \theta$. Now if $b = L_\beta \in L_\alpha$ then $L_\alpha \models \forall x \in a \theta^b$ as required.

Conversely, let $\alpha > \omega$ be admissible, and let φ be a Π_2^0 sentence of $\mathcal{L}_\in$ such that $L_\alpha \models \varphi$. We may assume that φ has the form $\forall x_1 ... x_n \exists y_1 ... y_m \Psi$ where Ψ is Σ_0^0. Hence $L_\alpha \models \forall x_1 ... x_n \exists y \theta$ where θ is the Σ_0^0 formula $\exists y_1 \in y ... \exists y_m \in y \Psi$. For simplicity we shall just consider the case when $n = 1$. If $\beta < \alpha$ and $a = L_\beta$ then $L_\alpha \models \forall x_1 \in a \exists y \theta$. Hence by Σ_1^0-collection there is a $b \in L_\alpha$ such that $L_\alpha \models \forall x_1 \in a \exists y \in b \theta$. But $b \subseteq L_\gamma$ for some $\gamma < \alpha$ so that $\models \forall x_1 \in a \exists y \in L_\gamma \theta$. Let $f(\beta)$ be the least such $\gamma < \alpha$. Then $f: \alpha \rightarrow \alpha$ is α-recursive. Let $\beta_0 < \alpha$ such that every constant of θ occurs in L_{β_0}. Then by the Lemma 2.3 choose a limit ordinal β such that $\beta_0 < \beta < \alpha$ and β is closed under f. Then we must have $L_\beta \models \forall x_1 \exists y \theta$ so that α reflects the Π_2^0 sentence φ.

In order to prove (iv) of Theorem 1.9 we shall need

2.4. Theorem. *There is a Π_3^{0-} sentence σ_0 of $\mathcal{L}_\in$ such that the transitive class*

M is admissible if and only if $M \models \sigma_0$.

Proof. By Lemma 6 of [6] there are binary rud functions $F_0, \ldots, F_8$ such that the class M is rud closed if and only if it is closed under $F_0, \ldots, F_8$.

By Lemma 2 of [6] there are Σ_0^{0-} formulae $\varphi_i(x,y,z)$ of $\mathcal{L}_\in$ that define the graphs of F_i for $i \leq 8$. So M is rud closed if and only if $M \models \theta_0$ where θ_0 is the Π_2^{0-} sentence $\bigwedge_{i\leq 8} \forall x \forall y \exists z \varphi_i(x,y,z)$.

By Lemma 9 of [6] we may prove:

2.5. Lemma. *There is a Σ_1^{0-} formula $\mathrm{Sat}(x,y)$ of $\mathcal{L}_\in$ such that if $\theta(x)$ is a Σ_1^0 formula of $\mathcal{L}_\in$ with x as only free variable and $a = \ulcorner \theta(x) \urcorner$ then for all rud closed M, if the constants of $\theta(x)$ are in M then $a \in M$ and*

$$M \models \forall x(\theta(x) \leftrightarrow \mathrm{Sat}(a,x)) .$$

Using this lemma we see that the transitive rud closed class M is admissible if and only if $M \models \theta_1$ where θ_1 is the Π_3^{0-} sentence $\forall u \forall v[\forall x \in u \ \mathrm{Sat}(v,x) \rightarrow \exists z \forall x \in u \ \mathrm{Sat}(v,x)^z]$. The theorem now follows if we let σ_0 be $\theta_0 \wedge \theta_1$.

Proof of Theorem 1.9.

(i) b) $\Rightarrow$ a) is trivial.

c) $\Rightarrow$ b). Let $\alpha = \mathrm{Sup}(X \cap \alpha)$ and let φ be a Π_1^0 sentence such that $L_\alpha \models \varphi$. If $a_1, \ldots, a_n$ are the individual constants occurring in φ then $a_1, \ldots, a_n \in L_\alpha$ so that there is a $\beta \in X \cap \alpha$ such that $a_1, \ldots, a_n \in L_\beta$ and hence $L_\beta \models \varphi$, as φ is Π_1^0. So α is Π_1^0-reflecting on X and hence Σ_2^0-reflecting on X, by (iii).

a) $\Rightarrow$ c). Let α be Π_0^0-reflecting on X and let $\beta < \alpha$. Let φ be the sentence $\neg(\beta \in \beta)$. Then $L_\alpha \models \varphi$, so that as φ is Π_0^0 there is a $\gamma \in X \cap \alpha$ such that $L_\gamma \models \varphi$ and hence $\beta < \gamma$. Hence $\alpha = \mathrm{Sup}(X \cap \alpha)$.

(ii) $\Rightarrow$. Let α be Π_2^0-reflecting on X. Then α is Π_2^0-reflecting and hence by Theorem 1.8 $\alpha \in \mathrm{Ad}$. Now let $f : \alpha \to \alpha$ be α-recursive. Let $\theta(x,y)$ be a Σ_1^0 formula of $\mathcal{L}_\in$ that defines the graph of f on L_α. Then $L_\alpha \models \forall x \exists y(\theta(x,y) \vee (\neg \mathrm{On}(x) \wedge y = 0))$. Hence there is a $\beta \in X \cap \alpha$ such that $L_\beta \models \forall x \exists y(\theta(x,y) \vee (\neg \mathrm{On}(x) \wedge y = 0))$. Hence $\beta > 0$ is closed under f. So α is recursively Mahlo on X.

(ii) $\Leftarrow$. Let α be recursively Mahlo on X and let φ be a Π_2^0 sentence of $\mathcal{L}_\in$

such that $L_\alpha \models \varphi$. As in the proof of Theorem 1.8 we will suppose that φ has the form $\forall x_1 \exists y \theta$ where θ is $\boldsymbol{\Sigma}_0^0$ and define the α-recursive function $f : \alpha \to \alpha$, and the ordinal $\beta_0 < \alpha$. Now let $g(x) = \mathrm{Max}(\beta_0, x+1, f(x))$. Then $g : \alpha \to \alpha$ is α-recursive so that there is a $\beta \in X \cap \alpha$ such that $\beta > 0$ is closed under g. From the definition of g it follows that $\beta > \beta_0$ is a limit ordinal and is closed under f so that $L_\beta \models \forall x_1 \exists y \theta$. Thus α reflects φ as required.

(iii) $\Leftarrow$ is trivial. So for the converse let α be $\boldsymbol{\Pi}_n^0$-reflecting on X and let φ be $\boldsymbol{\Sigma}_{n+1}^0$ such that $L_\alpha \models \varphi$. φ has the form $\exists x_1 \ldots \exists x_n \theta(x_1, \ldots, x_n)$ where $\theta(x_1, \ldots, x_n)$ is $\boldsymbol{\Pi}_n^0$. So there are $a_1, \ldots, a_n \in L_\alpha$ such that $L_\alpha \models \theta(a_1, \ldots, a_n)$. Hence there is a $\beta \in X \cap \alpha$ such that $L_\beta \models \theta(a_1, \ldots, a_n)$. But then $L_\beta \models \exists x_1 \ldots \exists x_n \theta(x_1, \ldots, x_n)$. So α is $\boldsymbol{\Sigma}_{n+1}^0$-reflecting on X.

(iv) $\Leftarrow$ is trivial as $X \cap \mathrm{Ad} \subseteq X$. For the converse we use the $\boldsymbol{\Pi}_3^{0-}$ sentence σ_0 given by Theorem 2.4. Let θ_0 be a $\boldsymbol{\Sigma}_1^{0-}$ sentence expressing the existence of an infinite set. Then $\alpha \in \mathrm{Ad} \Leftrightarrow L_\alpha \models \sigma_0 \wedge \theta_0$, and $\sigma_0 \wedge \theta_0$ is a $\boldsymbol{\Pi}_3^{0-}$ sentence of $\mathcal{L}_\in$. Now let α be $\boldsymbol{\Pi}_m^n$ $(\boldsymbol{\Sigma}_m^n)$-reflecting on X and let φ be a $\boldsymbol{\Pi}_m^n$ $(\boldsymbol{\Sigma}_m^n)$ sentence of $\mathcal{L}_\in$ such that $L_\alpha \models \varphi$. Then because of the restrictions on n and m $\varphi \wedge \sigma_0 \wedge \theta_0$ is a $\boldsymbol{\Pi}_m^n$ $(\boldsymbol{\Sigma}_m^n)$ sentence of $\mathcal{L}_\in$ such that $L_\alpha \models \varphi \wedge \sigma_0 \wedge \theta_0$. Hence there is a $\beta \in X \cap \alpha$ such that $L_\beta \models \varphi \wedge \sigma_0 \wedge \theta_0$. Hence $\beta \in \mathrm{Ad}$ so that α reflects φ on $X \cap \mathrm{Ad}$ showing that α is $\boldsymbol{\Pi}_m^n$ $(\boldsymbol{\Sigma}_m^n)$-reflecting on $X \cap \mathrm{Ad}$.

§3. Ordinal theoretic characterisations

Let us call α $\boldsymbol{\Pi}_m^n$ $(\boldsymbol{\Sigma}_m^n)$ **-reflecting* [*on X*] if α reflects [on X] every $\boldsymbol{\Pi}_m^n$ $(\boldsymbol{\Sigma}_m^n)$ sentence of $\mathcal{L}_p$. Our proof of Theorem 1.10 will be a little indirect, in that we first prove Theorems 1.8* and 1.9*, obtained from Theorems 1.8 and 1.9 by replacing 'reflecting' everywhere by '*-reflecting'. But first we need the following lemma.

3.1. Lemma. *There is a bijective primitive recursive function* $N : \mathrm{ON} \to \mathrm{L}$ *such that if* $(\forall x < \alpha)\, 2^x < \alpha$ *then* $L_\alpha = N''\alpha$.

Proof. In Lemma 3.2 of [8] a primitive recursive bijection $N : \mathrm{ON} \to \mathrm{L}$ is obtained from Gödel's primitive recursive surjection $F : \mathrm{ON} \to \mathrm{L}$ by successively removing repetions in the values of F. Examining their definition of N it is not hard to see that $N''\alpha = F''\alpha$ for all limit ordinals α. If $(\forall x < \alpha)\, 2^x < \alpha$ then either $\alpha = 0$, $\alpha = \omega$ or α has the form $\alpha = \epsilon_\beta$. Clearly $L_0 = \emptyset = N''0$. If

$\alpha = \omega$ or $\alpha = \epsilon_\beta$ then in [12] it is shown that $L_\alpha = F''\alpha$ and hence it follows that $L_\alpha = N''\alpha$ as α is a limit ordinal.

This result relativises to give a bijection $N_A : \mathrm{ON} \to L[A]$ which is primitive recursive in A, such that $L_\alpha[A] = N_A''\alpha$ if $(\forall x < \alpha) 2^x < \alpha$.

3.2. Lemma. *If α is Π_2^0 *-reflecting then*

(i) *α is a limit ordinal* $> \omega$.

(ii) $a, b < \alpha \Rightarrow a+b < \alpha$.

(iii) $b < \alpha \Rightarrow 2^b < \alpha$.

Proof. First note that the graphs of primitive recursive functions are primitive recursive relations and hence are allowed in the language $\mathcal{L}_p$.

(i) If $a < \alpha$ then $\alpha \models \exists x(x=a)$ so that $\beta \models \exists x(x=a)$ for some $\beta < \alpha$ and hence $a < \beta < \alpha$, thus α is a limit number. So $\alpha \models \forall x \exists y (x < y)$. By reflection there is a limit ordinal $< \alpha$; i.e. α is a limit ordinal $> \omega$.

(ii) Given $a < \alpha$ we prove that $a+b < \alpha$ by induction on $b < \alpha$. If $b = 0$ this is trivial. If $0 < b < \alpha$ and $\forall x < b\ a+x < \alpha$ then, as α is a limit ordinal, $\alpha \models \forall x \exists y [x < b \to a+x+1 = y]$. So by reflection $\forall x < b\ a+x+1 < \beta$ for some $\beta < \alpha$. Hence $a+b = \mathrm{Sup}_{x<b}\, a+x+1 \leq \beta < \alpha$.

(iii) We prove that $2^b < \alpha$ by induction on $b < \alpha$. If $b = 0$ then $2^b = 1 < \omega < \alpha$. If $0 < b < \alpha$ and $\forall x < b\ 2^x < \alpha$ then by (ii) $\forall x < b\ 2^{x+1} = 2^x + 2^x < \alpha$. Hence $\alpha \models \forall x \exists y [x < b \to 2^{x+1} = y]$. So by reflection $\forall x < b\ 2^{x+1} < \beta$ for some $\beta < \alpha$. Hence $2^b = \mathrm{Sup}_{x<b}\, 2^{x+1} \leq \beta < \alpha$.

3.3. Theorem 1.8*. *α is Π_2^0 *-reflecting* $\Leftrightarrow \alpha \in \mathrm{Ad}$.

Proof. Let α be Π_2^0 *-reflecting. Then by (i) of Lemma 3.2 $\alpha > \omega$ and α is a limit ordinal so that as we have already observed L_α is rud closed. Hence it suffices to show that L_α satisfies Σ_1^0-collection. So let $L_\alpha \models \forall x \in a \exists y \Psi(x,y,b)$ where $\Psi(x,y,z)$ is Σ_0^{0-}. (We can assume without loss that there is only one existential quantifier $\exists y$ and only one constant b.) We must find $c \in L_\alpha$ such that

$$(*) \qquad L_\alpha \models \forall x \in a \exists y \in c\ \Psi(x,y,b).$$

Let $R \subseteq^3 \mathrm{ON}$ be the primitive recursive relation given by

$$R(\alpha,\beta,\gamma) \Leftrightarrow \models \Psi(N(\alpha), N(\beta), N(\gamma)).$$

Let $a = N(\alpha_0)$, $b = N(\beta_0)$. By the previous lemmas $L_\alpha = N''\alpha$ so that

$$\alpha \models \forall x(R_\in(x,\alpha_0) \to \exists y R(x,y,\beta_0)) \wedge \forall x \exists y\, (2^x = y)$$

where $R_\in(\alpha,\beta) \Leftrightarrow N(\alpha) \in N(\beta)$. Hence, as α is $\mathbf{\Pi}_2^0$ *-reflecting, there is a $\beta < \alpha$ such that $\beta \models \forall x(R_\in(x,\alpha_0) \to \exists y R(x,y,\beta_0)) \wedge \forall x \exists y\, (2^x = y)$. As $\forall x < \beta\, (2^x < \beta), L_\beta = N''\beta$ so that

$$L_\beta \models \forall x \in a\, \exists y\, \Psi(x,y,b)\,.$$

(*) follows if we let $c = L_\beta \in L_\alpha$.

3.4. Theorem 1.9*.

Proof. This follows the same pattern as the proof of Theorem 1.9 and so will be omitted. In the proof of (iv) we need the next lemma, which replaces Theorem 2.4.

3.5. Lemma. *There is a* $\mathbf{\Pi}_3^{0-}$ *sentence* σ_1 *of* $\mathcal{L}_p$ *such that* α *is admissible if and only if* $\alpha \models \sigma_1$.

Proof. Let us assume that the $\mathbf{\Pi}_3^{0-}$ sentence σ_0 of $\mathcal{L}_\in$ given in Theorem 2.4 is in Prenex form with $\mathbf{\Sigma}_0^{0-}$ matrix $\Psi(x_1, ..., x_k)$. Now let $R(\alpha_1, ..., \alpha_k) \Leftrightarrow\ \models \Psi(N(\alpha_1), ..., N(\alpha_k))$ for $\alpha_1, ..., \alpha_k \in \mathrm{ON}$. Then R is a primitive recursive relation. Let θ_0 be obtained from σ_0 by replacing $\Psi(x_1, ..., x_k)$ by $R(x_1, ..., x_k)$. Then if $L_\alpha = N''\alpha$

$$\alpha \models \theta_0 \Leftrightarrow L_\alpha \models \sigma_0\,.$$

Hence by Lemma 3.1 we can let σ_1 be $\theta_0 \wedge \forall x \exists y (2^x = y)$.

We can now turn to the proof of Theorem 1.10. (i)–(iii) of Theorems 1.9 and 1.9* yield the theorem in the cases $\mathbf{\Pi}_m^0$ $(m \leq 2)$ and $\mathbf{\Sigma}_m^0$ $(m \leq 3)$. By Theorems 1.8 and 1.8* and (iv) of Theorems 1.9 and 1.9* the remaining cases need only be proved when $\alpha \in \mathrm{Ad}$ and $X \subseteq \mathrm{Ad}$. With these restrictions the remaining cases will follow from:

3.6. Lemma. *Let $n+m>0$*

(i) *For each* $\mathbf{\Pi}^n_m$ *sentence* θ *of* $\mathcal{L}_p$ *there is a* $\mathbf{\Pi}^n_m$ *sentence* $\theta_\in$ *of* $\mathcal{L}_\in$ *such that for admissible* α

$$\alpha \models \theta \Leftrightarrow \mathrm{L}_\alpha \models \theta_\in .$$

(ii) *For each* $\mathbf{\Pi}^n_m$ *sentence* θ *of* $\mathcal{L}_\in$ *there is a* $\mathbf{\Pi}^n_m$ *sentence* θ_p *of* $\mathcal{L}_p$ *such that for admissible* α

$$\mathrm{L}_\alpha \models \theta \Leftrightarrow \alpha \models \theta_p .$$

Using this lemma let us conclude the proof of Theorem 1.10. Let $n>0$ or $m>2$ and let α be $\mathbf{\Pi}^n_m$-reflecting on X. Let θ be a $\mathbf{\Pi}^n_m$ sentence of $\mathcal{L}_p$ such that $\alpha \models \theta$. Then $\theta_\in$ is a $\mathbf{\Pi}^n_m$ sentence of $\mathcal{L}_\in$ such that $\mathrm{L}_\alpha \models \theta_\in$ as α is admissible. Hence there is a $\beta \in X \cap \alpha$ such that $\mathrm{L}_\beta \models \theta_\in$. As $X \subseteq \mathrm{Ad}$, β is admissible so that $\beta \models \theta$. Hence L_α reflects θ. Similarly if ($n>0$ or $m>3$) and α is $\mathbf{\Sigma}^n_m$-reflecting on X and θ is a $\mathbf{\Sigma}^n_m$ sentence of $\mathcal{L}_p$ then $\neg\theta$ is a $\mathbf{\Pi}^n_m$ sentence of $\mathcal{L}_p$ so that $\neg(\neg\theta)_\in$ is a $\mathbf{\Sigma}^n_m$ sentence of $\mathcal{L}_\in$ and the argument is as above. The proof of the converse implications is exactly similar using (ii) of the lemma instead of (i).

Proof of Lemma 3.6.

(i) By theStability Theorem 2.5 of [8] we may easily associate with each primitive recursive relation R a $\mathbf{\Sigma}^{0-}_1$ formula $\varphi_R(x_1, ..., x_n)$ of $\mathcal{L}_\in$ such that for admissible α and $\alpha_1, ..., \alpha_n < \alpha$

$$R(\alpha_1, ..., \alpha_n) \Leftrightarrow \mathrm{L}_\alpha \models \varphi_R(N(\alpha_1), ..., N(\alpha_n)) .$$

Now let θ be a sentence of $\mathcal{L}_p$. If θ contains individual constants for sets that are not ordinals, then $\alpha \models \theta$ can never hold, so let $\theta_\in$ be $(1 \in 0)$. Otherwise define θ_* as follows. First replace each constant for an ordinal β by a constant for $N(\beta)$. Then replace each occurrence of a relation symbol $R(s_1, ..., s_n)$ in θ by $\varphi_R(s_1, ..., s_n)$. Then for admissible α it is clear that

$$\alpha \models \theta \Leftrightarrow \mathrm{L}_\alpha \models \theta_* .$$

Now if θ is $\mathbf{\Pi}_m^n$ and $n > 0$ then θ_* is also $\mathbf{\Pi}_m^n$ and so we can let $\theta_\in$ be θ_*.

If θ is $\mathbf{\Pi}_m^0$ $(m > 0)$ then we have to be more careful. We may assume that θ is in prenex form. So it has the form of an alternating sequence of m blocks of universal and existential type 0 quantifiers followed by a $\mathbf{\Pi}_0^0$ formula $\Psi(x_1, ..., x_k)$. Now $\Psi(x_1, ..., x_k)$ is built up from primitive recursive relations and ordinals using the boolean operations and restricted quantifiers. Hence there is a primitive recursive relation R and ordinals $\beta_1, ..., \beta_1$ such that for all α

$$\alpha \models \Psi(\alpha_1, ..., \alpha_k) \Longleftrightarrow \alpha \models R(\beta_1, ..., \beta_1, \alpha_1, ..., \alpha_k) .$$

Now define $\theta_\in$ as follows: If m is even, replace $\Psi(x_1, ..., x_k)$ in θ by $\varphi_R(N(\beta_1), ..., N(\beta_1), x_1, ..., x_k)$ and if m is odd, replace $\Psi(x_1, ..., x_k)$ in θ by $\neg\varphi_{\neg R}(N(\beta_1), ..., N(\beta_1), x_1, ..., x_k)$. Then $\theta_\in$ is $\mathbf{\Pi}_m^0$ and has the desired properties.

(ii) Let θ be a sentence of $\mathcal{L}_\in$. If θ contains constants for non-constructible sets, then $\mathrm{L}_\alpha \models \theta$ never holds so we can let θ_p be $(0 = 1)$. Otherwise define θ_0 as follows. First replace each individual constant for the set a by the constant for ordinal α such that $N(\alpha) = a$. Then replace each occurrence of $s \in t$ in θ by $R_\in(s, t)$, where $R_\in(\alpha, \beta) \Longleftrightarrow N(\alpha) \in N(\beta)$. (When proving the relativised version of 3.6 there may be occurrences of an atomic formula $A(s_1, ..., s_n)$. These must be replaced by $R_A(s_1, ..., s_n)$ where R_A is the relation primitive recursive in A such that $R_A(\alpha_1, ..., \alpha_n) \Longleftrightarrow A(N_A(\alpha_1), ..., N_A(\alpha_n))$.) Clearly for admissible ordinals α

$$\mathrm{L}_\alpha \models \theta \Longleftrightarrow \alpha \models \theta_0 .$$

Now if θ is $\mathbf{\Pi}_m^n$ with $n > 0$ then θ_0 is also $\mathbf{\Pi}_m^n$ and hence we can let θ_p be θ_0. If θ is $\mathbf{\Pi}_m^0$ with $m > 0$ then we must again be more careful. We can assume that θ is in prenex form with a sequence of quantifiers followed by a $\mathbf{\Pi}_0^0$ formula $\Psi(x_1, ..., x_k)$. Now Ψ determines a primitive recursive relation R and ordinals $\beta_1, ..., \beta_1$ such that for all α

$$\mathrm{L}_\alpha \models \Psi(N(\alpha_1), ..., N(\alpha_k) \Longleftrightarrow \alpha \models R(\beta_1, ..., \beta_1, \alpha_1, ..., \alpha_k) .$$

Now define θ_p by replacing $\Psi(x_1, ..., x_k)$ in θ by $R(\beta_1, ..., \beta_1, x_1, ..., x_k)$. Then θ_p is a $\mathbf{\Pi}_m^0$ sentence of $\mathcal{L}_p$ satisfying the lemma.

We conclude this section with a characterisation of admissible ordinals that will be useful in the appendix. We state it in relativised form.

3.7. Theorem. *Let A be a relation on ordinals. The ordinal β is admissible relative to $A \upharpoonright \beta$ if and only if for all $\alpha < \beta$ and all $R \subseteq^3$ ON that is primitive recursive in A if*

$$\forall x < \beta\, \exists y < \beta\, R(\alpha, x, y)$$

then there is $\alpha < \lambda < \beta$ such that

$$\forall x < \lambda\, \exists y < \lambda\, R(\alpha, x, y)\,.$$

Proof. Note that this characterisation uses a restricted form of $\mathbf{\Pi}_2^0$ *-reflection. Hence it is only necessary to observe that this special form is sufficient for the proofs of 3.2 and 3.3.

§4. The relative sizes of the first order reflecting ordinals

In this section we shall need some more results about ordinal recursion on an admissible ordinal. If f is a partial function on the admissible ordinal α then f is *α-partial recursive* if the graph of f is definable on L_α by a $\mathbf{\Sigma}_1^0$ formula of $\mathcal{L}_\in$.

As in Theorem 4.4 of [8] we may prove:

4.1. Normal Form Theorem. *For each $n \geq 0$ there is a primitive recursive relation T_n and there is a primitive recursive function U such that if α is admissible and f is an n-ary α-partial recursive function then there is an $e < \alpha$ such that for $\alpha_1, \ldots, \alpha_n < \alpha$*

$$f(\alpha_1, \ldots, \alpha_n) \simeq U(\mu_\alpha y\, T_n(e, \alpha_1, \ldots, \alpha_n, y))\,.$$

Moreover e depends only on a $\mathbf{\Sigma}_1^0$ formula of $\mathcal{L}_\in$ that defines the graph of f on L_α. If this formula contains no constants then $e < \omega$. e is called an α-index of f.

Note the uniformity in this theorem. For example it follows that if $F:{}^{n}\mathrm{ON}\to\mathrm{ON}$ is primitive recursive then there is an $e<\omega$ such that $F\restriction\alpha$ is α-recursive with α-index e for all admissible ordinals α.

Let us write $\{e\}_\alpha(\alpha_1,\ldots,\alpha_n)$ for $U(\mu_\alpha y\,T_n(e,\alpha_1,\ldots,\alpha_n,y))$. It will be useful to allow $n=0$.

The next result is a uniform generalisation of Kleene's $S-m-n$ theorem.

4.2. Theorem. *For each $m>0$ there is a primitive recursive function S_m such that for all admissible ordinals α if $e,a_1,\ldots,a_m,\alpha_1,\ldots,\alpha_n<\alpha$ then $\{e\}_\alpha(a_1,\ldots,a_m,\alpha_1,\ldots,\alpha_n)\simeq\{S_m(e,a_1,\ldots,a_m)\}_\alpha(\alpha_1,\ldots,\alpha_n)$.*

This theorem may be proved roughly as follows: If f is an $m+n$-ary α-partial recursive function whose graph is defined by the Σ_1^0 formula $\theta(x_1,\ldots,x_m,x_{m+1},\ldots,x_{m+n})$ on L_α then for $a_1\ldots a_m<\alpha$ $\lambda\alpha_1\ldots\alpha_n f(a_1,\ldots,a_m,\alpha_1,\ldots,\alpha_n)$ is also α-partial recursive, with graph defined by the Σ_1^0 formula $\theta(a_1,\ldots,a_m,x_1,\ldots,x_n)$ on L_α. Now S_m is chosen so that if e is the index of f determined by $\theta(x_1,\ldots,x_{m+n})$ then $S_m(e,a_1,\ldots,a_m)$ is the index of $\lambda\alpha_1,\ldots,\alpha_n f(a_1\ldots a_m,\alpha_1\ldots\alpha_n)$ determined by $\theta(a_1,\ldots,a_m,x_1,\ldots,x_n)$. We leave a detailed definition of S_m as a primitive recursive function independent of α to the imagination of the reader.

We now use Theorem 4.1 to define universal Π_{m+1}^0 and Σ_{m+1}^0 formulae of $\mathcal{L}_p$. For each $n\geq 0$ let $\Sigma_1(x_0,\ldots,x_n)$ be $\exists y T_n(x_0,\ldots,x_n,y)$ and let $\Sigma_{m+1}(x_0,\ldots,x_n)$ be $\exists y\Pi_m(x_0,\ldots,x_n,y)$ for $m>0$, where $\Pi_m(x_0,\ldots,x_k)$ is $\neg\Sigma_m(x_0,\ldots,x_k)$. Clearly $\Sigma_m(x_0,\ldots,x_n)$ is a Σ_m^0 formula of $\mathcal{L}_p$ and $\Pi_m(x_0,\ldots,x_n)$ is a Π_m^0 formula of $\mathcal{L}_p$ for each $m>0$, $n\geq 0$.

Let us call two formulae of $\mathcal{L}_p$ $\theta_1(x_1,\ldots,x_n)$, $\theta_2(x_1,\ldots,x_n)$ *equivalent on* α if for all $a_1,\ldots,a_n<\alpha$

$$\alpha\models\theta_1(a_1,\ldots,a_n)\Leftrightarrow\alpha\models\theta_2(a_1,\ldots,a_n)\,.$$

4.3. Lemma. *Let $m>0$. If $\varphi(x_1,\ldots,x_n)$ is a Σ_m^{0-} (Π_m^{0-}) formula of $\mathcal{L}_p$ then there is an $e<\omega$ such that $\varphi(x_1,\ldots,x_n)$ and $\Sigma_m(e,x_1,\ldots,x_n)(\Pi_m(e,x_1,\ldots,x_n))$ are equivalent on every admissible ordinal.*

Proof. This is by induction on m. Note that the Π_m^0 case follows from the

$\boldsymbol{\Sigma}_m^0$ case by taking negations. If $m = 1$ and $\theta(x_1, ..., x_n)$ is a $\boldsymbol{\Sigma}_1^{0-}$ formula of $\mathcal{L}_p$, then, using the stability Theorem 2.5 of [8], we may find a $\boldsymbol{\Sigma}_1^{0-}$ formula $\varphi(x_1, ..., x_n, x_{n+1})$ of $\mathcal{L}_\in$ such that for admissible α and $\alpha_1, ..., \alpha_n, \beta < \alpha$

$$\mathrm{L}_\alpha \models \varphi(\alpha_1, ..., \alpha_n, \beta) \Leftrightarrow \alpha \models \theta(\alpha_1, ..., \alpha_n) \;\&\; \beta = 0 .$$

But $\varphi(x_1, ..., x_{n+1})$ defines the graph of an α-partial recursive function on each admissible ordinal with index $e < \omega$ independent of α. Hence if α is admissible and $\alpha_1, ..., \alpha_n < \alpha$ then

$$\begin{aligned} \alpha \models \theta(\alpha_1, ..., \alpha_n) &\Leftrightarrow \mathrm{L}_\alpha \models \varphi(\alpha_1, ..., \alpha_n, \beta) \quad \text{for some } \beta \\ &\Leftrightarrow \{e\}_\alpha(\alpha_1, ..., \alpha_n) \quad \text{is defined} \\ &\Leftrightarrow \alpha \models \exists y T_n(e, \alpha_1, ..., \alpha_n, y) . \end{aligned}$$

Hence $\theta(x_1, ..., x_n)$ is equivalent to $\boldsymbol{\Sigma}_1(e, x_1, ..., x_n)$ on admissibles.

Now suppose that the result has been proved for $m > 0$ and let $\varphi(x_1, ..., x_n)$ be $\boldsymbol{\Sigma}_{m+1}^0$. Then we may assume that it has the form $\exists y_1 ... \exists y_k \theta(y_1, ..., y_k, x_1, ..., x_n)$ for some $\boldsymbol{\Pi}_m^0$ formula $\theta(y_1, ..., y_k, x_1, ..., x_n)$.

Now let G be the graph of a primitive recursive function mapping k-tuples of ordinals one-one onto the ordinals. Then $\varphi(x_1, ..., x_n)$ is equivalent on every admissible to $\exists y \theta'(x_1, ..., x_n, y)$ where $\theta'(x_1, ..., x_n, y)$ is the $\boldsymbol{\Pi}_m^0$ formula

$$\forall y_1 ... \forall y_k (G(y_1, ..., y_k, y) \to \theta(y_1, ..., y_k, x_1, ..., x_n)) .$$

By induction hypothesis there is an $e < \omega$ such that $\theta'(x_1, ..., x_n, y)$ is equivalent to $\boldsymbol{\Pi}_m(e, x_1, ..., x_n, y)$ on every admissible. Hence $\varphi(x_1, ..., x_n)$ is equivalent to $\boldsymbol{\Sigma}_{m+1}(e, x_1, ..., x_n)$ on every admissible.

4.4. Corollary. *If $X \subseteq \mathrm{Ad}$ then for $n > 0$*

$$\alpha \in M_n(X) \Leftrightarrow \alpha \in X \;\&\; \forall a < \alpha [\alpha \models \boldsymbol{\Pi}_n(a) \Rightarrow (\exists \beta \in X \cap \alpha) \beta \models \boldsymbol{\Pi}_n(a)] .$$

Proof. By Theorem 1.10 $\alpha \in M_n(X)$ if and only if $\alpha \in X$ and for every $\boldsymbol{\Pi}_n^0$ sentence φ of $\mathcal{L}_p$, $\alpha \models \varphi \Rightarrow (\exists \beta \in X \cap \alpha) \beta \models \varphi$. By Lemma 4.3 and Theorem

4.2 if φ is a $\mathbf{\Pi}_n^0$ sentence of $\mathcal{L}_p$ then there is an ordinal a such that φ is equivalent to $\mathbf{\Pi}_n(a)$ on every admissible. The corollary now follows when $X \subseteq \mathrm{Ad}$.

Below we shall be concerned with operators F on classes of ordinals that have the following properties.

4.5. (i) $F(X) \subseteq \mathrm{L}(X)$
(ii) $X \subseteq Y \Rightarrow F(X) \subseteq F(Y)$
(iii) $\lambda < \alpha \in F(X) \Rightarrow \alpha \in F(X \cap (\lambda, \alpha])$
where $(\lambda, \alpha] = \{\beta : \lambda < \beta \leq \alpha\}$.
It follows from (iii) that for all λ

$$F(X) \subseteq F(X \cap (\lambda, \infty]) \cup (\lambda + 1)$$

where $(\lambda, \infty] = \{\beta : \lambda < \beta\}$.

Examples of such F are $\mathrm{L}, M, H_n, RM, M_n$. Moreover, if F has these properties, then so does F^λ for $\lambda > 0$ and also F^Δ.

4.6. Definition. If F satisfies (i)–(iii) above and $n > 0$, then F *is* $\mathbf{\Pi}_n^0$*-preserving* if there is a primitive recursive function $f : \mathrm{ON} \to \mathrm{ON}$ such that if $X = \{\alpha \in \mathrm{Ad} : \alpha \models \mathbf{\Pi}_n(a)\}$ then
a) $F(X) = \{\alpha \in \mathrm{Ad} : \alpha \models \mathbf{\Pi}_n(f(a))\}$ and
b) $M_n(\mathrm{Ad}) \subseteq X \cup \mu \Rightarrow M_n(\mathrm{Ad}) \subseteq F(X) \cup \mu$ for all $\mu \in \mathrm{ON}$.

4.7. Lemma. *For* $n > 0$, M_n *is* $\mathbf{\Pi}_{n+1}^0$*-preserving.*

Proof. If $X = \{\alpha \in \mathrm{Ad} : \alpha \models \mathbf{\Pi}_{n+1}(a)\}$ then by 4.4 $\alpha \in M_n(X)$ if and only if $\alpha \in \mathrm{Ad}$ & $\alpha \models [\mathbf{\Pi}_{n+1}(a)$ & $\forall x \exists y (\mathbf{\Pi}_n(x) \to R(a,x,y))]$ where R is the primitive recursive relation such that $R(a,b,\beta) \Longleftrightarrow \beta \models \mathbf{\Pi}_n(b)$ & $\beta \in \mathrm{Ad}$ & $\beta \models \mathbf{\Pi}_{n+1}(a)$. So by 4.3 $M_n(X) = \{\alpha \in \mathrm{Ad} : \alpha \models \mathbf{\Pi}_{n+1}(e,a)\}$ for some $e < \omega$. Now if $f = \lambda x\, S_1(e,x)$ then f is primitive recursive and

$$M_n(X) = \{\alpha \in \mathrm{Ad} : \alpha \models \mathbf{\Pi}_{n+1}(f(a))\}.$$

Now let $M_{n+1}(\mathrm{Ad}) \subseteq X \cup \mu$ and let $\alpha \in M_{n+1}(\mathrm{Ad})$. We must show that $\alpha \in M_n(X) \cup \mu$. If $\alpha < \mu$, then we are done. Otherwise $\alpha \in X$ so that $\alpha \in \mathrm{Ad}$ and $\alpha \models \mathbf{\Pi}_{n+1}(a)$. Now suppose that $\alpha \models \mathbf{\Pi}_n(b)$. Then $\alpha \models \mathbf{\Pi}_n(b) \wedge \mathbf{\Pi}_{n+1}(a)$.

As α is $\mathbf{\Pi}^0_{n+1}$-reflecting on Ad there is a $\beta \in \mathrm{Ad} \cap \alpha$ such that $\beta \models \mathbf{\Pi}_n(b) \wedge \mathbf{\Pi}_{n+1}(a)$. Hence $\beta \in X \cap \alpha$ and $\beta \models \mathbf{\Pi}_n(b)$. Thus we have shown that $\alpha \in M_n(X)$.

4.8. Lemma. *If F is $\mathbf{\Pi}^0_n$-preserving, then so is F^Δ.*

Proof. Let F be $\mathbf{\Pi}^0_n$-preserving and let f be a primitive recursive function such that $F(\{\alpha \in \mathrm{Ad} : \alpha \models \mathbf{\Pi}_n(a)\}) = \{\alpha \in \mathrm{Ad} : \alpha \models \mathbf{\Pi}_n(f(a))\}$. Our first aim is to find a primitive recursive function g such that for admissible α and $a, c \in \mathrm{ON}$

$$(1) \quad \alpha \models \mathbf{\Pi}_n(g(a,c)) \Leftrightarrow c < \alpha \;\&\; (\forall b < c)\, \alpha \models \mathbf{\Pi}_n(f(g(a,b))) \;\&\; \alpha \models \mathbf{\Pi}_n(a)\,.$$

So let $\theta(x_1,x_2,x_3)$ be the formula

$$\mathbf{\Pi}_n(x_2) \wedge \forall y \forall u \forall v (y < x_3 \wedge T_2(x_1,x_2,y,u) \wedge R(u,v) \rightarrow \mathbf{\Pi}_n(v))$$

where $R = \{(u,v) : f(U(u)) = v\}$ is primitive recursive. Clearly this is $\mathbf{\Pi}^{0-}_n$ so that $\theta(x_1,x_2,x_3)$ is equivalent on admissibles to $\mathbf{\Pi}_n(e_0,x_1,x_2,x_3)$ for some $e_0 < \omega$. By a uniform version of the second recursion theorem on admissible ordinals there is an $e < \omega$ such that $\{e\}_\alpha(a,x) \simeq S_3(e_0,e,a,x)$ for $a, x < \alpha$ and admissible α. Now let $g = \lambda a, x S_3(e_0,e,a,x)$. Then on admissibles $\mathbf{\Pi}_n(g(a,c))$ is equivalent to $\mathbf{\Pi}_n(e_0,e,a,c)$ which is equivalent to $\theta(e,a,c)$. Hence for admissible α

$$\begin{aligned}
&\alpha \models \mathbf{\Pi}_n(g(a,c))\\
&\Leftrightarrow c < \alpha \;\&\; \alpha \models \mathbf{\Pi}_n(a) \;\&\; (\forall b<c)(\forall u,v<\alpha)(T_2(e,a,b,u) \;\&\; f(U(u)) = v\\
&\qquad\qquad \Rightarrow \alpha \models \mathbf{\Pi}_n(v))\\
&\Leftrightarrow c < \alpha \;\&\; \alpha \models \mathbf{\Pi}_n(a) \;\&\; (\forall b<c)\, \alpha \models \mathbf{\Pi}_n(f(g(a,b)))
\end{aligned}$$

so that (1) is proved.

Let

$$F^{(\beta)}(X) = \{\alpha > \beta : \alpha \in F^\beta(X)\}\,.$$

Our next aim is to show that for all $\beta \in \mathrm{ON}$

$$(2) \quad F^{(\beta)}(\{\alpha \in \mathrm{Ad}) : \alpha \models \mathbf{\Pi}_n(a)\}) = \{\alpha \in \mathrm{Ad} : \alpha \models \mathbf{\Pi}_n(g(a,\beta))\}\,.$$

This will be proved by induction on β. Let $X = \{\alpha \in \mathrm{Ad} : \alpha \models \mathbf{\Pi}_n(a)\}$. By induction hypothesis, for $b < \beta < \alpha$

$$\begin{aligned} \alpha \in F(F^b(X)) &\Longleftrightarrow \alpha \in F(F^{(b)}(X)) \qquad \text{by 4.5 (iii)} \\ &\Longleftrightarrow \alpha \in \mathrm{Ad} \ \& \ \alpha \models \mathbf{\Pi}_n(f(g(a,b))) \\ &\Longleftrightarrow \alpha \in X \ \& \ \alpha \models \mathbf{\Pi}_n(f(g(a,b))) . \end{aligned}$$

Hence by (1)

$$\begin{aligned} \alpha \in F^{(\beta)}(X) &\Longleftrightarrow \beta < \alpha \ \& \ \alpha \in F^{\beta}(X) \\ &\Longleftrightarrow \beta < \alpha \ \& \ (\forall b < \beta)\, \alpha \in F(F^b(X)) \\ &\Longleftrightarrow \beta < \alpha \ \& \ \alpha \in X \ \& \ (\forall b < \beta)\, \alpha \models \mathbf{\Pi}_n(f(g(a,b))) \\ &\Longleftrightarrow \beta < \alpha \ \& \ \alpha \in \mathrm{Ad} \ \& \ \alpha \models \mathbf{\Pi}_n(a) \ \& \ (\forall b < \beta)\, \alpha \models \mathbf{\Pi}_n(f(g(a,b))) \\ &\Longleftrightarrow \alpha \in \mathrm{Ad} \ \& \ \alpha \models \mathbf{\Pi}_n(g(a,\beta)) . \end{aligned}$$

So that (2) is proved.

Now we shall find a primitive recursive function f' such that

(3) $\quad F^{\Delta}(\{\alpha \in \mathrm{Ad} : \alpha \models \mathbf{\Pi}_n(a)\}) = \{\alpha \in \mathrm{Ad} : \alpha \models \mathbf{\Pi}_n(f'(a))\}$.

Let $X = \{\alpha \in \mathrm{Ad} : \alpha \models \mathbf{\Pi}_n(a)\}$. The formula $\forall x \forall y [g(z,x) = y \to \mathbf{\Pi}_n(y)]$ is a $\mathbf{\Pi}_n^{0-}$ formula so that there is an $e_1 < \omega$ such that for admissible α and $a \in \mathrm{ON}$

$$\begin{aligned} \alpha \models \mathbf{\Pi}_n(e_1, a) &\Longleftrightarrow \alpha \models \forall x \forall y [g(a,x) = y \to \mathbf{\Pi}_n(y)] \\ &\Longleftrightarrow (\forall b < \alpha)\, \alpha \models \mathbf{\Pi}_n(g(a,b)) \\ &\Longleftrightarrow (\forall b < \alpha)\, \alpha \in F^{(b)}(X) \qquad \text{by (2)} \\ &\Longleftrightarrow (\forall b < \alpha)\, \alpha \in F^b(X) \\ &\Longleftrightarrow \alpha \in F^{\Delta}(X) . \end{aligned}$$

But $F^{\Delta}(X) \subseteq X \subseteq \mathrm{Ad}$ so that $F^{\Delta}(X) = \{\alpha \in \mathrm{Ad} : \alpha \models \mathbf{\Pi}_n(f'(a))\}$ where $f' = \lambda x S_1(e_1, x)$. So (3) is proved.

It now remains to show that if $X = \{\alpha \in \mathrm{Ad} : \alpha \models \mathbf{\Pi}_n(a)\}$ and

$M_n(\mathrm{Ad}) \subseteq X \cup \mu$ then $M_n(\mathrm{Ad}) \subseteq F^{\Delta}(X) \cup \mu$. So let X, μ satisfy the above assumptions. We first show that for all $\beta \in \mathrm{ON}$:

$$(4) \qquad M_n(\mathrm{Ad}) \subseteq F^{\beta}(X) \cup \mathrm{Max}(\mu, \beta+1) .$$

This will be proved by induction on β. By induction hypothesis, if $b < \beta$, then

$$\begin{aligned} M_n(\mathrm{Ad}) &\subseteq F^{b}(X) \cup \mathrm{Max}(\mu, b+1) \\ &\subseteq F^{(b)}(X) \cup \mathrm{Max}(\mu, b+1) . \end{aligned}$$

But as F is $\mathbf{\Pi}^0_n$-preserving, by (2), if $b < \beta$, then

$$\begin{aligned} M_n(\mathrm{Ad}) &\subseteq F(F^{(b)}(X)) \cup \mathrm{Max}(\mu, b+1) \\ &\subseteq F(F^{b}(X)) \cup \mathrm{Max}(\mu, \beta+1) \qquad \text{by 4.5 (iii)} . \end{aligned}$$

Hence

$$\begin{aligned} M_n(\mathrm{Ad}) &\subseteq (X \cap \textstyle\bigcap_{b<\beta} F(F^{b}(X))) \cup \mathrm{Max}(\mu, \beta+1) \\ &\subseteq F^{\beta}(X) \cup \mathrm{Max}(\mu, \beta+1) . \end{aligned}$$

Hence (4) is proved and now if $\alpha \in M_n(\mathrm{Ad})$ then if $\alpha < \mu$ we are done. Otherwise, by (4) $\alpha \in \bigcap_{\beta<\alpha} F^{\beta}(X)$ so that $\alpha \in F^{\Delta}(X) \cup \mu$. Thus $M_n(\mathrm{Ad}) \subseteq F^{\Delta}(X) \cup \mu$.

We can now prove Theorem 1.11.

If F is $\mathbf{\Pi}^0_{n+1}$-preserving, then $M_{n+1}(\mathrm{Ad}) \subseteq F(\mathrm{Ad})$ as $\mathrm{Ad} = \{\alpha \in \mathrm{Ad} : \alpha \models \mathbf{\Pi}_n(e_0)\}$ for some $e_0 < \omega$. Hence Theorem 1.11 follows from the previous two lemmas.

4.9. *Remark.* If Y is a primitive recursive class of ordinals such that $Y \subseteq \mathrm{Ad}$ and Ad is replaced by Y in Definition 4.6, then the proofs of the previous two lemmas still hold so that we get that for $n > 0$:

$$M_{n+1}(Y) \subseteq M_n^{\Delta}(Y) , \qquad (M_n^{\Delta})^{\Delta}(Y) , \quad \text{etc.} .$$

§5. Reflecting ordinals and indescribable cardinals

In this section we will prove Theorems 1.14 and 1.16.

5.1. Lemma. *If κ is 2-regular then $\kappa > \omega$ and κ is regular.*

Proof. Let κ be 2-regular. It suffices to show that every $g : \kappa \to \kappa$ has a witness. For a given g, let $F : {}^{\kappa}\kappa \to {}^{\kappa}\kappa$ be defined by $F(f)(\xi) = g(f(0))$ for all $f : \kappa \to \kappa$ and all $\xi < \kappa$. F is clearly κ-bounded. Let α be a witness for F. We show α is a witness for g. Let $\beta < \kappa$ and $f : \kappa \to \kappa$ such that $f(\xi) = \beta$ for all $\xi < \kappa$. Then $f''\alpha \subseteq \alpha$ and hence $F(f)''\alpha \subseteq \alpha$. Thus

$$g(\beta) = g(f(0)) = F(f)(0) < \alpha .$$

Hence $g''\alpha \subseteq \alpha$.

5.2. Proof of Theorem 1.14. κ is 2-regular iff κ is strongly $\mathbf{\Pi}^1_1$-indescribable.
We show

$$\text{(a) } \kappa \text{ 2-regular} \Rightarrow \left\{ \begin{array}{l} \text{(b) } \kappa \text{ is strongly inaccessible} \\ \& \\ \text{(c) } \kappa \text{ is } \mathbf{\Pi}^1_1\text{-indescribable} \end{array} \right.$$

$$\Rightarrow \quad \text{(d) } \kappa \text{ is strongly } \mathbf{\Pi}^1_1\text{-indescribable}$$

$$\Rightarrow \quad \text{(a).}$$

We first show (a) $\Rightarrow$ (b). Let κ be 2-regular. Since κ is regular it remains to show $\lambda < \kappa \Rightarrow 2^{\lambda} < \kappa$. Suppose not. Let $\lambda < \kappa$ and $2^{\lambda} \geq \kappa$. Let r map 2^{λ} onto κ. Define $F : {}^{\kappa}\kappa \to {}^{\kappa}\kappa$ by

$$F(f)(\xi) = \begin{cases} r(f \upharpoonright \lambda) & \text{if } \ f''\lambda \subseteq 2\,, \\ 0, & \text{otherwise,} \end{cases}$$

for $\xi < \kappa$. F is κ-bounded since $F(f)(\xi)$ is determined by values of f on $\lambda < \kappa$. Let α be a witness for F. It is clear that α can be chosen ≥ 2. Let $g : \lambda \to 2$ such that $r(g) > \alpha$ and let $f : \kappa \to \kappa$ so that $f \upharpoonright \lambda = g$ and $f(\xi) = 0$ for $\xi \geq \lambda$. Then $f''\alpha \subseteq 2 \subseteq \alpha$. Since α is a witness for F, $F(f)(0) < \alpha$. But

$$F(f)(0) = r(f \upharpoonright \lambda) = r(g) > \alpha$$

which is a contradiction.

To show (a) ⇒ (c) let φ be a $\mathbf{\Pi}_1^1$ sentence of $\mathcal{L}$ such that $\kappa \models \varphi$. We must find a $0 < \alpha < \kappa$ such that $\alpha \models \varphi$. Let P be one of the standard bijective mappings of On × On onto On, and K, L be the associated pairing functions (cf. Lévy [11]). We first switch from set quantifiers to quantifiers of binary relations which are characteristic functions, and then switch to the language with unary function quantifiers instead of set quantifiers. In this language with the aid of P, K, L we can put $\mathbf{\Pi}_1^1$ formulas in a normal form (cf. Rogers [17], where this is done for formulas of second-order arithmetic). Thus there is a quantifier-free formula Q such that $\kappa \models \forall f \exists \xi Q(f, \xi)$ and for every $\alpha \leq \kappa$ which is closed under P,

$$\alpha \models \varphi \Leftrightarrow \alpha \models \forall f \exists \xi Q(f, \xi) .$$

Furthermore, Q can be chosen so that in Q there is no nesting of f (i.e. no terms of the form $f(f(...)))$. For a given f and ξ the truth or falsity of $Q(f, \xi)$ is determined by the values of f for finitely many arguments and the answers to finitely many questions about membership in the relations appearing in Q. Since there is no nesting of f the finite set of arguments of f needed depends only on ξ and the relations in Q but not on f itself. Thus,

$$\forall \xi < \kappa \, \exists \eta < \kappa B(\xi, \eta) ,$$

where

$$B(\xi, \eta) \Leftrightarrow \forall f \forall g [f \upharpoonright \eta = g \upharpoonright \eta \Rightarrow Q(f, \xi) \Leftrightarrow Q(g, \xi)] .$$

Hence, since κ is regular,

$$\forall \beta < \kappa \, \exists \eta < \kappa . \, C(\beta, \eta) ,$$

where

$$C(\beta, \eta) \Leftrightarrow \forall \xi \leq \beta . \, B(\xi, \eta) .$$

Let $h(\beta) = \mu\eta . \, C(\beta, \eta)$. Then $h : \kappa \to \kappa$ and for all $f, g : \kappa \to \kappa$,

(1) $f \restriction h(\beta) = g \restriction h(\beta) \Rightarrow [\forall \xi \leq \beta.\ Q(f,\xi) \Leftrightarrow Q(g,\xi)]$.
Let $G : {}^{\kappa}\kappa \to {}^{\kappa}\kappa$ so that
(2) $G(f) = \mu\sigma\,[P''\sigma \times \sigma \subseteq \sigma \;\&\; \exists \xi \leq \sigma.\ Q(f,\xi) \;\&\; h(\xi) \leq \sigma]$,
and let $F : {}^{\kappa}\kappa \to {}^{\kappa}\kappa$ so that $F(f)(\beta) = G(f)$ for all $\beta < \kappa$. F is κ-bounded since

$$\mu\xi.\,Q(f,\xi) = \mu\xi.\,Q(g,\xi) \Rightarrow F(f) = F(g)\,.$$

Let α be a witness for F and let $f : \alpha \to \alpha$. We show $\exists \xi < \alpha.\ Q(f,\xi)$. Let $g : \kappa \to \kappa$ so that $g \restriction \alpha = f$. Then $g''\alpha \subseteq \alpha$ and $F(g)''\alpha \subseteq \alpha$ since α is a witness. Let $\delta = \mu\xi.\,Q(g,\xi)$. Then $\delta, h(\delta) \leq G(g) \leq F(g)(0) < \alpha$ by definition of G. Thus from (1),

$$\forall \xi \leq \delta\ \ Q(f,\xi) \Leftrightarrow Q(g,\xi)\,,$$

and hence $\mu\xi.\,Q(f,\xi) = \delta < \alpha$. Thus $\alpha \models \forall f\, \exists \xi Q(f,\xi)$ and since α is closed under P (by (2)), $\alpha \models \varphi$.

A proof that (b) & (c) $\Rightarrow$ (d) appears in Lévy [11] p. 217. It remains to show (d) $\Rightarrow$ (a). Let κ be strongly $\mathbf{\Pi}^1_1$-indescribable and $F : {}^{\kappa}\kappa \to {}^{\kappa}\kappa$ be κ-bounded. We show that F has a witness. Let

$$X = \{\langle f \restriction \gamma, \xi, \eta\rangle : f \in {}^{\kappa}\kappa \;\&\; \xi, \eta, \gamma < \kappa \;\&\; \forall g \in {}^{\kappa}\kappa\,[g \restriction \gamma = f \restriction \gamma \Rightarrow F(g)(\xi) = \eta]\}.$$

Then $X \subseteq R(\kappa)$. Note that $\langle f \restriction \gamma, \xi, \eta\rangle \in X \Rightarrow F(f)(\xi) = \eta$. Since F is κ-bounded,

$$R(\kappa) \models \forall f\, \forall \xi\, \exists \gamma, \eta\,[\langle f \restriction \gamma, \xi, \eta\rangle \in X]\,.$$

Since κ is strongly $\mathbf{\Pi}^1_1$-indescribable there is an $\alpha < \kappa$ such that

$$R(\alpha) \models \forall f\, \forall \xi\, \exists \gamma, \eta\,[\langle f \restriction \gamma, \xi, \eta\rangle \in X]\,,$$

i.e. $0 < \alpha < \kappa$ and
(3) $\forall f \in {}^{\alpha}\alpha\, \forall \xi < \alpha\, \exists \gamma, \eta < \alpha\,[\langle f \restriction \gamma, \xi, \eta\rangle \in X \cap R(\alpha)]$.
We show α is a witness for F. Let $f : \kappa \to \kappa$ such that $f''\alpha \subseteq \alpha$. Since $f \restriction \alpha \in {}^{\alpha}\alpha$ and $f \restriction \alpha \restriction \gamma = f \restriction \gamma$ for $\gamma < \alpha$, we have from (3)

$$\forall \xi < \alpha\, \exists \eta < \alpha\, F(f)(\xi) = \eta, \quad \text{i.e. } F(f)''\alpha \subseteq \alpha\,.$$

5.3. *Remark.* In the proof of (d) ⇒ (a), the assumption that F is κ-bounded cannot be eliminated. For each κ it is easy to define an $F : {}^{\kappa}\kappa \to {}^{\kappa}\kappa$ which is not κ-bounded and has no witness.

5.4. Proof of Theorem 1.16. κ is 2-admissible iff κ is $\mathbf{\Pi}_3^0$-reflecting.

Suppose κ is $\mathbf{\Pi}_3^0$-reflecting. Let $\{\xi\}_\kappa$ map κ-recursive functions to κ-recursive functions. We show ξ has a witness. By hypothesis,

$$\forall\beta < \kappa[\{\beta\}_\kappa : \kappa \to \kappa \Rightarrow \{\{\xi\}_\kappa(\beta)\}_\kappa : \kappa \to \kappa] .$$

By using the T predicate this is equivalent to

$$\kappa \models \forall x[\forall y\, \exists z T(x,y,z) \to \forall y . T(\xi,x,y) \to \forall u\, \exists v T(U(y),u,v)] .$$

The sentence on the right is equivalent to a $\mathbf{\Pi}_3^0$ sentence $\varphi(\xi)$. Since κ is $\mathbf{\Pi}_3^0$-reflecting (and hence $\mathbf{\Pi}_3^0$-reflecting on Ad) there is some $\alpha \in \kappa \cap \mathrm{Ad}$ such that $\alpha \models \varphi(\xi)$. But by the definition of $\varphi(\xi)$ this implies $\{\xi\}_\alpha$ maps α-recursive functions to α-recursive functions and hence α is a witness for ξ.

Now suppose κ is 2-admissible and let φ be a $\mathbf{\Pi}_3^0$ sentence of $\mathcal{L}_p$ such that $\kappa \models \varphi$. We show that κ reflects φ. For simplicity we assume that φ is of the form $\forall x\, \exists y\, \forall z\, \psi(x,y,z)$ where ψ is a $\mathbf{\Sigma}_0^0$ formula with constants less than κ. Let α be admissible so that all constants in ψ are less than α. We introduce certain Gödel numbers of α-partial recursive functions which, by the uniformity of the Normal Form and $S-m-n$ theorems, can be chosen to be independent of the particular choice of α. First choose $a < \alpha$ so that

$$\{S_1(a,\beta)\}(\gamma) \simeq \{a\}_\alpha(\beta,\gamma) \simeq \mu_\alpha \delta . \neg\psi(\beta,\gamma,\delta) .$$

Then,

$$\begin{aligned}
(4) \quad \alpha \models \varphi &\Leftrightarrow \forall\beta < \alpha\, \exists\gamma < \alpha\, \forall\delta < \alpha\, \psi(\beta,\gamma,\delta) \\
&\Leftrightarrow \forall\beta < \alpha \neg \forall\gamma < \alpha\, \exists\delta < \alpha \neg\psi(\beta,\gamma,\delta) \\
&\Leftrightarrow \forall\beta < \alpha . \neg\{S(a,\beta)\}_\alpha : \alpha \to \alpha \\
&\Leftrightarrow \forall\beta < \alpha . \{\beta\}_\alpha : \alpha \to \alpha \Rightarrow \forall\gamma < \alpha\, \exists\delta < \alpha . S_1(a,\gamma) = \delta \;\&\; \delta \neq \beta .
\end{aligned}$$

Let g be a primitive recursive ordinal function such that

$$\{g(\beta)\}_\alpha(\gamma) \simeq \mu_\alpha \delta . S_1(a,\gamma) = \delta \;\&\; \delta \neq \beta$$

and let $\xi < \omega$ be a Gödel number (independent of α) of g. Then from (4),

$$(5) \quad \alpha \models \varphi \Leftrightarrow \forall \beta < \alpha . \{\beta\}_\alpha : \alpha \to \alpha \Rightarrow \{g(\beta)\}_\alpha : \alpha \to \alpha$$

$$\Leftrightarrow \forall \beta < \alpha . \{\beta\}_\alpha : \alpha \to \alpha \Rightarrow \{\{\xi\}_\alpha(\beta)\}_\alpha : \alpha \to \alpha$$

$$\Leftrightarrow \{\xi\}_\alpha \text{ maps } \alpha\text{-recursive functions to } \alpha\text{-recursive functions.}$$

Since $\kappa \models \varphi$, by (5) $\{\xi\}_\kappa$ maps κ-recursive functions to κ-recursive functions. Since κ is 2-admissible there is an $\alpha \in \kappa \cap \mathrm{Ad}$ which is a witness for ξ. But then by (5), $\alpha \models \varphi$.

5.5. *Remark.* The definition of 2-admissible given here is equivalent to the definition which appears in [2]. The full definition of n-admissible is given in [2].

§6. Stability

In this section we prove Theorems 1.18 and 1.19.

Note that if $A \prec_{\Sigma_1^0} B$ and $A \subseteq C \subseteq B$ then $A \prec_{\Sigma_1^0} C$. It follows that if α is β-stable and $\alpha < \gamma < \beta$ then α is γ-stable. Hence the weakest stability property for an ordinal α is that of being $\alpha+1$-stable. 1.18 implies that even this weakest stability property determines ordinals with rather strong reflecting properties. But first we need:

6.1. Lemma. *If α is $\alpha+1$-stable then α is admissible.*

Proof. Let $L_\alpha \models \forall x \in a \varphi$ where φ is a Σ_1^0 formula $\mathcal{L}_\in$. Then $L_{\alpha+1} \models \forall x \in a \varphi^b$ where $b = L_\alpha \in L_{\alpha+1}$. Hence $L_{\alpha+1} \models \exists z \forall x \in a \varphi^z$. If α is $\alpha+1$-stable then $L_\alpha \models \exists z \forall x \in a \varphi^z$. Hence L_α satisfies Σ_1^0-collection. The lemma now follows, as α is clearly a limit ordinal so that L_α is rud closed.

Proof of 1.18. α is Π_0^1-reflecting if and only if α is $\alpha+1$-stable.

Let α be Π_0^1-reflecting. Let φ be a Σ_1^0 sentence of $\mathcal{L}_\in$, with constants only for sets in $\mathcal{L}_\in$, such that $L_{\alpha+1} \models \varphi$. We may assume that φ has the form

$\exists x_1 \ldots \exists x_n \theta(x_1, \ldots, x_n)$ where $\theta(x_1, \ldots, x_n)$ is $\mathbf{\Pi}^0_0$. So let $a_1, \ldots, a_n \in L_{\alpha+1}$ such that $L_{\alpha+1} \models \theta(a_1, \ldots, a_n)$. As $L_{\alpha+1} = \mathrm{Def}(L_\alpha)$ there are $\mathbf{\Pi}^1_0$ formulae $\theta_1(v_0), \ldots, \theta_n(v_0)$ of $\mathcal{L}_\in$, with constants in L_α, such that $a_i = \{b \in L_\alpha : L_\alpha \models \theta_i(b)\}$ for $i = 1, \ldots, n$. Let θ' be obtained from $\theta(a_1, \ldots, a_n)$ by first replacing every occurrence of $a_i \in s$ by $\exists y \in s \forall z (z \in y \leftrightarrow z \in a_i)$ and then replacing every occurrence of $s \in a_i$ by $\theta_i(s)$ for $i = 1, \ldots, n$. Clearly θ' is a $\mathbf{\Pi}^1_0$ sentence such that $L_\alpha \models \theta'$. As α is $\mathbf{\Pi}^1_0$-reflecting there is a $\beta < \alpha$ such that $L_\beta \models \theta'$. Now if $a'_i = \{b \in L_\beta : L_\beta \models \theta_i(b)\}$ then $a'_i \in L_{\beta+1}$, as we may assume that x_1 actually occurs free in $\theta(x_1, \ldots, x_n)$ so that the constants of $\theta_i(v_0)$ are also constants of θ' and hence are constants for sets in L_β. It follows that $L_\alpha \models \theta(a'_1, \ldots, a'_n)$ and hence $L_\alpha \models \varphi$.

Conversely let α be $\alpha+1$-stable. Let σ_0 be the $\mathbf{\Pi}^{0-}_3$ sentence given in Theorem 2.4. Let $\Psi(x)$ be a $\mathbf{\Sigma}^0_1$ formula of $\mathcal{L}_\in$ that defines L inside each admissible class A. Hence $\forall x \Psi(x)$ expresses V = L and the transitive models of $\sigma_0 \wedge \forall x \Psi(x)$ all have the form L_β for some admissible β. Now let φ be a $\mathbf{\Pi}^1_0$ sentence of $\mathcal{L}_\in$ such that $L_\alpha \models \varphi$. Then $L_\alpha \models \varphi_1$ where φ_1 is $\varphi \wedge \sigma_0 \vee \forall x \Psi(x)$ is also a $\mathbf{\Pi}^1_0$ sentence of $\mathcal{L}_\in$. Hence $L_{\alpha+1} \models \exists x (\mathrm{trans}(x) \wedge \varphi_1^{(x)})$ where $\varphi_1^{(x)}$ is obtained from φ_1 by restricting all quantifiers to x, and trans(x) is $\forall y \in x \forall z \in y (z \in x)$. As α is $\alpha+1$-stable it follows that $L_\alpha \models \exists x (\mathrm{trans}(x) \wedge \varphi_1^{(x)})$. Hence there is a transitive set in L_α that satisfies φ_1. But this must have the form L_β for some admissible β and $L_\beta \models \varphi$. It follows that α reflects φ, so that α is $\mathbf{\Pi}^1_0$-reflecting.

We now turn to the proof of 1.19. In fact we shall prove a generalisation of that result in 6.4. Some of the ideas in [4] will be basic to our proof. For a transitive set A let A^+ be the smallest admissible set such that $A \in A^+$. If $S \subseteq A$ we say that S is $\mathbf{\Pi}^n_m$ ($\mathbf{\Sigma}^n_m$) *over* A if $S = \{a \in A : A \models \varphi(a)\}$ for some $\mathbf{\Pi}^n_m$ ($\mathbf{\Sigma}^n_m$) formula $\varphi(x)$ of $\mathcal{L}_\in$. Theorem 3.1(a) of [4] states that if A is a countable transitive set closed under unordered pairs then for $S \subseteq A$, S is $\mathbf{\Pi}^1_1$ over A if and only if S is $\mathbf{\Sigma}^0_1$ over A^+. The proof of this result in [4] may be made to yield the following formulation which gives us the extra information we shall need.

6.2. Theorem. (i) *If $\varphi(v_1, \ldots, v_n)$ is a $\mathbf{\Pi}^1_1$ formula of $\mathcal{L}_\in$ then there is a $\mathbf{\Sigma}^0_1$ formula $\varphi^+(v_0, v_1, \ldots, v_n)$ of $\mathcal{L}_\in$ having the same constants as $\varphi(v_1, \ldots, v_n)$ such that for every non-empty countable transitive set A and every admissible set B such that $A \in B$, if $a_1, \ldots, a_n \in A$ then*

$$A \models \varphi(a_1, ..., a_n) \ \ iff \ \ B \models \varphi^+(A, a_1, ..., a_n) .$$

(ii) *If* $\varphi(v_1, ..., v_n)$ *is a* $\mathbf{\Sigma}_1^0$ *formula of* $\mathcal{L}_\in$ *then there is a* $\mathbf{\Pi}_1^1$ *formula* $\varphi^-(v_1, ..., v_n)$ *having the same constants as* $\varphi(v_1, ..., v_n)$ *such that if* A *is an infinite transitive set containing the sets whose constants occur in* $\varphi(v_1, ..., v_n)$ *then for* $a_1, ..., a_n \in A$

$$A^+ \models \varphi(a_1, ..., a_n) \ \ iff \ \ A \models \varphi^-(a_1, ..., a_n) .$$

Proof. We shall require some familiarity with the infinitary languages $\mathcal{L}_B$ for admissible B. See for example [3].

(i) Let $\varphi(v_1, ..., v_n)$ be a $\mathbf{\Pi}_1^1$ formula of $\mathcal{L}_\in$. We may assume that it has the form $\forall X_1 ... \forall X_m \theta(v_1, ..., v_n)$ where $\theta(v_1, ..., v_n)$ is a $\mathbf{\Pi}_0^1$ formula of $\mathcal{L}_\in$ with extra relation symbols $X_1, ..., X_m$. Given a non-empty transitive set A we may define the infinitary sentences $\Psi_0(A)$ and $\Psi_1(A)$ as follows: $\Psi_0(A)$ is $\bigwedge_{a \in A} \forall y (y \in a \leftrightarrow \bigvee_{b \in a} (y = b))$ and $\Psi_1(A)$ is $\forall x \bigvee_{a \in A} (a = x)$. Then the models of $\Psi_0(A) \wedge \Psi_1(A)$ are all isomorphic to $\langle A, \in \restriction A, a \rangle_{a \in A}$. Hence if $a_1, ..., a_n \in A$ then

(1) $A \models \varphi(a_1, ..., a_n)$ iff $\Psi_0(A) \wedge \Psi_1(A) \rightarrow \theta(a_1, ..., a_n)$ is logically valid.

Note that if $A \in B$ where B is admissible then $(\Psi_0(A) \wedge \Psi_1(A) \rightarrow \theta(a_1, ..., a_n)) \in B$ i.e. it is a sentence of $\mathcal{L}_B$.

Now it is a routine matter, using [3] to find $\mathbf{\Sigma}_1^0$ formulae of $\mathcal{L}_\in$ $\Psi(v_0, v_1, ..., v_{n+1})$ and $\chi(v_0)$, such that if $A, B, a_1, ..., a_n$ are as above then if $b \in B$

(2) $B \models \Psi(A, a_1, ..., a_n, b)$ iff $b = (\Psi_0(A) \wedge \Psi_1(A) \rightarrow \theta(a_1, ..., a_n))$, and if b is countable then

(3) $B \models \chi(B)$ iff b is a logically valid sentence of $\mathcal{L}_B$. (3) follows trom the completeness theorem for countable infinitary sentences (see Theorem 2.7 of [3]).

$\Psi(v_0, ..., v_{n+1})$ may be chosen to have the same constants as $\varphi(v_1, ..., v_n)$, while $\chi(v_0)$ may be chosen to have no constants.

Now let $\varphi^+(v_0, ..., v_n)$ be $\exists v_{n+1}(\Psi(v_0, ..., v_{n+1}) \wedge \chi(v_{n+1}))$. The result follows from (1)–(3) using the fact that $(\Psi_0(A) \wedge \Psi_1(A) \rightarrow \theta(a_1, ..., a_n))$ is countable if A is countable.

(ii) Let $\varphi(v_1, ..., v_n)$ be a $\mathbf{\Sigma}_1^0$ formula of $\mathcal{L}_\in$. Let KP be the theory of admissible sets, as formulated in [3]. Then by 3.3 of [4], if A is a transitive set and $\mathfrak{B}$ is an end extension of $A \cup \{A\}$ that is a model of KP then $\mathfrak{B}$ is an end extension of A^+. Hence if $a_1, ..., a_n \in A$ then $A^+ \models \theta(a_1, ..., a_n)$ iff $\mathfrak{B} \models \theta(a_1, ..., a_n)$ for every A-model $\mathfrak{B}$ of KP where an *A-model of* KP is a model of KP that is an end-extension of $A \cup \{A\}$.

Now by the downward Löwenheim–Skolem theorem every A-model $\mathfrak{B}$ of KP has an elementary subsystem $\mathfrak{B}' \prec \mathfrak{B}$ that is an A-model of KP of the same cardinality as A, assuming that A is infinite. Every such A-model $\mathfrak{B}'$ is isomorphic to $\langle A, E\rangle$ for some $E \subseteq A \times A$. Then there is an $f: A \to A$ and $a \in A$ such that

(a) $\langle A, E\rangle$ is a model of KP

(b) $f: \langle A, \in \restriction A\rangle \cong \langle a_E, E \restriction a_E\rangle$ where $a_E = \{b \in A : bEa\}$

(c) $b_E \subseteq a_E$ for all $b \in a_E$.

It follows from the above that $A^+ \models \theta(a_1, ..., a_n)$ iff $\langle A, E\rangle \models \theta(f(a_1), ...$ $..., f(a_n))$ for all $E \subseteq A \times A$, $f: A \to A$ and $a \in A$ such that (a) & (b) & (c).

It is now a routine matter to find the required $\mathbf{\Pi}^1_1$ formula obtained by formalizing the right-hand side of the above equivalence.

6.3. Definition. An admissible set A is *$\mathbf{\Pi}^1_1$-reflecting* if $A \models \varphi \Rightarrow \exists a \in A$ ($a \models \varphi$ and a is transitive) for all $\mathbf{\Pi}^1_1$ sentences φ of $\mathcal{L}_\in$.

The following is a generalisation of 1.19.

6.4. Theorem. *The countable admissible set A is $\mathbf{\Pi}^1_1$-reflecting if and only if $A \prec_{\Sigma^0_1} A^+$.*

Proof. Let A be a countable admissible set that is $\mathbf{\Pi}^1_1$-reflecting. Let φ be a $\mathbf{\Sigma}^0_1$ sentence of $\mathcal{L}_\in$, with constants only for sets in A, such that $A^+ \models \varphi$. Let τ be $\forall x\, \exists y\, (x \in y)$. Then, by (6.2)(ii) with $n = 0$, $A \models \varphi^- \wedge \tau$. Hence $a \models \varphi^- \wedge \tau$ for some transitive $a \in A$. It follows that a is an infinite transitive set such that $a \models \varphi^-$. By (6.2)(ii) $a^+ \models \varphi$. But as $a^+ \subseteq A$ and φ is $\mathbf{\Sigma}^0_1$ it follows that $A \models \varphi$. Hence $A \prec_{\Sigma^0_1} A^+$.

Conversely, let $A \prec_{\Sigma^0_1} A^+$ and let φ be a $\mathbf{\Pi}^1_1$ sentence of $\mathcal{L}_\in$ such that $A \models \varphi$. Then by (6.2)(i) with $n = 0$, $A^+ \models \varphi^+(A)$. Hence $A^+ \models \varphi_1$ where φ_1 is the $\mathbf{\Sigma}^0_1$ sentence $\exists x\, (\text{trans}(x) \wedge \varphi^+(x))$. But φ_1 only has constants for sets in A, so that $A \models \varphi_1$ i.e. $A \models \varphi^+(a)$ for some transitive set $a \in A$. As A is countable so is a, so that by (6.2)(i) $a \models \varphi$. Thus A is $\mathbf{\Pi}^1_1$-reflecting.

In order to obtain 1.19 we need:

6.5. Lemma. L_α *is* $\mathbf{\Pi}^1_1$*-reflecting iff* α *is* $\mathbf{\Pi}^1_1$*-reflecting.*

Proof. Let α be $\mathbf{\Pi}^1_1$-reflecting. If φ is a $\mathbf{\Pi}^1_1$ sentence of $\mathcal{L}_\in$ such that $L_\alpha \models \varphi$ then $L_\beta \models \varphi$ for some $\beta < \alpha$. But now $a = L_\beta$ is a transitive element of A such that $a \models \varphi$. Hence L_α is $\mathbf{\Pi}^1_1$-reflecting.

Conversely, let L_α be $\mathbf{\Pi}^1_1$-reflecting. Let σ be the $\mathbf{\Pi}^0_3$ sentence $\sigma_0 \wedge \forall x \Psi(x)$ occurring in the proof of 1.18. If φ is a $\mathbf{\Pi}^1_1$ sentence such that $L_\alpha \models \varphi$ then $L_\alpha \models \varphi \wedge \sigma$. Hence there is a transitive set $a \in L_\alpha$ such that $a \models \varphi \wedge \sigma$. But $a = L_\beta$ for some $\beta < \alpha$. So $L_\beta \models \varphi$ for some $\beta < \alpha$. Hence α is $\mathbf{\Pi}^1_1$-reflecting.

Now 1.19 follows from 6.4 and 6.5 when we observe that $(L_\alpha)^+ = L_{\alpha^+}$ for every ordinal α.

PART II. INDUCTIVE DEFINITIONS

§7. First order inductive definitions, I

We begin by considering inductive definitions which are either recursive or closely related to recursive inductive definitions. These very simple cases illustrate some of the principles used in characterizing the closure ordinals of more complicated inductive definitions.

7.1. Definition. For any inductive definitions Γ_0, Γ_1 let

$$n \in [\Gamma_0, \Gamma_1](X) \Leftrightarrow n \in \Gamma_0(X) \vee [\Gamma_0(X) \subseteq X \;\&\; n \in \Gamma_1(X)] \,.$$

Let $\Gamma = [\Gamma_0, \Gamma_1]$. In constructing the transfinite sequence $\langle \Gamma^\alpha : \alpha \in \mathrm{On} \rangle$ one repeatedly applies Γ_0 until closure under Γ_0 is reached, (i.e. until a λ is reached such that $\Gamma_0(\Gamma^\lambda) \subseteq \Gamma^\lambda$); then Γ_1 is applied once; then Γ_0 is repeatedly applied until closure under Γ_0 is reached, etc. Γ_1 is applied only when closure under Γ_0 is reached. Note that if $\Gamma_0(\Gamma^\lambda) \subseteq \Gamma^\lambda$ then $\Gamma(\Gamma^\lambda) = \Gamma_1(\Gamma^\lambda)$. $|[\Gamma_0, \Gamma_1]|$ is the least λ such that both $\Gamma_0(\Gamma^\lambda) \subseteq \Gamma^\lambda$ and $\Gamma_1(\Gamma^\lambda) \subseteq \Gamma^\lambda$.

For any recursive relation R and inductive definition Γ, the truth or falsity of $R(n, \Gamma^\lambda)$ is determined by the answers to a finite number of questions about membership in Γ^λ. For limit λ, the answers to these questions are the same as the answers to the same questions about membership in Γ^ξ for suitable large $\xi < \lambda$. Hence for recursive R and limit λ,

(1) $R(n, \Gamma^{\lambda}) \Rightarrow \exists \xi < \lambda.\ R(n, \Gamma^{\xi})$,

(2) $$R(n, \Gamma^{\lambda}) \Leftrightarrow \exists \xi < \lambda \forall \delta.\ \xi \leq \delta < \lambda \Rightarrow R(n, \Gamma^{\delta})$$
$$\Leftrightarrow \forall \xi < \lambda \exists \delta.\ \xi \leq \delta < \lambda \,\&\, R(n, \Gamma^{\delta}).$$

Using (1) and (2) we can prove the following trivial result. Let $[\mathbf{\Pi}^0_0, \mathbf{\Pi}^0_0] = \{[\Gamma_0, \Gamma_1] : \Gamma_0, \Gamma_1 \in \mathbf{\Pi}^0_0\}$, $[\mathbf{\Pi}^0_0, \mathbf{\Pi}^0_0, \mathbf{\Pi}^0_0] = \{[\Gamma_0, [\Gamma_1, \Gamma_2]] : \Gamma_0, \Gamma_1, \Gamma_2 \in \mathbf{\Pi}^0_0\}$ etc..

7.2. Proposition. (i) $|\mathbf{\Pi}^0_0| = \omega$,
(ii) $|[\mathbf{\Pi}^0_0, \mathbf{\Pi}^0_0]| = \omega^2$, $|[\mathbf{\Pi}^0_0, \mathbf{\Pi}^0_0, \mathbf{\Pi}^0_0]| = \omega^3$, etc.

Proof. (i) Let $\Gamma \in \mathbf{\Pi}^0_0$. Then for some recursive R, $n \in \Gamma(X) \Leftrightarrow R(n, X)$. Hence,

$$n \in \Gamma(\Gamma^{\omega}) \Rightarrow R(n, \Gamma^{\omega})$$
$$\Rightarrow R(n, \Gamma^{\xi}) \qquad \text{for some } \xi < \omega, \text{ by (1)}$$
$$\Rightarrow n \in \Gamma(\Gamma^{\xi}) \subseteq \Gamma^{\xi+1} \subseteq \Gamma^{\omega}.$$

Thus $\Gamma(\Gamma^{\omega}) \subseteq \Gamma^{\omega}$ and hence $|\Gamma| \leq \omega$. To show $|\mathbf{\Pi}^0_0| \geq \omega$, let $\Gamma_0(X) = \{0\} \cup \{\langle 1, x\rangle : x \in X\}$. Then $\Gamma_0 \in \mathbf{\Pi}^0_0$. $0 \in \Gamma^{\infty}$ and $|0| = 0$; if $n \in \Gamma_0^{\infty}$ and $|n| = \xi$, then $\langle 1, n\rangle \in \Gamma_0^{\infty}$ and $|\langle 1, n\rangle| = \xi + 1$. Thus $|\Gamma_0| \geq \omega$.

(ii) Let $n \in \Gamma_0(X) \Leftrightarrow R_0(n, X)$ and $n \in \Gamma_1(X) \Leftrightarrow R_1(n, X)$ where R_0 and R_1 are recursive. Then as in the proof of (i) we have:

(3) *If limit* λ *then* $\Gamma_0(\Gamma^{\lambda}) \subseteq \Gamma^{\lambda}$ *and hence* $\Gamma(\Gamma^{\lambda}) = \Gamma_1(\Gamma^{\lambda})$.

Now let $n \in \Gamma(\Gamma^{\omega^2})$. We show $n \in \Gamma^{\omega^2}$. Since limit ω^2, $n \in \Gamma_1(\Gamma^{\omega^2})$ by (3), i.e. $R_1(n, \Gamma^{\omega^2})$. Then by (2) there is some $\xi < \omega^2$ such that $\forall \delta < \omega^2$. $\xi \leq \delta \Rightarrow R_1(n, \Gamma^{\delta})$. Since the limit ordinals are cofinal in ω^2 there is some limit δ, $\xi \leq \delta < \omega^2$, and hence $R_1(n, \Gamma^{\delta})$. Thus $n \in \Gamma_1(\Gamma^{\delta})$. But since limit δ, $\Gamma_1(\Gamma^{\delta}) = \Gamma(\Gamma^{\delta})$ and hence $n \in \Gamma(\Gamma^{\delta}) \subseteq \Gamma^{\delta+1} \subseteq \Gamma^{\omega^2}$. Let Γ_0 be as above and let $\Gamma_1(X) = \{\langle 2, x\rangle : x \in X\}$. Let $\Gamma = [\Gamma_0, \Gamma_1]$. It is easy to show that $|\Gamma| \geq \omega^2$.

7.3. *Remark.* If R is $\mathbf{\Sigma}^0_1$ then we still have (1) and the first equivalence in (2) so that 7.2 still holds if $\mathbf{\Sigma}^0_1$ replaces $\mathbf{\Pi}^0_0$ everywhere.

Note that $\mathbf{\Pi}^0_0 \subseteq [\mathbf{\Pi}^0_0, \mathbf{\Pi}^0_0] \subseteq [\mathbf{\Pi}^0_0, \mathbf{\Pi}^0_0, \mathbf{\Pi}^0_0] \subseteq \ldots \mathbf{\Pi}^0_1$. Thus $\omega < \omega^2 < \omega^3 < \ldots < |\mathbf{\Pi}^0_1|$. We have

7.4. Theorem (Gandy). $|\Pi_1^0| = |\Sigma_2^0| = \omega_1$.

As $|\Sigma_2^0| \geq |\Pi_1^0| \geq |\Pi_1^0\text{-mon}| \geq \omega_1$ we have one half of the theorem. For the other half we will use the next two lemmas. These will also be used for getting upper bounds for other classes of first order inductive definitions.

7.5. Lemma. *Let* $\Gamma \in \Pi_0^1$. *Then* $\langle \Gamma^\xi : \xi < \lambda \rangle$ *is uniformly* Σ_1^0 *on* L_λ *for* $\lambda \in \mathrm{Ad}$. *Hence for* $\lambda \in \mathrm{Ad}$ Γ^λ *is* Σ_1^0 *on* L_λ.

Proof. If $\lambda \in \mathrm{Ad}$ and $x \subseteq \omega$ such that $x \in L_\lambda$ then $\Gamma(x)$ is Π_0^0 on L_λ as it is defined by a formula with quantifiers restricted to $\omega < \lambda$. Hence $\Gamma(x) \in L_\lambda$ as $\Gamma(x) \subseteq \omega$. So if $G(x,y) = \bigcup\{\Gamma(y'z \cap \omega) : z \in x\}$ then $G \restriction L_\lambda : L_\lambda \times L_\lambda \to L_\lambda$. Moreover $G \restriction L_\lambda$ is uniformly Σ_1^0 on L_λ for $\lambda \in \mathrm{Ad}$. Let $F(x) = G(x, F \restriction x)$. Then by 2.2 $F \restriction L_\lambda : L_\lambda \to L_\lambda$ and is uniformly Σ_1^0 on L_λ. By an easy induction we see that $\Gamma^\xi = F(\xi)$ for all $\xi \in \mathrm{ON}$, so that $\langle \Gamma^\xi : \xi < \lambda \rangle = F \restriction \lambda$ is uniformly Σ_1^0 on L_λ for $\lambda \in \mathrm{Ad}$. Hence Γ^λ is Σ_1^0 on L_λ as $x \in \Gamma \Leftrightarrow (\exists \xi < \lambda)\, x \in \Gamma^\xi$.

7.6. Lemma. *Let* $\langle \Gamma^\xi : \xi < \lambda \rangle$ *be* Σ_1^0 *on* L_λ *where* $\lambda \in \mathrm{Ad}$. *Let* R *be recursive. Then*

$$\forall x R(n,x,\Gamma^\lambda) \Rightarrow \exists \xi < \lambda \forall x R(n,x,\Gamma^\xi)\,.$$

Proof. Suppose $\forall x R(n,x,\Gamma^\lambda)$ where $\lambda \in \mathrm{Ad}$. Then for each x, $\forall z < x R(n,z,\Gamma^\lambda)$. Since λ is a limit, by (2), $\forall z \leq x R(n,z,\Gamma^\xi)$ for some $\xi < \lambda$. Let $f(x) \simeq \mu\xi < \lambda \forall z \leq x R(n,z,\Gamma^\xi)$. Then $f : \omega \to \lambda$ is λ-recursive. As $\omega < \lambda$ $\alpha = \mathrm{Sup}_{n<\omega} f(n) < \lambda$. It remains to show that $\forall x R(n,x,\Gamma^\alpha)$.

Case 1. Limit α. Suppose that for some z_0, $\neg R(n,z_0,\Gamma^\alpha)$. Then there is some $\xi < \alpha$ such that $\neg R(n,z_0,\Gamma^\delta)$ whenever $\xi \leq \delta < \alpha$. Since limit α, there is some $x > z_0$ such that $\xi < f(x) < \alpha$ and hence $\neg R(n,z_0,\Gamma^{f(x)})$. But then by definition of f, $\forall z \leq x R(n,z,\Gamma^{f(x)})$. In particular $R(n,z_0,\Gamma^{f(x)})$ which is a contradiction.

Case 2. Not limit α. Then since f is non-decreasing there is some y such that for all $x \geq y$, $f(x) = \alpha$. But by definition of f this clearly implies $\forall x R(n,x,\Gamma^\alpha)$.

We can now complete the proof of Theorem 7.4. We must show:

7.7. Lemma. $|\Sigma_2^0| \leq \omega_1$.

Proof. Let Γ be Σ_2^0. Then

$$n \in \Gamma(X) \Leftrightarrow \exists y \forall x R(n,y,x,X)$$

for some recursive R. Then

$$\begin{aligned} n \in \Gamma(\Gamma^{\omega_1}) &\Rightarrow \forall x R(n,y,x,\Gamma^{\omega_1}) && \text{for some } y < \omega, \\ &\Rightarrow \forall x R(n,y,x,\Gamma^{\xi}) && \text{for some } \xi < \omega_1,\ y < \omega, \\ &\Rightarrow n \in \Gamma(\Gamma^{\xi}) \subseteq \Gamma^{\omega_1}. \end{aligned}$$

Hence $\Gamma(\Gamma^{\omega_1}) \subseteq \Gamma^{\omega_1}$ so that $|\Gamma| \leq \omega_1$. A different proof appeared in [2]. The present proof is due to Grilliot.

As in the definitions of $[\Pi_0^0, \Pi_0^0]$, $[\Pi_0^0, \Pi_0^0, \Pi_0^0]$ etc. we may define $[\mathcal{C}_0, \mathcal{C}_1]$, $[\mathcal{C}_0, \mathcal{C}_1, \mathcal{C}_2]$ etc. for any classes $\mathcal{C}_0, \mathcal{C}_1, \ldots$ of i.d.'s.

7.8. Lemma. (i) $|[\Pi_1^0, \Pi_n^0]| \leq |[\Sigma_2^0, \Sigma_{n+1}^0]| \leq$ *least element of* $M_{n+1}(\mathrm{Ad})$;
(ii) $|[\Pi_1^0, \Pi_n^0, \Pi_m^0]| \leq |[\Sigma_2^0, \Sigma_{n+1}^0, \Sigma_{m+1}^0]| \leq$ *least element of* $M_{m+1}(M_{n+1}(\mathrm{Ad}))$ etc.

Proof. The first inequalities in (i) and (ii) are trivial, so we turn to the last inequalities.

(i) Let $\Gamma = [\Gamma_0, \Gamma_0]$ where $\Gamma_0 \in \Sigma_2^0$ and $\Gamma_1 \in \Sigma_{n+1}^0$. Then $\Gamma \in \Pi_0^1$ and hence $\langle \Gamma^\xi : \xi < \lambda \rangle$ is Σ_1^0 on L_λ for $\lambda \in \mathrm{Ad}$, so that by the proof of Lemma 7.7

(4) If $\lambda \in \mathrm{Ad}$ then $\Gamma_0(\Gamma^\lambda) \subseteq \Gamma^\lambda$ and hence $\Gamma(\Gamma^\lambda) = \Gamma_1(\Gamma^\lambda)$. Suppose first that n is even. Then for some recursive R

$$a \in \dot{\Gamma}_1(X) \Leftrightarrow \exists x_0 \forall x_1 \ldots \exists x_n R(a, x_0, \ldots, x_n, X).$$

Hence by (2) and (4), if $\lambda \in \mathrm{Ad}$ then

$$\begin{aligned} (5)\quad a \in \Gamma(\Gamma^\lambda) &\Leftrightarrow a \in \Gamma_1(\Gamma^\lambda)) \\ &\Leftrightarrow (\exists x_0 \in \omega)(\forall x_1 \in \omega) \ldots (\exists x_n \in \omega) R(a, x_0, \ldots, x_n, \Gamma^\lambda) \\ &\Leftrightarrow (\exists x_0 \in \omega)(\forall x_1 \in \omega) \ldots (\exists x_n \in \omega)(\exists \xi < \lambda)(\forall \delta < \lambda) \\ &\qquad [\xi \leq \delta \Rightarrow R(a, x_0, \ldots, x_n, \Gamma^\delta)] \\ &\Leftrightarrow L_\lambda \models Q(a) \end{aligned}$$

for some Σ^0_{n+2} formula $Q(v)$ of $\mathcal{L}_\in$ that is independent of $\lambda \in \mathrm{Ad}$.

Now let κ be the least element of $M_{n+1}(\mathrm{Ad})$. Suppose $a \in \Gamma(\Gamma^\kappa)$. Then as $\kappa \in \mathrm{Ad}$ it follows from (5) that $L_\kappa \models Q(a)$. As κ is Σ^0_{n+2}-reflecting on Ad there is a $\lambda < \kappa$ such that $\lambda \in \mathrm{Ad}$ and $L_\lambda \models Q(a)$. Hence by (5) $a \in \Gamma(\Gamma^\lambda) \subseteq \Gamma^\kappa$. Thus $\Gamma(\Gamma^\kappa) \subseteq \Gamma^\kappa$ and hence $|\Gamma| \leq \kappa$.

If n is odd then for some recursive R

$$a \in \Gamma_1(X) \Leftrightarrow \exists x_0 \forall x_1 \ldots \forall x_n R(a, x_0, \ldots, x_n, X).$$

Hence using (2) and (4) again, if $\lambda \in \mathrm{Ad}$ then

$$\begin{aligned} a \in \Gamma(\Gamma^\lambda) &\Leftrightarrow (\exists x_0 \in \omega) \ldots (\forall x_n \in \omega)(\forall \xi < \lambda)(\exists \delta < \lambda) \\ &\qquad [\xi \leq \delta \;\&\; R(a, x_0, \ldots, x_n, \Gamma^\delta)] \\ &\Leftrightarrow L_\lambda \models Q(a) \end{aligned}$$

for some Σ^0_{n+2} formula $Q(v)$ of $\mathcal{L}_\in$ that is independent of $\lambda \in \mathrm{Ad}$. The rest of the proof is as before.

(ii) This follows the same pattern as the proof of (i). Let $\Gamma = [\Gamma_0, \Gamma_1, \Gamma_2]$ where $\Gamma_0 \in \Sigma^0_2$, $\Gamma_1 \in \Sigma^0_{n+1}$ and $\Gamma_2 \in \Sigma^0_{m+1}$. The proof of (i) shows that

(4′) If $\lambda \in M_{n+1}(\mathrm{Ad})$ then $[\Gamma_0, \Gamma_1](\Gamma^\lambda) \subseteq \Gamma^\lambda$ and hence $\Gamma(\Gamma^\lambda) = \Gamma_2(\Gamma^\lambda)$.

Then as in (5) if $\lambda \in M_{n+1}(\mathrm{Ad})$

(5′) $a \in \Gamma(\Gamma^\lambda) \Leftrightarrow L_\lambda \models Q(a)$

for some Σ^0_{m+2} formula $Q(v)$ of $\mathcal{L}_\in$ that is independent of $\lambda \in M_{n+1}(\mathrm{Ad})$. The rest of the proof is as in (i).

In the next section we will prove results which will enable us to reverse the inequalities in this lemma.

§8. Closed classes of inductive definitions

In this section we formulate the notion of a closed class $\mathcal{C}$. The results in this section will enable us to give characterisations of $|\mathcal{C}|$ and $\mathrm{Ind}(\mathcal{C})$ for many of these classes.

8.1. Definition. $f : \Delta \leq_m \Gamma$ if

(a) f is a recursive function and $\{f(e)\}$ is total for all e;

(b) if $\{e\} : X \leq_m Y$ then $\{f(e)\} : \Delta(X) \leq_m \Gamma(Y)$.

$\mathbf{\Delta} \leq_m \Gamma$ if $f : \mathbf{\Delta} \leq_m \Gamma$ for some f. $\mathbf{\Delta} \leq_1 \Gamma$ is defined similarly.

8.2. Theorem. *If* $\mathbf{\Delta} \leq_m \Gamma$ *then* $\mathbf{\Delta}^\infty \leq_m \Gamma^\infty$ *and* $|\mathbf{\Delta}| \leq |\Gamma|$. *Similarly, with* $\leq_m$ *replaced by* $\leq_1$.

This is an immediate consequence of the following:

8.3. Lemma. *If* $\mathbf{\Delta} \leq_m \Gamma$ *there is a recursive function g such that for all* α, $g : \mathbf{\Delta}^\alpha \leq_m \Gamma^\alpha$.

Proof. Let $f : \mathbf{\Delta} \leq_m \Gamma$. By the recursion theorem there is an e such that $\{e\} = \{f(e)\} = g$, say. g is total since $\{f(e)\}$ is. We show by induction on α that $g : \mathbf{\Delta}^\alpha \leq_m \Gamma^\alpha$. Suppose

$$\{e\} = g : \mathbf{\Delta}^\beta \leq_m \Gamma^\beta \text{ and hence } g = \{f(e)\} : \mathbf{\Delta}(\mathbf{\Delta}^\beta) \leq_m \Gamma(\Gamma^\beta)$$

for all $\beta < \alpha$. Then,

$$\begin{aligned} x \in \mathbf{\Delta}^\alpha &\Leftrightarrow \exists \beta < \alpha.\ \ x \in \mathbf{\Delta}(\mathbf{\Delta}^\beta) \\ &\Leftrightarrow \exists \beta < \alpha.\ \ g(x) \in \Gamma(\Gamma^\beta) \\ &\Leftrightarrow g(x) \in \Gamma^\alpha . \end{aligned}$$

8.4. Definition. Γ is $\mathcal{C}$*-complete* if $\Gamma \in \mathcal{C}$ and $\mathcal{C} = \{\mathbf{\Delta} : \mathbf{\Delta} \leq_m \Gamma\}$.

8.5. Theorem. *If* Γ *is* $\mathcal{C}$*-complete then* $|\mathcal{C}| = |\Gamma|$ *and* $\mathrm{Ind}(\mathcal{C}) = \{X : X \leq_m \Gamma^\infty\}$.

Proof. Let Γ be $\mathcal{C}$-complete. As $\Gamma \in \mathcal{C}$, $|\mathcal{C}| \geq |\Gamma|$ and $\mathrm{Ind}(\mathcal{C}) \supseteq \{X : X \leq_m \Gamma^\infty\}$. If $\mathbf{\Delta} \in \mathcal{C}$ then $\mathbf{\Delta} \leq_m \Gamma$ and hence by 8.2 $|\mathbf{\Delta}| \leq |\Gamma|$ and $\mathbf{\Delta}^\infty \leq_m \Gamma^\infty$. Hence $|\mathcal{C}| \leq |\Gamma|$ and $\mathrm{Ind}(\mathcal{C}) \subseteq \{X : X \leq_m \Gamma^\infty\}$.

8.6. Theorem. *There is a* $\mathbf{\Pi}^n_{m+1}$*-complete operator. Similarly for* $\mathbf{\Sigma}^n_{m+1}$.

Proof. We shall need the following folklore result, which is well-known when

$n = 0$ or $n = 1$, but is equally true for larger n.

8.7. Proposition. *There is a universal* $\mathbf{\Pi}^n_{m+1}$ *operator. Similarly for* $\mathbf{\Sigma}^n_{m+1}$.

By this we mean a $\mathbf{\Pi}^n_{m+1}$ operator Γ such that every $\mathbf{\Pi}^n_{m+1}$ operator $\mathbf{\Delta}$ has the form $\mathbf{\Delta}(X) = \Gamma_a(X) = \{x \mid \langle a,x\rangle \in \Gamma(X)\}$ for some $a \in \omega$.

We will show that Γ is $\mathbf{\Pi}^n_{m+1}$-complete. Let $\mathbf{\Delta}_1(X) = \{\langle e,x\rangle : x \in \mathbf{\Delta}(\{e\}^{-1}X)\}$. When $n > 0$, $\mathbf{\Delta}_1$ is easily seen to be $\mathbf{\Pi}^n_{m+1}$ and hence has the form Γ_a for some $a \in \omega$. Now let f be a recursive function such that $\{f(e)\}(x) = \langle a,\langle e,x\rangle\rangle$. Then $f : \mathbf{\Delta} \leq_m \Gamma$.

When $n = 0$ we must be more careful as $\mathbf{\Delta}_1$ may not be $\mathbf{\Pi}^0_{m+1}$. We will define a $\mathbf{\Pi}^0_{m+1}$ operator $\mathbf{\Delta}_2$ such that if $\{e\}$ is total then $\langle e,x\rangle \in \mathbf{\Delta}_1(X) \Leftrightarrow \langle e,x\rangle \in \mathbf{\Delta}_2(X)$. Then $f : \mathbf{\Delta} \leq_m \Gamma$ if we let $\mathbf{\Delta}_2 = \Gamma_a$. Let $\varphi(X,x)$ be a $\mathbf{\Pi}^0_{m+1}$ formula defining Γ. By separating out positive and negative occurrences of X in $\varphi(X,x)$ we may write the formula as $\theta(X, \omega - X, x)$ where $\theta(X, Y, x)$ contains only positive occurrences of X and Y. Then

$$\langle e,x\rangle \in \mathbf{\Delta}_1(X) \Leftrightarrow \theta(\{e\}^{-1}X,\ \omega - \{e\}^{-1}X, x)\,.$$

Now if m is *odd* let

$$\mathbf{\Delta}_2(X) = \{\langle e,x\rangle : \theta(\{e\}^{-1}X,\ \{e\}^{-1}(\omega - X), x)\}$$

and if m is *even* let

$$\mathbf{\Delta}_2(X) = \{\langle e,x\rangle : \theta(\omega - \{e\}^{-1}(\omega - X),\ \omega - \{e\}^{-1}X, x)\}\,.$$

Then in each case $\mathbf{\Delta}_2$ is $\mathbf{\Pi}^0_{m+1}$.

8.8. Definition. $\mathcal{C}$ is *closed* if

(a) There is a $\mathcal{C}$-complete operator;

(b) $\Gamma_1, \Gamma_2 \in \mathcal{C} \Rightarrow \Gamma_1 \cup \Gamma_2,\ \Gamma_1 \cap \Gamma_2 \in \mathcal{C}$;

(c) Every recursive operator is in $\mathcal{C}$.

The following result is now trivial.

8.9. Theorem. $\mathbf{\Pi}^n_{m+1}$ *and* $\mathbf{\Sigma}^n_{m+1}$ *are closed.*

In order to obtain characterisations of $|\mathcal{C}|$ and $\mathrm{Ind}(\mathcal{C})$ for closed classes we will need a method for constructing notation systems $\mathcal{M} = (M, \|)$ which is more general than that mentioned in the introduction. We shall first give an example which bears some resemblance to Kleene's systems of notations for the constructive ordinals. We define a transfinite sequence of sets $\langle M_\xi : \xi \in \mathrm{ON}\rangle$. In the definition $|a| = \mu\xi(a \in M_{\xi+1})$. $\lambda x[b](x, X)$ is the b'th function primitive recursive in X in a recursive enumeration, uniform in X, of the functions primitive recursive in X.

$$M_0 = \emptyset$$

$$M_{\alpha+1} = M_\alpha \cup \{0\} \cup \{\langle 1, a, b\rangle : a \in M_\alpha \ \&\ \forall x[b](x, M_{|a|}) \in M_\alpha\}$$

$$M_\lambda = \bigcup_{\xi<\lambda} M_\xi \text{ for limit } \lambda$$

$$M = \bigcup_{\xi \in \mathrm{ON}} M_\xi .$$

Note that the definition of M has the appearance of a set inductively defined by an i.d. But the situation is complicated by the fact that the definition of $M_{\alpha+1}$ depends not only on the previously defined M_α, but also on $\langle M_{|a|} : a \in M_\alpha\rangle$. Given any sequence $\langle M_\xi : \xi \in \mathrm{ON}\rangle$ we use the following notation.

$$M = \bigcup\{M_\xi : \xi \in \mathrm{ON}\}$$

$$|a| = \mu\xi(a \in M_{\xi+1}) \qquad \text{for } a \in M$$

$$|M| = \mathrm{Sup}\{|x| : x \in M\}$$

$$M_\alpha^* = \{\langle x, y\rangle : x, y \in M_\alpha \ \&\ |x| \leq |y|\} \qquad \text{for } \alpha \in \mathrm{ON}$$

$$M^* = \bigcup\{M_\alpha^* : \alpha \in \mathrm{ON}\} .$$

If $X \subseteq \omega$ let $\mathcal{F}(X) = \{x : \langle x, x\rangle \in X\}$ and $X_{<x} = \{y : \langle y, x\rangle \in X \ \&\ \langle x, y\rangle \notin X\}$. Then clearly $M_\alpha = \mathcal{F}(M_\alpha^*)$ and $M_{|a|} = (M_\alpha^*)_{<a}$ for $a \in M_\alpha$.

Hence the definition of $M_{\alpha+1}$ above may be written

$$M_{\alpha+1} = M_\alpha \cup \Theta(M_\alpha^*)$$

where

$$\Theta(X) = \{0\} \cup \{\langle 1, a, b\rangle : a \in \mathcal{F}(X) \ \&\ \forall x[b](x, X_{<a}) \in \mathcal{F}(X)\} .$$

Notice that Θ is $\mathbf{\Pi}_1^0$. We will see below that M^* is inductively defined by a $\mathbf{\Pi}_1^0$ i.d. We now generalize the above procedure to an arbitrary Θ.

8.10. Definition. For any i.d. Θ, $\mathcal{M}^\Theta = (M^\Theta, \|)$ is defined by:

$$M_0 = \emptyset$$

$$M_{\alpha+1} = M_\alpha \cup \Theta(M_\alpha^*)$$

$$M_\lambda = \bigcup\{M_\alpha : \alpha < \lambda\} \quad \text{if limit } \lambda$$

$$M^\Theta = \bigcup\{M_\alpha : \alpha \in \mathrm{ON}\}\,.$$

Although M^Θ does not have an obvious inductive definition we show that M^* does.

8.11. Definition. For any i.d. Θ, $\Theta_\leq$ is defined by $\Theta_\leq(X) = \{\langle x,y\rangle : x \in \mathcal{F}(X) \,\&\, y \in \Theta(X) \setminus \mathcal{F}(X)\} \cup \{\langle x,y\rangle : x,y \in \Theta(X) \setminus \mathcal{F}(X)\}$.

8.12. *Remark.* For closed $\mathcal{C}$ note that $\Theta \in \mathcal{C} \Rightarrow \Theta_\leq \in \mathcal{C}$.

8.13. Lemma. *For all α $\Theta_\leq^\alpha = M_\alpha^*$ and hence $|\Theta_\leq| = |M^\Theta|$ and $\Theta_\leq^\infty = M^*$.*

Proof. Note that $M_{\alpha+1}^* = M_\alpha^* \cup \Theta_\leq(M_\alpha^*)$. The result now follows by induction on α.

8.14. Equivalence Theorem. *If $\mathcal{C}$ is closed and Θ is $\mathcal{C}$-complete then $|\mathcal{C}| = |M^\Theta|$ and* $\mathrm{Ind}(\mathcal{C}) = \{X \subseteq \omega : X \leq_m M^\Theta\}$.

Proof. If $\mathcal{C}$ is closed and $\Theta \in \mathcal{C}$ it follows from 8.12 and 8.13 that $|\mathcal{C}| \geq |M^\Theta|$ and $\mathrm{Ind}(\mathcal{C}) \supseteq \{X \subseteq \omega : X \leq_m M^\Theta\}$ as $\Theta_\leq \in \mathcal{C}$ and $x \in M^\Theta \Leftrightarrow \langle x,x\rangle \in \Theta_\leq^\infty$.

For the converse it is sufficient to prove the following.

8.15. Lemma. *If $\Gamma \leq_m \Theta$ then there is a recursive function f such that for all α $f : \Gamma^\alpha \leq_m M_\alpha$; hence $\Gamma^\infty \leq_m M^\Theta$ and $|\Gamma| \leq |M^\Theta|$.*

Proof. Let f_0 be a recursive function such that $\{f_0(e)\}(x) \simeq \langle\{e\}(x), \{e\}(x)\rangle$. So $\{e\}^{-1}\mathcal{F} = \{f_0(e)\}^{-1}$. Let $f_1 : \Gamma \leq_m \Theta$. Then for total $\{a\}$, $\Gamma\{a\}^{-1} = \{f_1(a)\}^{-1}\Theta$. Then

$$\Gamma\{e\}^{-1}\mathcal{F} = \Gamma\{f_0(e)\}^{-1} = \{f_1 f_0(e)\}^{-1}\Theta .$$

Now choose e such that $\{e\} = \{f_1 f_0(e)\}$ and let $f = \{e\}$.

Then $\Gamma f^{-1}\mathcal{F} = f^{-1}\Theta$. Hence $\Gamma f^{-1}(M_\alpha) = \Gamma f^{-1}\mathcal{F}(M_\alpha^*) = f^{-1}\Theta(M_\alpha^*)$, so that $f^{-1}M_\alpha \cup \Gamma(f^{-1}M_\alpha) = f^{-1}M_\alpha \cup f^{-1}\Theta(M_\alpha^*) = f^{-1}(M_\alpha \cup \Theta(M_\alpha^*)) = f^{-1}M_{\alpha+1}$. It follows by induction on α that $\Gamma^\alpha = f^{-1}M_\alpha$; i.e. $f : \Gamma^\alpha \leq_m M_\alpha$.

In §7 we have seen how to prove, for certain $\mathcal{C}$, that $|\mathcal{C}| \leq \kappa$ where κ is the least reflecting ordinal of a certain kind. In order to show that $|\mathcal{C}| = \kappa$ we will choose a 'good' notation system $\mathcal{M} = \langle M, \|\rangle$ such that $|M| \leq |\mathcal{C}|$ and show that M has the required reflection property. As M is a set of notations for the ordinals $< |M|$, statements about ordinals $< |M|$ can be rewritten as statements about M. The reflection property for $|M|$ will then follow from closure properties of M. The Coding Lemma below gives a formulation of this rewriting process for Σ_1^0 statements.

8.16. Definition. A notation system $\mathcal{M} = \langle M, \|\rangle$ is *good* if $\mathcal{M} = \mathcal{M}^\Theta$ where $\Theta(X) = \Xi(X) \cup \Phi(X)$ and $\Xi(X) = \{0\} \cup \{\langle 1, a, b\rangle : a \in \mathcal{F}(X)$ & $\forall x [b](x, X_{<a}) \in \mathcal{F}(X)\} \cup \{\langle 2, a, b\rangle : a \in \mathcal{F}(X)$ or $b \in \mathcal{F}(X)\}$, and $\Phi(X)$ is always disjoint from $\{0\} \cup \{\langle 1, a, b\rangle : a, b \in \omega\} \cup \{\langle 2, a, b\rangle : a, b \in \omega\}$.

If $\mathcal{M}$ is good then an ordinal $\lambda \leq |M|$ is $\mathcal{M}$*-good* if $\Xi(M_\lambda^*) \subseteq M_\lambda$. Thus $|M|$ is $\mathcal{M}$-good, but usually there will be $\mathcal{M}$-good ordinals $< |M|$.

8.17. Coding Lemma. *Let* $\mathcal{M} = \langle M, \|\rangle$ *be a good notation system and let* $T_\mathcal{M} = \{(x, \alpha) : \alpha \in \mathrm{On}$ & $x \in M_\alpha\}$. *Then*

(i) *Every* $\mathcal{M}$*-good ordinal is in* $\mathrm{Ad}(T_\mathcal{M})$.

(ii) *For every* Σ_1^{0-} *formula* $\varphi(v_1, \ldots, v_n)$ *of* $\mathcal{L}_p(T_\mathcal{M})$ *there is a primitive recursive function h such that for every* $\mathcal{M}$*-good ordinal* λ

$$a_1, \ldots, a_n \in M_\lambda \;\&\; \lambda \models \varphi(|a_1|, \ldots, |a_n|) \Leftrightarrow h(a_1, \ldots, a_n) \in M_\lambda .$$

(iii) *If* λ *is* $\mathcal{M}$*-good then for* $X \subseteq \omega$ *X is* λ*-r.e. in* $T_\mathcal{M} \restriction \lambda \Leftrightarrow X \leq_m M_\lambda$.

This lemma will be proved in the appendix.

8.18. Corollary. *Let* $\mathcal{M}$, $T_{\mathcal{M}}$ *be as in the lemma, and let* $f(\alpha) = \mu n[n \in M \,\&\, |n| = \alpha]$ *for* $\alpha < |M|$. *Then for each* $\mathcal{M}$*-good ordinal* λ $f \restriction \lambda : \lambda \to \omega$ *is a* λ*-recursive in* $T_{\mathcal{M}} \restriction \lambda$ *injection.*

Proof. $f(\alpha) = \mu n[(n, \alpha+1) \in T_{\mathcal{M}} \,\&\, (n, \alpha) \notin T_{\mathcal{M}}]$. Hence $f \restriction \lambda$ is λ-recursive in $T_{\mathcal{M}} \restriction \lambda$ for $\mathcal{M}$-good λ. It is clearly an injection.

8.19 Theorem. *Let* $\mathcal{C} \supseteq \Pi^0_1$ *be closed and let* Γ *be* $\mathcal{C}$*-complete. Let* $\lambda = |\mathcal{C}|$ *and* $A_\Gamma = \{(n, \alpha) : \alpha < \lambda \,\&\, n \in \Gamma^\alpha\}$. *Then*

(i) λ *is admissible relative to* A_Γ;

(ii) λ *is projectible to* ω *relative to* A_Γ;

(iii) $\mathrm{Ind}(\mathcal{C}) = \{X \subseteq \omega : X$ *is* λ*-r.e. in* $A_\Gamma\}$.

Proof. Let $\Theta(X) = \Xi(X) \cup \Phi(X)$ where $\Phi(X) = \{\langle 3, a\rangle : a \in \Gamma(X)\}$. Then as $\Xi \in \Pi^0_1 \subseteq \mathcal{C}$ and $\mathcal{C}$ is closed it follows that $\Theta \in \mathcal{C}$. Also $\lambda x \langle 3, x\rangle :$ $\Gamma(X) \leq_m \Theta(X)$ for all X and hence $f' : \Gamma \leq_m \Theta$ where $f : \Gamma \leq_m \Gamma$ and f' is a recursive function such that $\{f'(e)\}(x) \simeq \langle 3, \{f(e)\}(x)\rangle$. Hence Θ is $\mathcal{C}$-complete. Now $\mathcal{M} = \mathcal{M}^\Theta$ is a good notation system and by the Equivalence Theorem $|\mathcal{C}| = |M|$ and $\mathrm{Ind}(\mathcal{C}) = \{X \subseteq \omega : X \leq_m M\}$. Hence by the Coding Lemma and its corollary the theorem follows as long as we replace A_Γ by $T_{\mathcal{M}} \restriction \lambda$. It only remains to show that A_Γ and $T_{\mathcal{M}} \restriction \lambda$ are λ-recursive in each other. But by 8.15 there is a recursive function h such that $h : \Gamma^\alpha \leq_m M_\alpha$ for all α. Hence $(n, \alpha) \in A_\Gamma \Leftrightarrow (h(n), \alpha) \in T_{\mathcal{M}} \restriction \lambda$, so that A_Γ is λ-recursive in $T_{\mathcal{M}} \restriction \lambda$. For the converse note that as Γ is $\mathcal{C}$-complete and $\Theta_{\leq} \in \mathcal{C}$ it follows that $g : \Theta_{\leq} \leq_m \Gamma$ for some g. Hence $g : M^*_\alpha \leq_m \Gamma^\alpha$ for all α, by 8.3 and 8.13. So

$$(n, \alpha) \in T_{\mathcal{M}} \restriction \lambda \Leftrightarrow (g(\langle n, n\rangle), \alpha) \in A_\Gamma\,,$$

showing that $T_{\mathcal{M}} \restriction \lambda$ is λ-recursive in A_Γ.

§9. First order inductive definitions, II

We are now ready to characterise the ordinals of first order inductive definitions.

9.1. Theorem. (i) $|\mathbf{\Pi}_1^0|$ *is the least element of* Ad;
(ii) $|[\mathbf{\Pi}_1^0, \mathbf{\Pi}_n^0]|$ *is the least element of* $M_{n+1}(\mathrm{Ad})$;
(iii) $|[\mathbf{\Pi}_1^0, \mathbf{\Pi}_m^0, \mathbf{\Pi}_n^0]|$ *is the least element of* $M_{n+1}(M_{m+1}(\mathrm{Ad}))$, etc.

9.2. *Remark.* By 7.8 this theorem is also true when each $\mathbf{\Pi}_k^0$ is replaced by $\mathbf{\Sigma}_{k+1}^0$.

Before proving the theorem we derive some immediate consequences.

9.3. Corollary. $|\mathbf{\Pi}_n^0| = |\mathbf{\Sigma}_{n+1}^0| = \pi_{n+1}^0 = \sigma_{n+2}^0$.

Proof. $|\mathbf{\Pi}_0^0| = |\mathbf{\Sigma}_1^0| = \omega = \pi_1^0$ by 1.9(i), 7.2 and 7.3. $|\mathbf{\Pi}_1^0| = |\mathbf{\Sigma}_2^0| = \omega_1 = \pi_2^0$ by 1.8 and 7.4. For $n > 1$ $[\mathbf{\Pi}_1^0, \mathbf{\Pi}_n^0] = \mathbf{\Pi}_n^0$ and by 1.9(iv) $M_{n+1}(\mathrm{Ad}) = M_{n+1}(\mathrm{ON})$. Hence $|\mathbf{\Pi}_n^0| = |\mathbf{\Sigma}_{n+1}^0| = \pi_{n+1}^0$ by 9.1 and 9.2. By 1.9(iii) $\pi_n^0 = \sigma_{n+1}^0$ for all n.

9.4. Corollary. (i) $|[\mathbf{\Pi}_1^0, \mathbf{\Pi}_0^0]|$ *is the least recursively inaccessible ordinal*; $|[\mathbf{\Pi}_1^0, \mathbf{\Pi}_0^0, \mathbf{\Pi}_0^0]|$ *is the least recursively inaccessible limit of recursively inaccessible ordinals, etc.*

(ii) $|[\mathbf{\Pi}_1^0, \mathbf{\Pi}_1^0]|$ *is the least recursively Mahlo ordinal*; $|[\mathbf{\Pi}_1^0, \mathbf{\Pi}_1^0, \mathbf{\Pi}_1^0]|$ *is the least recursively hyper-Mahlo ordinal, etc.*

We now turn to the proof of the theorem. By 7.8 it only remains to prove:

9.5. Lemma. (i) $\alpha \in \mathrm{Ad}$ *for some* $\alpha \leq |\mathbf{\Pi}_1^0|$;
(ii) $\alpha \in M_{n+1}(\mathrm{Ad})$ *for some* $\alpha \leq |[\mathbf{\Pi}_1^0, \mathbf{\Pi}_n^0]|$;
(iii) $\alpha \in M_{n+1}(M_{m+1}(\mathrm{Ad}))$ *for some* $\alpha \leq |[\ {}_1^0,\ {}_m^0,\ {}_n^0]|$; *etc.*

Proof. (i) This follows from Theorem 7.4, whose proof assumed the result $|\mathbf{\Pi}_1^0\text{-mon}| \geq \omega_1$. To give a direct proof let $\mathcal{M} = \mathcal{M}^\Theta$ where $\Theta = \Xi$, and let $\alpha = |M|$. Then $\mathcal{M}$ is a good notation system so that $\alpha \in \mathrm{Ad}$, by the Coding Lemma. As $\Theta \in \mathbf{\Pi}_1^0$, so is $\Theta_{\leq}$ so that $\alpha = |\Theta_{\leq}| \leq |\mathbf{\Pi}_1^0|$.

(ii) First assume that n is odd. Let $\mathcal{M} = \mathcal{M}^\Theta$ where $a \in \Theta(X) \Leftrightarrow a \in \Xi(X) \vee [\Xi(X) \subseteq \mathcal{F}(X) \,\&\, a \in \Phi_n^1(X)]$, $\Phi_n^1(X) = \{\langle 3, e\rangle : e \in \Phi_n(\mathcal{F}(X))\}$ and $a \in \Phi_n(X) \Leftrightarrow (\forall x_1 \in X)(\exists x_2 \in X) \ldots (\forall x_n \in X)[a](x_1, \ldots, x_n) \in X$. Here $\lambda x_1, \ldots, x_n [a](x_1, \ldots, x_n)$ is the a'th n-ary primitive recursive function in a recursive enumeration of them.

An easy argument shows that $\Theta_{\leq} = [\Xi_{\leq}, (\Phi_n^1)_{\leq}]$, so that $\Theta_{\leq} \in [\mathbf{\Pi}_1^0, \mathbf{\Pi}_n^0]$

as $(\Phi_n^1)_{\leq} \in \mathbf{\Pi}_n^0$. $\mathcal{M}$ is a good notation system so that by the Coding Lemma $\alpha = |M| \in \mathrm{Ad}$. Let φ be a $\mathbf{\Pi}_{n+1}^0$ sentence of $\mathcal{L}_p$ such that $\alpha \models \varphi$. We may assume that φ has the form

$$\forall x_1 \exists x_2 \ldots x_n \Psi(x_1, \ldots, x_n, |c_1|, \ldots, |c_k|)$$

where Ψ is a Σ_1^{0-} formula of $\mathcal{L}_p$ and $c_1, \ldots, c_k \in M$. By the Coding Lemma there is a primitive recursive function h such that for all $\mathcal{M}$-good ordinals λ

$$a_1, \ldots, a_{n+k} \in M_\lambda \;\&\; \lambda \models \Psi(|a_1|, \ldots, |a_{n+k}|) \Leftrightarrow h(a_1, \ldots, a_{n+k}) \in M_\lambda .$$

Now choose $e \in \omega$ such that

$$[e](a_1, \ldots, a_n) = h(a_1, \ldots, a_n, c_1, \ldots, c_k) .$$

Then it follows that for $\mathcal{M}$-good λ

$$\begin{aligned} \lambda \models \varphi &\Leftrightarrow e \in \Phi_n(M_\lambda) \\ &\Leftrightarrow \langle 3, e\rangle \in \Phi_n^1(M_\lambda^*) \subseteq M_{\lambda+1} . \end{aligned}$$

Hence as $\alpha \models \varphi$ and α is $\mathcal{M}$-good

$$\langle 3, e\rangle \in M .$$

Now if $\lambda = |\langle 3, e\rangle|$ then $\lambda < \alpha$, λ is $\mathcal{M}$-good, and hence admissible, and $\langle 3, e\rangle \in \Phi_n^1(M_\lambda^*)$ so that $\lambda \models \varphi$. Thus $\alpha \in M_{n+1}(\mathrm{Ad})$ and $\alpha = |\Theta_{\leq}| \leq |[\mathbf{\Pi}_1^0, \mathbf{\Pi}_n^0]|$, as required.

When $n > 0$ is even the proof is as above except that

$$a \in \Phi_n(X) \Leftrightarrow (\forall x_1 \in X)(\exists x_2 \in X) \ldots (\exists x_n \in X)[a](x_1, \ldots, x_n) \notin X ,$$

and the $\mathbf{\Pi}_{n+1}^0$ sentence φ now has the form

$$\forall x_1 \exists x_2 \ldots \exists x_n \neg \Psi(x_1, \ldots, x_n, |c_1|, \ldots, |c_k|)$$

with $\Psi, c_1, \ldots, c_k$ as before.

In case $n = 0$ let $\Phi_n(X) = X$. Then as before $\alpha = |M| \leq |[\mathbf{\Pi}_1^0, \mathbf{\Pi}_n^0]|$ and $\alpha \in \mathrm{Ad}$. In order to show that $\alpha \in M_1(\mathrm{Ad})$ we must show that α is a limit of admissibles. So let $\beta < \alpha$. Then $\beta = |a|$ for some $a \in M$. Then $\langle 3, a\rangle \in M$ as $\Xi(M^*) \subseteq M$. Let $\lambda = |\langle 3, a\rangle| < \alpha$. Then λ is $\mathcal{M}$-good and hence admissible, and $\beta = |a| < \lambda$. So $\alpha \in M_1(\mathrm{Ad})$.

(iii) Let $\mathcal{M} = \mathcal{M}^\Theta$ where

$$a \in \Theta(X) \Leftrightarrow a \in \Xi(X) \vee [\Xi(X) \subseteq \mathcal{F}(X) \;\&\; a \in \Phi_m^1(X)]$$
$$\vee [\Xi(X) \cup \Phi_m^1(X) \subseteq \mathcal{F}(X) \;\&\; a \in \Phi_n^{11}(X)]$$

where $\Phi_n^{11}(X) = \{\langle 5, e\rangle : e \in \Phi_n(\mathcal{F}(X))\}$ and Φ_m, Φ_m^1 are as in (ii). Then as in (ii) $\Theta_{\leq} = [\Xi_{\leq}, (\Phi_m^1)_{\leq}, (\Phi_n^{11})_{\leq}] \in [\mathbf{\Pi}_1^0, \mathbf{\Pi}_m^0, \mathbf{\Pi}_n^0]$ so that $\alpha = |M| = |\Theta_{\leq}| \leq |[\mathbf{\Pi}_1^0, \mathbf{\Pi}_m^0, \mathbf{\Pi}_n^0]|$.

As in the proof of (ii) we may show that $\alpha \in M_{m+1}(\mathrm{Ad})$. Moreover we may show that $|\langle 5, e\rangle| \in M_{m+1}(\mathrm{Ad})$ whenever $\langle 5, e\rangle \in M$. Hence using once more the argument in the proof of (ii) we can show that $\alpha \in M_{n+1}(M_{m+1}(\mathrm{Ad}))$.

The next result characterises $\mathrm{Ind}(\mathcal{C})$ for certain classes of first order inductive definitions.

9.6. Theorem. *If $\mathcal{C}$ is any of the classes $\mathbf{\Pi}_1^0$, $[\mathbf{\Pi}_1^0, \mathbf{\Pi}_n^0]$, $[\mathbf{\Pi}_1^0, \mathbf{\Pi}_m^0, \mathbf{\Pi}_n^0]$, etc. and $\lambda = |\mathcal{C}|$ then there is a $\Gamma \in \mathcal{C}$ such that $\lambda = |\Gamma|$ and* $\mathrm{Ind}(\mathcal{C}) = \{X \subseteq \omega : X \leq_m \Gamma^\infty\} = \{X \subseteq \omega : X$ *is* λ-*r.e.*$\}$.

9.7. *Remark.* This result also applies to the classes $\mathbf{\Pi}_{n+1}^0$ and to the classes obtained from the ones considered by replacing each $\mathbf{\Pi}_k^0$ by $\mathbf{\Sigma}_{k+1}^0$.

Proof. The proof has the same form in each case. We illustrate with $\mathcal{C} = [\mathbf{\Pi}_1^0, \mathbf{\Pi}_n^0]$. In the proof of 9.5 (ii) a good notation system $\mathcal{M} = \mathcal{M}^\Theta$ is defined such that $|M| \leq \lambda$ and $|M| \in M_{n+1}(\mathrm{Ad})$. But λ is the first element of $M_{n+1}(\mathrm{Ad})$ by 9.1. Hence $|M| = \lambda$. So if $\Gamma = \Theta_{\leq} \in \mathcal{C}$ then $\lambda = |\Gamma|$. By the Coding Lemma $\{X \subseteq \omega : X$ is λ-r.e.$\} \subseteq \{X \subseteq \omega : X \leq_m \Gamma^\infty\}$, as $M \leq_m M^* = \Gamma^\infty$. Hence $\{X \subseteq \omega : X$ is λ-r.e.$\} \subseteq \mathrm{Ind}(\mathcal{C})$ as $\Gamma \in \mathcal{C}$. $\Lambda \in \mathcal{C}$ implies that $\Lambda^\infty = \Lambda^\lambda$ is λ-r.e. by 7.5. Hence $\mathrm{Ind}(\mathcal{C}) \subseteq \{X \subseteq \omega : X$ is λ-r.e.$\}$, proving the theorem.

For completeness we conclude this section with the following easily proved result.

9.8. Theorem. (i) $\operatorname{Ind}(\mathbf{\Pi}_0^0) = \{X \subseteq \omega : X \textit{ is r.e.}\}$.
(ii) $\operatorname{Ind}(\mathbf{\Sigma}_1^0) = \{X \subseteq \omega : X$ *is a "recursive" union of arithmetical sets*$\}$.

§ 10. Higher order inductive definitions, I

In the previous section it was shown that the closure ordinals of certain classes of first-order inductive definitions are reflecting ordinals of a prescribed form. In this section we obtain corresponding results for higher type inductive definitions. The techniques are similar to those of §§7 and 9.

In Lemma 7.5 it was shown that for $\Gamma \in \mathbf{\Pi}_0^1$, $\langle \Gamma^\xi : \xi < \lambda \rangle$ is uniformly $\mathbf{\Sigma}_1^0$ on L_λ for $\lambda \in \mathrm{Ad}$. For other Γ this need not be the case and this makes the characterisation of $|\mathbf{\Pi}_n^m|$ for $m, n > 0$ somewhat more difficult.

In the following lemma the class A is some relation on ordinals.

10.1. Lemma. *Suppose $m, n > 0$ and $\Gamma \in \mathbf{\Pi}_n^m$. Let X be a class of limit ordinals greater than ω, and $\kappa \in X$ be $\mathbf{\Pi}_n^m(A)$-reflecting on X. If $\langle \Gamma^\xi : \xi < \lambda \rangle$ is uniformly $\mathbf{\Sigma}_0^1$ on $\mathrm{L}_\lambda[A]$ for $\lambda \in X$, then $|\Gamma| \leq \kappa$. Similarly with $\mathbf{\Pi}_n^m(A)$ and $\mathbf{\Pi}_n^m$ replaced by $\mathbf{\Sigma}_n^m(A)$ and $\mathbf{\Sigma}_n^m$, respectively.*

Proof. Let $\Gamma \in \mathbf{\Pi}_n^m$. Then for some $\mathbf{\Pi}_n^m$ formula $\varphi(y, Y)$ of $\mathcal{L}_\in$, for all $Y \subseteq \omega$

$$n \in \Gamma(Y) \Leftrightarrow \omega \models \varphi(n, Y)\,.$$

Let $\psi_0(z)$ be the formula $z \in \omega$, and $\psi_{k+1}(Z^{k+1})$ be $\forall Y^k [Z^{k+1}(Y^k) \to \psi_k(Y^k)]$. Then each ψ_{k+1} is a $\mathbf{\Pi}_1^k$ formula (in the constant ω). Let $\varphi^*(n, X)$ be obtained from φ by restricting each quantifier of type k to ψ_k. Then $\varphi^* \in \mathbf{\Pi}_n^m$ and for $\lambda > \omega$ and $Y \subseteq \omega$,

$$n \in \Gamma(Y) \Leftrightarrow \mathrm{L}_\lambda[A] \models \varphi^*(n, Y)\,.$$

Let $\varphi^*(y, Y)$ be $\boldsymbol{QZ}.\,\psi(\mathbf{Z}, y, Y)$ where $\boldsymbol{QZ}$ is a sequence of quantified variables of appropriate type for a prenex $\mathbf{\Pi}_n^m$ formula and ψ is $\mathbf{\Sigma}_0^0$. Then for $\lambda \in X$,

$$\begin{aligned} (1) \quad s \in \Gamma(\Gamma^\lambda) &\Leftrightarrow \mathrm{L}_\lambda[A] \models \boldsymbol{QZ}.\,\psi(\mathbf{Z}, s, \Gamma^\lambda) \\ &\Leftrightarrow \mathrm{L}_\lambda[A] \models \boldsymbol{QZ}\,\forall \xi \in \mathrm{On}\; \exists \delta \in \mathrm{On} [\xi \leq \delta \wedge \psi(\mathbf{Z}, s, \Gamma^\delta)] \\ &\Leftrightarrow \mathrm{L}_\lambda[A] \models \boldsymbol{QZ}\,\forall \xi \in \mathrm{On}\; \exists \delta \in \mathrm{On}\; \exists y [\xi \leq \delta \wedge R(\delta, y) \wedge \psi(\mathbf{Z}, s, y)] \\ &\Leftrightarrow \mathrm{L}_\lambda[A] \models \varphi_1(s)\,, \end{aligned}$$

where R is a Σ_0^1 formula of $\mathcal{L}_{\in}(A)$ (independent of $\lambda \in X$) such that for $\delta < \lambda$, $y \in \mathrm{L}_\lambda[A]$

$$\Gamma^\delta = y \Leftrightarrow \mathrm{L}_\lambda[A] \models R(\delta, y) ,$$

and $\varphi_1(s)$ is the sentence appearing immediately above it in (1). Clearly φ_1 is a $\mathbf{\Pi}_n^m$ formula of $\mathcal{L}_{\in}(A)$. Now let $\kappa \in X$ be $\mathbf{\Pi}_n^m(A)$-reflecting and $s \in \Gamma(\Gamma^\kappa)$. We show $s \in \Gamma^\kappa$. $\mathrm{L}_\kappa[A] \models \varphi_1(s)$ by (1). Since φ_1 is a $\mathbf{\Pi}_n^m$ formula of $\mathcal{L}_{\in}(A)$ there is some $\lambda \in X \cap \kappa$ such that $\mathrm{L}_\lambda[A] \models \varphi_1(s)$. Hence by (1), $s \in \Gamma(\Gamma^\lambda) \subseteq \Gamma^\kappa$.

10.2. Definition. For a given i.d. Γ let $A_\Gamma(x,y) \Leftrightarrow y \in \mathrm{On} \;\&\; x \in \Gamma^y$. Let $\pi_n^m(\Gamma)$ be the least $\mathbf{\Pi}_n^m(A_\Gamma)$-reflecting ordinal; similarly for $\sigma_n^m(\Gamma)$.

10.3. Theorem. *Let $m,n > 0$ and Γ be complete $\mathbf{\Pi}_m^n$. Then $|\mathbf{\Pi}_n^m| = \pi_n^m(\Gamma)$. Similarly $|\mathbf{\Sigma}_n^m| = \sigma_n^m(\Gamma)$ if Γ is complete $\mathbf{\Sigma}_n^m$.*

Proof. We prove this for $\mathbf{\Pi}_n^m$. The proof for $\mathbf{\Sigma}_n^m$ is similar. Let Γ be complete $\mathbf{\Pi}_n^m$. For $\lambda \in \mathrm{Ad}(A_\Gamma)$ and $\xi < \lambda$, $\Gamma^\xi = \{x \in \omega : A_\Gamma(x,\xi)\} \in \mathrm{L}_\lambda[A_\Gamma]$, by $\mathbf{\Pi}_0^0$-separation. And since for $y \in \mathrm{L}_\lambda[A_\Gamma]$,

$$\Gamma^\xi = y \Leftrightarrow \mathrm{L}_\lambda[A_\Gamma] \models [\forall x \in \omega A_\Gamma(x,\xi) \to x \in y] \wedge \forall x \in y.\, A_\Gamma(x,\xi) ,$$

it follows that $\langle \Gamma^\xi : \xi < \lambda \rangle$ is uniformly $\mathbf{\Sigma}_1^0$ on $\mathrm{L}_\lambda[A_\Gamma]$ for $\lambda \in \mathrm{Ad}(A_\Gamma)$. Letting $X = \mathrm{Ad}(A_\Gamma)$ and $\kappa = \pi_n^m(\Gamma)$ in Lemma 1, we see that $|\mathbf{\Pi}_n^m| \leq \pi_n^m(\Gamma)$.

To show $|\mathbf{\Pi}_n^m| \geq \pi_n^m(\Gamma)$ it suffices to show $|\mathbf{\Pi}_n^m|$ is $\mathbf{\Pi}_n^m(\Gamma)$-reflecting. Let Θ be as in the proof of 8.19 and let $\mathcal{M} = \langle M, \|\rangle = \mathcal{M}^\Theta$. By 8.14 $|M| = |\mathbf{\Pi}_n^m|$. If $T_{\mathcal{M}}$ and h are as in the proof of 8.19 then

$$\begin{aligned} A_\Gamma(x,\alpha) &\Leftrightarrow T_{\mathcal{M}}(h(x),\alpha) \\ &\Leftrightarrow \exists y \in \omega[h(x) = y \;\&\; T_{\mathcal{M}}(y,\alpha)] . \end{aligned}$$

Since h is recursive, the predicate $h(x) = y$ is $\mathbf{\Sigma}_1^0$ on L_ω and hence $\mathbf{\Sigma}_0^0$ on $\mathrm{L}_\lambda[T_{\mathcal{M}}]$ for $\lambda > \omega$. Hence A_Γ is $\mathbf{\Sigma}_0^0$ on $\mathrm{L}_\lambda[T_{\mathcal{M}}]$ for $\lambda > \omega$. So if φ is a $\mathbf{\Pi}_n^m$ sentence of $\mathcal{L}_{\in}(A_\Gamma)$ there is a $\mathbf{\Pi}_n^m$ sentence φ^* of $\mathcal{L}_{\in}(T_{\mathcal{M}})$ such that for $\lambda > \omega$ $\mathrm{L}_\lambda[A_\Gamma] \models \varphi \Leftrightarrow \mathrm{L}_\lambda[T_{\mathcal{M}}] \models \varphi^*$. Hence it suffices to show that $|M|$ is $\mathbf{\Pi}_n^m(T_{\mathcal{M}})$-reflecting. This will follow from the next lemma. Let $\varphi(v_1, ..., v_\ell)$ be a $\mathbf{\Pi}_n^{m-}$ formula of $\mathcal{L}_p(T_{\mathcal{M}})$ with the indicated free variables.

10.4. Lemma. *There is a* $\boldsymbol{\Pi}_n^m$ *i.d.* Ψ *such that for* $\mathcal{M}$*-good* λ *and* $c_1, \ldots, c_\ell \in M_\lambda$

$$\lambda \models \varphi(|c_1|, \ldots, |c_\ell|) \Leftrightarrow \langle c_1, \ldots, c_\ell \rangle \in \Psi(M_\lambda^*) .$$

Proof. This will be in five parts. Assume throughout that λ ranges over $\mathcal{M}$-good ordinals.

(1) If R is primitive recursive in $T_\mathcal{M}$ then by the Coding Lemma there is a primitive recursive function h_R, independent of λ, such that for $a_1, \ldots, a_n \in M$

$$\lambda \models R(|a_1|, \ldots, |a_k|) \Leftrightarrow h_R(a_1, \ldots, a_k) \in M_\lambda .$$

In particular

$$a, b \in M_\lambda \;\&\; |a| = |b| \Leftrightarrow h_=(a, b) \in M_\lambda .$$

(2) Let $\mathcal{G}(Y, X)$ if and only if $X \subseteq \omega$ and $Y \subseteq \omega \times \omega$ is the graph of a bijection $f : \omega \cong Q \subseteq X$ such that

(i) $x, y \in Q \;\&\; h_=(x, y) \in X \Rightarrow x = y$, and

(ii) $y \in X \Rightarrow \exists x \in Q h_=(x, y) \in X$.

It should be clear that $\mathcal{G}$ is arithmetical. Moreover $\mathcal{G}(Y, M_\lambda)$ holds if and only if Y is the graph of a bijection $f : \omega \cong Q \subseteq M_\lambda$ such that $Y^* = \lambda x |f(x)|$ is a bijection: $\omega \cong \lambda$. Hence $\exists Y \mathcal{G}(Y, M_\lambda)$.

(3) If R is primitive recursive in $T_\mathcal{M}$ let $\theta_R(Y, X, x_1, \ldots, x_k)$ be the $\boldsymbol{\Sigma}_1^{0-}$ formula of $\mathcal{L}_p$

$$\exists y_1 \ldots \exists y_k \exists y [\bigwedge_{1 \le i \le k} Y(x_i, y_i) \wedge X(y) \wedge (h_R(y_1, \ldots, y_k) = y)] .$$

Then if $\mathcal{G}(Y, M_\lambda)$ and $a_1, \ldots, a_k \in \omega$

$$\lambda \models R(Y^*(a_1), \ldots, Y^*(a_k)) \Leftrightarrow \omega \models \theta_R(Y, M_\lambda, a_1, \ldots, a_k) .$$

(4) Let $\varphi^*(Y, X, v_1, \ldots, v_\ell)$ be obtained from $\varphi(v_1, \ldots, v_\ell)$ by replacing every atomic formula $R(x_1, \ldots, x_k)$ by $\theta_R(Y, X, x_1, \ldots, x_k)$. Then φ^* is a $\boldsymbol{\Pi}_n^{m-}$ formula of $\mathcal{L}_p$, and if $\mathcal{G}(Y, M_\lambda)$ and $a_1, \ldots, a_\ell \in \omega$ then

$$\lambda \models \varphi(Y^*(a_1), \ldots, Y^*(a_\ell)) \Leftrightarrow \omega \models \varphi^*(Y, M_\lambda, a_1, \ldots, a_\ell) .$$

(5) $\Psi(X)$ may now be defined to be the set of $\langle c_1, ..., c_\ell \rangle$ such that $c_1, ..., c_\ell \in \mathcal{F}(X)$ and for all Y such that $\mathcal{G}(Y, \mathcal{F}(X))$ and all $a_1, ..., a_\ell$, $b_1, ..., b_\ell$, if $\bigwedge_{1 \leq i \leq l} (Y(a_i, b_i) \wedge h_=(b_i, c_i) \in \mathcal{F}(X))$ then

$$\omega \models \varphi^*(Y, \mathcal{F}(X), a_1, ..., a_\ell) .$$

Then Ψ is a $\mathbf{\Pi}_n^m$ i.d. that satisfies the lemma.

We can now complete the proof of the theorem. Let $|M| \models \varphi$ where φ is a $\mathbf{\Pi}_n^m$ sentence of $\mathcal{L}_p(T_{\mathcal{M}})$. We must find $\lambda < |M|$ such that $\lambda \models \varphi$. φ must have the form $\varphi(|a_1|, ..., |a_\ell|)$ for some $\mathbf{\Pi}_n^{m-}$ formula of $\mathcal{L}_p(T_{\mathcal{M}})$ and $a_1, ..., a_\ell \in M$. Let Ψ be the i.d. given by Lemma 10.4. Then as Ψ is $\mathbf{\Pi}_n^m$ and Γ is complete $\mathbf{\Pi}_n^m$ there is a $g : \Psi(X) \leq_m \Gamma(X)$ for all X. Let $a = g(\langle a_1, ..., a_\ell \rangle)$. Then for $\mathcal{M}$-good λ

$$\begin{aligned} \lambda \models \varphi &\Leftrightarrow \langle a_1, ..., a_\ell \rangle \in \Psi(M_\lambda^*) \\ &\Leftrightarrow a \in \Gamma(M_\lambda^*) \\ &\Leftrightarrow \langle 3, a \rangle \in \Theta(M_\lambda^*) . \end{aligned}$$

But $|M| \models \varphi$. Hence $\langle 3, a \rangle \in \Theta(M^*) \subseteq M$. Let $\lambda = |\langle 3, a \rangle|$. Then $\lambda < |M|$ is $\mathcal{M}$-good and $\langle 3, a \rangle \in \Theta(M_\lambda^*)$. Hence $\lambda \models \varphi$.

10.5. Corollary. *For $m, n > 0$ $|\mathbf{\Pi}_n^m|$ is $\mathbf{\Pi}_n^m$-reflecting and $|\mathbf{\Sigma}_n^m|$ is $\mathbf{\Sigma}_n^m$-reflecting. Hence $|\mathbf{\Pi}_n^m| \geq \pi_n^m$ and $|\mathbf{\Sigma}_n^m| \geq \sigma_n^m$.*

Proof. By Theorem 10.3 $|\mathbf{\Pi}_n^m|$ is $\mathbf{\Pi}_n^m(\Gamma)$-reflecting and hence $\mathbf{\Pi}_n^m$-reflecting. Similarly for $\mathbf{\Sigma}_n^m$.

In general we cannot expect that $|\mathbf{\Pi}_n^m| = \pi_n^m$ or $|\mathbf{\Pi}_n^m| = \sigma_n^m$ for $m, n > 0$. In order to use Lemma 10.1 to show that $|\mathbf{\Pi}_3^1| \leq \pi_3^1$, for example, we would want to show that for $\Gamma \in \mathbf{\Pi}_3^1$, $X \in \mathrm{L}_{\pi_3^1} \cap P(\omega) \Rightarrow \Gamma(X) \in \mathrm{L}_{\pi_3^1}$. But there is no guarantee that $\Gamma^1 = \Gamma(\emptyset)$ belongs to $\mathrm{L}_{\pi_3^1}$ or even to L. In the case of $\mathbf{\Pi}_1^1$ and $\mathbf{\Sigma}_1^1$, however, we can do better by making use of a result due to Barwise, Gandy and Moschovakis, formulated here in Theorem 6.2.

Let In be the class of recursively inaccessible ordinals.

10.6. Lemma. *If Γ is $\mathbf{\Pi}_1^1$ or $\mathbf{\Sigma}_1^1$ then $\langle \Gamma^\xi : \xi < \lambda \rangle$ is uniformly $\mathbf{\Sigma}_1^0$ on L_λ for $\lambda \in \mathrm{In}$.*

Proof. It is sufficient to show that if $\lambda \in \mathrm{In}$ then

(i) $x \in \mathrm{L}_\lambda \Rightarrow \Gamma(x \cap \omega) \in \mathrm{L}_\lambda$, and

(ii) $\{(x,y) \in \mathrm{L}_\lambda \times \mathrm{L}_\lambda : \Gamma(x \cap \omega) = y\}$ is $\mathbf{\Sigma}_1^0$ on L_λ uniformly for $\lambda \in \mathrm{In}$, as we may then carry through the proof of Lemma 7.5.

Let Γ be $\mathbf{\Pi}_1^1$. Then by 6.2(i) there is a $\mathbf{\Sigma}_1^{0-}$ formula φ^+ of $\mathcal{L}_\in$ such that if A, B are admissible sets and $x \in A \in B$ then

$$n \in \Gamma(x \cap \omega) \Leftrightarrow B \models \varphi^+(A, n, x)\,.$$

But if $\lambda \in \mathrm{In}$ and $x \in \mathrm{L}_\lambda$ then $x \in \mathrm{L}_\mu$ for some admissible $\mu < \lambda$ and also $\mu^+ < \lambda$. So

$$n \in \Gamma(x \cap \omega) \Leftrightarrow \mathrm{L}_{\mu^+} \models \varphi^+(\mathrm{L}_\mu, n, x)\,.$$

Hence $\Gamma(x \cap \omega)$ is $\mathbf{\Sigma}_1^0$ on L_{μ^+} so that $\Gamma(x \cap \omega) \in \mathrm{L}_{\mu^+ + 1} \subseteq \mathrm{L}_\lambda$, proving (i).

To prove (ii) let σ_0 be the $\mathbf{\Pi}_3^{0-}$ sentence of 2.4. Then if $x, y \in \mathrm{L}_\lambda$ and $\lambda \in \mathrm{In}$, $\Gamma(x \cap \omega) = y$ if and only if there are transitive sets $A, B \in \mathrm{L}_\lambda$ such that $[A, B$ are transitive & $x \in A \in B$ & $A \models \sigma_0$ & $B \models \sigma_0$ & $y \subseteq \omega$ & $\forall n \in \omega$ $(B \models \varphi^+(A, n, x) \Leftrightarrow n \in y)]$. The expression [...] can be defined by a $\mathbf{\Sigma}_0^{0-}$ formula $\Psi(A, B, x, y)$ of $\mathcal{L}_\in$, independent of λ, so that $\Gamma(x \cap \omega) = y \Leftrightarrow \mathrm{L}_\lambda \models \exists a \exists b \Psi(a, b, x, y)$, proving (ii).

If Γ is $\mathbf{\Sigma}_1^1$ then the proof is as above except that φ^+ is now $\mathbf{\Pi}_1^{0-}$.

10.7. Theorem. $|\mathbf{\Pi}_1^1| = \pi_1^1$ *and* $|\mathbf{\Sigma}_1^1| = \sigma_1^1$.

Proof. $|\mathbf{\Pi}_1^1| \geq \pi_1^1$ and $|\mathbf{\Sigma}_1^1| \geq \sigma_1^1$ follows from 10.5. Note that by Theorem 1.9 $\pi_1^1, \sigma_1^1 \in \mathrm{In}$, π_1^1 is $\mathbf{\Pi}_1^1$-reflecting on In and σ_1^1 is $\mathbf{\Sigma}_1^1$-reflecting on In. Hence by Lemma 10.6 we may use Lemma 10.1 with $n = m = 1$, $X = \mathrm{In}$ and $A = \emptyset$ to get $|\mathbf{\Pi}_1^1| \leq \pi_1^1$ and $|\mathbf{\Sigma}_1^1| \leq \sigma_1^1$.

§11. Higher order inductive definitions, II

Recall from the introduction that $\omega(\mathcal{R}) = \mathrm{Sup}\{|\prec| : \prec \in \mathcal{R}$ well-orders a subset of $\omega\}$, where $|\prec|$ is the order type of the well-ordering $\prec$. In this section we will be concerned with comparing the ordinals $|\mathbf{\Pi}^n_m|$, $|\mathbf{\Sigma}^n_m|$, $|\mathbf{\Delta}^n_m|$, π^n_m, σ^n_m, $\omega(\mathbf{\Pi}^n_m)$, $\omega(\mathbf{\Sigma}^n_m)$ and $\omega(\mathbf{\Delta}^n_m)$ when $n, m > 0$. We will prove Theorems E, F and G of the introduction.

11.1. Theorem. *Let* $m, n > 0$.

(i) *If* $m+n > 2$ *then* $|\mathbf{\Delta}^n_m| = \omega(\mathbf{\Delta}^n_m)$.

(ii)
$$\begin{array}{cccccccccc} & > & |\mathbf{\Pi}^n_m| & \geq & \pi^n_m & \geq & \omega(\mathbf{\Sigma}^n_m) & \geq & \\ |\mathbf{\Delta}^n_{m+1}| & & & & & & & & \omega(\mathbf{\Delta}^n_m) \\ & > & |\mathbf{\Sigma}^n_m| & \geq & \sigma^n_m & \geq & \omega(\mathbf{\Pi}^n_m) & \geq & \\ & \text{(a)} & & \text{(b)} & & \text{(c)} & & \text{(d)} & \end{array}$$

Proof. (i) We first show that $|\mathbf{\Delta}^n_m| \geq \omega(\mathbf{\Delta}^n_m)$. Let $\prec$ be a $\mathbf{\Delta}^n_m$ well-ordering of $A \subseteq \omega$. It suffices to find $\Gamma \in \mathbf{\Delta}^n_m$ such that $|\Gamma| = |\prec|$. Let

$$\Gamma(X) = \{x \in A : \forall z (z \prec x \Rightarrow z \in X)\}\,.$$

As $n > 0$ Γ is clearly $\mathbf{\Delta}^n_m$. By induction on α, $\Gamma^\alpha = \{x \in A : |x|_\prec < \alpha\}$ where $|x|_\prec$ is the order type of x in the well-ordering $\prec$. Hence $|\Gamma| = |\prec|$.

We next show that $\omega(\mathbf{\Delta}^n_m) \geq |\mathbf{\Delta}^n_m|$. The technique here is implicit in [14]. Let $\Gamma \in \mathbf{\Delta}^n_m$ and

$$x \in Q \Leftrightarrow x \in \Gamma^\infty \;\&\; \forall y < x (y \in \Gamma^\infty \Rightarrow |y| \neq |x|)\,.$$

Then Q is a univalent system of notations for the ordinals less than $|\Gamma|$. Let

$$x \prec y \Leftrightarrow x, y \in Q \;\&\; |x| < |y|\,.$$

The $\prec$ is a well-ordering and $|\prec| = |\Gamma|$. It suffices to show that $\prec$ is $\mathbf{\Delta}^n_m$.

For $X \subseteq \omega \times \omega$ let $X_k = \{y : yXk\}$. If $S \subseteq \omega \times \omega$ and $Y \subseteq \omega$ let $\Phi(S, X, Y) \Leftrightarrow S$ is a (strict) well-ordering of Y & $\forall k \notin Y (X_k = \emptyset)$ & $\forall k \in Y (X_k = \bigcup\{\Gamma(X_l) : lSk\})$ & $\forall k, l \in Y (k \neq l \Rightarrow X_k \neq X_l)$ & $\Gamma(\bigcup_{k<\omega} X_k) \subseteq \bigcup_{k<\omega} X_k$.

Then as $m,n>0$ and $m+n>2$ Φ is $\mathbf{\Delta}^n_m$. Clearly $\Phi(S,X,Y)$ if and only if S is a well-ordering of Y of order type $|\Gamma|$ such that $X_k = \Gamma^{|k|_S}$ for $k \in Y$, and $X_k = \emptyset$ for $k \notin Y$. Hence if

$$T(x,y) \Leftrightarrow x,y \in \Gamma^\infty \;\&\; |x| \leq |y| ,$$

then

$$T(x,y) \Leftrightarrow \exists S \exists X \exists Y [\Phi(S,X,Y) \;\&\; \psi(X,x,y)]$$

$$\Leftrightarrow \forall S \forall X \forall Y [\Phi(S,X,Y) \Rightarrow \psi(X,x,y)]$$

where $\psi(X,x,y) \Leftrightarrow [x,y \in \bigcup_{k\in\omega} X_k \;\&\; \forall k(y \in X_k \Rightarrow x \in X_k)]$. Hence T is $\mathbf{\Delta}^n_m$. But

$$x \in Q \Leftrightarrow T(x,x) \;\&\; \forall y < x(T(y,y) \Rightarrow \neg(T(y,x) \;\&\; T(x,y))) ,$$

so that Q is $\mathbf{\Delta}^n_m$. As

$$x \prec y \Leftrightarrow x,y \in Q \;\&\; \neg T(y,x) ,$$

it follows that $\prec$ is $\mathbf{\Delta}^n_m$.

(ii) (a) Let Γ_1 be complete $\mathbf{\Sigma}^n_m$ such that $a \notin \Gamma_1^\infty$. Let $\Gamma_2(X) = \{a\}$ for all X and let $\Gamma = [\Gamma_1, \Gamma_2]$. Then $\Gamma \in \mathbf{\Delta}^n_{m+1}$ and $|\Gamma| = |\Gamma_1| + 1 = |\mathbf{\Sigma}^n_m| + 1$. Hence $|\mathbf{\Delta}^n_{m+1}| > |\mathbf{\Sigma}^n_m|$; $|\mathbf{\Delta}^n_{m+1}| > |\mathbf{\Pi}^n_m|$ is proved in the same way.

(b) is just 10.5. For (c) let $\prec$ be a $\mathbf{\Pi}^n_m$ well-ordering. Then there is a $\mathbf{\Sigma}^n_m$ sentence φ of $\mathcal{L}_P$ logically equivalent to

(1) $$\exists f \forall k \forall l [k \prec l \Rightarrow f(k) < f(l)] .$$

Then $\lambda \models \varphi \Leftrightarrow \lambda \geq |\prec|$ and hence $|\prec|$ is not $\mathbf{\Sigma}^n_m$-reflecting The proof that $\pi^n_m \geq \omega(\mathbf{\Sigma}^n_m)$ is similar to the above, interchanging $\mathbf{\Pi}^n_m$ and $\mathbf{\Sigma}^n_m$ throughout and replacing (1) by

$$\neg \exists f \exists k \forall \alpha \forall \beta [\alpha < \beta \Rightarrow f(\alpha) \prec f(\beta) \prec k] .$$

(d) is trivial.

Remark. We do not know of any cases where equality holds in (c). Note that

$\pi^n_m \leq |\Pi^n_m| < \pi^n_{m+1}$. Thus π^n_m and $|\Pi^n_m|$ are not too far apart. Similarly for σ^n_m and $|\Sigma^n_m|$.

11.2. Theorem. $|\Delta^1_1|$ *is not admissible.*

Proof. We shall use the following fact extracted from §7.10 of [19].

Proposition. There is a Π^1_1 relation $J \subseteq \omega \times \omega \times \mathcal{P}(\omega)$ such that Γ is Δ^1_1 if and only if $\Gamma(X) = \Gamma_n(X) = \{y : J(n,y,X)\}$ for some $n < \omega$.

11.3. Lemma. *If* λ *is recursively inaccessible then* $\langle \Gamma^\xi_n : n < \omega \;\&\; \xi < \lambda \rangle$ *is* Σ^0_1 *on* L_λ.

Proof. By 10.6, as each Γ_n is Π^1_1, $\langle \Gamma^\xi_n : \xi < \lambda \rangle$ is Σ^0_1 on L_λ for all $n < \omega$. But as Γ_n is Π^1_1 uniformly in n, and the proof of 10.6 is uniform, $\langle \Gamma^\xi_n : \xi < \lambda \rangle$ may be seen to be Σ^0_1 on L_λ uniformly in n giving the lemma.

11.4. Lemma. $|\Delta^1_1|$ *is a limit of admissibles.*

Proof. Let $\alpha < |\Delta^1_1|$. Then there is a $\Gamma \in \Delta^1_1$ such that $\alpha \leq |\Gamma|$. Let $\Theta(X) = \Xi(X) \cup \{\langle 3, a \rangle : a \in \Gamma(X)\}$, where Ξ is as in 8.16. Then $\mathcal{M} = \mathcal{M}^\Theta$ is a good notation system so that by the Coding Lemma $|M|$ is admissible. But as $\Gamma \leq_m \Theta$ it follows from 8.15 that $|\Gamma| \leq |M|$. But $|M| = |\Theta_{\leq}|$ and, as Γ is Δ^1_1, so is Θ and hence $\Theta_{\leq}$. Now let

$$\Lambda(X) = \Theta_{\leq}(X) \cup \{x : \Theta_{\leq}(X) \subseteq X \;\&\; x \in \omega\}\,.$$

Then Λ is Δ^1_1 and $|\Lambda| = |\Theta_{\leq}| + 1$. Hence $|M| < |\Lambda| \leq |\Delta^1_1|$, so that $\alpha \leq |M| < |\Delta^1_1|$ and $|M|$ is admissible proving the lemma. Note that we have shown that $|\Gamma| < |\Delta^1_1|$ for all $\Gamma \in \Delta^1_1$.

We can now prove the theorem. Suppose that $\lambda = |\Delta^1_1|$ is admissible. Then by the previous lemma it is recursively inaccessible and hence by Lemma 11.3 $\langle \Gamma^\xi_n : n < \omega \;\&\; \xi < \lambda \rangle$ is Σ^0_1 on L_λ. Let $f(n) = |\Gamma_n|$ for $n < \omega$. Then $f : \omega \to \lambda$ is λ-recursive, as

$$f(n) = \mu\xi\,[\Gamma^{\xi+1}_n = \Gamma^\xi_n]\,.$$

But $\lambda = |\Delta_1^1| = \mathrm{Sup}_{n<\omega}|\Gamma_n| = \mathrm{Sup}_{n<\omega} f(n)$, contradicting the admissibility of λ.

We conclude this section by showing that under very general conditions $|\mathcal{C}| \neq |\neg\mathcal{C}|$. We also obtain a related spectrum result.

11.5. Definition. If $\mathcal{C}$ is a class of inductive definitions then the *spectrum* $\mathrm{sp}(\mathcal{C})$ of $\mathcal{C}$ is $\{|\Gamma| : \Gamma \in \mathcal{C}\}$.

11.6. Definition. A closed class $\mathcal{C}$ is $\forall$-*closed* if $\Sigma_1^0 \subseteq \mathcal{C}$ and $\Gamma \in \mathcal{C}$ implies $\Gamma_1 \in \mathcal{C}$, where

$$\Gamma_1(X) = \{e : \forall x \{e\}(x) \in \Gamma(X)\} .$$

11.7. Theorem. *If $\mathcal{C}$ is $\forall$-closed then $|\mathcal{C}| \notin \mathrm{sp}(\neg\mathcal{C})$ and in particular $|\mathcal{C}| \neq |\neg\mathcal{C}|$.*

Note that the last inequality follows because $|\neg\mathcal{C}| \in \mathrm{sp}(\neg\mathcal{C})$. If $|\mathcal{C}| < |\neg\mathcal{C}|$ the theorem implies $\mathrm{sp}(\neg\mathcal{C})$ is not an initial segment of ordinals. Σ_m^0 is not $\forall$-closed. However, Π_{m+1}^0, Π_m^n, and Σ_m^n are $\forall$-closed for $m, n > 0$. In particular we get

11.8. Corollary. *If $m, n > 0$ then $|\Pi_m^n| \neq |\Sigma_m^n|$.*

We turn to the proof of the theorem. Let $\mathcal{C}$ be $\forall$-closed. Let Γ be $\mathcal{C}$-complete and $\Delta \in \neg\mathcal{C}$. It is sufficient to show that $|\Delta| \neq |\Gamma|$ as $|\Gamma| = |\mathcal{C}|$. Let $\Theta(X) =$

$$\begin{aligned}
&\{\langle 1, a\rangle : a \in \Gamma(X)\} \\
&\cup \{\langle 2, a, b\rangle : a \in \mathcal{F}(X) \;\&\; \langle 4, b\rangle \in X_{<a}\} \\
&\cup \{\langle 3, a, b\rangle : \langle 6, a\rangle \in \mathcal{F}(X) \;\&\; b \notin \Delta(\{x : \langle 2, a, x\rangle \in \mathcal{F}(X)\})\} \\
&\cup \{\langle 4, a\rangle : \exists x\, [\langle 6, \langle 6, x\rangle\rangle \in \mathcal{F}(X) \;\&\; \langle 3, x, a\rangle \notin \mathcal{F}(X)]\} \\
&\cup \{a : a = \langle 5\rangle \;\&\; \forall x\, [x \in \Delta(\{y : \langle 4, y\rangle \in \mathcal{F}(X)\}) \Rightarrow \langle 4, x\rangle \in \mathcal{F}(X)]\} \\
&\cup \{\langle 6, a\rangle : a \in \mathcal{F}(X)\} .
\end{aligned}$$

As $\mathcal{C}$ is first order closed $\Theta \in \mathcal{C}$. As $\Gamma \leq_m \Theta$ it follows that Θ is $\mathcal{C}$-complete and hence by Theorem 8.14

Lemma 1. $|\Gamma| = |M|$.

Lemma 2. *For* $\alpha \leq |M|$,

$$\langle 4,a\rangle \in M_\alpha \Leftrightarrow \exists\nu\,[\nu+3<\alpha\ \&\ a\in \mathbf{\Delta}(\{y:\langle 4,y\rangle \in M_\nu\})]\ .$$

Proof.

$$\langle 2,x,y\rangle \in M_\nu \Leftrightarrow \exists\lambda<\nu.\ \langle 2,x,y\rangle \in \Theta(M^*_\lambda)$$

$$\Leftrightarrow \exists\lambda<\nu.\ \langle 6,x\rangle \in M_{\lambda+1}\ \&\ \langle 4,y\rangle \in M_{|x|}$$

$$\Leftrightarrow \langle 6,x\rangle \in M_\nu\ \&\ \langle 4,y\rangle \in M_{|x|}\ . \tag{2}$$

Now

$$\langle 6,x\rangle \in M_\lambda \Leftrightarrow \exists\sigma<\lambda.\ \langle 6,x\rangle \in \Theta(M^*_\sigma)$$

$$\Leftrightarrow \exists\sigma<\lambda.\ x\in M_\sigma$$

$$\Leftrightarrow |x|+1<\lambda\ .$$

Hence, letting

$$Q(a,\sigma) \Leftrightarrow a\in \mathbf{\Delta}(\{y:\langle 4,y\rangle \in M_\sigma\})\ ,$$

we have by (2),

$$\langle 3,x,a\rangle \in M_\nu \Leftrightarrow \exists\lambda<\nu.\ \langle 6,x\rangle \in M_\lambda\ \&\ a\notin \mathbf{\Delta}(\{y:\langle 2,x,y\rangle \in M_\lambda\})$$

$$\Leftrightarrow \exists\lambda<\nu.\ \langle 6,x\rangle \in M_\lambda\ \&\ a\notin \mathbf{\Delta}(\{y:\langle 6,x\rangle \in M_\lambda\ \&\ \langle 4,y\rangle \in M_{|x|}\})$$

$$\Leftrightarrow \langle 6,\langle 6,x\rangle\rangle \in M_\nu\ \&\ \neg Q(a,|x|)\ .$$

Hence

$$\langle 4,a\rangle \in M_\alpha \Leftrightarrow \exists\lambda<\alpha\exists x[\langle 6,\langle 6,x\rangle\rangle \in M_\lambda\ \&\ \langle 3,x,a\rangle \notin M_\lambda]$$

$$\Leftrightarrow \exists\lambda<\alpha\exists x[\langle 6,\langle 6,x\rangle\rangle \in M_\lambda\ \&\ [\langle 6,\langle 6,x\rangle\rangle \notin M_\lambda \vee Q(a,|x|)]]$$

$$\Leftrightarrow \exists\lambda<\alpha\exists x[|x|+2<\lambda\ \&\ Q(a,|x|)]$$

$$\Leftrightarrow \exists\nu[\nu+3<\alpha\ \&\ Q(a,\nu)]\quad (\text{since } \alpha\leq|M|)\ .$$

Lemma 3. *For* $\alpha \leq |M|$ *and* $i < 4$, $\langle 4, a\rangle \in M_{4\alpha+i} \Leftrightarrow a \in \Delta^{\alpha}$. *In particular,* $\Delta^{\alpha} \leq_m M_{4\alpha}$.

Proof. We use induction on α.

$$\begin{aligned}\langle 4,a\rangle \in M_{4\alpha+i} &\Leftrightarrow \exists \nu[\nu+3<4\alpha+i \;\&\; \bar{Q}(a,\nu)]\\ &\Leftrightarrow \exists \beta<\alpha.\ \exists j<4\, \bar{Q}(a,4\beta+j)\\ &\Leftrightarrow \exists \beta<\alpha.\ a \in \Delta\Delta^{\beta}\\ &\Leftrightarrow a \in \Delta^{\alpha}.\end{aligned}$$

Since $|\Gamma| = |M|$ it suffices to prove:

Lemma 4. $|\Delta| \neq |M|$.

Proof. Suppose $|\Delta| = |M| = \alpha$. We get a contradiction by showing $|\Delta| < \alpha$. Since $a \in M \Rightarrow \langle 6,a\rangle \in M \;\&\; |a|+1 = |\langle 6,a\rangle|$, α must be a limit ordinal and hence $\alpha = 4\alpha$. By definition of α, $\Delta\Delta^{\alpha} \subseteq \Delta^{\alpha}$. Hence by Lemma 3,

$$\forall x[x \in \Delta(\{a : \langle 4,a\rangle \in M_{\alpha}\}) \Rightarrow \langle 4,x\rangle \in M_{\alpha}]$$

i.e. $\langle 5\rangle \in M_{\alpha+1} \subseteq M$. Let $4\beta + i = |\langle 5\rangle| < \alpha$. Then by definition of M,

$$\forall x[x \in \Delta(\{a : \langle 4,a\rangle \in M_{4\beta+i}\}) \Rightarrow \langle 4,x\rangle \in M_{4\beta+i}]\,.$$

By Lemma 3,

$$\forall x[x \in \Delta(\Delta^{\beta}) \Rightarrow x \in \Delta^{\beta}]\,,$$

and hence $|\Delta| \leq \beta < \alpha$.

Appendix. Proof of the Coding Lemma

§A.1. Acceptable ordinal systems.

We begin by discussing certain closure properties on systems $\mathcal{M}$ and show that if $\mathcal{M}$ is a good notation system, then $\mathcal{M}$ has these closure properties.

A.1. Definition. Let $\mathcal{M} = \langle M, \|\rangle$ be any ordinal system and $B \subseteq \omega$. $\mathcal{M}$ is *B-restricted productive* if there is a primitive recursive function p such that for every $n \in \omega$,

(i) $\forall x[\{n\}(x,B) \in M] \Rightarrow p(n) \in M \;\&\; |p(n)| \geq \sup\{|\{n\}(x,B)| + 1 : x \in \omega\}$;

(ii) $p(n) \in M \Rightarrow \forall x[\{n\}(x,B) \in M]$.

p is called a B-restricted productive function for $\mathcal{M}$.

The closure condition (i) is analogous to the closure of infinite regular cardinals with respect to mappings from smaller ordinals. (ii) is a technical requirement which ensures that there are not extraneous notations in M.

A.2. Definition. $\mathcal{M}$ is *acceptable* if there are recursive functions j, $\circledV$ and a primitive recursive function p such that

(i) $j : M \leq_m M$ and for $a \in M$, if $|j(a)| \leq \alpha$ then $J(M_{|a|})$ is recursive in M_α uniformly in a, where J is the complete Σ_1^0 i.d. $J(X) = \{x : \exists y T^X(x,x,y)\}$;

(ii) if $a \in M$ then $\lambda n.\, p(a,n)$ is an $M_{|a|}$-restricted productive function for $\mathcal{M}$.

(iii) $a \in M \vee b \in M \Rightarrow a \circledV b \in M \;\&\; \inf\{|a|, |b|\} \leq |a \vee b|$, where $|x| = |M|$ if $x \notin M$. We say that $\mathcal{M}$ is acceptable *in terms of* $j, p, \circledV$.

We next show that there are functions $j, p, \circledV$ such that if $\mathcal{M}$ is a good notation system and λ is $\mathcal{M}$-good then $\mathcal{M}_\lambda$ is acceptable in terms of $j, p, \circledV$.

A.3. Lemma. *Let $\mathcal{M} = \langle M, \|\rangle$ be a good notation system. If $a \in M$ then $J(M_{|a|})$ is recursive in M_α for all $\alpha \geq |a| + 2$, uniformly in a.*

Proof. Let $\alpha \geq |a| + 2$ and let e be a recursive function such that

$$[e(a,x)](t, M_{|a|}) = \begin{cases} a & \text{if } \neg T^{M_{|a|}}(x,x,t) \\ 1 & \text{otherwise}. \end{cases}$$

Note that $1 \notin M$. It suffices to show there is a recursive function f such that for all x, $x \notin J(M_{|a|}) \Longleftrightarrow f(a,x) \in M_\alpha$. Now

$$\begin{aligned} x \notin J(M_{|a|}) &\Longleftrightarrow \forall t \neg T^{M_{|a|}}(x,x,t) \\ &\Longleftrightarrow \forall t\,[e(a,x)](t, M_{|a|}) = a \in M_{|a|+1} \\ &\Longleftrightarrow \langle 1, a, e(a,x)\rangle \in M_\alpha\,. \end{aligned}$$

Thus let $f(a,x) = \langle 1, a, e(a,x)\rangle$.

To show that $\mathcal{M}$ satisfies A.2 (i) it remains to find a recursive function j so that for all $\mathcal{M}$-good λ, $j : M_\lambda \leq_m M_\lambda$ and for $a \in M_\lambda$, $|(a)| = |a| + 2$. Let e_1 be a recursive function such that for all a, t $[e_1(a)](t, M_{|0|}) = a$. Then for $\mathcal{M}$-good λ,

$$a \in M_\lambda \Rightarrow \forall t [e_1(a)](t, M_{|0|}) = a \in M_\lambda$$
$$\Rightarrow \langle 1, 0, e_1(a)\rangle \in M_\lambda \;\&\; |\langle 1, 0, e_1(a)\rangle| = |a| + 1\,.$$

Also $a \notin M_\lambda \Rightarrow \langle 1, 0, e_1(a)\rangle \notin M_\lambda$. Thus let $j(a) = \langle 1, 0, e_1(\langle 1, 0, e_1(a)\rangle)\rangle$.

A.4. Lemma. *There are functions j, p, $\circledvee$ such that if $\mathcal{M}$ is a good notation system and λ is $\mathcal{M}$-good, then $\mathcal{M}_\lambda$ is acceptable in terms of j, p, $\circledvee$.*

Proof. It remains to find $\vee$ and p. It is easy to see that $a \circledvee b = \langle 2, a, b\rangle$ has the desired property. To find p, let e be a primitive recursive function such that for any n and $X \subseteq \omega$, the range of $\lambda t[e(n)](t, X)$ equals $\{0\} \cup$ the range of $\lambda t.\{n\}(t, X)$. Then for $a \in M_\lambda$,

$$\forall x\{n\}(x, M_{|a|}) \in M_\lambda \Leftrightarrow \forall x[e(n)](x, M_{|a|}) \in M_\lambda$$

and if $\forall x\{n\}(x, M_{|a|}) \in M_\lambda$ then

$$\mathrm{Sup}\{|\{n\}(x, M_{|a|})| : x \in \omega\} = \mathrm{Sup}\,\{|\,[e(n)](x, M_{|a|})| : x \in \omega\}\,.$$

Thus it is easy to see that we can choose $p(a,n) = \langle 1, a, e(n)\rangle$.

In view of Lemma A.4, to prove the Coding Lemma it suffices to prove the following:

A.5. Theorem. *Let $\mathcal{M} = \langle M, ||\rangle$ be acceptable in terms of j, p, $\circledvee$.*

(i) $|M| \in \mathrm{Ad}(T_{\mathcal{M}})$;

(ii) *For every Σ_1^{0-} formula $\varphi(v_1, ..., v_n)$ of $\mathcal{L}_p(T_{\mathcal{M}})$, there is a primitive recursive function h such that*

$$a_1, ..., a_n \in M \;\&\; |M| \models \varphi(|a_1|, ..., |a_n|) \quad \textit{iff} \quad h(a_1, ..., a_n) \in M\,.$$

Furthermore, h is completely determined by the functions j, p, Ⓥ, *a member* $u_0 \in M$, *and the formula* φ.

(iii) *Let F be an ordinal function which is* $|M|$*-partial recursive in* $T_{\mathfrak{M}}$. *Then there is a recursive function k such that for* $a_1, ..., a_n \in M$, *if* $F(|a_1|, ..., |a_n|)$ *is defined then* $k(a_1, ..., a_n) \in M$ *and* $F(|a_1|, ..., |a_n|) \leq |k(a_1, ..., a_n)|$.

(iv) $X \subseteq \omega$ *is* $|M|$-r.e. *in* $T_{\mathfrak{M}}$ *iff* $X \leq_m M$.

The remainder of the appendix is devoted to the proof of Theorem A.5. Suppose $\mathfrak{M} = \langle M, || \rangle$ is acceptable in terms of j, p, Ⓥ. Let $u_0 \in M$ and $|u_0| = 0$. In Lemmas A.6–16 below the reader should observe that the functions described are either independent of the particular acceptable system or are completely determined by j, p, Ⓥ and u_0. (In Lemma A.8 the choice of e is independent of $\mathfrak{M}$; in Lemma A.9 an index of h can be found as the value of a recursive function of the indices of the f_i, g_i which is independent of $\mathfrak{M}$.)

A.6. Lemma. *There is a recursive function* $+_M$ *such that*:

(i) $a \in M \;\&\; b \in M \Rightarrow a +_M b \in M \;\&\; |a +_M b| > \max\{|a|, |b|\}$;

(ii) $a +_M b \in M \Rightarrow a \in M \;\&\; b \in M$.

Proof. Let e be a recursive function such that

$$\{e(a,b)\}(t, M_0) = \begin{cases} a & \text{if } t = 0 \\ b & \text{otherwise,} \end{cases}$$

and let $a +_M b = p(u_0, e(a,b))$.

§A.2. $\mathfrak{M}$-recursion.

We next define a class of partial number-theoretic functions based on $\mathfrak{M}$. These functions behave very much like the functions partial recursive in the type 2 functional E of Kleene [9], where for $f \in {}^{\omega}\omega$,

$$E(f) = \begin{cases} 0 & \text{if } \exists t[f(t) = 0], \\ 1 & \text{otherwise.} \end{cases}$$

Using these functions we are able to carry out computations involving $\mathcal{M}$ which are needed to show that $|M|$ is admissible. The following definition by schemata of the predicate $\{z\}^{\mathcal{M}}(\boldsymbol{x}) \simeq y$ parallels the corresponding definition by Kleene [9] of the partial recursive functionals of finite types. As described in [9] this definition by schemata may be viewed as a transfinite inductive definition. The essential difference here is that there are an infinite number of starting functions in the case S0. Thus the characteristic function of each M_α for $\alpha < |M|$ is given outright.

In the following, $\boldsymbol{x}$ and $\boldsymbol{y}$ are abbreviations for $x_1, \ldots, x_n$ and $y_1, \ldots, y_m$, respectively.

S0.a $\{\langle 0, 1, a\rangle\}^{\mathcal{M}}(x) = \begin{cases} 0 & \text{if } x \in M_{|a|}, \\ 1 & \text{otherwise,} \end{cases}$ for each $a \in M$;

S1. $\{\langle 1, n\rangle\}^{\mathcal{M}}(\boldsymbol{x}) = x_1 + 1$;

S2. $\{\langle 2, n, q\rangle\}^{\mathcal{M}}(\boldsymbol{x}) = q$;

S3. $\{\langle 3, n\rangle\}^{\mathcal{M}}(\boldsymbol{x}) = x_1$;

S4. $\{\langle 4, n, a, b\rangle\}^{\mathcal{M}}(\boldsymbol{x}) \simeq \{a\}^{\mathcal{M}}(\{b\}^{\mathcal{M}}(\boldsymbol{x}), \boldsymbol{x})$;

S5. $\begin{cases} \{\langle 5, n{+}1, a, b\rangle\}^{\mathcal{M}}(0, \boldsymbol{x}) \simeq \{a\}^{\mathcal{M}}(\boldsymbol{x}) \\ \{\langle 5, n{+}1, a, b\rangle\}^{\mathcal{M}}(y{+}1, \boldsymbol{x}) \simeq \{b\}^{\mathcal{M}}(y, \{\langle 5, n{+}1, a, b\rangle\}^{\mathcal{M}}(y, \boldsymbol{x}), \boldsymbol{x}) \end{cases}$;

S6. $\{\langle 6, n, k, a\rangle\}^{\mathcal{M}}(\boldsymbol{x}) \simeq \{a\}^{\mathcal{M}}(\boldsymbol{x}_1)$;

where $\boldsymbol{x}_1$ is obtained from $\boldsymbol{x}$ by moving x_{k+1} to the front.

S8. $\{\langle 8, n, a\rangle\}^{\mathcal{M}}(\boldsymbol{x}) \simeq E(\lambda t . \{a\}^{\mathcal{M}}(t, \boldsymbol{x}))$;

where both sides are undefined if for the given $\boldsymbol{x}$, $\lambda t . \{a\}^{\mathcal{M}}(t, \boldsymbol{x})$ is not total.

S9. $\{\langle 9, n{+}m{+}1, m\rangle\}^{\mathcal{M}}(z, \boldsymbol{x}, \boldsymbol{y}) \simeq \{z\}^{\mathcal{M}}(\boldsymbol{x})$;

Since we are defining only partial *number-theoretic* functions there is no S7 clause. $\{z\}^{\mathcal{M}}$ is called the *$\mathcal{M}$-partial recursive function with index z*. $\{z\}^{\mathcal{M}}$ is *$\mathcal{M}$-recursive* if it is everywhere defined. Note that if z is the index of an $\mathcal{M}$-partial recursive function, then $(z)_1$ is the number of variables of the function. It is easy to prove the standard theorems of recursive function theory with the exception of the normal form theorem. In particular the Kleene S–m–n theorem, the Kleene second recursion theorem and the theorem on definition by cases are proved exactly as in [9]. Thus we have:

A.7. Lemma. *For each $m \geq 1$: there is a primitive recursive function $S^m(z, y_1, ..., y_m)$ such that if $f(y_1, ..., y_m, \boldsymbol{x})$ is an $\mathcal{M}$-partial recursive function with index z then for each fixed $y_1, ..., y_m$, $S^m(z, y_1, ..., y_m)$ is an index of $f(y_1, ..., y_m, \boldsymbol{x})$ as a function of $\boldsymbol{x}$.*

A.8. Lemma (Second recursion theorem). *Given any $\mathcal{M}$-partial recursive function $f(z, \boldsymbol{x})$ an integer e can be found such that $\{e\}^{\mathcal{M}}(\boldsymbol{x}) \simeq f(e, \boldsymbol{x})$.*

A.9. Lemma. *If f_0, f_1, g are $\mathcal{M}$-partial recursive then the function*

$$h(\boldsymbol{x}) \simeq \begin{cases} f_0(\boldsymbol{x}) & \text{if } g(\boldsymbol{x}) \simeq 0, \\ f_1(\boldsymbol{x}) & \text{if } g(\boldsymbol{x}) \simeq 1, \end{cases}$$

is $\mathcal{M}$-partial recursive.

A.10. Lemma. *If f is $\mathcal{M}$-partial recursive, then so is $\mu y [f(\boldsymbol{x}, y) \simeq 0]$.*

A.11. *Remark.* It is clear from scheme S0 and the fact that the $\mathcal{M}$-partial recursive functions are closed under composition, primitive recursion and the μ-scheme, that each function partial recursive in some M_α, where $\alpha < |M|$, is $\mathcal{M}$-partial recursive.

It is convenient to deal with functions of just one variable. Let $\{z\}^{\mathcal{M}}[a] \simeq \{z\}^{\mathcal{M}}((a_0), ..., (a)_{(z)_1 \dot{-} 1})$ and let $D = \{\langle z, a\rangle : \{z\}^{\mathcal{M}}[a] \text{ is defined}\}$. The inductive definition described by schemata S0–S9 associates with each $\langle z, a\rangle \in D$ an ordinal as follows. $|\langle z, a\rangle|^{\mathcal{M}} = 0$ if $(z)_0$ is 0, 1, 2, or 3, that is if $\{z\}^{\mathcal{M}}((a)_0, ..., (a)_{(z)_1 \dot{-} 1})$ is defined by one of S0–S3. In case S4, letting $\boldsymbol{a} = (a)_0, ..., (a)_{(z)_1 \dot{-} 1}$,

$$\{z\}^{\mathcal{M}}[a] = \{z\}^{\mathcal{M}}(\boldsymbol{a}) = \{b\}^{\mathcal{M}}(\{c\}^{\mathcal{M}}(\boldsymbol{a}), \boldsymbol{a}) = \{b\}^{\mathcal{M}}[\langle\{c\}^{\mathcal{M}}[a], \boldsymbol{a}\rangle]$$

for some b, c. Thus let

$$|\langle z,a\rangle|^{\mathfrak{M}} = \max\{|\langle c,a\rangle|^{\mathfrak{M}}, |\langle b,\langle\{c\}^{\mathfrak{M}}[a],\boldsymbol{a}\rangle\rangle|^{\mathfrak{M}}\} + 1 .$$

The cases S5, S6, and S9 are similar. For example in case S9,

$$\{u\}^{\mathfrak{M}}[\langle z,\boldsymbol{a},\boldsymbol{y}\rangle] = \{u\}^{\mathfrak{M}}(z,\boldsymbol{a},\boldsymbol{y}) = \{z\}^{\mathfrak{M}}(\boldsymbol{a}) = \{z\}^{\mathfrak{M}}[a] .$$

Thus let $|\langle u,\langle z,\boldsymbol{a},\boldsymbol{y}\rangle\rangle|^{\mathfrak{M}} = |\langle z,a\rangle|^{\mathfrak{M}} + 1$. In case S8 we have

$$\{z\}^{\mathfrak{M}}[a] = \{z\}^{\mathfrak{M}}(\boldsymbol{a}) = E(\lambda t.\{b\}^{\mathfrak{M}}(t,\boldsymbol{a})) = E(\lambda t.\{b\}^{\mathfrak{M}}[\langle t,\boldsymbol{a}\rangle])$$

for some b. In this case we define

$$|\langle z,a\rangle|^{\mathfrak{M}} = \sup\{|\langle b,\langle t,\boldsymbol{a}\rangle\rangle|^{\mathfrak{M}} + 1 : t\in\omega\} .$$

The ordinal function $|\ |^{\mathfrak{M}}$ makes possible proofs by induction. The following lemma and corollary are a generalization of the fact that a function recursive in E is actually recursive in O_α for some $\alpha < \omega_1$ (where O is from Kleene [22]).

A.12. Lemma. *There are recursive functions f and g such that*

(i) $\langle z,a\rangle \in D \Longleftrightarrow g(z,a)\in M$;

(ii) *If $\langle z,a\rangle\in D$ then for all x,*

$$\{z\}^{\mathfrak{M}}[a] = \{f(z,a)\}(x, M_{|g(z,a)|}) .$$

Proof. The recursive functions f and g are defined simultaneously by the Kleene second recursion theorem of ordinary recursive function theory. (ii) and (i) in the direction $\Rightarrow$ are then proven by induction on $|\langle z,a\rangle|^{\mathfrak{M}}$. (i) in the direction $\Leftarrow$ is proven by induction on $|g(z,a)|$. A rigorous proof would require an elaborate definition of f and g involving a number of auxiliary functions arising from applications of the S–m–n theorem of ordinary recursive function theory. Instead of this we give an informal description suppressing explicit reference to most of the auxiliary functions. We begin by assuming $\langle z,a\rangle\in D$ and show in case Si how $f(z,a)$ and $g(z,a)$ must be defined in this case so that they satisfy (ii) and (i) in the direction $\Rightarrow$. Then we show in Si that if $f(z,a)$ and $g(z,a)$ are defined as in Si then $g(z,a)\in M$ implies $\langle z,a\rangle\in D$. The reader familiar with the Second Recursion Theorem will have no trouble in showing, if desired, that there actually exist such recursive functions f and g.

Case S0. $\langle z,a\rangle \in D$ and

$$\{z\}^{\mathcal{M}}[a] = \{\langle 0,1,(z)_2\rangle\}^{\mathcal{M}}((a)_0) = \begin{cases} 0 & \text{if } (a)_0 \in M_{|(z)_2|}\,, \\ 1 & \text{otherwise}\,. \end{cases}$$

Thus let $g(z,a) = (z)_2$ and choose $f(z,a)$ so that for all x,

$$\{f(z,a)\}(x, M_{|(z)_2|}) = \begin{cases} 0 & \text{if } (a)_0 \in M_{|(z)_2|}, \\ 1 & \text{otherwise}\,. \end{cases}$$

Case S'0. $z = \langle 0,1,(z)_2\rangle$ and $f(z,a) = (z)_2 \in M$. Since $(z)_2 \in M$, $\{z\}^{\mathcal{M}}[a] \simeq \{\langle 0, 1,(z)_2\rangle\}^{\mathcal{M}}((a)_0)$ is defined by clause S0 in the definition of $\{\ \}^{\mathcal{M}}$.

The definition of f and g is trivial in cases S1–S3, and easy in cases S5, S9. We consider in detail cases S8 and S4.

Case S8. $\langle z,a\rangle \in D$ and

$$\{z\}^{\mathcal{M}}[a] = E(\lambda t.\{b\}^{\mathcal{M}}[\langle t,\boldsymbol{a}\rangle]) = \begin{cases} 0 & \text{if } \exists t.\{b\}^{\mathcal{M}}[\langle t,\boldsymbol{a}\rangle] = 0\,, \\ 1 & \text{otherwise}\,, \end{cases}$$

where $b = (z)_2$. By the Induction Hypothesis,

(1) $$\forall t[g(b,\langle t,\boldsymbol{a}\rangle) \in M]$$

and for all t, x,

(2) $$\{b\}^{\mathcal{M}}[\langle t,\boldsymbol{a}\rangle] = \{f(b,\langle t,\boldsymbol{a}\rangle)\}(x, M_{|g(b,\langle t,\boldsymbol{a}\rangle)|})\,.$$

Since $\mathcal{M}$ is ϕ-restricted productive we can find $u \in M$ such that for all t, $|jg(b,\langle t,\boldsymbol{a}\rangle)| < |u|$. (More precisely $u = p(u_0, e)$ where for all t, $\{e\}(t, M_{|u_0|}) = jg(b,\langle t,\boldsymbol{a}\rangle)$. e of course depends on g, a and b.) Then by A.2. (i),

(3) $$M_{|g(b,\langle t,\boldsymbol{a}\rangle)|} \leq_t J(M_{|g(b,\langle t,\boldsymbol{a}\rangle)|})$$
$$\leq_t M_{|u|} \leq_t J(M_{|u|})$$
$$\leq_t M_{|j(u)|}\,,$$

where these reducibilities are uniform in a, b, t. Then from (2), (3) we can find a v (depending on a, b) so that for all t, $\{b\}^{\mathfrak{M}}[\langle t,\boldsymbol{a}\rangle] = \{v\}(t,M_{|u|})$. Then choose w so that

(4) $$E(\lambda t.\{b\}^{\mathfrak{M}}[\langle t,\boldsymbol{a}\rangle]) = 0 \Leftrightarrow \exists t.\{b\}^{\mathfrak{M}}[\langle t,\boldsymbol{a}\rangle] = 0$$
$$\Leftrightarrow \exists t.\{v\}(t,M_{|u|}) = 0$$
$$\Leftrightarrow w \in J(M_{|u|})\,.$$

Let $g(z,a) = j(u)$. Then choose $f(z,a)$ so that for all x,

$$\{f(z,a)\}(x,M_{|j(u)|}) = \begin{cases} 0 & \text{if } w \in J(M_{|u|})\,,\\ 1 & \text{otherwise}\,. \end{cases}$$

It follows from (4) that $f(z,a)$ satisfies the desired equation.
Case S'8. $z = \langle 8,n,b\rangle$ and $g(z,a) \in M$. Since $g(z,a) = j(u)$, $j(u) \in M$; since $j: M \leq_m M$, $u \in M$; since $u = p(u_0,e) \in M$ we have by A.2(ii), for all t, $\{e\}(t,M_{|u_0|}) = jg(b,\langle t,\boldsymbol{a}\rangle) \in M$ and hence $g(b,\langle t,\boldsymbol{a}\rangle) \in M$. Also by A.2(i), (ii), for all t,

$$|g(b,\langle t,\boldsymbol{a}\rangle)| < |jg(b,\langle t,\boldsymbol{a}\rangle)| < |u| < |j(u)| = |g(z,a)|\,.$$

Hence by the Induction Hypothesis, for all t, $\langle b,\langle t,\boldsymbol{a}\rangle\rangle \in D$, i.e. $\{b\}^{\mathfrak{M}}[\langle t,\boldsymbol{a}\rangle]$ is defined. Hence $\{z\}^{\mathfrak{M}}[a] \simeq E(\lambda t.\{b\}^{\mathfrak{M}}[\langle t,\boldsymbol{a}\rangle])$ is defined, i.e. $\langle z,a\rangle \in D$.
Case S4. $\langle x,a\rangle \in D$ and $\{z\}^{\mathfrak{M}}[a] = \{b\}^{\mathfrak{M}}[\langle\{c\}^{\mathfrak{M}}[a],\boldsymbol{a}\rangle]$. Let $d = \langle\{c\}^{\mathfrak{M}}[a],\boldsymbol{a}\rangle$. Then $\langle c,a\rangle \in D$ and $\langle b,d\rangle \in D$, and $|\langle c,a\rangle|^{\mathfrak{M}}, |\langle b,d\rangle|^{\mathfrak{M}} < |\langle z,a\rangle|^{\mathfrak{M}}$. By the induction hypothesis,

$$g(c,a) \in M \quad \text{and} \quad g(b,d) \in M\,,$$

and for all x,

$$(5) \qquad \{c\}^{\mathfrak{M}}[a] = \{f(c,a)\}(x, M_{|g(c,a)|}),$$

and

$$\{z\}^{\mathfrak{M}}[a] = \{b\}^{\mathfrak{M}}[d] = \{f(b,d)\}(x, M_{|g(b,d)|}).$$

We begin by showing how to choose $g(z,a)$ so that $g(z,a) \in M$ and

$$(6) \qquad |jg(c,a)|,\ |jg(b,d)| < |g(z,a)|.$$

By (5) we can find a u (depending on a, b, c) such that for every x,

$$(7) \qquad \begin{aligned} jg(b,d) &= jg(b, \langle \{f(c,a)\}(x, M_{|g(c,a)|}), \boldsymbol{a} \rangle) \\ &= \{u\}(x, M_{|g(c,a)|}). \end{aligned}$$

By A.2 (ii), $p(g(c,a),u) \in M$ and

$$(8) \qquad \begin{aligned} |p(g(c,a),u)| &\geq \operatorname{Sup}\{|\{u\}(x, M_{|g(c,a)|})| + 1 : x \in \omega\} \\ &= |jg(b,d)| + 1. \end{aligned}$$

Let $g(z,a) = p(g(c,a),u) +_M jg(c,a)$. It is clear from (8) that $g(z,a)$ satisfies (6). To find $f(z,a)$ observe that by (6) and A.2 (i),

$$(9) \qquad M_{|g(b,d)|} \leq_t J(M_{|g(b,d)|}) \leq_t M_{|g(z,a)|},$$

uniformly in b, d; and

$$(10) \qquad M_{|g(c,a)|} \leq_t J(M_{|g(c,a)|}) \leq_t M_{|g(z,a)|},$$

uniformly in c, a. From (5), (9) we can find v, w, y so that for all x,

$$\begin{aligned} \{z\}^{\mathfrak{M}}[a] &= \{f(b,d)\}(x, M_{|g(b,d)|}) \\ &= \{v\}(x, b, d, M_{|g(z,a)|}) \\ &= \{w\}(x, b, \{f(c,a)\}(x, M_{|g(c,a)|}), M_{|g(z,a)|}) \\ &= \{y\}(x, M_{|g(z,a)|}), \end{aligned}$$

where w is obtained by eliminating d in the previous equation by referring back to the definition of d and then using (10), and y is obtained by using (10). Thus let $f(z,a) = y$.

Case S'4. $z = \langle 4,n,a,b\rangle$ and $g(z,a) \in M$. Since

$$g(z,a) = p(g(c,a),u) +_M jg(c,a) \in M$$

we have $g(c,a)$, $p(g(c,a),u) \in M$ and $|g(c,a)|, |p(g(c,a),u)| < |g(z,a)|$. By the induction hypothesis, $\{c\}^{\mathfrak{M}}[a]$ is defined. Also by (7),

$$jg(b,d) = \{u\}(0,M_{|g(c,a)|}) \in M\,,$$

and hence $g(b,d) \in M$ and $|g(b,d)| < |p(g(c,a),u)| < |g(z,a)|$. Again by the induction hypothesis, $\{b\}^{\mathfrak{M}}[d]$ is defined. Since $\{z\}^{\mathfrak{M}}[a] \simeq \{b\}^{\mathfrak{M}}[d]$, $\{z\}^{\mathfrak{M}}[a]$ is defined.

$X \subseteq {}^n\omega$ is said to be $\mathfrak{M}$*-r.e.* if X is the domain of an $\mathfrak{M}$-partial recursive function; X is $\mathfrak{M}$*-recursive* if the representing function of X is $\mathfrak{M}$-recursive.

A.13. Corollary. *Let* $X \subseteq \omega$.

(i) *X is $\mathfrak{M}$-r.e. iff* $X \leq_m M$;

(ii) *X is $\mathfrak{M}$-recursive iff $X \leq_t M_\alpha$ for some* $\alpha < |M|$;

(iii) *If h is $\mathfrak{M}$-partial recursive there is a recursive function k such that for all* $\boldsymbol{a}$,

$$h(\boldsymbol{a}) \in M \Rightarrow k(\boldsymbol{a}) \in M \;\&\; |h(\boldsymbol{a})| \leq |k(\boldsymbol{a})|\,;$$

(iv) *If $h : \omega \to M$ and h is $\mathfrak{M}$-recursive, then* $\operatorname{Sup}\{|h(x)| : x \in \omega\} < |M|$.

Proof. (i) If X is $\mathfrak{M}$-r.e. then there is a $z \in \omega$ such that $n \in X \Leftrightarrow \{z\}^{\mathfrak{M}}(n)$ is defined. So if g is as in Lemma A.12 then $n \in X \Leftrightarrow g(z,\langle n\rangle) \in M$. Thus $X \leq_m M$. Now suppose h is a recursive function such that $h : X \leq_m M$ and choose e so that $\{e\}^{\mathfrak{M}}(n) \simeq \{\langle 0,1,h(n)\rangle\}^{\mathfrak{M}}(0)$. Then,

$$\begin{aligned} n \in X &\Leftrightarrow h(n) \in M \\ &\Leftrightarrow \{\langle 0,1,h(n)\rangle\}^{\mathfrak{M}}(0) \quad \text{is defined} \\ &\Leftrightarrow \{e\}^{\mathfrak{M}}(n) \quad \text{is defined.} \end{aligned}$$

Thus X is $\mathcal{M}$-r.e.

(ii) It follows from Remark A.11 that if X is recursive in M_α for some $\alpha < |M|$, then X is $\mathcal{M}$-recursive. For the other direction it suffices to show that each total function $\{z\}^{\mathcal{M}}$ is recursive in M_α for some $\alpha < |M|$. By A.12, for all n,

$$\text{(11)} \qquad \{z\}^{\mathcal{M}}(n) = \{z\}^{\mathcal{M}}[\langle n\rangle] = \{f(z,\langle n\rangle)\}(0, M_{|g(z,\langle n\rangle)|}) .$$

Choose e so that for all n, $\{e\}(n, M_{|u_0|}) = jg(z,n)$ and let $c = p(u_0, e)$. Then $c \in M$ and $|c| > |jg(z,n)|$ for all n. Hence for all n, $M_{|g(z,n)|} \leq_t M_{|c|}$ uniformly in n. Thus there is an e_1 such that for all n,

$$\{e_1\}(n, M_{|c|}) = \{f(z,\langle n\rangle)\}(0, M_{|g(z,\langle n\rangle)|}) .$$

Then from (11), $\{z\}^{\mathcal{M}}$ is recursive in $M_{|c|}$.

(iii) Let $h = \{z\}^{\mathcal{M}}$, $a = \langle \boldsymbol{a}\rangle$, and $k(\boldsymbol{a}) = p(g(z,a), f(\hat{z},a))$. For $h(\boldsymbol{a}) \in M$, $\{z\}^{\mathcal{M}}[a] = h(\boldsymbol{a}) \in M$; hence by A.12, $g(z,a) \in M$ and for all x,

$$\{z\}^{\mathcal{M}}[a] = \{f(z,a)\}(x, M_{|g(z,a)|}) \in M .$$

Thus $k(\boldsymbol{a}) \in M$ and

$$|k(\boldsymbol{a})| \geq \sup\{|\{f(z,a)\}(x, M_{|g(z,a)|})| + 1 : x \in \omega\} = |h(\boldsymbol{a})| + 1 .$$

(iv) Let k be as in (iii). Then for all x, $k(x) \in M$ and $|h(x)| < |k(x)|$. Choose e so that for all x, $k(x) = \{e\}(x, M_{|u_0|})$. Then $p(u_0, e) \in M$ and for all x, $|h(x)| < |k(x)| < |p(u_0, e)|$.

§A.3. Selection.

Using Ⓥ and the fact that by S0 every M_α is $\mathcal{M}$-recursive uniformly in a notation for α, given $a \in M \vee b \in M$ it is possible to decide $\mathcal{M}$-recursively whether $|a| \leq |b|$ or $|b| < |a|$. (Recall that if $a \notin M$ then $|a| = |M|$.) More precisely:

A.14. Lemma. *There is an $\mathcal{M}$-partial recursive function d such that:*

(i) $a \in M \;\&\; |a| \leq |b| \Rightarrow d(a,b) = 0$;

(ii) $|b| < |a| \Rightarrow d(a,b) = 1$.

Proof. Let $k(x) = \langle 0, 1, x\rangle$, $h(a,b) = a \circledV b +_M u_0$ and

$$d(a,b) \simeq \begin{cases} 0 \text{ if } \{kh(a,b)\}^{\mathfrak{M}}(a) \simeq 0 \,\&\, \{k(a)\}^{\mathfrak{M}}(b) \simeq 1\,; \\ 1 \text{ if } [\{kh(a,b)\}^{\mathfrak{M}}(a) \simeq 0 \,\&\, \{k(a)\}^{\mathfrak{M}}(b) \simeq 0] \vee \{kh(a,b)\}^{\mathfrak{M}}(a) \simeq 1\,. \end{cases}$$

Using A.9 it is easy to see that d is $\mathfrak{M}$-partial recursive. If $a \in M \,\&\, |a| \leq |b|$ then $h(a,b) \in M$ and $|a| \leq |a \vee b| < |h(a,b)|$; Hence $a \in M_{|h(a,b)|} \,\&\, b \notin M_{|a|}$ and so $\{kh(a,b)\}^{\mathfrak{M}}(a) = 0 \,\&\, \{k(a)\}^{\mathfrak{M}}(b) = 1$. Thus $d(a,b) = 0$. Similarly, if $|b| < |a|$ then $h(a,b) \in M$ and either $a \in M_{|h(a,b)|} \,\&\, b \in M_{|a|}$ or $a \notin M_{|h(a,b)|}$ and hence by the definition of d, $d(a,b) = 1$.

The following argument is similar to Gandy's unpublished proof of the existence of selection functions associated with functionals of type 2.

A.15. Lemma. *There is an $\mathfrak{M}$-partial recursive function υ such that if $\lambda t.\{z\}(t)$ is total then*

$$\exists t\,\{z\}(t) \in M \Rightarrow \upsilon(z) \text{ is defined } \,\&\, \{z\}(\upsilon(z)) \in M\,.$$

Proof. Let g be the recursive function of A.12 and let e be obtained from the second recursion theorem (Lemma A.8) so that

$$(12)\quad \{e\}^{\mathfrak{M}}(t,z) \simeq \begin{cases} 0 \text{ if } \{z\}(t) \in M \\ \{e\}^{\mathfrak{M}}(t+1,z)+1 \text{ if } d(\{z\}(t), g(e,\langle t+1,z\rangle)) = 1\,. \end{cases}$$

e is found by using the recursion theorem in a manner similar to the proof of Kleene [9, XVI]. Let y be the least t such that $\{z\}(t) \in M$; equivalently, y is the least t such that $\{e\}^{\mathfrak{M}}(t,z) = 0$. We show by induction on x that $\{e\}^{\mathfrak{M}}(y-x,z) = x$ for $0 \leq x \leq y$. This is true if $x = 0$ since in this case $\{e\}^{\mathfrak{M}}(y-x,z) = \{e\}^{\mathfrak{M}}(y,z) = 0 = x$. Suppose $x > 0$. Then by the induction hypothesis $\{e\}^{\mathfrak{M}}(y-(x-1),z) = x-1$. In particular, since $\{e\}^{\mathfrak{M}}(y-(x-1),z)$ is defined, $g(e,\langle y-(x-1),z\rangle) \in M$. Since also $\{z\}(y-x) \notin M$, we have $d(\{z\}(y-x), g(e,\langle y-(x-1),z\rangle)) = 1$. This implies by (12),

$$\{e\}^{\mathfrak{M}}(y-x,z) = \{e\}^{\mathfrak{M}}(y-(x-1),z) + 1 = (x-1) + 1 = x\,.$$

Setting $x = y$ this gives $\{e\}^{\mathcal{M}}(0,z) = y$. Thus it suffices to let $v(z) \simeq \{e\}^{\mathcal{M}}(0,z)$.

A.16. Corollary. *Let $Q \subseteq {}^{n+1}\omega$ be $\mathcal{M}$-r.e. Then there is an $\mathcal{M}$-partial recursive function $\lambda \boldsymbol{x} \nu y Q(\boldsymbol{x},y)$ of n variables such that for all $\boldsymbol{x}$,*

$$\exists y Q(\boldsymbol{x},y) \Rightarrow \nu y Q(\boldsymbol{x},y) \text{ is defined } \& \; Q(\boldsymbol{x}, \nu y Q(\boldsymbol{x},y)) .$$

Proof. Let $Q(\boldsymbol{x},y) \Leftrightarrow \{z\}^{\mathcal{M}}(\boldsymbol{x},y)$ is defined. Then

$$\begin{aligned} Q(\boldsymbol{x},y) &\Leftrightarrow \{S^m(z,\boldsymbol{x})\}^{\mathcal{M}}(y) \qquad \text{is defined} \\ &\Leftrightarrow g(S^m(z,\boldsymbol{x}), \langle y \rangle) \in M . \end{aligned}$$

Let e be a recursive function such that $\{e(\boldsymbol{x})\}(y) = g(S^m(z,\boldsymbol{x}), \langle y \rangle)$. Then

$$\exists y Q(\boldsymbol{x},y) \Leftrightarrow \exists y \{e(\boldsymbol{x})\}(y) \in M .$$

Thus let $\nu y Q(\boldsymbol{x},y) \simeq \nu e(\boldsymbol{x})$.

The following lemma summarises some of the properties of $\mathcal{M}$-r.e. relations.

A.17. Lemma. (i) *If f is $\mathcal{M}$-partial recursive, then the relation $f(\boldsymbol{x}) = z$ is* $\mathcal{M}$-r.e.

(ii) *If Q is $\mathcal{M}$-r.e., then the function*

$$f(\boldsymbol{x}) \simeq \begin{cases} z \text{ if } Q(\boldsymbol{x}) , \\ \text{undefined otherwise} , \end{cases}$$

is $\mathcal{M}$-partial recursive.

(iii) *The relations $y \in M \;\&\; |x| < |y|$; $y \in M \;\&\; |x| \leq |y|$, $y \in M \;\&\; |y| \leq |x|$ are* $\mathcal{M}$-r.e.

(iv) *The $\mathcal{M}$-r.e. relations are closed under conjunction, disjunction, universal and existential number quantification, and inverse images by $\mathcal{M}$-recursive functions.*

Proof. Suppose f is $\mathcal{M}$-partial recursive. Let

$$e(x,z) = \begin{cases} \langle 2, 1, 0\rangle & \text{if } x = z\,, \\ 0 & \text{otherwise}\,, \end{cases}$$

and $g(\boldsymbol{x}, z) \simeq \{e(f(\boldsymbol{x}), z)\}^{\mathcal{M}}(0)$. Then g is $\mathcal{M}$-partial recursive and

$$\begin{aligned} g(\boldsymbol{x}, z) \text{ is defined } &\Leftrightarrow e(f(\boldsymbol{x}), z)) = \langle 2, 1, 0\rangle \\ &\Leftrightarrow f(\boldsymbol{x}) = z\,. \end{aligned}$$

To prove (ii) let $Q(\boldsymbol{x}) \Leftrightarrow g(\boldsymbol{x})$ is defined, and let $f(\boldsymbol{x}) \simeq 0 \cdot g(\boldsymbol{x}) + z$. To prove (iii) we have $y \in M \;\&\; |x| < |y| \Leftrightarrow \{\langle 0, 1, y\rangle\}^{\mathcal{M}}(x) = 0$; then use (i). The other relations in (iii) are handled similarly. To prove (iv) we consider just the cases of universal and existential quantification. Suppose $Q(\boldsymbol{x}, y)$ is $\mathcal{M}$-r.e. By (ii) the function

$$f(\boldsymbol{x}, y) \simeq \begin{cases} 0 & \text{if } Q(\boldsymbol{x}, y) \\ \text{undefined otherwise} \end{cases}$$

is $\mathcal{M}$-partial recursive. Then

$$\begin{aligned} \forall y Q(\boldsymbol{x}, y) &\Leftrightarrow \forall y\,[f(\boldsymbol{x}, y) \text{ is defined}] \\ &\Leftrightarrow E(\lambda y f(\boldsymbol{x}, y)) \text{ is defined.} \end{aligned}$$

To treat existential quantification, let $Q(\boldsymbol{x}, y) \Leftrightarrow \{z\}^{\mathcal{M}}(\boldsymbol{x}, y)$ is defined. Then by A.16,

$$\begin{aligned} \exists y Q(\boldsymbol{x}, y) &\Leftrightarrow Q(\boldsymbol{x}, \nu y Q(\boldsymbol{x}, y)) \\ &\Leftrightarrow \{z\}^{\mathcal{M}}(\boldsymbol{x}, \nu y Q(\boldsymbol{x}, y)) \quad \text{is defined.} \end{aligned}$$

§A.4. Proof of Theorem A5.

A.18. Lemma. *There is an $\mathcal{M}$-partial recursive function g such that if $z \in M$ and $\{e\}^{\mathcal{M}}(u, \boldsymbol{x})$ is defined for each $u \in M_{|z|}$ then*

$$|g(e, z, \boldsymbol{x})| = \operatorname{Sup}\,\{|\{e\}^{\mathcal{M}}(u, \boldsymbol{x})| : u \in M_{|z|}\}\,.$$

Proof. Let $g(e,z,\boldsymbol{x}) \simeq \nu y Q(e,z,\boldsymbol{x},y)$ where

$$Q(e,z,\boldsymbol{x},y) \Leftrightarrow y,z \in M \;\&\; \forall u \in M_{|z|}[|\{e\}^{\mathcal{M}}(u,\boldsymbol{x})| \le |y|]$$
$$\& \; \forall v \in M_{|y|} \exists u \in M_{|z|}[|v| \le |\{e\}^{\mathcal{M}}(u,\boldsymbol{x})|] \;.$$

A.19. Lemma. *For each ordinal function F primitive recursive in $T_{\mathcal{M}}$ there is an $\mathcal{M}$-partial recursive function h_F such that if $a_1, ..., a_n \in M$ then $h_F(a_1, ..., a_n) \in M$ and*

$$F(|a_1|, ..., |a_n|) = |h_F(a_1, ..., a_n)| \,.$$

Proof. We shall use the characterisation of the ordinal functions primitive recursive in $T_{\mathcal{M}}$ given by the following schemata (see [8]):

(i) $F(x,y) = \begin{cases} 0 & \text{if } x \in M_y \\ 1 & \text{otherwise} \end{cases}$

(ii) $F(\boldsymbol{x}) = x_i$

(iii) $F(x) = 0$

(iv) $F(x) = x + 1$

(v) $F(x,y,u,v) = \begin{cases} x & \text{if } u < v\,, \\ y & \text{otherwise}\,. \end{cases}$

(vi) $F(\boldsymbol{x},\boldsymbol{y}) = G(\boldsymbol{x},H(\boldsymbol{x}),\boldsymbol{y})$

(vii) $F(\boldsymbol{x},\boldsymbol{y}) = G(H(\boldsymbol{x}),\boldsymbol{y})$

(viii) $F(z,\boldsymbol{x}) = G(\sup_{u<z} F(u,\boldsymbol{x}),z,\boldsymbol{x})\,.$

For (i) we first need an $\mathcal{M}$-recursive function k such that $|k(n)| = n$ for all n. Let $k(0) = u_0$ and

$$k(n+1) \simeq \nu y[y \in M \;\&\; |k(n)| < |y| \;\&\; \forall z[|z| < |y| \Rightarrow |z| \le |k(n)|]] \,.$$

Then k has the desired property. Now in case (i) let $|u_1| = 1$ and

$$h_F(a,b) \simeq \begin{cases} u_0 & \text{if } a,b \in M \;\&\; \exists n[|k(n)| = |a| \;\&\; n \in M_{|b|}] \\ u_1 & \text{if } a,b \in M \;\&\; \exists n[|k(n)| = |a| \;\&\; n \notin M_{|b|}]. \end{cases}$$

Now define h_F in each remaining case as follows:

(ii) $h_F(\boldsymbol{a}) = a_i$

(iii) $h_F(a) = u_0$

(iv) $h_F(a) \simeq \nu y[a,y \in M \;\&\; |a| < |y| \;\&\; \forall z[|z| < |y| \Rightarrow |z| \leq |a|]]$

(v) $h_F(a,b,c,d) \simeq \nu y[c,d \in M \;\&\; (|c| < |d| \;\&\; y = a) \vee (|d| \leq |c| \;\&\; y = b)]$

(vi) $h_F(\boldsymbol{a},\boldsymbol{b}) \simeq h_G(\boldsymbol{a},h_H(\boldsymbol{a}),\boldsymbol{b})$

(vii) $h_F(\boldsymbol{a},\boldsymbol{b}) \simeq h_G(h_H(\boldsymbol{a}),\boldsymbol{b})$

(viii) $h_F(a,\boldsymbol{b}) \simeq h_G(g(e,a,\boldsymbol{b}),a,\boldsymbol{b})$,

where g is from A.18 and e is chosen by the Second Recursion Theorem for $\mathcal{M}$-recursion so that $h_F = \{e\}^{\mathcal{M}}$.

A.20. Lemma. (i) *$|M|$ is closed under functions primitive recursive in $T_{\mathcal{M}}$.*
(ii) *Let $R \subseteq$ On be primitive recursive in $T_{\mathcal{M}}$. Then*

$$\{(a_1, ..., a_n) : a_1, ..., a_n \in M \;\&\; R(|a_1|, |a_2|, ..., |a_n|)\}$$

is $\mathcal{M}$-r.e.

Proof. (i) is an immediate consequence of A.19. Let $A = \{(a_1, ..., a_n)$. $a_1, ..., a_n \in M \;\&\; R(|a_1|, ..., |a_n|)\}$ and F be the representing function of R. Then using A.19,

$$\begin{aligned}(a_1,...,a_n) \in A &\Leftrightarrow a_1,...,a_n \in M \;\&\; F(|a_1|,...,|a_n|) = 0 \\ &\Leftrightarrow a_1,...,a_n \in M \;\&\; h_F(a_1,...,a_n) \in M_1 .\end{aligned}$$

Thus $A = {}^nM \cap h_F^{-1}M_1$ which is $\mathcal{M}$-r.e. using A.17.

A.21. Lemma (proof of A.5.1). $|M| \in \mathrm{Ad}(T_{\mathcal{M}})$.

Proof. By Theorem 3.7 it remains to show that if $R \subseteq {}^3\mathrm{On}$ is primitive recursive in $T_{\mathcal{M}}$, $\alpha < |M|$ and

(13) $$\forall x < |M| \, \exists y < |M| \, R(\alpha, x, y)$$

then there is $\alpha < \lambda < |M|$ such that

(14) $$\forall x < \lambda \, \exists y < \lambda \, R(\alpha, x, y) .$$

Suppose (13) holds. Let $|c| = \alpha$ and $f(x) \simeq \nu y [y \in M \,\&\, R(|c|, |x|, |y|)]$. Then f is $\mathcal{M}$-partial recursive by A.20. Let g be the $\mathcal{M}$-recursive function defined by: $g(0) = c$ and

$$g(n+1) = \nu y [y \in M \,\&\, |g(n)| < |y| \,\&\, \forall x \in M_{|g(n)|} |f(x)| \leq |y|] .$$

Then $|g(n)| < |g(n+1)|$ and $|x| < |g(n)| \Rightarrow |f(x)| \leq |g(n+1)|$. Let $\lambda = \mathrm{Sup}_{n<\omega} |g(n)|$. $\alpha < \lambda < |M|$ by A.13 (iv). We show that λ satisfies (14).

(15) $$\begin{aligned} |x| < \lambda &\Rightarrow \exists n \, |x| < |g(n)| \\ &\Rightarrow \exists n \, |f(x)| < |g(n)| \\ &\Rightarrow |f(x)| < \lambda . \end{aligned}$$

(14) then follows from (15) and the definition of f.

A.22. Lemma (proof of A.5 (ii)). *If $R \subseteq {}^n|M|$ is $|M|$-r.e. in $T_{\mathcal{M}}$ then there is a primitive recursive function h such that*

$$a_1, \ldots, a_n \in M \,\&\, R(|a_1|, \ldots, |a_n|) \Longleftrightarrow h(a_1, \ldots, a_n) \in M .$$

Proof. Let R be $|M|$-r.e. in $T_{\mathcal{M}}$. Then there is a primitive recursive relation S such that

$$R(\alpha_1, \ldots, \alpha_n) \Longleftrightarrow \exists \beta < |M| . \, S(\beta, \alpha_1, \ldots, \alpha_n) .$$

Let

$$A = \{(a_1, ..., a_n) : a_1, ..., a_n \in M \;\&\; R(|a_1|, ..., |a_n|)\} .$$

Then

$$(a_1, ..., a_n) \in A \Leftrightarrow a_1, ..., a_n \in M \;\&\; \exists b[b \in M \;\&\; S(|b|, |a_1|, ..., |a_n|)] .$$

It follows that A is $\mathcal{M}$-r.e. Hence by A.13 there is a recursive function h_1 such that $A = h_1^{-1}({}^nM)$. It remains to find a primitive recursive h. By the S–m–n theorem for ordinary recursive function theory there is a primitive recursive function such that, letting $\boldsymbol{a} = a_1, ..., a_n$, $\{g(\boldsymbol{a})\}(x, M_{|u_0|}) = h_1(\boldsymbol{a})$ for all x. Let $h(\boldsymbol{a}) = p(u_0, g(\boldsymbol{a}))$; then h is primitive recursive and

$$(\boldsymbol{a}) \in A \Leftrightarrow h_1(\boldsymbol{a}) \in M$$
$$\Leftrightarrow \forall x[\{g(\boldsymbol{a})\}(x, M_{|u_0|}) \in M$$
$$\Leftrightarrow h(\boldsymbol{a}) \in M .$$

Remark. The above proof is the only place where we use the fact that p is *primitive* recursive instead of just recursive.

A.23. Lemma (proof of A.5 (iii)). *Let F be $|M|$-partial recursive in $T_{\mathcal{M}}$. Then there is a recursive function k such that for $a_1, ..., a_n \in M$, if $F(|a_1|, ..., |a_n|)$ is defined then $k(a_1, ..., a_n) \in M$ and $F(|a_1|, ..., |a_n|) \leq |k(a_1, ..., a_n)|$.*

Proof. By the normal form theorem relativised to $T_{\mathcal{M}}$, the graph of F is $|M|$-r.e. in $T_{\mathcal{M}}$; hence by A.22 there is a recursive function h such that

$$\boldsymbol{a}, b \in M \;\&\; F(|a_1|, ..., |a_n|) = |b| \Leftrightarrow h(a_1, ..., a_n, b) \in M .$$

Let $f(\boldsymbol{a}) \simeq \nu y[y \in M \;\&\; h(a_1, ..., a_n, y) \in M]$. Then f is $\mathcal{M}$-partial recursive. Hence by A.13 there is a recursive function k such that

$$f(\boldsymbol{a}) \in M \Rightarrow k(\boldsymbol{a}) \in M \;\&\; |f(\boldsymbol{a})| \leq |k(\boldsymbol{a})| .$$

Then for $\boldsymbol{a} \in M$, if $F(|a_1|, ..., |a_n|)$ is defined

$$F(|a_1|, ..., |a_n|) = |f(\boldsymbol{a})| \leq |k(\boldsymbol{a})| .$$

A.24. Lemma (proof of A.5 (iv). *$X \subseteq \omega$ is $|M|$-r.e. in $T_{\mathcal{M}}$ iff $X \leq_m M$.*

Proof. M is $|M|$-r.e. in $T_{\mathcal{M}}$ since $n \in M \Leftrightarrow \exists \alpha < |M|.\, T_{\mathcal{M}}(n, \alpha)$. Hence if $X \leq_m M$, X is also $|M|$-r.e. in $T_{\mathcal{M}}$. Now suppose X is $|M|$-r.e. in $T_{\mathcal{M}}$, and let k be an $\mathcal{M}$-recursive function such that $|k(n)| = n$ for all n (see the proof of A.19). Then using A.22 there is a recursive function h such that

$$n \in X \Leftrightarrow |k(n)| \in X$$
$$\Leftrightarrow hk(n) \in M\,.$$

X is the inverse image of the $\mathcal{M}$-r.e. set M under the $\mathcal{M}$-recursive function hk and hence is $\mathcal{M}$-r.e. by A.17 (iv). Hence $X \leq_m M$ by A.13 (i).

Remarks. (i) The definition of an acceptable ordinal system differs slightly from that given in [16]. The requirement that j (called there g) be a many-one reduction of M to M is necessary for the proof of A.12 and its omission was an oversight in [16]. The other change is the requirement that p be primitive recursive instead of recursive, and as mentioned above this is only to ensure that the function h of A.5 is *primitive* recursive.

(ii) A.5 (iii) is not used elsewhere but it appears to be of interest in its own right and its proof comes naturally from our construction. It was used in [16] in an earlier proof of some of our results but is not needed in our present formulation.

(iii) The method we have used in proving the Coding Lemma is to utilize techniques from the well-developed theory of recursive functionals of type 2. In particular, the crucial results needed about $\mathcal{M}$-recursion, namely the Boundeness Theorem (A.13 (iv)), and Theorem A.15 on the existence of selection functions are proved by standard methods from the theory of recursive functionals of type 2. On the other hand the theory of recursive functionals may be regarded as a part of the theory of inductive definitions. This suggests that an ultimately simpler and more elegant proof of the Coding Lemma in a more general setting can be provided within the "pure" theory of inductive definitions.

References

[1] S. Aanderaa, Inductive definitions and their closure ordinals, this volume.

[2] P. Aczel and W. Richter, Inductive definitions and analogues of large cardinals, in: Conference in Mathematical Logic, London '70 (Springer, Berlin, 1971) 1–10.

[3] J. Barwise, Infinitary logic and admissible sets, J. Symb. Logic 34 (1969) 226–251.

[4] J. Barwise, R.O. Gandy and Y.N. Moschovakis, The next admissible set, J. Symb. Logic 36 (1971) 108–120.

[5] D. Cenzer, Analytic inductive definitions, Abstract, Notices, A.M.S. 703-E2, Vol. 20 (1973) pA376.

[6] K. Devlin, An introduction to the fine structure of the constructible hierarchy, this volume.

[7] W.P. Hanf and D. Scott, Classifying inaccessible cardinals, Notices of the A.M.S. 8 (1961) 445.

[8] R.B. Jensen and C. Karp, Primitive recursive set functions, in: D. Scott (ed.) Axiomatic Set Theory, Proceedings Pure Math. 13 (Amer. Math. Soc. Providence, R.I., 1971) 143–176.

[9] S.C. Kleene, Recursive functionals and quantifiers of finite types I, Trans Amer. Math. Soc. 91 (1959) 1–52.

[10] S. Kripke, Transfinite recursion, constructible sets and analogues of cardinals, in Lecture Notes prepared in connection with the Summer Institute on Axiomatic Set Theory, held at Los Angeles (1967).

[11] A. Lévy, The sizes of the indescribable cardinals, in: D. Scott (ed.) Axiomatic Set Theory, Proceedings Pure Math. 13 (Amer. Math. Soc., Providence, R.I., 1971) 143–176.

[12] Th.A. Linden, Equivalences between Gödel's definitions of constructibility, in: J.N. Crossley (ed.) Sets, Models and Recursion Theory (North-Holland, Amsterdam, 1967) 33–43.

[13] Y.N. Moschovakis, Elementary Induction on Abstract Structures (North-Holland, Amsterdam, 1973).

[14] H. Putnam, On hierarchies and systems of notations, Proc. Amer. Math. Soc. 15 (1964) 44–50.

[15] W. Richter, Constructive transfinite number classes, Bull. Amer. Math. Soc. 73 (1967) 261–265.

[16] W. Richter, Recursively Mahlo ordinals and inductive definitions, in: R.O. Gandy and C.E.M. Yates (eds.) Logic Colloquium '69 (North-Holland, Amsterdam, 1971) 273–288.

[17] H. Rogers, Jr., Theory of recursive functions and effective computability (McGraw-Hill, 1967).

[18] G. Sacks, The 1-section of a type *n* object, this volume.

[19] J.R. Shoenfield, Mathematical logic (Addison Wesley, Reading, Mass., 1967).

[20] C. Spector, Inductively defined sets of natural numbers, in: Infinitistic Methods (Pergamon Press, Oxford, 1961) 97–102.

[21] H. Tanaka, On analytic well-orderings, J. Symb. Logic 35 (1970) 198–204.

[22] S.C. Kleene, On the forms of the predicates in the theory of constructive ordinals, II, Amer. J. of Math. 77 (1955) 405–428.

PART IV

AXIOMATIC APPROACHES AND GENERAL DISCUSSION

J.E.Fenstad, P.G.Hinman (eds.), *Generalized Recursion Theory*

ON AXIOMATIZING RECURSION THEORY

Jens Erik FENSTAD
University of Oslo

Generalized recursion theory can be many different things. Starting from ordinary recursion theory one may e.g. move up in types over ω, or look to more general domains such as ordinals, admissible sets and acceptable structures. Alternatively, one may want to study in a more general setting one particular approach to ordinary recursion theory, thus e.g. try to develop a general theory based on schemes or fixed point operators, or work out a general theory of inductive definability, or develop in a suitable abstract setting the various model theoretic approaches such as representability in formal systems or invariant and implicit definability.

The approach of this paper is axiomatic. This is nothing new. Of previous axiomatic studies of recursion theory we mention Strong [14], Wagner [15], and Friedman [4]. Our interest in the axiomatics of generalized recursion theory was more directly inspired by Moschovakis [10], and any one familiar with his "Axioms for Computation Theories" will soon see our dependence upon his work.

Our objective is two-fold: First to contribute to the discussion and choice of the "correct" primitives for axiomatic recursion theory. Second to indicate new results, partly proved, partly conjectural, within the (modified) Moschovakis framework.

First one general remark on axiomatizing recursion theory. This may in itself be a worthy objective. Through an axiomatic analysis one may hope to get a satisfying classification and comparison of existing generalizations (technically through "representation" theorems and "imbedding" results). And one may, perhaps, also obtain a better insight into the "concrete" examples on which the axiomatization is based. But it is not clear – and some disagree – that the field is at present ripe for axiomatization. Hence we are approaching our topic in a tentative manner.

As Moschovakis in [10] we take as our basic relation

$$\{a\}(\sigma) \simeq z ,$$

which asserts that the "computing device" named or coded by a acting on the input sequence $\sigma = (x_1, ..., x_n)$ gives z as output.

Let Θ denote the set of all computation tuples (a, σ, z) such that the relation $\{a\}(\sigma) \simeq z$ obtains. It is possible to write down axioms for a computation set Θ which suffice to derive the most basic results of recursion theory, say up through the fixed-point or second recursion theorem.

However, many arguments seem to require an analysis not only of the computation tuples, but of the whole structure of "subcomputations" of a given computation tuple. Now computations, and hence subcomputations, can be many different things. And in an axiomatic analysis of the variety of approaches hinted at in the opening paragraph of the paper it would be rash to commit oneself at the outset to one specific idea of 'computation'.

In [10] Moschovakis emphasized the fact that whatever computations may be, they have a well-defined length, which always is an ordinal, finite or infinite. Thus he proposed to add as a further primitive a map from the set Θ of computation tuples to the ordinals, denoting by $|a, \sigma, z|_\Theta$ the ordinal associated with the tuple $(a, \sigma, z) \in \Theta$.

In this paper we shall abstract another aspect of the notion of computation. We shall add as a further primitive a relation between computation tuples

$$(a', \sigma', z') < (a, \sigma, z) ,$$

which is intended to express that (a', σ', z') is a subcomputation of (a, σ, z), or, in other words, that the computation (a, σ, z) depends upon (a', σ', z'). The basic axioms will state that the relation is transitive and wellfounded.

Remark. In our approach we have chosen functions and computations rather than sets and inductive definitions as basic notions. We have also chosen to exhibit the codes for the computations directly in the axioms rather than tried to develop the theory in a more "coordinate-free" or invariant manner. It is, perhaps, still an open question which will be the most "useful" way to organize generalized recursion theory into a theory.

The rest of the paper will be divided in four sections. In Section 1 we give the basic definition of a computation theory. In this we follow Moschovakis closely, making the modifications necessary due to our use of the subcomputation relation $<$ instead of the length concept.

In Section 2 we list some basic facts about computation theories. This is in all essentials a repetition of material from [10], but is included for the convenience of the reader. In this part of the theory there does not seem to be much difference between the length concept and the subcomputation relation. The important thing is that both allows us to carry over to the abstract setting certain results proved in the "concrete" examples by transfinite induction on associated ordinals, or, alternatively, by a course-of-value induction on "subcomputations". The basic result here, due to Moschovakis in the axiomatic setting, is the first recursion theorem. The section concludes with the definition of *regular* computation theories. Such theories have selection operators, hence we have a reasonable theory for the computable and semicomputable relations. And for this class of theories we can introduce an adequate notion of finiteness, a set being finite if we can computably quantify over it.

In Section 3 we discuss the problem how to strengthen regularity. Of several possibilities we have chosen to emphasize two: One is the idea that a "computation" should be a finite object in the sense of the theory; the other is that the theory should satisfy the prewellordering property. The first we can express by requiring that for every $(a,\sigma,z)\in\Theta$, the set

$$\boldsymbol{S}_{(a,\sigma,z)} = \{(a',\sigma',z') \mid (a',\sigma',z') < (a,\sigma,z)\}$$

is finite in the theory. One formulation of the other says that the set

$$\{(a',\sigma',z') \mid (a',\sigma',z')\in\Theta \wedge \|a',\sigma',z'\| \leq \|a,\sigma,z\|\}$$

is computable in the theory (where $\|a,\sigma,z\|$ is the ordinal of $\boldsymbol{S}_{(a,\sigma,z)}$).

In Section 4 we move beyond the "normality" conditions discussed above, representation theorems and imbedding results being the main themes. Our exposition will be sketchy for several reasons: One is that a complete development would be too long, – another and more important one is that several of the results are still in a preliminary stage.

In conclusion I would like to acknowledge my great debts to Peter Aczel and Peter Hinman, who patiently have explained many results and methods of general recursion theory to me, and to Johan Moldestad and Dag Normann, who have with great enthusiasm participated in the investigations reported on in this paper.

1. Computation theories: basic definitions

In this section we give the basic definitions using the subcomputation relation as primitive notion.

Definition 1. A computation domain is a structure

$$\mathfrak{A} = \langle A, C, N, s, M, K, L \rangle ,$$

where A is the universe, $N \subseteq C \subseteq A$ and $\langle N, s \restriction N \rangle$ is isomorphic to the non-negative integers. C is called the set of codes. M is a pairing function on C, i.e.

$$a, b \in C \quad \text{iff } M(a,b) \in C ,$$

and

$$M(a,b) = M(a',b') \in C \quad \text{implies} \quad a = a' \wedge b = b' .$$

K and L are inverses to M, i.e. they map C into C and

$$c = M(a,b) \in C \quad \text{iff } a = K(c) \wedge b = L(c) .$$

To facilitate the presentation we introduce some notational convention (following Moschovakis [10]). We use
$x, y, z, \ldots$ for elements in A.
$a, b, c, \ldots$ for elements in C.
$i, j, k, \ldots$ for elements in N.
$\sigma, \tau, \ldots$ for finite sequences from A.
σ, τ or (σ, τ) denotes the concatenation of sequences. And as usual $\mathrm{lh}(\sigma)$ = the length of the sequence σ. A computation tuple is any sequence (a, σ, z) such that $a \in C$ and $\mathrm{lh}(a, \sigma, z) \geq 2$.

Definition 2. The system $\langle \Theta, < \rangle$ is called a computation structure on the domain $\mathfrak{A}$ if $<$ is a transitive relation on the set of computation tuples and Θ is the wellfounded part of $<$.

Thus $(a, \sigma, z) \in \Theta$ iff the set

$$S_{(a,\sigma,z)} = \{(a', \sigma', z') \mid (a', \sigma', z') < (a, \sigma, z)\}$$

is wellfounded with respect to the relation $<$. Note that if $(a, \sigma, z) \in \Theta$ and $(a', \sigma', z') < (a, \sigma, z)$, then $(a', \sigma', z') \in \Theta$.

Note: We have built into our definition the convention that something which looks like a computation, i.e. an arbitrary computation tuple (a, σ, z), is not a computation if and only if its "subcomputation tree" contains an infinite descending path. In practice this may not always be so, but if an attempt at a computation stops after a finite number of steps without giving a *bona fide* computation, we can always start repeating ourselves in some suitable way so as to obtain an infinite descending path.

As in [10] we shall make use of the notions of partial multiple-valued (pmv) function and functional. We recall some notations:

$f(\sigma) \to z$	iff	$z \in f(\sigma)$.
	iff	z is one value of f at σ.
$f(\sigma) = g(\sigma)$	iff	$\forall z [f(\sigma) \to z$ iff $g(\sigma) \to z]$.
$f(\sigma) = z$	iff	$f(\sigma) = \{z\}$.
	iff	$f(\sigma) \to z \wedge \forall u [f(\sigma) \to u \Rightarrow u = z]$.
$f \subseteq g$	iff	$\forall \sigma \forall z [f(\sigma) \to z \Rightarrow g(\sigma) \to z]$.

A mapping is a total, single-valued function.

Let $\langle \Theta, < \rangle$ be a computation structure on $\mathfrak{A}$. To every $a \in C$ and every natural number n we can associated a pmv function $\{a\}^n_\Theta$ in the following way:

$$\{a\}^n_\Theta(\sigma) \to z \quad \text{iff} \quad \mathrm{lh}(\sigma) = n \wedge (a, \sigma, z) \in \Theta .$$

Definition 3. Let $\langle \Theta, < \rangle$ be a computation structure on $\mathfrak{A}$. A pmv function f on A is Θ-computable if for some $\hat{f} \in C$

$$f(\sigma) \to z \quad \text{iff} \quad (\hat{f}, \sigma, z) \in \Theta .$$

We call $\hat{f}$ a Θ-code of f and write $f = \{\hat{f}\}^n_\Theta$, where n is the number of arguments of f.

A pmv functional on A

$$\varphi(f,\sigma) = \varphi(f_1, ..., f_1, x_1, ..., x_n)$$

maps pmv functions on A and elements of A into subsets of A (including the empty subset, $\emptyset$). φ is called continuous if

$$f_1 \subseteq g_1, ..., f_1 \subseteq g_1, \varphi(f,\sigma) \to z \Rightarrow \varphi(g,\sigma) \to z \,.$$

Definition 4. Let $\langle\Theta,<\rangle$ be a computation structure on the domain $\mathfrak{A}$. A pmv continuous functional φ on A is called Θ-computable if there exists a $\hat{\varphi} \in C$ such that for all $e_1, ..., e_\ell \in C$ and all $\sigma = (x_1, ..., x_n)$ from A, we have:
a) $\varphi(\{e_1\}_\Theta^{n_1}, ..., \{e_\ell\}_\Theta^{n_\ell}, \sigma) \to z$ if $\{\hat{\varphi}\}_\Theta^{\ell+n}(e_1, ..., e_\ell, \sigma) \to z$.
b) If $\varphi(\{e_1\}_\Theta^{n_1}, ..., \{e_\ell\}_\Theta^{n_\ell}, \sigma) \to z$, then there exist pmv functions $g_1, ..., g_\ell$ such that

i. $g_1 \subseteq \{e_1\}_\Theta^{n_1}, ..., g_\ell \subseteq \{e_\ell\}_\Theta^{n_\ell}$ and $\varphi(g_1, ..., g_\ell, \sigma) \to z$.
ii. For all $i = 1, ..., \ell$, if $g_i(t_1, ..., t_{n_i}) \to u$, then

$$(e_i, t_1, ..., t_{n_i}, u) < (\hat{\varphi}, e_1, ..., e_\ell, \sigma, z) \,.$$

Note: This is the first essential use of the suncomputation relation. For a motivation of the definition of Θ-computable functional, see [10, p. 209].

Definition 5. Let $\langle\Theta,<\rangle$ be a computation structure on the domain $\mathfrak{A}$. $\langle\Theta,<\rangle$ is called a *computation theory* on $\mathfrak{A}$ if there exist Θ-computable mappings $p_1, ..., p_{13}$ such that the following functions and functionals are Θ-computable with Θ-codes as indicated and such that the iteration property holds:

I. $$f(\sigma) = y, y \in C \qquad f = \{p_1(n,y)\}_\Theta^n,\ n = \mathrm{lh}(\sigma) \,.$$

II–VIII. Similar to I and state that the following functions are Θ-computable: Identity function, the successor function s, the characteristic functions of C and N, the pairing function M and the inverses K and L.

IX. $$\varphi(f,g,\sigma) = f(g(\sigma),\sigma) \qquad \hat{\varphi} = p_9(n),\ n = \mathrm{lh}(\sigma) \,.$$

X–XII. Similar to IX and state that the following functionals are Θ-computable: Primitive recursion, permutation of arguments, point evaluation.

XIII. Iteration property: For all n,m $p_{13}(n,m)$ is a Θ-code for a mapping $S^n_m(a,x_1,...,x_n)$ such that for all $a,x_1,...,x_n \in C$ and all $y_1,...,y_m \in A$:
(i) $\{a\}^{n+m}_{\Theta}(x_1,...,x_n,y_1,...,y_m) = \{S^n_m(a,x_1,...,x_m)\}(y_1,...,y_m)$.
(ii) If $\{a\}^{n+m}_{\Theta}(x_1,...,x_n,y_1,...,y_m) \to z$, then

$$(a,x_1,...,x_n,y_1,...,y_m) < (S^n_m(a,x_1,...,x_n),y_1,...,y_m,z) .$$

Remarks. 1. The missing parts of the definition can be found in [10, pp. 205–206].

2. If we drop the primitive $<$, the rest can be stated using Θ alone, and we arrive at Moschovakis' notion of a *pre-computation theory*. This part seems to contain the basic core (i.e. "pre-Post" theory) of any systematization, including the fixed-point or second recursion theorem.

3. To every tuple $(a,\sigma,z) \in \Theta$ there is associated an ordinal $\|a,\sigma,z\|$ = the ordinal of the set $\boldsymbol{S}_{(a,\sigma,z)}$. Θ with this ordinal assignment is a computation theory in the sense of Moschovakis.

2. Computation theories: basic facts

2.1. *Inductive generation of theories and equivalence.* Let $\mathfrak{A}$ be a computation domain and let

$$\boldsymbol{f} = f_1,...,f_\ell \quad \text{and} \quad \boldsymbol{\varphi} = \varphi_1,...,\varphi_k$$

be sequences of pmv functions and continuous pmv functionals on A. It is possible to construct a theory PR $[\boldsymbol{f},\boldsymbol{\varphi}]$, the prime computation theory generated by $\boldsymbol{f}$ and $\boldsymbol{\varphi}$, which in the following precise sense is the least computation theory which makes all the functions $\boldsymbol{f}$ and functionals $\boldsymbol{\varphi}$ (uniformly) computable.

Definition 6. Let $\langle\Theta,<\rangle$ and $\langle\Theta',<'\rangle$ be computation theories on the same domain $\mathfrak{A}$. We say that Θ' *extends* Θ, in symbols,

$$\Theta \leq \Theta'$$

(dropping, as we usual do, the explicit reference to $<$ and $<'$) if there exists a Θ'-computable mapping $p(a,n)$ such that $p(C,N) \subseteq N$ and such that for all n-tuples σ, all $a \in C$ and $z \in A$

i. $(a,\sigma,z) \in \Theta$ iff $(p(a,n),\sigma,z) \in \Theta'$.

ii. If $(a,\sigma,z), (a',\sigma',z') \in \Theta$ and $(a,\sigma,z) < (a',\sigma',z')$, then $(p(a,n),\sigma,z) <' (p(a',n'),\sigma',z')$.

If $\Theta' \leq \Theta$ and $\Theta \leq \Theta'$, we say that Θ and Θ' are equivalent and write $\Theta \sim \Theta'$.

Remark. It seems that Moschovakis' motivation for his version of the notion of equivalence (see [10, pp. 217–218]) is even more appropriate for the present version.

As in [10, p. 218] we have the following result which justifies the claim made above.

(i) Let $\langle\Theta,<\rangle$ be a computation theory on $\mathfrak{A}$, and let f and $\boldsymbol{\varphi}$ be sequences of pmv functions and continuous functionals on $\mathfrak{A}$. Then

$$\Theta \leq \Theta[f,\boldsymbol{\varphi}],$$

and if H is any other computation theory on $\mathfrak{A}$ such that $\Theta \leq H$ and f are H-computable and $\boldsymbol{\varphi}$ are uniformly H-computable, then

$$\Theta[f,\boldsymbol{\varphi}] \leq H.$$

Remark. There are some difficulties in carrying over (iii) [10, p. 219] to the present frame. We mention this since this is the only example of a result in [10, §§1–8] which has not had an immediate counterpart. (The difficulty is that if we pass from Θ to H via a map p and then back to Θ via a map q, the ordinal of (a,σ,z) is less than the ordinal of $(q(p(a,n),n),\sigma,z)$, but the former is not necessarily a subcomputation of the latter.

2.2. *The first recursion theorem.* The theorem was proved by Moschovakis in the axiomatic setting. The proof carries immediately over to the present set-up.

Theorem. *Let* $\langle\Theta,<\rangle$ *be a computation theory on* $\mathfrak{A}$. *Let* $\varphi(f,x)$ *be a*

Θ-computable continuous pmv *functional over A. Let f^* be the least solution of*

$$\forall x \in A : \varphi(f,x) = f(x) .$$

Then f^* is Θ-computable.

The theorem is particularly important in discussing the relationship between recursion theory and inductive definability. It corresponds to the fact, and can sometimes be used to show, that Σ_1 inductive definitions has a Σ_1 minimal solution.

2.3. *Selection operators.* We first give the definition of Θ-semicomputable and Θ-computable relations.

Definition 7. The relation $R(\sigma)$ is Θ-*semicomputable* if there is a Θ-computable pmv function f such that

$$R(\sigma) \text{ iff. } f(\sigma) \to 0 .$$

The relation $R(\sigma)$ is Θ-*computable* if there is a Θ-computable mapping f such that

$$R(\sigma) \text{ iff. } f(\sigma) = 0 .$$

The existence of a selection operator seems to be necessary in order to prove some of the basic facts about Θ-semicomputable and Θ-computable relations, such as closure of Θ-semicomputable relations under $\exists$-quantification and disjunction, and also to prove that a relation R is Θ-computable iff R and $\neg R$ are Θ-semicomputable.

Definition 8. Let $\langle\Theta,<\rangle$ be a computation theory on $\mathfrak{A}$. An n-ary *selection operator* for $\langle\Theta,<\rangle$ is an n+1-ary Θ-computable pmv function $q(a,\sigma)$ with Θ-code $\hat{q}$ such that

(i) If there is an x such that $\{a\}_\Theta(x,\sigma) \to 0$, then $q(a,\sigma)$ is defined and $\forall x[q(a,\sigma) \to x \Rightarrow \{a\}(x,\sigma) \to 0]$.

(ii) If $\{a\}(x,\sigma) \to 0$ and $q(a,\sigma) \to x$, then

$$(a,x,\sigma,0) < (\hat{q},a,\sigma,x) .$$

Remarks. 1. For a motivation of (ii), equally valid in the present case, see [10, p. 225].

2. By an oversight Moschovakis [10, p. 255] only required $\{a\}(q(a,\sigma),\sigma) \to 0$ rather than $\forall x[q(a,\sigma) \to x \Rightarrow \{a\}(x,\sigma) \to 0]$, which is necessary when working with pmv objects.

2.4. *The notion of finiteness in general computation theories.* Any good general approach to recursion theory must embody a suitable notion of "finiteness". We repeat the basic definitions from [10, pp. 230–233].

Definition 9. A computation theory $\langle \Theta, < \rangle$ on $\mathfrak{A}$ is called *regular* if
(i) $C = A$.
(ii) Equality "$x = y$" is Θ-computable.
(iii) $\langle \Theta, < \rangle$ has selection operators.

Definition 10. Let $\langle \Theta, < \rangle$ be a regular theory on $\mathfrak{A}$, and let $B \subseteq A$. By the B-quantifier we understand the continuous pmv functional $\boldsymbol{E}_B(f)$ defined by

$$\boldsymbol{E}_B(f) \to \begin{cases} 0 & \text{if } \exists x \in B\,[f(x) \to 0]\,. \\ 1 & \text{if } \forall x \in B\,[f(x) \to 1]\,. \end{cases}$$

The set B is called Θ-*finite* with Θ-canonical code e, if the B-quantifier $\boldsymbol{E}_B$ is Θ-computable with Θ-code e.

We refer the reader to [10] for a list of five properties of this particular notion of finiteness which may justify the claim that it is "natural". But it would be premature to conclude from this that the present version gives all the properties of ordinary (i.e. true) finiteness necessary for the combinatorial arguments of e.g. degree theory.

3. Extending regularity

In this section we will discuss the problem of how to extend regularity. Of several possibilities we have chosen to emphasize two.

A. The idea that a "computation" is a finite object in the sense of the theory seems to play an important role in many arguments of recursion theory. "Computation" is not a primitive of our system, but a perhaps satisfactory approximation consists in requiring that a computation tuple depends on only a finite number of other tuples, i.e. for every $(a, \sigma, z) \in \Theta$, the set

$$\boldsymbol{S}_{(a,\sigma,z)} = \{(a', \sigma', z') \mid (a', \sigma', z') < (a, \sigma, z)\}$$

is finite (uniformly in (a, σ, z)) in the sense for Section 2.4.

Note: A more complete technical statement would require that there is a Θ-computable mapping $p(n)$ such that for each $(a, \sigma, z) \in \Theta$ $\{p(n)\}_\Theta(a, \sigma, z)$ is a Θ-canonical code for $\boldsymbol{S}_{(a,\sigma,z)}$. And since we are dealing with sequences of arbitrary finite length, we assume some suitable coding convention.

Remark. A computation theory in the sense of Moschovakis is a pair $\langle \Theta, ||_\Theta \rangle$, where $||_\Theta$ is a map from Θ into the ordinals. Using a length-function it seems to be difficult to capture the idea that a "computation" should be a finite object in the sense of the theory.

As a first approximation one could consider the set

$$\{(a', \sigma', z') \| a', \sigma', z' |_\Theta \leq |a, \sigma, z|_\Theta\} .$$

But one cannot outright say that this set, which is the set of *all* computations with smaller length, is a finite object in the theory. Indeed, if this requirement is made, the set of natural numbers necessarily will be finite in Θ.

Some restriction must therefore be added and Moschovakis proposed to compare computations of equal length, i.e. he required the finiteness of the set

(∗) $$\{(a', \sigma', z') \mid \mathrm{lh}(\sigma') = \mathrm{lh}(\sigma) \wedge |a', \sigma', z'|_\Theta \leq |a, \sigma, z|_\Theta\} .$$

It is possible to proceed on this basis, and it may have some advantage later on (see Section 4), but it is not in my opinion a satisfactory conceptual analysis of the motivation behind normality (see [10, p. 233]).

B. The prewellordering property is another important tool in general recursion theory. One formulation is as follows. Let $\|a, \sigma, z\|$ denote the ordinal of the set $\boldsymbol{S}_{(a,\sigma,z)}$, $(a,\sigma,z) \in \Theta$. The theory $\langle \Theta, < \rangle$ has the prewellordering propery if there exists a Θ-computable function $p(x,y)$ such that if either $x \in \Theta$ or $y \in \Theta$ then $p(x,y)$ is defined and single-valued, and whenever $y \in \Theta$, then

$$p(x,y) = 0 \quad \text{iff} \quad x \in \Theta \wedge \|x\| \leq \|y\| \,,$$

in other words, the set $\{x \mid x \in \Theta \wedge \|x\| \leq \|y\|\}$ is Θ-computable, uniformly in y.

We shall make some remarks on the relationship between the finiteness of the sets $\boldsymbol{S}_{(a,\sigma,z)}$ and the prewellordering property. It is convenient to introduce some terminology.

Definition 11. Let $\langle \Theta, < \rangle$ be a computation theory on a domain $\mathfrak{A}$.
1. $\langle \Theta, < \rangle$ is called *p-normal* if it is regular and has the prewellordering property (see **B** above).
2. $\langle \Theta, < \rangle$ is called *s-normal* if it is regular and the sets $\boldsymbol{S}_{(a,\sigma,z)}$ are (uniformly) Θ-finite for $(a,\sigma,z) \in \Theta$ (see **A** above).
3. $\langle \Theta, < \rangle$ is called *strongly normal* if it is both p-normal and s-normal.

Remarks. 1. "Normality" in the sense of Moschovakis [10] requires that the sets in (∗) above are uniformly Θ-finite.

2. If the domain $\mathfrak{A}$ is Θ-finite, normality in the sense of Moschovakis, p-normality, and strong normality lead to essentially the same class of theories, viz. the so-called "Spector theories" (see Section 4.1).

(i) S-normality implies a "weak" form of the prewellordering property: We can define a Θ-computable pmv function $q(x,y)$ such that if

$$x,y \in \Theta, \qquad \text{then } q(x,y) = 0 \quad \text{iff} \quad \|x\| \leq \|y\| \,.$$

This is so since we can computably quantify over finite sets, hence have the following recursion equation

$$\|x\| \leq \|y\| \quad \text{iff} \quad \forall x' \in \boldsymbol{S}_x \; \exists y' \in \boldsymbol{S}_y \; \|x'\| \leq \|y'\| \,.$$

(ii) If we in addition assume that the relation $<$ is Θ-semicomputable, then we have the prewellordering property. In this case we have the following recursion equations for the function $p(x,y)$:
(a) $p(x,y) = 0$ if $\forall x' \in \boldsymbol{S}_x\ \exists y'[y' < y \wedge p(x',y') = 0]$.
(b) $p(x,y) = 1$ if $\exists x'[x' < y \wedge \forall y' \in \boldsymbol{S}_y\ p(x',y') = 1]$.

Note: Since $<$ is defined for all tuples of the form (a,σ,z) and Θ is the well-founded part of $<$, the assumption that $<$ is Θ-semicomputable is rather problematic. However, the following argument may add some plausibility. An arbitrary computation tuple (a,σ,z) may or may not represent a "true" computation. But as soon as we are given a set of instructions a and an input sequence σ we should be able to start "generating" the "subcomputations", and this is what the Θ-semicomputability of $<$ is intended to express. This argument seems to carry force in the case of single-valued computations. It is in the case of multiple-valued computations that we would have difficulties in extending a relation $<$ *on* Θ to a relation $<$ on *all* tuples (a,σ,z) such that the extended relation is Θ-semicomputable and such that Θ is the well-founded part of the extended $<$.

(iii) If $\langle\Theta,<\rangle$ is an s-normal theory over a Θ-finite domain $\mathfrak{A}$, we have the prewellordering property in several important cases, e.g. if Θ is the recursion theory associated with some total type-2 functional F (assuming ${}^2E \leq F$), or the recursion theory associated with the type of partial type-2 functionals considered by Hinman [6], or the "quantifiers" of Aczel [1]. Since (see 4.1.1) every s-normal $\langle\Theta,<\rangle$ is equivalent to a theory PR $[F]$ for some continuous partial F, it remains an interesting problem to determine those theories PR $[F]$ for which we have the prewellordering property.

(iv) For infinite s-normal theories $\langle\Theta,<\rangle$ on a partially ordered domain $\langle\mathfrak{A},\precsim\rangle$ we have the prewellordering property under the assumptions 4.2.3–4.2.5. If we drop the finiteness assumption 4.2.5, a counterexample can probably be constructed.

(v) P-normality does not imply s-normality. "Ordinary" recursion theory over ω can be constructed in such a way as to provide a counterexample.

4. Beyond normality

Moving beyond normality there is one fundamental distinction to make: Either a theory $\langle\Theta,<\rangle$ has a domain which is finite in the sense of the theory, or its domain is infinite.

4.1. *Theories over finite domains*. In analogy with Moschovakis [10] we call a p-normal (or, which amounts to essentially the same – see the remark following Definition 11 in Section 3 – strongly normal) theory $\langle\Theta,<\rangle$ on a domain $\mathfrak{A}$ a *Spector theory* if the domain A is Θ-finite. For such theories one can prove a great number of results which were originally established for hyperarithmetic and hyperprojective theory (see Moschovakis [9] and [10], – a detailed exposition within the axiomatic set up can be found in Vegem [16]).

Hyperarithmetic theory over ω is the theory of recursion in ${}^2E$, hyperprojective theory is a generalization of this to more general domains. How different is an arbitrary Spector theory from recursion in some functional over the domain, i.e. what kind of representation theorems do we have for Spector theories?

We state some results. But first a few terminological remarks. In many cases it is convenient to assume that the search operator ν is computable in Θ, where $\nu(f) = \{x \mid f(x) \to 0\}$. It is known from [10, p. 266] that if Θ has a selection operator, then Θ is weakly equivalent to $\Theta[\nu]$, hence, for reasons detailed there, we may as well work with $\Theta[\nu]$, which allows us greater freedom in defining pmv objects.

For convenience we also introduce the following notations:

sc(Θ) = the set of all Θ-computable relations on $\mathfrak{A}$.

en(Θ) = the set of all Θ-semicomputable relations on $\mathfrak{A}$.

(The notation en(Θ), the *envelope* of Θ, is taken from Moschovakis [11].)

4.1.1. *For any s-normal $\langle\Theta,<\rangle$ on a domain $\mathfrak{A}$, there exists a continuous partial functional F on $\mathfrak{A}$ such that*

$$\Theta \sim \mathrm{PR}\,[F]\,.$$

This result is jointly due to J. Moldestad and D. Normann, and the proof uses the uniform finiteness of the sets $\boldsymbol{S}_{(a,\sigma,z)}$. D. Normann has also adapted the main result of Sacks [12] to show

4.1.2. *For any Spector theory over* ω *there exists a total functional F such that*

$$\mathrm{sc}(\Theta) = {}_1\mathrm{sc}(F) \qquad (\text{and } {}_1\mathrm{en}(F) \subseteq \mathrm{en}(\Theta)) .$$

Remark. D. Normann actually proves a more general result: Let M be a countable admissible set which satisfies local countability and $\mathbf{\Delta}_0$-dependent choice, then: (i) $M = L^K_{0(M)}$, for some generic class $K \subseteq 0(M) = M \cap \mathrm{On}$; (ii) M is K-admissible, and (iii) $0(M)$ is the least β such that L^K_β is K-admissible. – 4.1.2 is an immediate corollary. (Note that the somewhat complicated hierarchy for type-2 recursion used by Sacks [12] is avoided in this approach. It may be essential for the k-section result.) We expect that Normann's result can be adapted to admissible sets with urelements, yielding a generalization of 4.1.2 to arbitrary countable domains.

In Moschovakis [11] we find a counterexample to show that 4.1.2 cannot be lifted from sections to envelopes:

4.1.3. *There exists a Spector theory* Θ *over* ω *such that* $\mathrm{en}(\Theta) \neq {}_1\mathrm{en}(F)$, *for all total type-2 functionals over* ω.

The problem remains to characterize those Spector theories which are equivalent to prime recursion is some total type-2 functional over the domain. This is, of course, only one step toward a full classification of Spector theories.

In this connection a Gandy–Spector theorem or, better, a normal form theorem for the class $\mathrm{en}(\Theta)$ may be of interest. It is not at all clear how to formulate such theorems in a sufficiently general form. Our proposal for a "weak" version:

Definition 12. Let $P(X, \sigma)$ be a second order relation over $\mathfrak{A}$. $P(X, \sigma)$ is called Θ-computable with index p if whenever $\{e\}_\Theta$ is the characteristic function of some set $X_e \in \mathrm{sc}(\Theta)$, then

$$\{p\}_\Theta(e, \sigma) \to \begin{cases} 0, & \text{if } P(X_e, \sigma) \\ 1, & ow . \end{cases}$$

We say that the theory Θ has the weak Gandy–Spector property if whenever $R \in \mathrm{en}(\Theta)$, there is a Θ-computable P such that

$$R(\sigma) \quad \text{iff.} \quad (\exists X \in \mathrm{sc}(\Theta))\, P(X, \sigma)\,.$$

4.1.4. *Every Spector theory has the weak Gandy–Spector property.*

This has been proved by J. Moldestad. It remains an interesting problem to find other "natural" kinds of normal form theorems which can be used to characterize certain classes of Spector theories.

Remark. The original or "strong" Gandy–Spector theorem for hyperarithmetical theory provides a P which is first order with respect to the language adequate to describe the domain $\mathfrak{A}$.

We shall comment on one more topic. The connection between inductive definability and hyperprojectivity was investigated by Grilliot [5]. His results was adapted to the present frame by Moldestad [8]. Let $\mathrm{Ind}(\Delta)$ denote the class of relations which are inductive in some operator of class Δ, i.e. reducible to some Γ^∞, where $\Gamma \in \Delta$.

4.1.5. *Let Θ be a Spector theory on $\mathfrak{A}$ and $\boldsymbol{R}$ a sequence of Θ-computable relations on $\mathfrak{A}$.*

(i) $\mathrm{Ind}(\Sigma_2(\boldsymbol{R})) \subseteq \mathrm{en}(\Theta)$.

(ii) $\Theta \sim \mathrm{PR}\,[\boldsymbol{R}, =, \nu, E]$, *if and only if* $\mathrm{Ind}(\Sigma_2(\boldsymbol{R})) = \mathrm{en}(\Theta)$.

Here the implication from right to left in (ii) goes beyond Grilliot [5], and gives a certain characterization of the "minimal" Spector theory on a domain $\mathfrak{A}$. A general result would give a necessary and sufficient condition on Θ for the validity of the equation $\mathrm{Ind}(\Sigma_n(\boldsymbol{R})) = \mathrm{en}(\Theta)$.

4.2. *Theories over infinite domains.* In Moschovakis [10] the case of infinite domains is particularly satisfying. Any Friedberg theory in his sense is a recursion theory generated in a "natural" way from an admissible prewellordering of the domain.

In our case the situation is more complicated. Our definition of normality does not automatically give us a prewellordering of the domain. In order for us to proceed we must add extra assumptions on the domain. The goal will be to abstract the "natural" or "minimal" recursion theory associated with an admissible set (possibly with urelements, – for this concept see Barwise [2]).

Remark. The more complicated situation in our set up is perhaps not a too serious disadvantage. There seems to be recursion theories over infinite domains (i.e. infinite in the sense of the theory) on which there is no natural associated prewellordering (see [7]). And there seems to be admissible sets where the prewellordering associated with the rank function is not admissible in the sense of [10]. (Let A be an admissible set such that A is uncountable but $A \cap \mathrm{On}$ is countable.) Such examples should not be denied their proper existence in an axiomatic analysis of computation theories.

We admit at once that we do not yet claim to have a "good" definition of computation theories over infinite domains. We shall make some preliminary suggestions, and hope that further work will lead to a "correct" analysis.

Let $\langle \Theta, < \rangle$ be an s-normal theory on the domain $\mathfrak{A}$. The basic intention is to express that there is a suitable correspondence between the complexity of the domain $\mathfrak{A}$ and the complexity of the computations in Θ. Some partial ordering of the domain seems to be necessary in order to code whole computations in a "natural", i.e. order preserving way into the domain. We make the following proposal:

4.2.1. *There is a Θ-computable partial ordering $\preceq$ of A such that the initial segments of $\preceq$ are well-founded and (uniformly) Θ-finite.*

A weak way of expressing the correspondence between Θ and the domain is to assume:

4.2.2. $|\preceq| = \sup\{\|a, \sigma, z\| \mid (a, \sigma, z) \in \Theta\}$.

Remark. A more refined analysis would probably postulate the existence of some kind of "coding"-function:

4.2.3. *There is a Θ-computable mapping $\kappa : \Theta \to A$ such that if we set*

$\kappa^*(x) = \{w \in A \mid w \precsim \kappa(x)\}$, *then*

(i) $\kappa^*(x)$ *is* Θ*-finite for each* $x \in \Theta$;

(ii) $x \in \mathbf{S}_y$ *implies* $\kappa^*(x) \subseteq \kappa^*(y)$;

(iii) $A = \bigcup_{x \in \Theta} \kappa^*(x)$.

Note that 4.2.2 follows from (iii) in 4.2.3 and that 4.2.3(i) is already implied by 4.2.1.

In addition to the coding process we seem to need some sort of "decoding" assumption. As a first approximation we propose:

4.2.4. *There is a* Θ*-computable mapping* $p(n)$ *such that* $\{p(n)\}_\Theta(a, \sigma, z, w) = 0$ *iff* $(a, \sigma, z) \in \Theta \wedge \|a, \sigma, z\| = |w|$.

In other words, the relation $(a, \sigma, z) \in \Theta^{|w|}$ is Θ-computable.

The prewellordering associated with $\precsim$ is easily seen to be uniformly Θ-computable. If we strengthen this to

4.2.5. $\{w \in A \mid |w| \leq |w_0|\}$ *is uniformly* Θ*-finite in* $w_0 \in A$,

we arrive at the class of "Friedberg theories" in the sense of Moschovakis [10, § 10]. And as he shows these are exactly the class of computation theories associated with admissible prewellorderings. It remains to isolate the properties which characterizes the recursion theories associated with admissible partial orderings, i.e. with arbitrary admissible sets.

A topic of central importance is the relationship between theories over finite and infinite domains. The basic example here is the relationship between hyperarithmetic and meta-recursion (or L_{ω_1}-recursion) theory. This was generalized in Barwise, Gandy, Moschovakis [3] (– see also Barwise [2]).

In our context we are looking for a theorem which states that finite theories can be imbedded into infinite ones. The idea behind is simply this. A computation theory on an infinite domain can be a "good" recursion theory, in particular, if we have a suitable correspondence (coding/decoding) between the domain and computations over the domain, and if the semi-computable relations are exactly the Σ_1-definable relations over the domain. The imbedding theorem should say that we can "enlarge" a finite theory to a "good" infinite theory. And, as a possible application, we would expect that fine structure results for, say, the semi-computable relations of the given finite theory could be obtained by "pull-back" from the enlargement. The motivating example is again various results for Π_1^1 sets obtained *via* meta-recursion theory.

We state the following preliminary version of the imbedding theorem:

4.2.6. *Let* $\langle \Theta, < \rangle$ *be a Spector theory on a domain* $\mathfrak{A}$. *It is then possible to construct*

(i) *a domain* $\langle \mathfrak{A}^*, \preceq^* \rangle$ *where* $\mathfrak{A}^*$ *extends* $\mathfrak{A}$ *and* $\preceq^*$ *is a well-founded partial ordering on* $\mathfrak{A}^*$,

(ii) *a relation* R *on* $\langle \mathfrak{A}^*, \preceq^* \rangle$, *and*

(iii) *an "infinite" theory* $\langle \Theta^*, <^* \rangle$ *on* $\langle \mathfrak{A}^*, \preceq^* \rangle$, *such that*

(a) $\preceq^*$ *is* Θ^**-computable and initial segments of* $\preceq^*$ *are uniformly* Θ^**-finite,*

(b) R *is* Θ^**-computable and* Θ^**-semicomputability equals* $\Sigma_1(R)$*-definability over* $\mathfrak{A}^*$, *and*

(c) *a subset* $R \subseteq A$, *where* A *is the domain of* $\mathfrak{A}$, *is* Θ*-semicomputable if and only if* R *is* Θ^**-semicomputable.*

Remark. The result and method of proof is clearly inspired by Barwise, Gandy, Moschovakis [3] and it uses in the construction of $\mathfrak{A}^*$ the recent developed theory of admissible sets with urelements, see Barwise [2] and his forthcoming lecture notes. There is also some recent work of P. Aczel and Y. Moschovakis which overlaps with the present result. Both Aczel and Moschovakis work in the context of a Spector class (which is equivalent to being the envelope of a Spector theory) and their goal is to relate these classes to the "next admissible ordinal" (Aczel) or "next admissible set" (Moschovakis). Both Aczel and Moschovakis carry their analysis a step further than the above result, showing the existence of a unique, minimal "next" structure. From this also follows the existence of a unique minimal "infinite" extension Θ^* of a given Spector theory Θ.– We have deliberately used the word "infinite theory" to describe Θ^*, since it is a bit unclear at the moment how strong properties we can enforce on Θ^*; it will satisfy the properties 4.2.1–4.2.5. We also expect that *c* can be strengthened to assert the equivalence between Θ and $\Theta^* \upharpoonright \mathfrak{A}$, but some details remain to be sorted out.

4.3. One further goal is to push the analysis of "computation" so far as to establish the domain of validity for the priority arguments. Only then can we claim to have a reasonably complete axiomatic analysis of generalized recursion theory. This is very much an open field.

Remark. The concept of a Friedberg theory does not seem to be entirely adequate, see Simpson's paper "Post's problem for admissible sets" in this volume. On the positive side see the appendix to that paper and also Simpson [13].

References

[1] P. Aczel, Representability in some systems of second order arithmetic, Israel Jour. Math. 8 (1970) 309–328.
[2] K.J. Barwise, Admissible sets over models of set theory, this volume.
[3] K.J. Barwise, R. Gandy and Y.N. Moschovakis, The next admissible set, J. Symbolic Logic 36 (1971) 108–120.
[4] H. Friedman, Axiomatic recursive function theory, in: R.O. Gandy and C.E.M. Yates (eds.) Logic Colloquium '69 (North-Holland, Amsterdam, 1971) 113–137.
[5] T. Grilliot, Inductive definitions and computability, Trans. Amer. Math. Soc. 158 (1971) 309–317.
[6] P. Hinman, Hierarchies of effective descriptive set theory, Trans. Amer. Math. Soc. 131 (1968) 526–543.
[7] P. Hinman and Y.N. Moschovakis, Computability over the continuum, in: R.O. Gandy and C.E.M. Yates (eds.) Logic Colloquium '69 (North-Holland, Amsterdam, 1971) 77–105.
[8] J. Moldestad, cand. real. thesis, Oslo 1972 (in Norwegian).
[9] Y.N. Moschovakis, Abstract first order computability I, Trans. Amer. Math. Soc. 138 (1969) 427–464; and II, 138 (1969) 465–504.
[10] Y.N. Moschovakis, Axioms for computation theories – first draft, in: R.O. Gandy and C.E.M. Yates (eds.) Logic Colloquium '69 (North-Holland, Amsterdam, 1971) 199–255.
[11] Y.N. Moschovakis, Structural characterizations of classes of relations, this volume.
[12] G.E. Sacks, The 1-section of a type n object, this volume.
[13] S. Simpson, Admissible selection operators, Notices Amer. Math. Soc. 19 (1972) A-599.
[14] H.R. Strong, Algebraically generalized recursive function theory, IBM Jour. Res. Devel. 12 (1968) 465–475.
[15] E.G. Wagner, Uniform reflexive structures: on the nature of Gödelizations and relative computability, Trans. Amer. Math. Soc. 144 (1969) 1–41.
[16] M. Vegem, cand. real. thesis, Oslo 1972 (in Norwegian).

J.E.Fenstad, P.G.Hinman (eds.), Generalized Recursion Theory

DISSECTING ABSTRACT RECURSION

Thomas J. GRILLIOT
The Pennsylvania State University

1. Introduction

We have two prototypes for this discussion: recursiveness on the natural numbers and hyperarithmeticalness on the natural numbers. The latter differs from the former in that a quantification-over-ω scheme is admitted. Crudely speaking, to say that F is recursive in $G_1, ..., G_n$ means that F is computable or can be combinatorially generated from the structure $\langle\omega; 0, s, =, G_1, ..., G_n\rangle$, where 0 and s are the usual zero and successor function of ω. (Note: $0, s, =$ are sufficient to specify the structure of ω as we know from the investigations of Peano.) Similarly, to say that F is hyperarithmetical in $G_1, ..., G_n$ means that F can be generated combinatorially from the structure $\langle\omega; 0, s, =, G_1, ..., G_n\rangle$ with the aid of an oracle that can test quantification over ω. It is natural to replace the structure $\langle\omega; 0, s, =, G_1, ..., G_n\rangle$ by an arbitrary structure $\langle A; G_1, ..., G_n\rangle$ where A is a set and $G_1, ..., G_n$ are functions or predicates on A. This is precisely what our investigation is all about: to find out what recrusiveness on an arbitrary structure means. Unfortunately, the matter is not so easy in that the structure of the natural numbers is fairly unique with respect to other structures. Consequently the abstract study of recursiveness and hyperarithmeticalness may tend to be prejudiced by preconceptions based on usual recursiveness and hyperarithmeticalness on the natural numbers. However, with a little investigation one can isolate these potential prejudices and discover that they seem to be five in number, which we denote by C (constant),

E (equality), S (search), A (for all), I (for infinitely many).

C-scheme: constant functions should be recursive. This scheme is true for ω because it is finitely generated by 0, s. In general, however, an arbitrary set need not be so combinatorially generated, and hence one cannot argue convincingly that constant functions are *ipso facto* recursive.

E-scheme: equality relation should be recursive. This scheme holds for ω because 0, s generate ω, s is one-one and 0 can be distinguished from successor numbers. The general situation is quite different. One cannot, for example, convincingly argue that equality on real numbers is *ipso facto* recursive.

S-scheme: unordered search operator should be recursive. The situation of ω is particularly nice in that one has not only an unordered search operator but even an ordered one (the so-called μ-operator). This is because ω is finitely generated by 0, s. An alternative formulation of the S-scheme is that semi-recursive predicates are closed under existential quantification. Again one cannot argue on purely combinatorial grounds that an unordered search operator is *ipso facto* recursive.

A-scheme: the universal-quantifier operator should be "recursive". This certainly holds for hyperarithmeticity on ω. In fact, one of the more elegant formulations of "hyperarithmetical" is based on this idea (see Kleene [5]). The A-scheme is the most natural candidate for formulating an abstract notion of hyperarithmeticalness, but compare with the following scheme.

I-scheme: the operator introducing the quantifer "for infinitely many" should be "recursive". This certainly holds for hyperarithmeticity on ω. In fact, the A-scheme and the I-scheme are equivalent on ω (in the presence of bounded quantification):

$$IxP(x) \leftrightarrow \forall y \exists x (P(x) \,\&\, y < x)$$

$$\forall x P(x) \leftrightarrow Iy \forall x (P(x) \,\&\, x < y) \,.$$

In some structures the two schemes are independent.

Since the five schemes listed above are independent, we have $32 (= 2^5)$ different formulations of recursion (of varying degrees of interest). To label them properly, we will use the terminology (C,E,S,A,I)-recursive, (C,E,A)-recursive, etc. Thus to say that F (on A) is (C,E,S)-recursive in $\langle A;G \rangle$ means that F can be generated combinatorially from the structure $\langle A;G \rangle$ with the aid of schemes C,E,S. An exact formulation of what this means is given in the next section.

An inquisitive reader will certainly question whether the five schemes listed above are exhaustive. In Section 5 we will show that they are (at least for countable structures) in the following sense: any scheme that holds for recursiveness (respectively, hyperarithmeticalness) on the natural numbers and can be formulated for arbitrary structures is derivable in (C,E,S)-recursion (respectively, (C,E,S,A,I)-recursion).

Historical note. Perhaps the earliest formulation of abstract recursion is due to Fraïssé [1]. However, he did not distinguish between the combinatorial and noncombinatorial (schemes C,E,S,A,I) aspects of recursion. Moschovakis [10] seems to have been first to isolate the schemes C,E,S and a combined A/I scheme. The completeness of the C,E,S schemes was asserted by Lacombe [8] in a surreptitious way. The completeness of the C,E,S,A,I schemes was proved by Grilliot [2].

2. Syntactic description of recursion

Our choice of formulation of recursion is proof-oriented. Besides being fairly versatile, it has the advantage of bridging the computational and the model-theoretic aspects of recursion.

Definition. We define $\Psi \vdash \theta$, where θ is a sentence and Ψ a set of sentences, inductively as follows. $\Psi \vdash \theta$ if:

(a) $\theta \in \Psi$ or $\varphi, \neg\varphi \in \Psi$ for some φ [initial]; or
(b) $\varphi \,\&\, \psi \in \Psi$ and $\Psi \cup \{\varphi, \psi\} \vdash \theta$ [&-elimination]; or
(c) $\varphi \vee \psi \in \Psi$, $\Psi \cup \{\varphi\} \vdash \theta$ and $\Psi \cup \{\psi\} \vdash \theta$ [$\vee$-elimination]; or
(d) $\forall x \varphi(x) \in \Psi$ and $\Psi \cup \{\varphi(t)\} \vdash \theta$ for some term t [$\forall$-elimination]; or
(e) $\exists x \varphi(x) \in \Psi$ and $\Psi \cup \{\varphi(y)\} \vdash \theta$ where y is a constant not in Ψ, θ [weak $\exists$-elimination]; or
(f) $\exists x \varphi(x) \in \Psi$ and $\Psi \cup \{\varphi(b)\} \vdash \theta$ for all $b \in B$ [strong $\exists$-elimination]; or
(g) $Ix\varphi(x) \in \Psi$ and $\Psi \cup \{\exists x_1 \ldots x_n(\varphi(x_1) \,\&\, \ldots \,\&\, \varphi(x_n) \,\&\, x_1 \neq x_2 \,\&\, x_1 \neq x_3 \,\&\, \ldots \,\&\, x_{n-1} \neq x_n)\} \vdash \theta$ for some $n \in \omega$ [I-elimination]; or
(h) $Cx\varphi(x) \in \Psi$ and $\Psi \cup \{\exists x_1 \ldots x_n \forall y(\varphi(y) \vee y = x_1 \vee \ldots \vee y = x_n)\} \vdash \theta$ for all $n \in \omega$ [C-elimination].

Clearly Ix represents the quantifier, for infinitely many x, and Cx represents the quantifier, for cofinitely many x; thus Ix and $\neg Cx \neg$ are expected to represent the same thing. Because of our specialized purpose, there is no need to have a $\neg$-elimination scheme (since $\neg$'s can be driven through the other connectives) nor introduction schemes (since θ will always be atomic or negated atomic) nor equality schemes (since they can be incorporated into Ψ as desired).

To say that Q is recursive in $\langle B; P_1, ..., P_n \rangle$ will mean that there is a formal description of Q in terms of $P_1, ..., P_n$ such that

(information about $P_1, ..., P_n$) $\cup$ (description) $\vdash$ (information about Q).

The description must be both complete (*all true* information about Q is derivable) and consistent (*no false* information about Q is derivable). The description may have intermediary predicate or function symbols in it. For example, the usual description of $\cdot$ (on ω) in terms of $0, s =$ utilizes $+$ as an intermediary. Let us formulate this definition of recursion precisely.

Definition. Let $\langle B; P_1, ..., P_n \rangle$ be a structure, where B is a set and $P_1, ..., P_n$ are relations (possibly partial) on B. Let Q be another relation (possibly partial) on B. For a language that includes one or more names (constant symbols) for each element of B and one or more names (predicate symbols) for each of $P_1, ..., P_n, Q$, define Δ_Q to be the set of formal sentences

$$\{Q(b_1, ..., b_n) : Q(b_1, ..., b_n) \text{ is true}\}$$

$$\cup \{\neg Q(b_1, ..., b_n) : Q(b_1, ..., b_n) \text{ is false}\}$$

and similarly for $\Delta_{P_1}, ..., \Delta_{P_n}$, where we use the same letter for an object and its formal name(s) so long as confusion is avoided. *Q is (C,E,S,A,I)-recursive in the structure* $\langle B; P_1, ..., P_n \rangle$ if there is a sentence φ such that

$$\Delta_= \cup \Delta_{P_1} \cup ... \cup \Delta_{P_n} \cup \{\varphi\} \vdash \theta \quad \text{for all } \theta \in \Delta_Q$$

and some expansion of $\langle B; P_1, ..., P_n \rangle$ is a model of φ. In this definition, some of the relations $P_1, ..., P_n, Q$ may be replaced by (partial) functions where, if f is a function, Δ_f is defined to be

$$\{f(b_1, ..., b_k) = c : f(b_1, ..., b_k) \simeq c\}.$$

With regard to the formation of Δ_f, if multiple names are used to denote one element of B, all names must appear as arguments of the function symbol but only one name need appear as value. Also multivalued functions are acceptable, only in this case, the symbol = occuring in Δ_f may have to be replaced by another symbol to avoid inconsistency with $\Delta_=$. An n-ary multivalued function amounts to the same as an $(n+1)$-ary relation in the presence of the E-scheme and S-scheme; but in the absence of the S-scheme the value-place of the function plays a role different from the argument-places as we shall see later.

As part of the definition of Q being (C,E,S,A,I)-recursive in $\langle B;P_1, ..., P_n\rangle$, we said that some expansion of $\langle B;P_1, ..., P_n\rangle$ must be a model of φ. This condition can be relaxed by saying some extension of $\langle B;P_1, ..., P_n\rangle$ must be a model of φ. In other words, it does not matter if the universe of the model is larger than B. This can be seen by the following little trick. Given φ with a model with universe larger than B, form a new description

$$\varphi^* \;\&\; \forall x_1 ... x_{k_1}(P_1(x_1, ..., x_{k_1}) \rightarrow P_1'(fx_1, ..., fx_{k_1}))$$

$$\& \;...\; \& \;\forall x_1 ... x_{k_n}(P_n(x_1, ..., x_{k_n}) \rightarrow P_n'(fx_1, ..., fx_{k_n}))$$

$$\& \;\forall x_1 ... x_k(Q'(fx_1, ..., fx_k) \rightarrow Q(x_1, ..., x_k))$$

where $P_1', ..., P_n', Q', f$ are new symbols and φ^* is made from φ by replacing each $P_1, ..., P_n, Q$ by $P_1', ..., P_n', Q'$ and each variable or constant x by fx. ($\psi \rightarrow \theta$ is regarded as an abbreviation for $\neg\psi \vee \theta$.)

Suppose Q is (C,E,S,A,I)-recursive in $\langle B;P\rangle$ with description φ. Thus

$$\Delta_= \cup \Delta_P \cup \{\varphi\} \vdash \psi \quad \text{for} \quad \psi \in \Delta_Q \, .$$

We may think of φ as a program for computing Q from P as follows. Assume that all $\neg$'s in φ are driven through the other connectives so that they occur only immediately preceding atomic formulas. To determine whether or not $Q(a)$ is true, one begins with a computation node $\{\varphi\} \vdash$. The connectives in φ are then systematically stripped in accord with clauses (b) through (h) so that one gets a tree of intermediate computation nodes of the form $\Psi \vdash$ where Ψ is a finite set. If at some node, $P(b) \in \Psi$ or $\neg P(b) \in \Psi$ or $b = c \in \Psi$ or $b \neq c \in \Psi$ for some $b, c \in B$, then one asks the oracle of P or = whether or not $P(b)$ or $b = c$ is true. If there is a conflict, clause (a) establishes that $\Delta_= \cup \Delta_P \cup \Psi \vdash Q(a)$ and $\Delta_= \cup \Delta_P \cup \Psi \vdash \neg Q(a)$; thus computation node

$\Psi \vdash$ is certified. Similarly if $\theta, \neg\theta \in \Psi$ for some θ, node $\Psi \vdash$ is certified. Also if $Q(a) \in \Psi$ (respectively, $\neg Q(a) \in \Psi$), clause (a) establishes $\Psi \vdash Q(a)$ (respectively, $\Psi \vdash \neg Q(a)$); thus node $\Psi \vdash$ is certified for $Q(a)$ (respectively, $\neg Q(a)$). As soon as all terminal nodes are certified for $Q(a)$ (respectively, $\neg Q(a)$), the computation procedure ceases with the answer being $Q(a)$ is true (respectively, false).

This computational procedure for (C,E,S,A,I)-recursion has five processes within it that are not purely combinatorial. We label these schemes by C,E,S,A,I. By removing one or more of these schemes as outlined below from the computational procedure, one gets 31 other forms of recursion ((C,E,S,A)-recursion, (C,E,S,I)-recursion, ..., ( )-recursion).

Role and irredundance of the I-scheme. C-elimination is an infinitistic rule of proof and hence is not combinatorial. Its role is that of introducing the Ix quantifier. Thus the I-scheme is retained or removed by retaining or removing the C-elimination rule from the inductive definition of $\vdash$. This turns out to be equivalent to allowing or disallowing the quantifiers Ix, Cx in the description φ. To verify the irredundance of the I-scheme, we can apply a result of II.C of [4] that there is a nonstandard model of first-order arithmetic in which every (C,E,S,A)-recursive set of standard numbers is finite. In particular, the set of all standard numbers (denoted by ω) is not (C,E,S,A)-recursive in this model. By contrast, ω is (C,E,S,A,I)-recursive in this model since $x \in \omega \leftrightarrow \neg Iy\ (y<x)$. In other words, we have the following description of ω:

$$\forall x[(Iy(y<x) \vee \omega(x)) \,\&\, (Cy(x<y) \vee \neg\omega(x))]$$

$$\& \text{ some equality axioms.}$$

Role and irredundance of the A-scheme. The other infinitistic rule is strong $\exists$-elimination. Its role is to introduce universal quantification over universe B. Thus the A-scheme is retained or removed by retaining or removing the strong $\exists$-elimination rule from the definition of $\vdash$. Consider the structure $\langle \omega \times \omega; < \rangle$ where $(a,b) < (c,d)$ iff $a < c$. The set $\{0\} \times \omega$ is (C,E,S,A,I)-recursive in $\langle \omega \times \omega; < \rangle$ since $x \in \{0\} \times \omega \leftrightarrow \neg\exists y(y<x)$; that is, the following is a description of $\{0\} \times \omega$:

$$\forall x((\forall y(y \not< x) \vee \neg P(x)) \,\&\, (\exists y(y<x) \vee P(x))) .$$

However, the set cannot be (C,E,S,I)-recursive in this structure. For suppose φ is a description of $\{0\} \times \omega$. Pick $n \in \omega$ so that no name for any (m,n) occurs in φ. Then φ is still a description of $\{0\} \times \omega$ even when a new element $(-1,n)$ is added to the structure and $<$ is extended so that $(-1,n) < (m,n)$ for all $m \in \omega$. On the other hand, φ must be a description of the transplant of $\{0\} \times \omega$ under the canonical isomorphism from $\langle \omega \times \omega; < \rangle$ to $\langle \omega \times \omega \cup \{(-1,n)\}; < \rangle$. Thus φ is a description of two different sets, a contradiction.

Role and irredundance of the C-scheme. This scheme allows all constant functions as automatically "recursive". The C-scheme is retained or removed by allowing or not allowing names for elements of B (other than those listed in $\langle B; P_1, ..., P_n \rangle$) in descriptions. At first glance, allowing names for arbitrary elements of B in a description φ is a purely combinatorial feature. However, in the course of a $\vdash$-tree such a name would have to be matched with its replicas in $\Delta_{P_1} \cup ... \cup \Delta_{P_n}$. Such matching cannot be regarded as automatically combinatorial. For example, let ω_1 be the set of countable ordinals. The predicate $P(x) \leftrightarrow (x$ is finite ordinal$)$ is (C,E)-recursive in the structure $\langle \omega_1; < \rangle$ since it has the description:

$$\forall x [(\omega < x \to \neg Px) \,\&\, (\omega = x \to \neg Px) \,\&\, (x < \omega \to Px)] .$$

The use of a name (ω) for the first limit ordinal is crucial to this description. With a compactness-like argument one sees that P cannot be generated combinatorially from $<$ and $=$; that is, P is not (E)-recursive in $\langle \omega_1; < \rangle$. With slightly more care, one can find a predicate P that is (C)-recursive in $\langle \omega_1; < \rangle$ but not (E,S,A,I)-recursive in $\langle \omega_1; < \rangle$. It should be noted, however, that in those structures $\langle B; P_1, ..., P_n \rangle$ in which B is generated by $P_1, ..., P_n$ (e.g., $\langle \omega; 0, s, = \rangle$) the C-scheme is redundant.

Role and irredundance of the E-scheme. This scheme says that the equality relation is automatically "recursive". At first glance, the scheme can be retained or removed by retaining or removing $\Delta_=$ in the definition. However, the removal of $\Delta_=$ may not be enough in some instances since subtle comparisons between objects of B may be made in the course of determining $\vdash$. For example, $\{\psi\} \vdash \theta$ is established when ψ is θ. But to verify that ψ is θ, one must make a character-for-character comparison which often includes a comparison between two names for elements of B which in turn insinuates a comparison of elements of B themselves. To avoid such surreptitious uses of equality, one must allow infinitely many names for each element of B. In this way, the combinatorial comparison of two names in no way establishes a non-combinatorial comparison of two elements of B. (Conversely, if one accepts the E-scheme, then one might as well use only one name for each element of B.) Consider the structure $\langle\omega; 0, s\rangle$. The predicate $Z(a) \leftrightarrow a = 0$ is (E)-recursive in $\langle\omega; 0, s\rangle$ (or, equivalently, is ( )-recursive in $\langle\omega; 0, s, =\rangle$). However, Z is not ( )-recursive in $\langle\omega; 0, s\rangle$. For suppose there is a description φ of Z. Let a be a name for 0 different from any name occuring in φ. Then one cannot combinatorially determine that $\Delta_s \cup \{\varphi\} \vdash Z(a)$. It is of interest to note that Z and $=$ are of equal strength in the presence of the predecessor function: $p(a) = a - 1$ when $a \neq 0$ and undefined otherwise. That is, the structure $\langle\omega; 0, s, =\rangle$ and $\langle\omega; s, p, Z\rangle$ are equivalent, and, in fact, the functions ( )-recursive in either structure are exactly the usual recursive functions, and the functions (A)-recursive (alternatively, (I)-recursive) in either structure are exactly the usual hyperarithmetical functions.

Role and irredundance of the S-scheme. This scheme allows one to search through the set B for some object that satisfies a "recursive" predicate. It is the abstract analogue of the μ-operator scheme. This search capability appears in the $\forall$-elimination rule in the definition of $\vdash$. To determine whether $\Psi \cup \{\forall x \varphi(x)\} \vdash \theta$, one must check whether $\Psi \cup \{\forall x \varphi(x), \varphi(t)\} \vdash \theta$ for some term t. Thus one must search through all terms t, which is tantamount to searching through B. To remove the S-scheme, one must restrict $\forall$-elimination to instances where the terms involved can be generated combinatorially from information already given and the oracles of the functions in the structure. To exemplify this restriction, consider the structure $\langle R; +, \cdot\rangle$ where R is the set of real numbers. Suppose that in the course of some computation one encounters

$$\Delta_+ \cup \Delta_\cdot \cup \{\forall x(x \cdot x \neq 2 \vee P(\sqrt{3}))\} \vdash P(\sqrt{3}) . \tag{1}$$

In the presence of the S-scheme, this can be established by replacing x by $\sqrt{2}$:

$$\Delta_+ \cup \Delta_\cdot \cup \{\sqrt{2} \cdot \sqrt{2} \neq 2 \vee P(\sqrt{3})\} \vdash P(\sqrt{3}) .$$

However, the new number $\sqrt{2}$ introduced cannot be found combinatorially from $2, \sqrt{3}, +, \cdot$; so one could not establish (1) in the absence of the S-scheme. The restricted $\forall$-elimination rule must be stated thus:

$$\Psi \cup \{\forall x \varphi(x)\} \vdash \theta \text{ if } \Psi \cup \{\forall x \varphi(x), \varphi(t)\} \vdash \theta$$

for some term t in which no name for an element of B occurs except those generated from the constant and function symbols occuring in $\Psi \cup \{\forall x \varphi(x), \theta\}$. (If θ is of the form $f(b_1, \ldots, b_n) = c$, the constant symbol c must be excluded from those "occuring" in $\Psi \cup \{\forall x \varphi(x), \theta\}$ unless it occurs elsewhere. This exception is made because the value(s) of a function being computed cannot be assumed to be given *a priori*. In this regard, n-ary multivalued functions are computationally different from $(n+1)$-ary relations.)

It should be noted that in those structures $\langle B; P_1, \ldots, P_n\rangle$ in which B is generated by $P_1, \ldots, P_n$ (e.g., $\langle \omega; 0, s, =\rangle$), the S-scheme is redundant. To further exemplify the irredundance of the S-scheme, consider the structure ⟨plane; compass, straightedge⟩. The functions "midpointing" and "angle-bisecting" are ( )-recursive in this structure; but "angle-trisecting" is not (C, E)-recursive in this structure, though it is (E, S)-recursive in this structure. More precisely, let A be the set of points in the plane, and let F, G, H be the following 0-, 1- or 2-valued functions on A:

$F(a, b, c, d)$ = point of intersection of line *ab* and line *cd*

$G(a, b, c, d)$ = point(s) of intersection of line *ab* and circle *cd*

$H(a, b, c, d)$ = point(s) of intersection of circle *ab* and circle *cd*

(Circle *ab* denotes the circle with center a passing through b.) Typical functions ( )-recursive in $\langle A; F, G, H\rangle$ are:

$M(a, b)$ = midpoint of segment *ab*

$B(a, b, c)$ = point d on line *ab* with the property that angle *acd* = angle *dcb*.

It can be shown that if J is (E)-recursive in $\langle A; F, G, H\rangle$, then $J(a_1, \ldots, a_n)$ takes on only finitely many values each of which is constructable from $a_1, \ldots, a_n$ using only compass and straightedge. It follows that the following function is not (E)-recursive or even (C, E)-recursive in $\langle A; F, G, H\rangle$:

$T(a, b, c)$ = point d on line *ab* with property angle *acb* = 3 angle *acd*.

However, T is (E, S)-recursive in $\langle A; F, G, H\rangle$ with description

$$\forall abcde\,(B(a,e,c) \neq d \vee B(d,b,c) \neq e \vee T(a,b,c) = d)$$

$$\&\ (\text{description of function } B)\ \&\ (\text{some equality axioms}).$$

In the absence of the S-scheme but in the presence of the A-scheme, one may allow a $\forall$-elimination rule of another sort:

$$\Psi \cup \{\forall x \varphi(x)\} \vdash \theta \ \text{ if, for all terms } t\,,$$

$$\Psi \cup \{\forall x \varphi(x), \varphi(t)\} \vdash \theta \ \text{ or } \ \Psi \cup \{\forall x \varphi(x), \neg\varphi(t)\} \vdash \theta$$

$$\text{and, for at least one term } t,\ \Psi \cup \{\forall x \varphi(x), \varphi(t)\} \vdash \theta.$$

Definition. Q is semi-(C,E,S,A,I)-recursive in $\langle B;P_1, ..., P_n\rangle$ if, for some φ,

$$\Delta_= \cup \Delta_{P_1} \cup ... \cup \Delta_{P_n} \cup \{\varphi\} \vdash Q(a) \ \text{ iff } \ Q(a)\,.$$

Normally we will talk of only *total* predicates as being semi-recursive whereas we allow *partial* predicates that are recursive. The removal of the schemes C,E,S,A,I in semi-recursion is just as in recursion. It should be fairly clear that a total predicate is (...)-recursive iff it and its negation are semi-(...)-recursive.

3. Semantic description of recursion: implicit definability

Each of the 32 variations of recursion that we discussed had something of the following form:

Q is recursive in $\langle B;P_1, ..., P_n\rangle$ if there is a description φ such that

$$\Delta_= \cup \Delta_{P_1} \cup ... \cup \Delta_{P_n} \cup \{\varphi\} \vdash \psi \ \text{ for } \ \psi \in \Delta_Q$$

and an expansion of $\langle B;P_1, ..., P_n\rangle$ is a model of φ.

These variations differed in that the meaning of $\vdash$ was varied. By I.C and I.E of [4], we see that $\vdash$ is complete at least in the case B is countable. Thus we can replace the syntactic $\vdash$ by the semantic $\models$. However,

$$\Delta_= \cup \Delta_{P_1} \cup ... \cup \Delta_{P_n} \cup \{\varphi\} \models \psi \ \text{ for } \ \psi \in \Delta_Q$$

simply means that every model of φ that interprets the symbols $=, P_1, \ldots, P_n$ as the relations (functions) $=, P_1, \ldots, P_n$ must interpret the symbol Q as the relation (function) Q. In other words, by replacing $\vdash$ by $\models$ we get the following semantic formulation of recursion:

Q is recursive in $\langle B; P_1, \ldots, P_n \rangle$ if there is a description φ such that some model of φ interprets $P_1, \ldots, P_n$ correctly and every such model interprets Q correctly. In effect, φ is *an implicit definition* of Q in terms of $P_1, \ldots, P_n$.

Each of the 32 variations of recursion has such an implicit definability form. They differ only in the kind of description allowed and in the kind of models allowed. These variations can be outlined very easily as follows.

Full (C,E,S,A,I)-recursion. φ may have $\forall, \exists, I, C$ quantifiers. Universes of models must be B.

Removal of the I-scheme. I and C quantifiers are not allowed in φ.

Removal of the A-scheme. Universes of models need not be B.

Removal of the S-scheme. Models that expand substructures of $\langle B; P_1, \ldots, P_n \rangle$ are allowed. One must then say that every model with universe C that interprets $P_1 \upharpoonright C, \ldots, P_n \upharpoonright C$ correctly interprets $Q \upharpoonright C$ correctly.

Removal of the C-scheme. φ may not have names of elements of B in it other than those in the list $P_1, \ldots, P_n$.

Removal of the E-scheme. Somehow one must allow models that interpret each element of B as many objects.

Historical note: Fraïssé's [1] definition of abstract recursion is essentially the implicit definability one. Kreisel [6] lays great emphasis on implicit definability. Indeed model-theoretic formulations of recursion have an air of completeness lacking in their combinatorial counterparts and thus add great weight to the validity of Church's Thesis. The equivalence of implicit definability with Moschovakis' scheme formulation of recursion is proved by Moschovakis [11], and also in the hyperarithmetical case by Grilliot [3].

4. Semantic description of recursion: quantifier form

We wish to generalize the classical results that semi-recursive predicates have $\mathbf{\Sigma}_1^0$ quantifier form and semi-hyperarithmetical predicates have $\mathbf{\Pi}_1^1$ quantifier form. The latter result is virtually equivalent to the result that the hyperarithmetical functions are precisely the unique solutions of $\mathbf{\Sigma}_1^1$ relationals. However, we must be more careful in the abstract case of describing quantifier forms than just by $\mathbf{\Pi}_n^m$ and $\mathbf{\Sigma}_n^m$, because we have the I and C quantifiers and also the contents of the matrix must be specified. For instance, if only $0, s, =$ are allowed in the matrix (and hence not $+, \cdot$, bounded quantifiers), then not every semi-recursive predicate is $\mathbf{\Sigma}_1^0$ but rather strict $-\mathbf{\Pi}_1^1$. Let bold-faced quantifiers ($\boldsymbol{\exists}, \boldsymbol{\forall}$) denote a list of second-order quantifiers (quantifiers over relations of objects) of that kind. Thus every $\mathbf{\Sigma}_1^1$ predicate has the form $\boldsymbol{\exists}\forall\exists$ over $\langle\omega; 0, s, =\rangle$ and every strict $-\mathbf{\Sigma}_1^1$ predicate has the form $\boldsymbol{\exists}\forall$ over $\langle\omega; 0, s, =\rangle$. The general results we would like are:

(1) Every predicate $\boldsymbol{\forall}\exists$ over $\langle B; P_1, ..., P_n\rangle$ is semi-(C,E,S)-recursive in $\langle B; P_1, ..., P_n\rangle$;

(2) every predicate $\boldsymbol{\forall}\exists I$ over $\langle B; P_1, ..., P_n\rangle$ is semi-(C,E,S,I)-recursive in $\langle B; P_1, ..., P_n\rangle$;

(3) every predicate $\boldsymbol{\forall}\exists\forall$ (also $\boldsymbol{\forall}\exists\forall\exists$, $\boldsymbol{\forall}\exists\forall\exists\forall$, etc.) over $\langle B; P_1, ..., P_n\rangle$ is semi-(C,E,S,A)-recursive in $\langle B; P_1, ..., P_n\rangle$;

(4) every predicate $\boldsymbol{\forall}\exists I\forall$ (also $\boldsymbol{\forall}$ (any list of first-order quantifiers)) over $\langle B; P_1, ..., P_n\rangle$ is semi-(C,E,S,A,I)-recursive in $\langle B; P_1, ..., P_n\rangle$.

We must assume that B is countable. The converses of these assertions will be considered shortly. To see (2), suppose $Q(a) \leftrightarrow \forall R_1 ... \forall R_m \varphi(a)$ where φ in prenex form is $\exists I$ and $R_1, ..., R_m$ are the relation symbols other than $P_1, ..., P_n, Q$ in φ. It follows that Q is (C,E,S,I)-recursive in $\langle B; P_1, ..., P_n\rangle$ with description $\forall x(\varphi(x) \to Q(x))$, which is a $\forall C$ sentence. For let $\mathfrak{A}$ be a model of this description. Then $\mathfrak{A} \mid B$ is also a model of it because it has quantifier form $\forall C$. If $Q(a)$ is true, then $\varphi(a)$ is true in every model with universe B and so $Q(a)$ is true in $\mathfrak{A} \mid B$ and hence in $\mathfrak{A}$.

The converses of (3) and (4) hold as is seen as follows. Suppose Q is semi-(C,E,S,A,I)-recursive in $\langle B; P_1, ..., P_n\rangle$ with description φ. First note that φ can be reduced to another description with quantifier form $\forall C$ by adding new intermediary relations. Indeed, any I-quantifier can be eliminated by noting that

$$\forall x\, \exists y\, (x < y \,\&\, \psi(y)) \,\&\, (< \text{ is irreflexive, transitive})$$

is a conservative extension of $Ix\,\psi(x)$, where $<$ is a new relation symbol. Next a description $\forall x\, \exists y\, \forall z\, Cu\, \exists v\, \Psi$ has the following conservative extension:

$$\forall x R_1(x) \,\&\, \forall x (R_1(x) \rightarrow \exists y R_2(x,y)) \,\&$$

$$\forall x \forall y (R_2(x,y) \rightarrow \forall z R_3(x,y,z)) \,\&$$

$$\forall x \forall y \forall z (R_3(x,y,z) \rightarrow Cu R_4(x,y,z,u)) \,\&$$

$$\forall x \forall y \forall z \forall u (R_4(x,y,z,u) \rightarrow \exists v \psi)\,.$$

Drawing quantifiers forward judiciously we get $\forall C \exists$ (or $\forall \exists C$) quantifier form. In this manner, any description can be reduced to one in quantifier form $\forall C \exists$. Finally note that if φ is such a description of Q in terms of $P_1, \ldots, P_n$, then $Q(a) \leftrightarrow \forall R_1 \ldots \forall R_m \forall Q (\varphi \rightarrow Q(a))$, where $R_1, \ldots, R_m$ are the predicate symbols in φ other than $Q, P_1, \ldots, P_n$. Thus Q is $\underset{\sim}{\forall} \exists I \forall$ over $\langle B; P_1, \ldots, P_n \rangle$.

We do not know if the converses of (1) and (2) are generally true, but they seem nearly to be true in the following sense. The converse of (1) is equivalent to the statement that every description φ for (C,E,S)-recursion can be made into one that is in $\forall$-form. We know from the discussion above that each description can be made into one that is in $\forall \exists$-form. However, $\exists$-quantifiers are fairly inert in the absence of strong $\exists$-elimination.

Historical note. Montague's [9] formulation of abstract recursion lays heavy emphasis on quantifier form. Moschovakis [12] first proved the abstract version of the Kleene–Suslin Theorem (HYP = $\mathbf{\Delta}_1^1$).

5. Completeness of C, E, S, A, I schemes

We will show that (C,E,S)-recursion is the strongest abstract recursion that generalizes usual recursiveness on the natural numbers, and that (C,E,S,A,I)-recursion is the strongest abstract recursion that generalizes usual hyperarithmeticalness on the natural numbers. To see this, suppose Q is recursive, in some sense, in $\langle B; P_1, ..., P_n \rangle$ where B is countable. Then Q is recursive in that sense in $\langle B; P_1, ..., P_n, 0, s, = \rangle$ for any $0 \in B$ and any successor function s on B (s is a successor function on B if s is injective and $B = \{0, s0, ss0, sss0, ...\}$). It follows that, for any bijection from B to ω, the transplant of Q is recursive in the usual sense in the transplants of $P_1, ..., P_n$. By the following assertion, Q is (C,E,S)-recursive in $\langle B; P_1, ..., P_n \rangle$. Let B be countable. Then

(1) Q is (C,E,S)-recursive in $\langle B; P_1, ..., P_n \rangle$ iff, for every bijection from B to ω, the transplant of Q is recursive (in the usual sense) in the transplants of $P_1, ..., P_n$ (bijection may be changed to injection);

(2) Q is (C,E,S,I)-recursive in $\langle B; P_1, ..., P_n \rangle$ iff, for every injection from B to ω, the transplant of Q is hyperarithmetical in the transplants of $P_1, ..., P_n$;

(3) Q is (C,E,S,A,I)-recursive in $\langle B; P_1, ..., P_n \rangle$ iff, for every bijection from B to ω, the transplant of Q is hyperarithmetical in the transplants of $P_1, ..., P_n$.

The proofs of (2) and (3) are nearly identical, so let us consider (2). The necessity is straightforward: if Q is (C,E,S,I)-recursive in $\langle B; P_1, ..., P_n \rangle$, then Q is (C,E,S,I)-recursive in $\langle C; P_1, ..., P_n \rangle$ for all countable $C \supseteq B$, and so Q is (C,E,S,I)-recursive in $\langle C; P_1, ..., P_n, 0, s, = \rangle$ for all countable $C \supseteq B, 0 \in C$ and successor functions s on C; this is equivalent to saying that, for each injection from B to ω, the transplant of Q is hyperarithmetical in the transplants of $P_1, ..., P_n$. Conversely suppose that, for each injection from B to ω, the transplant of Q is hyperarithmetical in the transplants of $P_1, ..., P_n$. Consider the theory $\mathbf{\Delta}$ whose symbols include names for elements of B and $0, s, <$ and whose axioms are $\mathbf{\Delta}_= \cup \mathbf{\Delta}_{P_1} \cup ... \cup \mathbf{\Delta}_{P_n}$ plus the usual axioms of arithmetic concerning $0, s, <$ plus the axiom $\forall x \neg \exists y (y < x)$. $\mathbf{\Delta}$ is clearly (C,E,S,I)-recursive in $\langle B; P_1, ..., P_n \rangle$. Also Q is (C,E,S,A,I)-recursive in every model of $\mathbf{\Delta}$. It follows from II.B of [4] that Q is (C,E,S,I)-recursive in $\langle B; P_1, ..., P_n \rangle$.

Historical note. Lacombe's [8] notion of $\forall$-recursiveness is that Q (on B) is $\forall$-recursive in P (on B) if, for every bijection from B to ω, the transplant of Q is recursive in the transplant of P. Lacombe asserted that $\forall$-recursiveness is equivalent to Fraïssé's invariant definability notion of recursiveness. Moschovakis [11] proved that $\forall$-recursiveness is equivalent to his schematic notion of recursiveness. Grilliot [2] proved the analogue for hyperarithmeticalness.

6. Extending the notion of finiteness

For simplicity, let us ignore the C, E, S, I schemes. The difference between recursion with and recursion without the A-scheme may be regarded as a difference in the idea of finiteness. The A-scheme says that (for some super-being) the universe can be comprehended with ease and so in an extended sense of the word the universe is "finite". There may be structures with sets other than the universe that are to be regarded as "finite" in an extended sense, e.g., admissible sets. It is of interest to define recursion for such a structure. Let $\langle B; P_1, ..., P_n\rangle$ be a countable structure and let $\mathcal{C}$ be a countable collection of subsets of B that are to be regarded as "finite". The definition of Q being recursive in $\langle B; P_1, ..., P_n\rangle$ is just as usual except that the A-scheme is dropped and the inductive definition of $\vdash$ allows for clauses that comprehend the elements of $\mathcal{C}$; namely, for $C \in \mathcal{C}$, $\Psi \vdash \theta$ if $\Psi \cup \{\varphi(c)\} \vdash \theta$ for all $c \in C$ and $\exists x \in C \varphi(x) \in \Psi$.

However, a problem of greater interest is the following. Given a structure $\langle B; P_1, ..., P_n\rangle$ and $\mathcal{C}$ a collection of "finite" subsets of B, what other subsets of B must necessarily be regarded as "finite"? In other words, how does one effect closure under the idea of "finiteness"? For example, if ω is regarded as "finite" for the structure $\langle \omega; =\rangle$, then any finite or cofinite subset of ω must be regarded as "finite", but there is no reason to regard the set of even numbers as "finite". On the other hand, if the structure is expanded to $\langle \omega; 0, s, =\rangle$, then the set of even numbers – in fact, any hyperarithmetical set – is readily derivable from the universe and hence must be regarded as "finite" though there is no reason to regard a nonhyperarithmetical set as "finite".

To repeat the question: given $\langle B; P_1, ..., P_n\rangle$ and a collection $\mathcal{C}$ of "finite" sets, characterize all the sets that are thereby "finite". We cite three possible answers to this question. (1) Any subset of B that is both $\boldsymbol{\Sigma}^1_1$-definable and $\boldsymbol{\Pi}^1_1$-definable-on-B in every model of

$$\boldsymbol{\Delta}_{P_1} \cup ... \cup \boldsymbol{\Delta}_{P_n} \cup \{\bigwedge_{C \in \mathcal{C}} \forall x (x \notin C \vee \bigvee_{c \in C} x = c)\}\,.$$

The definitions of definable and definable-on-B are given on pp. 122–122 of [7]. The definitions are easily adapted to allow second-order quantifiers.
(2) Any subset of B that belongs to every model of the set displayed above plus axioms that more or less state that two-element sets are "finite", "finite" sets are closed under subset, and "finite" unions of "finite" sets are "finite".

These axioms are chosen because they are virtually complete for the notion of "countable" (see I.F of [4]) and countability is completely degenerated finiteness. (3) Any subset of B that can be built up from $\mathcal{C}$ using effectivized versions of the axioms: two-element sets are "finite", "finite" sets are closed under subset, and "finite" unions of "finite" sets are "finite". For arbitrary structures we do not know how these three answers compare.

References

[1] R. Fraïssé, Une notion de récursivité relative, Infinitistic methods (Pergamon Press, Oxford, 1961) 323–328.

[2] T.J. Grilliot, Omitting types: application to recursion theory, J. Symbolic Logic, vol. 37 (1972) 81–89.

[3] T.J. Grilliot, Implicit definability and hyperprojectivity, Scripta Mathematica, to appear.

[4] T.J. Grilliot, Model theory for dissecting recursion theory, This volume.

[5] S.C. Kleene, Recursive functionals and quantifiers of finite types, I, Trans. Amer. Math. Soc., vol. 91 (1959) 1–52.

[6] G. Kreisel, Model theoretic invariants: application to recursive and hyperarithmetic operations, in: J. Addison et al. (eds.) The Theory of Models (North-Holland, Amsterdam, 1965) 190–205.

[7] G. Kreisel and J.L. Krivine, Elements of mathematical logic (North-Holland, Amsterdam, 1967).

[8] D. Lacombe, Deux généralizations de la notion de récursivité relative, Comptes Rendus de l'Academie des Sciences de Paris, vol. 258 (1964) 3410–3413.

[9] R. Montague, Recursion theory as a branch of model theory, in: B. van Rootselaar and J.F. Staal (eds.) Logic, methodology and philosophy of science III (North-Holland, Amsterdam, 1968) 63–68.

[10] Y.N. Moschovakis, Abstract first order computability, Trans. Amer. Math. Soc., vol. 138 (1969) 427–504.

[11] Y.N. Moschovakis, Abstract computability and invariant definability, J. Symbolic Logic, vol. 34 (1969) 605–633.

[12] Y.N. Moschovakis, The Suslin–Kleene Theorem for countable structures, Duke Math. J., vol. 37 (1970) 341–352.

J.E.Fenstad, P.G.Hinman (eds.), Generalized Recursion Theory

MODEL THEORY FOR DISSECTING RECURSION THEORY

Thomas J. GRILLIOT
The Pennsylvania State University

This paper summarizes some results of model theory that have an impact in examining recursion theory. No attempt has been made to document the results.

I. Completeness Theorems

A. Usual predicate calculus with $\&, \vee, \forall, \exists, \neg$.

First of all we make a blanket assumption that any set of axioms/rules is strong enough to drive $\neg$'s through other connectives so that, if needed, we may assume that $\neg$'s only appear before atoms. We show how a consistent set Φ of sentences has a model. The key to this demonstration as well as most of those following is the formation of a set Φ_∞ of sentences with the following properties: (a) $\Phi \subseteq \Phi_\infty$; (b) if $\varphi \;\&\; \psi \in \Phi_\infty$, then $\varphi, \psi \in \Phi_\infty$; (c) if $\varphi \vee \psi \in \Phi_\infty$, then $\varphi \in \Phi_\infty$ or $\psi \in \Phi_\infty$; (d) if $\forall x \varphi(x) \in \Phi_\infty$ and t is a term formed from the function symbols of Φ_∞, then $\varphi(t) \in \Phi_\infty$; (e) if $\exists x \varphi(x) \in \Phi_\infty$, then $\varphi(c) \in \Phi_\infty$ for some constant symbol c. Assuming that we have such a Φ_∞, let Φ^o_∞ be the set of all atomic or negated atomic sentences in Φ_∞. We can make the following observations: (1) since $\Phi \subseteq \Phi_\infty$ and $\Phi^o_\infty \subseteq \Phi_\infty$, any model of Φ_∞ is also a model of both Φ and Φ^o_∞, and the consistency of Φ_∞ implies the consistency of both Φ and Φ^o_∞; (2) since Φ^o_∞ consists of only atoms or negated atoms, its consistency implies that it has a model; (3) by induction on the number of connectives in a sentence of Φ_∞, one readily sees that a model $\mathfrak{A}$ of Φ^o_∞ with universe $\{t_{\mathfrak{A}}: t$ is a term formed from the function symbols of $\Phi_\infty\}$ is a model for all of Φ_∞. From these three observations it follows that the consistency of Φ implies that it has a model provided we can form Φ_∞ in such a way that the consistency of Φ implies the consistency of Φ_∞. We form Φ_∞ as a union of sets $\Phi_0, \Phi_1, \Phi_2, \ldots$ as follows. For simplicity assume Φ is countable. Let Φ_0 be Φ. Pick a sentence from Φ_0 with a connective other than $\neg$. If the sentence is $\varphi \,\&\, \psi$, then $\Phi_0 \cup \{\varphi, \psi\}$ must be consistent, so let Φ_1 be this set. If the sentence is $\varphi \vee \psi$, then $\Phi_0 \cup \{\varphi\}$ or $\Phi_0 \cup \{\psi\}$ is consistent, so let Φ_1 be the one that is consistent. If the sentence is $\forall x\, \varphi(x)$, then $\Phi_0 \cup \{\varphi(t)\}$ must be consistent for any term t, so let Φ_1 be $\Phi_0 \cup \{\varphi(t)\}$ where t is a variableless term. If the sentence is $\exists x\, \varphi(x)$, then $\Phi_0 \cup \{\varphi(c)\}$ must be consistent where c is a new constant symbol, so let Φ_1 be this set. In a similar manner, form $\Phi_2, \Phi_3, \Phi_4, \ldots$ in such a way that every connective (and every variableless term in connection with $\forall$) is acted upon at some stage. Then $\Phi_\infty = \cup\, \Phi_n$ is the desired set.

B. Predicate calculus with equality.

The following axioms when added to the usual predicate calculus axioms/rules are complete for =: $\forall x(x = x)$; $x = y \rightarrow \theta(x) \leftrightarrow \theta(y)$. This is proved just as in the preceding section except that the additional conditions are placed on Φ_∞: (f) $t = t \in \Phi_\infty$ for all t formed from function symbols of Φ_∞; (g) if $\theta(t) \in \Phi_\infty$ and $t = u \in \Phi_\infty$ or $u = t \in \Phi_\infty$, then $\theta(u) \in \Phi_\infty$. The axioms mentioned above are just enough to permit the formation of Φ_∞ with these two new conditions.

C. ω-logic.

Let A be a countable set whose elements have names in some countable language such that $a \neq b$ is an axiom when $a, b \in A$ and $a \neq b$. If one is interested in restricting the meaning of $\forall$ and $\exists$ to quantification over A, then the following infinitary rules when added to usual axioms/rules are complete: $\{\theta(a) : a \in A\}/\forall x\, \theta(x)$. The proof is as in the preceding section except that condition (e) concerning Φ_∞ must be revised to read: (e) if $\exists x \varphi(x) \in \Phi_\infty$, then $\varphi(a) \in \Phi_\infty$ for some $a \in A$. (One may have to add the axiom $\forall x \exists y\,(x = y)$ to Φ to make the details work out readily.) Several adaptations can be made. For example, add A as a unary predicate symbol to the language. To make $A(x)$ have the meaning $x \in A$ in any model, the following infinitary rules are complete: $\{\theta(a) : a \in A\} / \forall x(\neg A(x) \vee \theta(x))$. One can achieve the analogous result using countably many sets simultaneously.

D. Infinitary &'s and ∨'s.

One may wish to incorporate countably many infinitary conjunctions $\bigwedge$ and disjunctions $\bigvee$ in a countable language. For this modification, the following axioms/rules are complete: rules allowing $\neg$'s to be driven through $\bigwedge$'s and $\bigvee$'s according to deMorgan's Laws; $\theta_n \to \bigvee \theta_i$ for any n; $\{\theta_0, \theta_1, \ldots\} / \bigwedge \theta_i$. The proof is as in the case of the completeness of usual predicate calculus except that conditions (b) and (c) concerning Φ_∞ are revised to read: (b) if $\bigwedge \varphi_i \in \Phi_\infty$, then $\varphi_i \in \Phi_\infty$ for all i; (c) if $\bigvee \varphi_i \in \Phi_\infty$, then $\varphi_i \in \Phi_\infty$ for some i.

E. The infinity quantifier.

One may wish to add two new quantifiers Ix and Cx to a countable language with intended meanings of "for infinitely many x" and "for co-finitely many x". (Note: Ix and $\neg Cx \neg$ have same intended meaning.) For this modification, the following axioms/rules when added to the usual axioms/rules are complete: rules allowing $\neg$'s to be driven through I's and C's according to deMorgan's Laws; $\{\exists^{\geq 1} x\theta(x), \exists^{\geq 2} x\theta(x), \ldots\} / Ix\theta(x)$; $\forall^{\geq n} x\theta(x) \to Cx\theta(x)$ for any n, where $\exists^{\geq n} x\theta(x)$ is an abbreviation for $\exists x_1 \ldots \exists x_n(\theta(x_1) \,\&\, \ldots \,\&\, \theta(x_n) \,\&\, x_1 \neq x_2 \,\&\, \ldots)$ and $\forall^{\geq n} x\theta(x)$ is an abbreviation for $\neg\exists^{\geq n} x \neg\theta(x)$. The proof is as in the case of completeness of usual predicate calculus except that the following two additional conditions must be added concerning Φ_∞: (h) if $Ix\varphi(x) \in \Phi_\infty$, then $\exists^{\geq n} x\varphi(x) \in \Phi_\infty$ for every n; (i) if $Cx\varphi(x) \in \Phi_\infty$, then $\forall^{\geq n} x\varphi(x) \in \Phi_\infty$ for some n.

F. The uncountability quantifier.

One may wish to add two new quantifiers UCx and CCx to a countable language with the intended meanings of "for uncountably many x" and "for cocountably many x". Surprisingly, the additional rules needed for these quantifiers are finitary, and the completeness proof is quite different from those preceding. To simplify matters, use a two-sorted language with x, y, z

denoting type-0 variables and X, Y, Z type-1 variables and $\in$ a binary relation relation between type-0 and type-1. Let $UCx\theta(x)$ be an abbreviation for $\neg\exists X\forall y(\theta(y) \leftrightarrow y \in X)$ and let $CCx\theta(x)$ be an abbreviation for $\neg UCx\neg\theta(x)$. The following axioms are sufficient to make UC and CC have their intended meanings: $\exists X(y \in X \,\&\, z \in X)$; $\exists Y\forall z(z \in Y \leftrightarrow z \in X \,\&\, \theta(z))$; $\forall y \in X\, CCz\theta(y,z) \rightarrow CCz\forall y \in X\,\theta(y,z)$. Informally, these axioms state that all two-element sets are countable, countable sets are closed under subsets and countable unions of countable sets are countable. The middle axioms (subset axioms) can be dispensed with if $UCx\theta(x)$ is used as an abbreviation for $\neg\exists X\forall y(\theta(y) \rightarrow y \in X)$ instead of for $\neg\exists X\,\forall y(\theta(y) \leftrightarrow y \in X)$. Suppose Φ is a set of sentences that includes the above axioms and the axiom of extensionality. We want to show that its consistency implies that it has a model in which UC and CC have their intended meanings. By the standard completeness theorem, Φ has a countable model $\mathfrak{A}_0$, say with type-0 universe A_0 and type-1 universe $\bar{A}_0$. $\bar{A}_0$ may be regarded as a subset of the power set of A_0. Let $\{\varphi_i(x)\}$ be the collection of all formulas in the language of $\mathfrak{A}_0$ in one free variable x such that $\mathfrak{A}_0 \models CCx\,\varphi_i(x)$, and let $\psi(x)$ be any formula such that $\mathfrak{A}_0 \models UCx\,\psi(x)$. By the compactness theorem, $\mathfrak{A}_0$ clearly has an "extension" $\mathfrak{A}_1$ in which $\mathfrak{A}_1 \models \exists x[\psi(x) \,\&\, \bigwedge\varphi_i(x)]$. The trouble is that the compactness theorem only insures that the type-0 universe A_1 of $\mathfrak{A}_1$ includes A_0 and not that the type-1 universe $\bar{A}_1$ of $\mathfrak{A}_1$ includes $\bar{A}_0$. In order for $\bar{A}_0 \subseteq \bar{A}_1$, some types omitted in $\mathfrak{A}_0$ must remain omitted in $\mathfrak{A}_1$. We are thus faced with the problem of superimposing compactness upon omitting-types. This is possible thanks mainly to the countable union axioms, but the details will be left to a more appropriate section (section III.A). Thus $\mathfrak{A}_0$ has a *bone fide* elementary extension $\mathfrak{A}_1$ with the property that $\mathfrak{A}_1 \models \exists x[\psi(x) \,\&\, \bigwedge\varphi_i(x)]$. In like manner one forms $\mathfrak{A}_2, \mathfrak{A}_3, \ldots$ and indeed $\mathfrak{A}_\sigma$ for any countable successor ordinal σ. If σ is a limit ordinal, one lets $\mathfrak{A}_\sigma$ be $\bigcup_{\tau<\sigma}\mathfrak{A}_\tau$. $\mathfrak{A}_{\omega_1}$ is the desired model.

II. Satisfying an infinite sentence

A. Compactness ($\exists\bigwedge$).

Let Φ be a set of sentences, for simplicity, countable. The problem of compactness is to find a sufficient condition for when Φ has a model of an infinite conjunction or, what is virtually equivalent, an existentially quantified infinite conjunction. The answer is that $\Phi \cup \{\exists x \bigwedge_{i\in\omega}\varphi_i(x)\}$ has a model if, for each n, $\Phi \cup \{\exists x \bigwedge_{i\leq n}\varphi_i(x)\}$ has a model. (We have simplified the situation by considering only countable conjuncts and one existential quantifier.) To see this, let $\Sigma = \Phi \cup \{\varphi_i(c) : i \in \omega\}$ where c is a new constant symbol. Clearly every finite subset of Σ has a model. Form Σ_∞ with the usual closure conditions (see section I.A) plus the condition that every finite subset of Σ_∞ has a model. Since Σ^0_∞ consists only of atoms and negated atoms, it has a model because every finite subset of it has a model. Therefore, Σ_∞ and hence Σ and hence $\Phi \cup \{\exists x \bigwedge_{i\in\omega} \varphi_i(x)\}$ has a model.

B. Omitting types ($\forall\bigvee$).

Let Φ be a countable set of sentences. The problem of omitting types is to find a sufficient condition for when Φ has a model that is also a model of a universally quantified infinite disjunction. The usual answer is that $\Phi \cup \{\forall x \bigvee_{i\in\omega} \varphi_i(x)\}$ has a model if Φ has a model and, for all $\psi(x)$ such that $\Phi \cup \{\exists x \psi(x)\}$ has a model, $\Phi \cup \{\exists x(\psi(x) \;\&\; \varphi_i(x))\}$ has a model for some i. (We have simplified the situation by considering only one universal quantifier.) To see this, form Φ_∞ from Φ so that Φ_∞ has a model and satisfies the usual closure conditions plus the condition: if t is formed from the function symbols of Φ_∞, then $\varphi_i(t) \in \Phi_\infty$ for some i. The hypothesis above is adequate to insure that this new condition on Φ_∞ can be achieved. As is noted in section I.A, if Φ_∞ has a model $\mathfrak{A}$, then it has a submodel whose universe is $\{t\mathfrak{A} : t$ is formed from the function symbols of $\Phi_\infty\}$. This submodel is a model of $\Phi \cup \{\forall x \bigvee_{i\in\omega} \varphi_i(x)\}$. Variations can be made for languages with ω-rules, infinitary conjuncts and disjuncts and quantifiers Ix, Cx. Also, adding an infinite conjunct before the universally quantified infinite disjunction is no problem. Thus $\Phi \cup \{\bigwedge_{j\in\omega} \forall x \bigvee_{i\in\omega} \varphi_{ij}(x)\}$ has a model if Φ has a model and, for all j, for all $\psi(x)$ such that $\Phi \cup \{\exists x\, \psi(x)\}$ has a model, $\Phi \cup \{\exists x(\psi(x) \;\&\; \varphi_{ij}(x))\}$ has a model for some i. Another variation can be made for second-order quantifiers. Let the quantifiers $\exists p$, $\forall p$ vary through all relations of a specified number of arguments. Assume that p does not occur in Φ. $\Phi \cup \{\forall x(\bigvee_{i\in\omega} \exists p \varphi_i(x) \vee \bigvee_{i\in\omega} \forall p \psi_i(x))\}$ has a model if Φ has a model and, for all $\theta(x)$ in which p does not occur – though relation symbols other than those in Φ may – with the property that $\Phi \cup \{\exists x \theta(x)\}$ has a model, $\Phi \cup \{\exists x(\theta(x) \;\&\; \varphi_i(x))\}$ has a model for some i or $\Phi \cup \{\exists x(\theta(x) \;\&\; \neg\psi_i(x))\}$ has no model for some i. An important application is the following. Let Φ include in its language names for the elements of a countable set A. Then any subset of A that is $\mathbf{\Pi}^1_1$-definable-on-A in every model of Φ is semi-representable in some finite extension of Φ.

C. Omitting compactifiable types ($\forall\bigvee\exists\bigwedge$).

Let Φ be a countable set of sentences. $\Phi \cup \{\forall x \bigvee_{i\in\omega} \exists y \bigwedge_{j\in\omega} \varphi_{ij}(x,y)\}$ has a model if Φ has a model and, for all $\psi_0(x)$, $\psi_1(x)$, $\psi_2(x)$, ... such that $\Phi \cup \{\exists x \bigwedge_{n\leq j} \psi_n(x)\}$ has a model for each j, $\Phi \cup \{\exists x \exists y \bigwedge_{n\leq j}(\psi_n(x) \,\&\, \varphi_{in}(x,y))\}$ has a model for some i and all j. The proof is quite similar to the one outlined in the preceding section except that the formation of Φ_∞ is made so that the following condition holds: if t is formed from the function symbols of Φ_∞, then $\varphi_{ij}(t,c) \in \Phi_\infty$ for some i and all j, where c is a constant symbol. Variations of this result exist also. One important application is the following. Let Φ include in its language names for the elements of a countable set A. Then any subset of A that is Σ_1^1-definable in every model of Φ is finite. In fact, there exists one model of Φ in which every Σ_1^1-definable subset of A is finite.

III. Compactness after omitting types

A. Application to uncountability quantifier.

In section I.F we were confronted with the following situation. Given a second-order model $\mathfrak{A}$ in which $\mathfrak{A} \models CCx\,\varphi_i(x)$ for each i and $\mathfrak{A} \models UCx\,\psi(x)$, find an elementary extension $\mathfrak{B}$ in which $\exists x(\psi(x) \,\&\, \bigwedge_{i\in\omega}\varphi_i(x))$ is true. Since $\exists x(\psi(x) \,\&\, \bigwedge_{i\leq n}\varphi_i(x))$ is true in $\mathfrak{A}$ for all n, this would follow trivially from the compactness theorem if it were not that we insist that the type-1 universe of $\mathfrak{B}$ include the type-1 universe of $\mathfrak{A}$. In other words, if W is a type-1 object of $\mathfrak{A}$ and $a_0, a_1, \ldots$ are the type-0 objects of $\mathfrak{A}$ for which $\mathfrak{A} \models a_i \in W$, then the universally quantified infinite disjunction $\forall x(x \notin W \vee \bigvee_{i\in\omega} x = a_i)$ must hold in $\mathfrak{B}$. Since we are making infinitary requirements on $\mathfrak{B}$ of the sort $\forall\bigvee$, the usual compactness theorem is of no value by itself. Compactness must be achieved by using the axioms satisfied by $\mathfrak{A}$, especially the countable union axioms. Let Φ be the set of θ for which $\mathfrak{A} \models \theta$ together with $\psi(c)$, $\varphi_i(c)$ for $i \in \omega$, where c is a new constant symbol. We assume that $\varphi_0(x)$, $\varphi_1(x)$, ... is a complete list of those one-place formulas such that $CCx\,\varphi_i(x)$ is true in $\mathfrak{A}$. We need to show that $\Phi \cup \{\bigwedge_{W\in\mathfrak{A}} \forall x(x \notin W \vee \bigvee_{a\in W} x = a)\}$ has a model. If it does not, then by section II.B there exists $\theta(x,c)$ such that $\Phi \cup \{\exists x\theta(x,c)\}$ has a model, but that, for some $W \in \mathfrak{A}$, $\Phi \vdash \forall x(\theta(x,c) \to x \in W)$ and $\Phi \vdash \forall x(\theta(x,c) \to x \neq a)$ for all $a \in W$. Therefore, we have that $CCy[\psi(y) \to \forall x(\theta(x,y) \to x \in W)]$ and $\forall z \in W\, CCy[\psi(y) \to \forall x(\theta(x,y) \to x \neq z)]$ are true in $\mathfrak{A}$. Since $\mathfrak{A}$ satisfies the countable union axioms, the bounded quantifier $\forall z \in W$ can be drawn through the CCy quantifier so that

$CCy[\psi(y) \to \forall x(\theta(x,y) \to x \notin W)]$ is true in $\mathfrak{A}$. Combining this with the one above we get that $CCy[\psi(y) \to \forall x \neg\theta(x,y)]$ is true in $\mathfrak{A}$. This means that one of the $\varphi_i(c)$ is $\psi(c) \to \forall x \neg\theta(x,c)$, contradicting the fact that $\Phi \cup \{\exists x\theta(x,c)\}$ has a model.

B. Barwise compactness.

Let Φ be a countable collection of sentences including, possibly, some with infinite conjunctions and disjunctions. Assume that Φ is closed under components, the finite propositional connectives, and the distribution of $\neg$'s. In other words, $\varphi \in \Phi$ iff $\neg\varphi \in \Phi$, $\varphi \mathbin{\&} \psi \in \Phi$ iff $\varphi, \psi \in \Phi$, $\varphi \vee \psi \in \Phi$ iff $\varphi, \psi \in \Phi$, $\forall x\varphi(x) \in \Phi$ implies $\varphi(t) \in \Phi$ for all variableless terms t formed from a set of function symbols that includes an infinite number of constants, $\exists x\varphi(x) \in \Phi$ implies $\varphi(t) \in \Phi$ for those same terms, $\bigwedge_i \varphi_i \in \Phi$ implies $\varphi_i \in \Phi$ for all i, $\bigvee_i \varphi_i \in \Phi$ implies $\varphi_i \in \Phi$ for all i, $\neg\bigwedge_i \varphi_i \in \Phi$ iff $\bigvee_i \neg\varphi_i \in \Phi$, $\neg\bigvee_i \varphi_i \in \Phi$ iff $\bigwedge_i \neg\varphi_i \in \Phi$, $\neg\forall x\varphi(x) \in \Phi$ iff $\exists x \neg\varphi(x) \in \Phi$, $\neg\exists x\varphi(x) \in \Phi$ iff $\forall x \neg\varphi(x) \in \Phi$. Barwise compactness gives a sufficient condition for when a subset of Φ has a model. Let the notation $\varphi \in \psi$ denote that ψ is a conjunction and that φ is one of its components, and for a subset Σ of Φ let $\psi \subseteq \Sigma$ denote that every $\varphi \in \psi$ is an element of Σ. Let φ, ψ, θ vary through elements of Φ. A subset Σ of Φ has a model if:

(1) every $\varphi \subseteq \Sigma$ has a model;

and, for any ψ, ψ', ψ a conjunction,

(2) if, for every $\varphi \in \psi$, there is a $\theta \subseteq \Sigma$ such that $\theta \models \psi' \vee \varphi$, then there is $\theta \subseteq \Sigma$ such that $\theta \models \psi' \vee \psi$.

Note that if there are no infinite connectives in Φ, then condition (2) is automatic and so Barwise compactness reduces to classical compactness. Condition (2) has the form of a union axiom or a replacement axiom. Indeed condition (2) is automatically satisfied when Φ can be represented by an admissible set in such a way that Σ becomes a Σ_1 set. (In an admissible set, $\models$ is a Σ_1 relation.) To prove Barwise compactness, suppose that Σ satisfies the two conditions above. We form Σ_∞ in the usual manner as outlined in section I.D except that each stage Σ_n must satisfy condition (1). This is obvious for Σ_0 because Σ_0 is Σ. To see that it is true for Σ_1, recall the ways that Σ_1 may be constructed from Σ_0. The difficult one is when $\bigvee_i \varphi_i \in \Sigma_0$ and Σ_1 must be $\Sigma_0 \cup \{\varphi_i\}$ for some i. Such an i can be chosen; for otherwise, for each i, some

$\psi_i \subseteq \Sigma_0 \cup \{\varphi_i\}$ has no model which means that $\psi_i - \{\varphi_i\} \models \neg\varphi_i$; by condition (2), for some $\psi \subseteq \Sigma_0$, $\psi \models \bigwedge_i \neg\varphi_i$; thus $\psi \wedge \bigvee_i \varphi_i \subseteq \Sigma_0$ has no model, a contradiction. One proceeds in this manner to show that every Σ_n satisfies condition (1). (One must form the Σ_n's so that each is only a finite extension of Σ.) It follows that $\Sigma^{\text{o}}_{\infty}$ has a model and hence that Σ has a model.

J.E.Fenstad, P.G.Hinman (eds.), Generalized Recursion Theory
© North-Holland Publ. Comp., 1974

AXIOMATIC THEORY OF ENUMERATION

Andrzej GRZEGORCZYK
University of Warsaw

Sets may be considered to be simpler than functions. Hence I propose to study first an axiomatic theory with a fundamental epsilon-like notion E. (xEy means: x belongs to the set with number parameter y.) The stronger theory of enumeration of functions may be developed later. Besides the relation E we need some other individual constants or functions (primitive recursive in the standard model) as primitive notions. First the pairing function and its inverses:

A0. $\quad K\langle x,y\rangle = x\,, \quad L\langle x,y\rangle = y\,, \quad \langle Kx, Lx\rangle = x\,;$

then the shifting function S and its axiom:

A1. $\quad xES(y,z) \Leftrightarrow \langle x,z\rangle Ey\,.$

The next axiom A2 is a collection of six comprehension schemas (C1–C6). I shall use them in two parallel forms as existential formulas and as definitions of new individual constants (which is the same in the model):

C1	$\bigvee u \bigwedge x(xEu \Leftrightarrow \varphi = \psi)$	C′1	$xEc \Leftrightarrow \varphi = \psi$
C2	$\bigvee u \bigwedge x(xEu \Leftrightarrow \varphi \neq \psi)$	C′2	$xEc \Leftrightarrow \varphi \neq \psi$
C3	$\bigvee u \bigwedge x(xEu \Leftrightarrow \varphi E \psi)$	C′3	$xEc \Leftrightarrow \varphi E \psi$
C4	$\bigvee u \bigwedge x(xEu \Leftrightarrow \bigvee y \langle x,y\rangle Ea)$	C′4	$xEc \Leftrightarrow \bigvee y \langle x,y\rangle Ea$
C5	$\bigvee u \bigwedge x(xEu \Leftrightarrow (xEa \wedge xEb))$	C′5	$xEc \Leftrightarrow (xEa \wedge xEb)$
C6	$\bigvee u \bigwedge x(xEu \Leftrightarrow (xEa \vee xEb)$	C′6	$xEc \Leftrightarrow (xEa \vee xEb)\,.$

In C1–C6 φ, ψ, a and b are terms containing no occurrences of the variable u, but possibly containing some variables as parameters. In C′1–C′6, φ and ψ may contain only $\boldsymbol{x}$ as variable, a and b must be constant terms, and c is a new individual constant.

In order to pass to the enumeration of functions we must postulate the existence of a universal function $\mathcal{U}$ and the Lachlan function $\mathcal{E}$:

A3 a. $\langle\langle x,y\rangle ,z\rangle E\,\mathcal{U} \wedge \langle\langle x,v\rangle ,z\rangle E\,\mathcal{U} \rightarrow y = v$

A3 b. $\bigwedge xyv(\langle x,y\rangle Ez \wedge \langle x,v\rangle Ez \rightarrow y = v) \rightarrow$

$$\bigwedge x(xEz \Leftrightarrow \langle x, \mathcal{E}(z)\rangle E\,\mathcal{U}) .$$

To close the theory we postulate that:

A4 $\bigwedge x \bigvee w,u(\bigwedge z(zEx \Leftrightarrow \bigvee y\langle z,y\rangle Ew) \wedge \bigwedge z(zEu \Leftrightarrow \neg zEw)) .$

(Every element is the projection of a dual element.)

The shifting function S allows us to consider some elements as combinators with respect to the equivalence. If $P(x_1, ..., x_n)$ is a polynomial built of S and $x_1, ..., x_n$, then the combinator associated to P is an element a_P such that the following formula is a theorem:

(1) $xES(... S(a_P, x_1), ..., x_n) \Leftrightarrow xEP(x_1, ..., x_n) .$

Combinatory Property. *For every $P(x_1, ..., x_n)$ there is a combinator associated to P.*

Proof. By C3 we can define a_P to satisfy the formula:

$$\langle ... \langle x,x_n\rangle , ..., x_1\rangle_E\, a_P \Leftrightarrow xEP(x_1, ..., x_n) .$$

Then applying n times A1, we get (1).

I shall mention some other properties.

Fixed point theorem. *There is a function π which produces fixed points:*

$$(2) \qquad zES(x,\pi(x)) \Leftrightarrow zE\pi(x) .$$

Proof. Accordingly to A2 there is an element Ω such that:

$$\langle\langle z,y\rangle,x\rangle E\Omega \Leftrightarrow zES(x,S(y,y)) .$$

Hence by A1:

$$zES(S(\Omega,x),y) \Leftrightarrow zES(x,S(y,y)) .$$

Putting $y = S(\Omega,x)$ we get:

$$zES(S(\Omega,x),S(\Omega,x)) \Leftrightarrow zES(x,S(S(\Omega,x),S(\Omega,x))) .$$

Hence the function: $\pi(x) = S(S(\Omega,x),S(\Omega,x))$ satisfies the formula (2).

Definition. x is *dual* $\Leftrightarrow \bigvee y \bigwedge z(zEy \Leftrightarrow \neg zEx)$.

There are dual elements. There are also elements which are not dual, e.g. the element c satisfying the equivalence:

$$zEc \Leftrightarrow zEz .$$

The supposition that c is dual leads to a contradiction by Russell's argument: if for some $y, \bigwedge z(zEy \Leftrightarrow \neg zEz)$, then: $yEy \Leftrightarrow \neg yEy$.

Definition. x is *finite* $\Leftrightarrow \bigwedge y(\bigwedge z(zEy \rightarrow zEx) \rightarrow y$ is dual).

The intuition is that every infinite set contains a non dual set. Every finite element is of course dual. Every element defined by alternation of identities with constants is finite:

$$xEc \Leftrightarrow (x=a_1 \vee \ldots \vee x=a_n) .$$

Considering the Boolean operations $\cup$, $\cap$, and / as defined by means of E as epsilon, we get that the dual elements constitute a Boolean algebra. On the other hand, having non dual elements we can prove that the complement of c defined above (or defined by: $xEc \Leftrightarrow x \neq x$) is infinite because it contains a non dual element.

There is a sequence of infinitely many different infinite elements:

$$xEa_0 \Leftrightarrow x = x$$

(3)

$$xEa_{n+1} \Leftrightarrow x \neq a_0 \wedge \ldots \wedge x \neq a_n .$$

Proof. a_0Ea_0. Suppose that a_iEa_i for every $i \leqslant n$. Accordingly to the definition $\neg a_iEa_{n+1}$. Hence $a_i \neq a_{n+1}$, and by the definition $a_{n+1}Ea_{n+1}$.

Definition. x is *closed* $\Leftrightarrow \bigwedge y,u(\bigwedge z(zEy \Leftrightarrow zEu) \rightarrow (uEx \Leftrightarrow yEx))$.

Rice's Theorem. *x is closed $\wedge\, a_0Ex \wedge \neg(b_0Ex) \rightarrow x$ is not dual.*

Proof. Suppose x to be dual. Hence for some x':

(4) $$zEx' \Leftrightarrow \neg(zEx) .$$

By A2 there is an h such that:

(5) $$\langle n,y\rangle Eh \Leftrightarrow ((nEx \wedge y = b_0) \vee (nEx' \wedge y = a_0)) .$$

The element h considered as a set of pairs is a function which maps $\{n\colon nEx\}$ to b_0 and $\{n\colon \neg nEx\}$ to a_0. By A2 for h there is an m such that:

(6) $$\langle z,n\rangle Em \Leftrightarrow \bigwedge y(zEy \wedge \langle n,y\rangle Eh) .$$

Using the fixed point theorem we get $n_0 = \pi(m)$ such that:

(7) $$zES(m,n_0) \Leftrightarrow zEn_0 .$$

The element h as a function is total (by (4) and (5)). Hence for n_0 there is a y_0 such that:

(8) $$\langle n_0,y_0\rangle Eh$$

(9) $$\langle n_0,y\rangle Eh \rightarrow y = y_0 .$$

By (6)–(9) and A1 we get that:

(10) $zEn_0 \Leftrightarrow zEy_0$.

When x is closed, (10) implies that:

(11) $n_0Ex \Leftrightarrow y_0Ex$.

On the other hand (4) and (5) imply that:

(12) $\langle n,y\rangle Eh \rightarrow (nEx \Leftrightarrow \neg yEx)$,

and (8) and (12) imply that:

(13) $n_0Ex \Leftrightarrow \neg y_0Ex$.

(11) and (13) give a contradiction.

Non-extensionality theorem. *For every element x and for every n there exist more than n elements which are extensional with x.*

Proof. Suppose that there are only n elements $x_1, ..., x_n$ which are extensional with x. This means that:

(14) $\bigwedge z(zEu \Leftrightarrow zEx) \Leftrightarrow (u=x_1 \vee ... \vee u=x_n)$.

According to C′1 and C′6 there is an element y such that:

$$uEy \Leftrightarrow (u=x_1 \vee ... \vee u=x_n) .$$

By (14) y is closed. It is not empty because xEy, and by (3) there is some v such that $\neg vEy$. According to C′2 and C′5, y is dual. But this contradicts Rice's theorem.

Notice that for the above argument we need Rice's theorem in a uniform formulation. Instead of h take: $S(S(S(H,x), a_0), b_0)$ for suitable H, and instead of m take $S(M,h)$ for suitable M.

The axiom A3 enables us to consider the enumeration of functions. Accordingly to A3 the element U may be considered as a partial function and we can write:

$$U(z,x) = y \text{ instead of } \langle\langle x,y\rangle,z\rangle E\mathcal{U}.$$

We shall use the abbreviations $\cong$ and $\downarrow$:

$$U(z,x)\downarrow \Leftrightarrow \bigvee y \langle\langle x,y\rangle,z\rangle E\mathcal{U}.$$

Kleene's S_n^m-theorem. *There is a shifting function S' such that*:

$$U(S'(a,b),x) \simeq U(a,\langle b,x\rangle).$$

Proof. By A2 there is an element U_0 such that:

(15) $$\langle\langle\langle x,y\rangle,a\rangle,b\rangle E\, U_0 \Leftrightarrow \langle\langle\langle b,x\rangle,y\rangle,a\rangle E\mathcal{U}.$$

According to A1:

(16) $$\langle x,y\rangle ES(S(U_0,b),a) \Leftrightarrow \langle\langle\langle x,y\rangle,a\rangle,b\rangle E\, U_0.$$

By A3a, the element y is unique. Hence, applying A3b, we get that:

(17) $$\langle x,y\rangle ES(S(U_0,b),a) \Leftrightarrow \langle\langle x,y\rangle,\&(S(S(U_0,b),a))\rangle E\mathcal{U}.$$

Putting $S'(a,b) = \&(S(S(a_0,b),a))$ and applying (15)–(17) we get our theorem.

Wagner's URS theorem. *There are two elements α and ψ such that*:
a. $U(\alpha,x)\downarrow \wedge U(U(\alpha,x),y)\downarrow$
b. $U(U(U(\alpha,x),y),z) \simeq U(U(x,z),U(y,z))$
c. $U(\psi,x)\downarrow \wedge U(U(\psi,x),y)\downarrow \wedge U(U(U(\psi,x),y),z)\downarrow$
d. $U(U(U(U(\psi,x),y),z),z) = x$
e. $z \neq v \rightarrow U(U(U(U(\psi,x),y),z),v) = y$.

Proof. By A2 there are elements a_1, a_2, a_3 such that:

$$\langle\langle\langle b,c\rangle,x\rangle,y\rangle E a_1 \Leftrightarrow y = U(U(b,x),U(c,x))$$

$$\langle\langle b,c\rangle,y\rangle E a_2 \Leftrightarrow y = S'(\&(a_1),\langle b,c\rangle)$$

$$\langle b,y\rangle E a_3 \Leftrightarrow y = S'(\&(a_2),b)\,.$$

Putting $\alpha = \&(a_3)$, by A3 and Kleene's S^m_n-theorem we can deduce:

$$U(\alpha,b) = S'(\&(a_2),b)\,,$$

$$U(U(\alpha,b),c) \simeq U(S'(\&(a_2),b),c)$$

$$\simeq U(\&(a_2),\langle b,c\rangle) = S'(\&(a_1),\langle b,c\rangle)\,,$$

and

$$U(U(U(\alpha,b),c),x) \simeq U(S'(\&(a_1),\langle b,c\rangle),x)$$

$$\simeq U(\&(a_1),\langle\langle b,c\rangle,x\rangle) \simeq U(U(b,x),U(c,x))\,.$$

Similarly by A2 there are elements $e_1 - e_4$ such that:

$$\langle\langle\langle\langle a,b\rangle,c\rangle,x\rangle,y\rangle E e_1 \Leftrightarrow ((y=a \wedge c=x) \vee (y=b \wedge c \neq x))$$

$$\langle\langle\langle a,b\rangle,c\rangle,y\rangle E e_2 \Leftrightarrow y = S'(\&(e_1),\langle\langle a,b\rangle,c\rangle)$$

$$\langle\langle a,b\rangle,y\rangle E e_3 \Leftrightarrow y = S'(\&(e_2),\langle a,b\rangle)$$

$$\langle a,y\rangle E e_4 \Leftrightarrow y = S'(\&(e_3),a)\,.$$

Putting $\psi = \&(e_4)$ we easily verify conditions: c, d, and e.

There is of course a standard arithmetical model for A0–A4 in which the universe consists of natural numbers and "xEy" means: the number x belongs to recursively enumerable set having the number y. The other model consists of the recursive ordinals and metarecursive enumeration.

Is it a natural theory, or is it perhaps too weak to give more involved interesting theorems? One can trye to develop the hierarchies in it, but perhaps it may be more appropriate to add the weak second order logic.

Another question which seems to be interesting is: taking A0, A1, how many instances of A2 can one take and still get a theory compatible with extensionality? If it were possible to define two combinators of the λ-calculus, we would have a model for the λ-calculus, if the definitions could be compatible with extensionality.

J.E.Fenstad, P.G.Hinman (eds.), Generalized Recursion Theory
©North-Holland Publ. Comp., 1974

POST'S PROBLEM FOR ADMISSIBLE SETS [1]

S.G. SIMPSON
The University of California, Berkeley

In 1944 Post proved that there exists a recursively enumerable subset of ω having intermediate many-one degree. Post then asked whether there exists a recursively enumerable subset of ω having intermediate degree of unsolvability. In 1956 Friedberg and Muchnik solved Post's problem affirmatively by proving that there exist two recursively enumerable subsets of ω having incomparable degrees of unsolvability.

Recently, Sacks and Simpson [5] generalized the Friedberg–Muchnik theorem to the context of recursion theory on admissible sets of the form L_α. Admissible sets of this special form retain the following basic property of ω:

(W) { the universe is well-ordered by a recursive relation.

Kreisel [2: p. 173] asked whether [2] property (W) is in any way essential or "significant" for generalizations of the Friedberg–Muchnik theorem.

In the present paper we partially answer Kreisel's question. Namely we prove: there exists an admissible set M for which both (W) and the Friedberg–Muchnik theorem fail. However, our proof has one serious defect: it uses AD, the so-called axiom of determinacy, which is actually not an axiom but rather an unsupported (though pragmatically interesting) hypothesis. We conjecture that this defect can be eliminated. In the meantime, for background material on AD, the reader may consult [3].

[1] Research partially supported by NSF Contract GP-24352.

[2] See also Kreisel's 1973 Zentralblatt review of [4] in which Kreisel suggests that this question must be answered before it is reasonable to start thinking about axiomatics for post–Friedberg recursion theory.

Our main theorem, Theorem 1 below, is stronger than what was stated above, in two ways. First, while our admissible set M will not have property (W), it *will* have the following property:

$$\text{(PW)} \quad \left\{ \begin{array}{l} \text{the universe is prewellordered by an } M\text{-recursive} \\ \text{relation whose initial segments are uniformly} \\ M\text{-finite.} \end{array} \right.$$

Second, not only will the M-analog of the Friedberg–Muchnik theorem fail, but so will the M-analog of Post's weaker theorem mentioned in the first sentence of this paper.

Definition. Let M be an admissible set. $B \subseteq M$ is *complete*-$\Sigma(M)$ if (i) B is $\Sigma(M)$; and (ii) for each $\Sigma(M)$ set $A \subseteq M$ there is a $\Sigma(M)$ relation $C \subseteq M \times M$ such that

(a) $\forall x \exists y\, C(x,y)$
(b) $\forall x \forall y\, (C(x,y) \to (x \in A \leftrightarrow y \in B))$.

Remark. If M is Σ-uniformizable then (ii) is equivalent to every $\Sigma(M)$ set being many-one reducible to B. In any case, (ii) implies that every $\Sigma(M)$ set is $\Delta(\langle M, B\rangle)$.

Theorem 1. *Assume* AD. *Let* $M = R^+$, *the next admissible set after the continuum. Then every* $\Sigma(M)$ *set is either* $\Delta(M)$ *or complete.*

In particular, the Friedberg–Muchnik theorem fails for R^+. Note that R^+ is a "Friedberg theory" in the sense of Moschovakis [4].

Before proving Theorem 1, we establish some notation. Let $R = \omega^\omega$, the real continuum. Let $M = R^+$, the smallest admissible set such that $R \in M$. Put $\kappa = \mathrm{On} \cap M$. Clearly $M = \mathrm{L}_\kappa(R)$ where the constructible hierarchy over R is defined by

$$\begin{aligned} \mathrm{L}_0(R) &= \text{transitive closure of } R; \\ \mathrm{L}_{\alpha+1}(R) &= \{X \subseteq \mathrm{L}_\alpha(R) \mid X \text{ is first-order definable over } \langle \mathrm{L}_\alpha(R), \in \rangle \\ &\qquad \text{alowing parameters from } \mathrm{L}_\alpha(R)\}; \end{aligned}$$

$$L_\lambda(R) = \bigcup\{L_\alpha(R) \mid \alpha < \lambda\} \text{ for limit } \lambda;$$

$$L(R) = \bigcup\{L_\alpha(R) \mid \alpha \text{ an ordinal}\}.$$

Some Berkeley set theorists have conjectured that AD "holds" in $L(R)$ in the sense that plausible large cardinal axioms may be found which imply this. [3] Note that our theorem and proof take place entirely within $L(R)$.

Let H be the Moschovakis system of notations for the ordinals less than κ. Let $\|$ be the corresponding norm. Thus H is a subset of R and $\|$ maps H onto κ. For each $\alpha < \kappa$ put $M_\alpha = L_\alpha(R)$ and $H_\alpha = \{x \in H \mid |x| < \alpha\}$.

Facts. 1. H is complete $\Sigma(M)$.
2. For each $\alpha < \kappa$, $M_\alpha \in M$ and $H_\alpha \in M$.
3. The sequences $\langle M_\alpha \mid \alpha < \kappa \rangle$ and $\langle H_\alpha \mid \alpha < \kappa \rangle$ are $\Sigma(M)$. Hence $\|$ is $\Sigma(M)$.
4. For each $\alpha < \kappa$ there is $i \in M$ such that i maps R onto M_α.

The proofs of the above facts do not use AD and are buried in the writings of Moschovakis (see for example [1]).

Let I and J be subsets of R. We write $I \leq J$ if there exists a continuous function $f: R \to R$ such that $x \in I \leftrightarrow f(x) \in J$ for all $x \in R$. We shall not actually use AD but only the following consequence of it due to Wadge [7].

Lemma 1. *For every $I, J \subseteq R$ either $I \leq J$ or $J \leq R - I$.*

Lemma 2. *Let S be a subset of κ such that $\forall \alpha < \kappa (S \cap \alpha \in M)$. Then S is $\Delta(M)$.*

Proof. Assume hypothesis. We shall show that S is $\Sigma(M)$. Put $K = \{x \in H \mid |x| \in S\}$. It suffices to show that K is $\Sigma(M)$. We shall do this by showing that $K \leq H$. By Lemma 1 it suffices to show that $H \nleq R - K$. So suppose $H \leq R - K$ via f. Then for all $x \in R$ we have

$$x \in H \leftrightarrow \exists y \in R(y \notin H \wedge x \in H_{|f(y)|})$$

[3] I personally do not subscribe to this conjecture. However, I am impressed by the fact that a number of people have tried and failed to deduce a contradiction from ZF + AD.

whence H is $\mathbf{\Pi}(M)$ a contradiction.

Proof of Theorem 1. Let $A \subseteq M$ be $\mathbf{\Sigma}(M)$. *Case I*: $A \cap M_\alpha \in M$ for all $\alpha < \kappa$. In this Case we shall show that A is $\mathbf{\Delta}(M)$. Let A be defined over M by

$$x \in A \leftrightarrow \exists y\, D(x, y)$$

where D is $\mathbf{\Delta}(M)$. For each $x \in M$ let $h(x)$ be the least η such that $D(x, y)$ holds for some $y \in M_\eta$. Thus dom$(h) = A$ and h is $\mathbf{\Sigma}(M)$. By the Case hypothesis and the admissibility of M, $h[M_\alpha]$ is bounded below κ for each $\alpha < \kappa$. Let $g(\alpha)$ be the least upper bound of $h[M_\alpha]$. Thus $g : \kappa \to \kappa$ and for each $x \in M$ we have

$$x \in A \leftrightarrow \forall \alpha < \kappa\, (x \in M_\alpha \to \exists y \in M_{g(\alpha)} D(x, y))\,.$$

By Lemma 2 g is $\mathbf{\Delta}(M)$ hence A is $\mathbf{\Pi}(M)$ q.e.d.

Case II: negation of Case I. Let $\alpha < \kappa$ be such that $A \cap M_\alpha \notin M$. Let $i \in M$ map R onto M_α. Put $I = \{r \in R \mid i(r) \in A\}$. Thus $I \subseteq R$ and I is $\mathbf{\Sigma}(M)$ but not $\mathbf{\Delta}(M)$. In particular $I \not\leq R - H$ hence by Lemma 1 $H \leq I$ hence I is complete $\mathbf{\Sigma}(M)$. From this it is immediate that A is complete $\mathbf{\Sigma}(M)$.

The proof of Theorem 1 is complete.

The following theorem is an immediate consequence of Wadge's lemma [7].

Theorem 2. *Assume* PD, *projective determinacy. Let* $M = H_{\omega_1}$, *the herditarily countable sets. Then every* $\mathbf{\Sigma}(M)$ *set is either* $\mathbf{\Delta}(M)$ *or complete.* [4]

Addendum.

The reader should not conclude from Theorems 1 and 2 that wellorderings are indispensable for priority arguments. Consider the following slight strengthening of property (PW):

[4] Hence there exists a *countable* admissible set M_0 for which the same conclusion holds. The proof that M_0 exists does not require the assumption of PD outright but only the assumption that PD has an admissible model.

$$(\mathrm{T})\quad \begin{cases} M \text{ has } \mathbf{\Delta}(M) \text{ prewellorderings } <_1 \text{ and } <_2 \\ \text{such that the initial segments of } <_1 \text{ are} \\ \text{uniformly } M\text{-finite, and the initial segments} \\ \text{of } <_2 \text{ are } M\text{-small.} \end{cases}$$

where $Y \subseteq M$ is said to be *M-finite* if $Y \in M$, and *M-small* if $Y \cap A \in M$ whenever A is $\mathbf{\Sigma}(M)$. We tentatively propose that an admissible set be called *thin* if it has property (T). Admissible sets of the form L_α are thin via the (pre)wellorderings $x < y$ and $f(x) < f(y)$ where $f: \alpha \to \alpha^*$ is α-recursive and one-one. Not every thin admissible set has property (W). For thin admissible sets one can imitate the proof of Theorem 4.1(i) in [6] yielding a version of the Friedberg–Muchnik theorem.

Bibliography

[1] K.J. Barwise, R.O. Gandy and Y.N. Moschovakis, The next admissible set, J. Symbolic Logic 36 (1971) 108–120.

[2] G. Kreisel, Some reasons for generalizing recursion theory, in: R.O. Gandy and C.E.M. Yates (eds.) Logic Colloquium '69 (North-Holland, Amsterdam, 1971) 139–198.

[3] J.E. Fenstad, The axiom of determinateness, in: J.E. Fenstad (ed.) Proceedings of the Second Scandinavian Logic Symposium (North-Holland, Amsterdam, 1971) 41–61.

[4] Y.N. Moschovakis, Axioms for computation theories – first draft, in: R.O. Gandy and C.E.M. Yates (eds.) Logic Colloquium '69 (North-Holland, Amsterdam, 1971) 199–255.

[5] G.E. Sacks and S.G. Simpson, The α-finite injury method, Annals of Math. Logic 4 (1972) 343–367.

[6] S.G. Simpson, Degree theory on admissible ordinals, this volume.

[7] W. Wadge, Degrees of complexity of subsets of the Baire space, Notices Amer. Math. Soc. 19 (1972) p. A-714.

PART V

BIBLIOGRAPHY OF GENERALIZED RECURSION THEORY

J.E.Fenstad, P.G.Hinman (eds.), Generalized Recursion Theory

SOME PAPERS ON GENERALIZED RECURSION THEORY ARRANGED ACCORDING TO SUBJECT MATTER

A number after an author's name singles out an item in the Uncritical Bibliography. Thus Grilliot (32) refers to: [32] T. Grilliot, Selection functions for recursive functionals, Notre Dame Jour. Formal Log. X (1969) 225–234. TV after an author's name refers to his paper in this volume.

Recursion in objects of finite type: Aczel and Hinman (TV). Gandy (26, 27). Grilliot (31, 32, 33). Harrington (TV). Kleene (48). MacQueen (75). Moschovakis (80, TV). Platek (92). Sacks (103, TV). Shoenfield (109).

Recursion on ordinals: Jensen and Karp (42). Kino and Takeuti (43). Kreisel and Sacks (61). Kripke (62). Lerman (68). Lerman and Sacks (69). Owings (88). Platek (92). Sacks (101, 102). Sacks and Simpson (105). Shore (110, 111). Simpson (TV). Takeuti (123, 124). Tugué (127).

Admissible sets: Barwise (8, TV). Barwise, Gandy and Moschovakis (11). Platek (92).

Inductive definability and hyperprojective sets: Aanderaa (TV). Aczel and Richter (TV). Cenzer (TV). Gandy (TV). Grilliot (34). Harrington (TV). Hinman and Moschovakis (40). Moschovakis (81, 82). Richter (99). Spector (117).

Model theoretic, axiomatic and other views of generalized recursion theory: Fenstad (TV). Fraissé (21). Friedman (22, 23). Gordon (30). Grilliot (TV). Kreisel (58, 59). Kunen (63). Lacombe (65, 66). Lambert (67). Montague (76, 77). Moschovakis (83). Strong (118, 119). Wagner (128, 129).

AN UNCRITICAL BIBLIOGRAPHY OF PAPERS ON GENERALIZED RECURSION THEORY

[1] P. Aczel, Representability in some systems of second order arithmetic, Israel Jour. Math. 8 (1970) 309–328.

[2] P. Aczel (Abstract) Implicit and inductive definability, Jour. Symb. Log. 35 (1970) 599.

[3] P. Aczel and W. Richter, Inductive definitions and analogues of large cardinals, in: Conference in Math. Log. – London '70 (Springer, Berlin, 1972) 1–9.

[4] J.W. Addison, Some consequences of the axiom of constructibility, Fund. Math. 46 (1959) 337–357.

[5] J.W. Addison, Some problems in hierarchy theory, Proc. Symp. Pure Math. vol. V (Amer. Math. Soc., Providence, R.I., 1962) 123–130.

[6] J.W. Addison and S.C. Kleene, A note on function quantification, Proc. Amer. Math. Soc. 8 (1957) 1002–1006.

[7] V.I. Amstislavskii, Extensions of recursive hierarchies and R-operations (Russian) Dokl. Ackad. Nauk SSSR 180 (1968) 1023–1026.

[8] J. Barwise, Infinitary logic and admissible sets, Jour. Symb. Log. 34 (1969) 226–252.

[9] J. Barwise, Applications of strict Π_1^1 predicates to infinitary Logic, Jour. Symb. Log. 34 (1969) 409–423.

[10] J. Barwise and E. Fisher, The Shoenfield Absoluteness Lemma, Israel Jour. Math. 8 (1970) 329–339.

[11] J. Barwise, R.O. Gandy and Y.N. Moschovakis, The next admissible set, Jour. Symb. Log. 36 (1971) 108–120.

[12] S. Bloom, The hyperprojective hierarchy, Zeit. Math. Log. Grund. Math. 16 (1970) 149–164.

[13] G. Boolos and H. Putnam, Degrees of unsolvability of constructible sets of integers, Jour. Symb. Log. 33 (1968) 497–513.

[14] R. Boyd, G. Hensel and H. Putnam, A recursion-theoretic characterisation of the ramified analytic hierarchy, Trans. Amer. Math. Soc. 141 (1969) 37–62.

[15] C.C. Chang and Y.N. Moschovakis, The Suslin–Kleene theorem for V_κ with cofinality $(\kappa) = \omega$, Pacif. Jour. Math. 35 (1970) 565–569.

[16] D.A. Clarke, Hierarchies of predicates of finite types, Memoir Amer. Math. Soc. No. 51 (1964) 1–95.

[17] G.C. Driscoll, Jr., Metarecursively enumerable sets and their metadegrees, Jour. Symb. Log. 33 (1968) 389–411.

[18] H.B. Enderton, The unique existential quantifier, Arch. Math. Log. Grund. 13 (1970) 52–54.

[19] H.B. Enderton and H. Putnam, A note on the hyperarithmetic hierarchy, Jour. Symb. Log. 35 (1970) 429–430.

[20] S. Feferman and G. Kreisel, Persistent and invariant formulas relative to theories of higher order, Bull. Amer. Math. Soc. 22 (1966) 480–485.

[21] R. Fraissé,Une notion de récursivité relative, in: Infinitistic Methods (Proceedings of the Warsaw Symposium 1959) (Pergamon, Oxford, 1961) 323–328.

[22] H. Friedman, Axiomatic recursive function theory, in: R.O. Gandy and C.E.M. Yates (eds.) Logic Colloquium '69 (North-Holland, Amsterdam, 1971) 113–137.

[23] H. Friedman, Algorithmic procedures, generalized Turing Algorithms and elementary recursion theories, in: R.O. Gandy and C.E.M. Yates (eds.) Logic Colloquium '69 (North-Holland, Amsterdam, 1971) 361–390.

[24] R.O. Gandy, On a problem of Kleene's, Bull. Amer. Math. Soc. 66 (1960) 501–502.

[25] R.O. Gandy, Proof of Mostowski's Conjecture, Bull. Acad. Polon. Sci. 8 (1960) 571–575.

[26] R.O. Gandy, General recursive functionals of finite type and hierarchies of functionals, Ann. Fac. Sci. Univ. Clermont-Ferrand No. 35 (1967) 5–24.

[27] R.O. Gandy, Computable functionals of finite type I, in: J. Crossley (ed.) Sets, Models and Recursion Theory (North-Holland, Amsterdam, 1967) 202–242.

[28] R.O. Gandy, G. Kreisel and W.W. Tait, Set existence I, Bull. Acad. Polon. Sci. 8 (1960) 577–583; and II 9 (1961) 881–882.

[29] R.O. Gandy and G.E. Sacks, A minimal hyperdegree, Fund. Math. 61 (1967) 215–223.

[30] C. Gordon, Comparisons between some generalisations of recursion theory, Compositio Math. 22 (1970) 333–346.

[31] T. Grilliot, Hierarchies based on objects of finite type, Jour. Symb. Log. 34 (1969) 177–182.

[32] T. Grilliot, Selection functions for recursive functionals, Notre Dame Jour. Formal Log. X (1969) 225–234.

[33] T. Grilliot, On effectively discontinuous type-2 objects, Jour. Symb. Log. 36 (1971) 245–248.

[34] T. Grilliot, Inductive definitions and computability, Trans. Amer. Math. Soc. 158 (1971) 309–317.

[35] T. Grilliot, Omitting types; applications to recursion theory, Jour. Symb. Log. 37 (1972) 81–89.

[36] A. Grzegorczyk, A Mostowski and C. Ryll-Nardzewski, Definability of sets in models of axiomatic theories, Bull. Acad. Polon. Sci. 9 (1961) 163–167.

[37] L. Harrington, Contributions to recursion theory in higher types, Ph.D. Thesis, Massachusetts Institute of Technology (1973).

[38] J. Harrison, Recursive pseudo-well-orderings, Trans. Amer. Math. Soc. 131 (1968) 526–543.

[39] P. Hinman, Hierarchies of effective descriptive set theory, Trans. Amer. Math. Soc. 142 (1969) 111–140.

[40] P. Hinman and Y.N. Moschovakis, Computability over the continuum, in: R.O. Gandy and C.E.M. Yates (eds.) Logic Colloquium '69 (North-Holland, Amsterdam, 1971) 77–105.

[41] M. Hirano, Some definitions for recursive functions of ordinal numbers, Sci. Rep. Tokyo Kyoiku Daigaku Sect. A 10 (1969) 135–141.

[42] R.B. Jensen and C. Karp, Primitive recursive set functions, Proceedings of Symposia in Pure Mathematics XIII Part I (Amer. Math. Soc., Providence, R.I., 1971) 143–176.

[43] A. Kino and G. Takeuti, On hierarchies of predicates of ordinal numbers, Jour. Math. Soc. Japan 14 (1962) 199–232.
[44] A. Kino and G. Takeuti, On predicates with constructive infinitely long expressions, Jour. Math. Soc. Japan 15 (1963) 176–190.
[45] S.C. Kleene, Hierarchies of number-theoretic predicates, Bull. Amer. Soc. 61 (1955) 193–213.
[46] S.C. Kleene, Arithmetic predicates and function quantifiers, Trans. Amer. Math. Soc. 79 (1955) 312–340.
[47] S.C. Kleene, On the forms of the predicates in the theory of constructive ordinals II, Amer. Jour. Math. 77 (1955) 405–428.
[48] S.C. Kleene, Recursive functionals and quantifiers of finite type I, Trans. Amer. Math. Soc. 91 (1959) 1–52; and II 108 (1963) 106–142.
[49] S.C. Kleene, Quantification of number-theoretic functions, Compos. Math. 14 (1959) 23–41.
[50] S.C. Kleene, Countable functionals, in: A. Heyting (ed.), Constructivity in Mathematics (Proceedings of the 1957 Amsterdam Colloquium) (North-Holland, Amsterdam, 1958) 81–100.
[51] S.C. Kleene, Herbrand–Gödel style recursive functionals of finite types, Proc. Symp. Pure Math. vol. V (Amer. Math. Soc., Providence, R.I., 1962) 49–75.
[52] S.C. Kleene, Lambda definable functionals of finite type, Fund. Math. 50 (1961) 281–303.
[53] S.C. Kleene, Turing machine computable functionals of finite type I, in: Logic, Methodology and Philosophy of Science (Proceedings of the 1960 Congress) (Stanford Univ. Press, Stanford, 1962) 38–45; and II, Proc. Lond. Math. Soc. 12 (1962) 245–258.
[54] D.L. Kreider and H. Rogers, Jr., Constructive versions of ordinal number classes, Trans. Amer. Math. Soc. 100 (1961) 325–369.
[55] G. Kreisel, Set theoretic methods suggested by the notion of infinite totality, in: Infinitistic Methods (Pergamon, Oxford, 1961) 97–102.
[56] G. Kreisel, La prédicative, Bull. Soc. Math. France 88 (1960) 371–391.
[57] G. Kreisel, The axiom of choice and the class of hyperarithmetic functions. Indag. Math. 24 (1962) 307–319.
[58] G. Kreisel, Model theoretic invariants: applications to recursive and hyperarithmetic operators, in: J. Addison et al. (eds.) Theory of Models (Proceedings of the 1963 Berkeley Symposium) (North-Holland, Amsterdam, 1965) 190–205.
[59] G. Kreisel, Some reasons for generalizing recursion theory, in: R.O. Gandy and C.E.M. Yates (eds.) Logic Colloquium '69 (North-Holland, Amsterdam, 1971) 139–198.
[60] G. Kreisel, J. Shoenfield and H. Wang, Number theoretic concepts and recursive wellorderings, Arch. Math. Log. Grund. 5 (1960) 42–64.
[61] G. Kreisel and G.E. Sacks, Metarecursive sets, Jour. Symb. Log. 30 (1965) 318–338.
[62] S. Kripke (Abstracts) Transfinite recursion on admissible ordinals I and II, Jour. Symb. Log. 29 (1964) 161–162.
[63] K. Kunen, Implicit definability and infinitary languages, Jour. Symb. Log. 33 (1968) 446–451.
[64] D. Lacombe, Deux généralisations de la notion de récursivité, C.R. Acad. Sci. Paris 258 (1964) 3141–3143.

[65] D. Lacombe, Deux généralisations de la notion de récursivité relative, C.R. Acad. Sci. Paris 258 (1964) 3410–3413.
[66] D. Lacombe, Recursion theoretic structures for relational systems, in: R.O. Gandy and C.E.M. Yates (eds.) Logic Colloquium '69 (North-Holland, Amsterdam, 1971) 3–17.
[67] W. Lambert, Jr., A notion of effectiveness in arbitrary structures, Jour. Symb. Log. 33 (1968) 577–602.
[68] M. Lerman On the suborderings of the α-recursively enumerable α-degrees, Ann. Math. Log. 4 (1972) 369–392.
[69] M. Lerman and G.E. Sacks, Some minimal pairs of α-recursively enumerable degrees, Ann. Math. Log. 4 (1972) 415–442.
[70] A. Levy, A hierarchy of formulas in set theory, Memoir Amer. Math. Soc. no. 57 (1965) 1–76.
[71] S.C. Liu, Recursive linear orderings and hyperarithmetic functions, Notre Dame Jour. Formal Log. 3 (1962) 129–132.
[72] P. Lorenzen and J. Myhill, Constructive definition of certain analytic sets of numbers, Jour. Symb. Log. 24 (1959) 37–49.
[73] M. Machtey, Admissible ordinals and the lattice of α-recursively enumerable sets, Ann. Math. Log. 2 (1970–71) 379–417.
[74] M. Machtey, Admissible ordinals and intrinsic consistency, Jour. Symb. Log. 35 (1970) 389–400.
[75] D. MacQueen, Post's problem for recursion in higher types, Ph.D. Thesis, Massachusetts Institute of Technology, 1972.
[76] R. Montague, Towards a general theory of computability, Synthese 12 (1960) 429–438.
[77] R. Montague, Recursion theory as a branch of model theory, in: B. van Rootselaar et al. (eds.) Logic Methodology and Philosophy of Science III (Proceedings of the 1967 Congress) (North-Holland, Amsterdam, 1968) 63–86.
[78] Y.N. Moschovakis, Many-one degrees of the $H_a(x)$ predicates, Pacif. Jour. Math. 18 (1966) 329–342.
[79] Y.N. Moschovakis, Predicative classes, in: Axiomatic Set Theory (Proceedings of Symposia in Pure Math. XIII, Part I, 1967) (Amer. Math. Soc., Providence, R.I., 1971) 247–264.
[80] Y.N. Moschovakis, Hyperanalytic predicates, Trans. Amer. Math. Soc. 129 (1967) 249–282.
[81] Y.N. Moschovakis, Abstract first order computability I, Trans. Amer. Math. Soc. 138 (1969) 427–464; and II 138 (1969) 465–504.
[82] Y.N. Moschovakis, Abstract computability and invariant definability, Jour. Symb. Log. 34 (1969) 605–633.
[83] Y.N. Moschovakis, Axioms for computation theories – first draft, in: R.O. Gandy and C.E.M. Yates (eds.) Logic Colloquium '69 (North-Holland, Amsterdam, 1971) 199–255.
[84] Y.N. Moschovakis, The Suslin–Kleene theorem for countable structures, Duke Math. Jour. 37 (1970) 341–352.
[85] A. Mostowski, Development and applications of the projective classification of sets of integers, in: Proc. Inter. Cong. Math. (1965) Amsterdam vol. III (E.P. Noordhoff, Groningen) 280–288.
[86] K. Ohashi, On a question of G.E. Sacks, Jour. Symb. Log. 35 (1970) 46–50.

[87] J.C. Owings, Jr., Recursion, metarecursion and inclusion, Jour. Symb. Log. 32 (1967) 173–179.
[88] J.C. Owings, Jr., Π_1^1 sets, ω-sets and metacompleteness, Jour. Symb. Log. 34 (1969) 194–204.
[89] J.C. Owings, Jr., The metarecursively enumerable sets, but not the Π_1^1 sets, can be enumerated without repetitions, Jour. Symb. Log. 35 (1970) 223–229.
[90] J.C. Owings, Jr., A splitting theorem for simple Π_1^1 sets, Jour. Symb. Log. 36 (1971) 433–438.
[91] R. Parikh, On the nonuniqueness in transfinite progressions, Jour. Indian Math. Soc. (N.S.) 31 (1967) 23–32.
[92] R. Platek, Foundations of Recursion Theory, Ph.D. Thesis, Stanford University, 1966.
[93] R. Platek, A countable hierarchy for the superjump, in: R.O. Gandy and C.E.M. Yates (eds.) Logic Colloquium '69 (North-Holland, Amsterdam, 1971) 257–271.
[94] H. Putnam, Uniqueness ordinals in higher constructive number classes, in: Essays on the Foundations of Mathematics (North-Holland, Amsterdam, 1961) 190–206.
[95] H. Putnam, On hierarchies and systems of notations, Proc. Amer. Math. Soc. 15 (1964) 44–50.
[96] W. Richter, Extensions of the constructive ordinals, Jour. Symb. Log. 30 (1965) 193–211.
[97] W. Richter, Constructive transfinite number classes, Bull. Amer. Math. Soc. 73 (1967) 261–265.
[98] W. Richter, Constructively accessible ordinal numbers, Jour. Symb. Log. 33 (1968) 43–55.
[99] W. Richter, Recursively Mahlo ordinals and inductive definitions, in: R.O. Gandy and C.E.M. Yates (eds.) Logic Colloquium '69 (North-Holland, Amsterdam, 1971) 273–288.
[100] J. Robinson, An introduction to hyperarithmetic functions, Jour. Symb. Log. 32 (1967) 325–342.
[101] G.E. Sacks, Post's problem, admissible ordinals and regularity, Trans. Amer. Math. Soc. 124 (1966) 1–23.
[102] G.E. Sacks, Metarecursion theory, in: J. Crossley (ed.) Sets, Models and Recursion (North-Holland, Amsterdam, 1967) 243–263.
[103] G.E. Sacks, Recursion in objects of finite type, in: Proceedings of the 1970 International Congress of Mathematicians (Gauthiers–Villars, Paris, 1971) 251–254.
[104] G.E. Sacks, On the reducibility of Π_1^1 sets, Advances in Math. 7 (1971) 57–82.
[105] G.E. Sacks and S.G. Simpson, The α-finite injury method, Ann. Math. Log. 4 (1972) 343–367.
[106] B. Scarpellini, A characterization of Δ_2^1 sets, Trans. Amer. Math. Soc. 117 (1965) 441–450.
[107] J.R. Shoenfield, The form of the negation of a predicate, Proc. Symp. Pure Math. vol. V (Amer. Math. Soc., Providence, R.I., 1962) 131–134.
[108] J.R. Shoenfield, The problem of predicativity, in: Essays on the Foundations of Mathematics (North-Holland, Amsterdam, 1961) 132–139.
[109] J.R. Shoenfield, A hierarchy based on a type-2 object, Trans. Amer. Math. Soc. 134 (1968) 103–108.
[110] R.A. Shore, Minimal α-degrees, Ann. Math. Log. 4 (1972) 393–414.
[111] R.A. Shore, Priority arguments in α-recursion theory, Ph.D. Thesis, Massachusetts Institute of Technology, 1972.

[112] S.G. Simpson, Admissible ordinals and recursion theory, Ph.D. Thesis, Massachusetts Institute of Technology, 1971).
[113] C. Spector, Recursive wellorderings, Jour. Symb. Log. 20 (1955) 151–163.
[114] C. Spector, On degrees of recursive unsolvability, Ann. of Math. 64 (1956) 581–592.
[115] C. Spector, Measure theoretic construction of incomparable hyperdegrees, Jour. Symb. Log. 23 (1958) 280–288.
[116] C. Spector, Hyperarithmetic quantifiers, Fund. Math. 48 (1959) 313–320.
[117] C. Spector, Inductively defined sets of natural numbers, in: Infinitistic Methods (Proceedings of Warsaw symposium 1959) (Pergamon Press, Oxford, 1971) 97–102.
[118] H.R. Strong, Algebraically generalized recursive function theory, IBM Jour. Res. Devel. 12 (1968) 465–475.
[119] H.R. Strong, Construction of models for algebraically generalized recursive function theory, Jour. Symb. Log. 35 (1970) 401–409.
[120] Y. Suzuki, A complete classification of the $\boldsymbol{\Delta}^1_2$ functions, Bull. Amer. Math. Soc. 70 (1964) 246–253.
[121] M. Takahashi, Recursive functions of ordinal numbers and Levy's hierarchy, Comment. Math. Univ. St. Paul 17 (1968) 21–29.
[122] H. Tanaka, On analytic wellorderings, Jour. Symb. Log. 35 (1970) 198–204.
[123] G. Takeuti, On the recursive functions of ordinal numbers, Jour. Math. Soc. Japan 12 (1960) 119–128.
[124] G. Takeuti, Recursive functions and arithmetic functions of ordinal numbers, in: Y. Bar-Hillel (ed.) Logic, Methodology and Philosophy of Science II (Proceedings of the 1964 Congress) (North-Holland, Amsterdam, 1965) 179–196.
[125] S.K. Thomason, On initial segments of hyperdegrees, Jour. Symb. Log. 35 (1970) 189–197.
[126] T. Tugué, Predicates recursive in a type-2 object and Kleene hierarchies, Comment. Math. Univ. St. Paul 8 (1959) 97–117.
[127] T. Tugué, On the partial recursive functions of ordinal numbers, Jour. Math. Soc. Japan 16 (1964) 1–31.
[128] E.G. Wagner, Uniform reflexive structures: on the nature of Gödelizations and relative computability, Trans. Amer. Math. Soc. 144 (1969) 1–41.
[129] E.G. Wagner, Uniformly reflexive structures: an axiomatic approach to computability, Informat. Sci. 1 (1968) 343–362.
[130] H. Wang, Ordinal numbers and predicative set theory, Zeit. Math. Log. Grund. Math. 5 (1959) 216–239.

INDEX *

* Boldface numbers refer to title pages of authors' chapters in this volume.